Neural Network Technologies and Brain-Computer Interfaces:

Innovations and Applications

Mohammed Saleh Al Ansari
University of Bahrain, Bahrain

Kapil Joshi
Uttaranchal University, India

IGI Global
Scientific Publishing
Publishing Tomorrow's Research Today

Vice President of Editorial — Melissa Wagner
Director of Acquisitions — Mikaela Felty
Director of Book Development — Jocelynn Hessler
Production Manager — Mike Brehm
Cover Design — Jose Rosado

Published in the United States of America by
IGI Global Scientific Publishing
701 East Chocolate Avenue
Hershey, PA, 17033, USA
Tel: 717-533-8845
Fax: 717-533-7115
Website: https://www.igi-global.com E-mail: cust@igi-global.com

Library of Congress Cataloging-in-Publication Data

Names: Al Ansari, Mohammed Saleh, 1966- editor | Joshi, Kapil, 1985- editor
Title: Neural network technologies and brain-computer interfaces : innovations and applications / edited by: Mohammed Saleh Al Ansari, Kapil Joshi.
Description: Hershey PA : IGI Global Scientific Publishing, [2025] | Includes bibliographical references. | Summary: "This book aims to provide an overview of the fundamental concepts, principles, and applications of neural network technologies and brain-computer interfaces. By converging artificial intelligence, neuroscience, and engineering, these innovations have transformative potential across various disciplines"-- Provided by publisher.
Identifiers: LCCN 2024060459 (print) | LCCN 2024060460 (ebook) | ISBN 9798369394458 hardcover | ISBN 9798369394465 paperback | ISBN 9798369394472 ebook
Subjects: LCSH: Brain-computer interfaces--Case studies | Neural networks (Computer science)--Case studies
Classification: LCC QP360.7 .N45 2025 (print) | LCC QP360.7 (ebook) | DDC 612.80285--dc23/eng/20250320
LC record available at https://lccn.loc.gov/2024060459
LC ebook record available at https://lccn.loc.gov/2024060460
British Cataloguing in Publication Data
A Cataloguing in Publication record for this book is available from the British Library.

Table of Contents

Detailed Table of Contents

The human brain, human-generated ideas and communications are non-stationary and nonlinear. Therefore, developing a technology that can draw greater insight from human thought processes is the challenging part. Applications of BCI will work better as these greater understandings are acquired. Brain-computer interfaces are gadgets that let individuals interact with computers solely by using their brain activity, which is frequently picked up by electroencephalography (EEG). Cutting edge in terms of brain-computer interfaces (BCI). An overview of BCIs is given at the outset, along with information on their most popular platforms and key working principles. The many parts of a BCI arrangement, including software, hardware, and digital signal processing methods, are then examined in the study.

Brain-Computer Interface (BCI)-based assistive technology offers promising solutions to enhance the independence and quality of life for people with disabilities. By allowing direct communication between the brain and external devices, BCIs can empower individuals to perform tasks that would otherwise be challenging or impossible. This study proposes a novel BCI system that integrates advanced

positively correlated with improved BCI performance, whereas increased work stress negatively affects accuracy.

A neural network is essentially an attempt to replicate how the brain works. Neural system theory is based on the idea that several of the most significant features of biological neurons can be extracted and applied to computations in computer order to produce a more simplified model of the brain. An ANN is configured for a specific use, such as pattern recognition or data classification, through a learning process. Learning involves altering the synaptic interactions between neurons in biological systems. Along with several ANN trainings, the fundamental MATLAB processes are presented. The training's goal is to reduce mean square error. T

Neural network organization on any website, we typically see the advertisements for the products we have previously searched for. Get you ever wondered how it's possible that, after searching for something on Flipkart, however, we miraculously get an advertisement for the same thing on Instagram? This is accomplished by employing the neural network concept. The neural network serves as a sophisticated data structure that links all of our online activities, processes the data, and uses the results for commercial objectives. The input set and the output are connected by a finite number of cells, nodes, units, or neuronal cells that make up the network. It is a component of an automated system that simulates the way the human brain processes and evaluates information. Neural networks that use artificial intelligence are used in a wide range of applications, including emergency room testing, character recognition, picture compression, is the process prediction of stock markets.

Mental health in migrants is often influenced by challenges related to healthcare access, including difficulties in understanding medical information and uncertainty in seeking appropriate assistance. This study analyzes survey responses from 82 adults (50 women and 32 men) to examine healthcare access barriers and their impact on mental health, vaccination status, and dental hygiene. Principal Component Analysis (PCA) was employed to extract significant features from the dataset, highlighting patterns such as healthcare literacy gaps and demographic influences. These features were used to train an Artificial Neural Network (ANN) for predictive modeling. The ANN demonstrated high accuracy, with F1 scores exceeding 85%, effectively predicting mental health conditions, vaccination rates, and dental hygiene practices. The results underscore the importance of addressing healthcare access barriers to improve overall well-being among migrants.

A comprehensive work promoting assistive technology that is meant to make life better for those who are visually impaired. They emphasise how machine learning methods and technologies may be used to customize cutting-edge apps to meet this user base's unique requirements. One such example, uses machine learning techniques to help people who are partially blind see better by improving how they see their surroundings. The literature-based approach helps visually impaired people recognise obstacles, improving mobility and safety. More recently, researchers have been working on making detection tools and visual aids that use artificial intelligence

BCIs offer a direct communication pathway between the brain and external devices, enabling individuals with neurological impairments to regain motor and cognitive functions. When combined with AI, BCIs can analyze neural signals more effectively, enabling personalized therapeutic interventions and adaptive learning models. This synergy facilitates real-time adjustments to rehabilitation protocols, improving patient outcomes and recovery rates. This chapter emphasizes the technological and clinical advancements in BCIs and AI, identifies barriers to implementation, and highlights their transformative potential for reshaping neurorehabilitation. Future research directions focus on improving device accessibility, ethical considerations, and expanding applications to broader neurological and cognitive disorders.

The integration of Neural Network Technologies (NNTs) and Brain-Computer Interfaces (BCIs) has transformed human-machine interaction, enabling direct brain-device communication. BCIs empower individuals with neurological disorders, while neural networks enhance signal processing, pattern recognition, and adaptive learning. Advances in deep learning, including CNNs, RNNs, GANs, and Federated Learning, have improved BCI accuracy in motor imagery, authentication, and neurofeedback. Hybrid BCIs combining multiple neural signals enhance adaptability and real-world application. However, challenges remain in ethics, data privacy, hardware, and personalized AI. Future AI-powered BCIs, brain-to-brain communication, and non-invasive neurostimulation promise breakthroughs in cognition, rehabilitation, and augmentation. This chapter explores key principles, innovations, applications, challenges, and future directions in AI-driven neuroscience.

Brain-Computer Interfaces (BCIs) are innovative systems that allow the brain to interact directly with external devices, eliminating the need for muscle-based communication. This technology is creating new possibilities in medical treatment, cognitive enhancement, and assistive tools. With the help of deep learning methods like CNNs and RNNs, BCIs can now decode brain signals more effectively and accurately. Applications range from helping individuals recover from strokes to improving communication for people with severe physical impairments. BCIs also show potential in treating neurological disorders and supporting mental health. Despite their promise, challenges like real-time processing and ethical considerations remain. Continued progress could transform both healthcare and human-machine interaction.

Brain-Computer Interface (BCI) technology enables direct communication between the brain and external devices, offering significant potential for individuals with disabilities. BCIs capture neural signals, interpret them, and convert them into commands to control devices like prosthetics, wheelchairs, or communication aids. This chapter explores the evolution of BCIs, from early brainwave control research to advanced systems that restore lost functions for those with physical, cognitive, and sensory disabilities. It also examines the design of user interfaces for accessibility and ease of use. Despite the promise, challenges such as signal accuracy, invasiveness, and usability persist, requiring improvement for better BCI-based assistive technologies. The chapter evaluates BCI effectiveness in enhancing user mobility, communication, and quality of life, emphasizing the need for ongoing research to overcome limitations and enhance BCI potential in assistive tech.

Early stroke detection by doctors improves treatment possibilities and leads to better patient results. Our research with Recurrent Neural Networks shows how CT image time series help hospitals detect stroke patients. StrokeTimeNet combines LSTMs and 3D-Convolutional neural networks to analyze spatial patterns in CT image sequences taken over time. The model showed better diagnosis results when it processed multiple CT scan datasets and trained itself to detect stroke changes. Our diagnostic tests checked how well the method detected the disease and measured its performance against 3D CNNs and ML while determining the AUC-ROC results and

Preface

Neural Network Technologies and Brain-Computer Interfaces: Innovations and Applications explores the convergence of two transformative areas in modern science and engineering: artificial neural networks (ANNs) and brain-computer interfaces (BCIs). As artificial intelligence generally evolves toward more human-like cognition and biological systems become increasingly integrated with machines, the synergies between neural computation and brain-machine communication are unlocking revolutionary possibilities. In terms of neural network technologies, neural networks are computational models inspired by the human brain's structure and function. They have rapidly advanced from shallow architectures to deep learning and transformer-based models, enabling significant breakthroughs in image recognition, natural language processing, and autonomous systems. Key innovations include: Deep Learning Architectures, lie CNNs, RNNs and Transformers, Neuro-symbolic systems integrating reasoning with perception, Edge and neuromorphic computing for low-power, real-time processing, Explainable AI (XAI) for transparency in decision-making. Brain-Computer Interfaces facilitate direct communication between the brain and external devices, bypassing traditional neuromuscular pathways. Initially developed for assistive technologies, BCIs now span neurorehabilitation, gaming, mental health, and cognitive augmentation. Innovations include: Non-invasive BCIs (e.g., EEG, fNIRS) and emerging wearable technologies, Invasive BCIs (e.g., ECoG, microelectrode arrays) offering high fidelity, AI-driven signal processing for decoding brain activity patterns, Closed-loop systems for adaptive feedback and real-time control. The discussion about integration and applications then the integration of neural networks and BCIs enables intelligent systems that can interpret, learn from, and respond to neural signals. Applications including medical and therapeutic systems for motor recovery, speech generation, and prosthetics, neuroadaptive interfaces in VR/AR and gaming, cognitive monitoring for mental health diagnostics, brain-to-brain communication and collaborative neural interfaces. Challenges and Future Directions are explained the field faces critical challenges such as ethical considerations, data privacy, long-term biocompatibility, and interpretability of neural models. Nonetheless, future directions point toward hybrid neuro-AI systems with biologically plausible learning, brain-inspired hardware for scalable and efficient computation, human-AI symbiosis in cognition, creativity, and decision-making.

Neural Network Technologies and Brain-Computer Interfaces: Innovations and Applications is a highly relevant and timely subject in today's world, where rapid advancements in artificial intelligence, neuroscience, and human-computer interaction are transforming how we understand and interact with machines. In today's landscape, neural network technologies form the foundation of modern AI systems, powering everything from voice assistants and recommendation engines to autonomous vehicles and medical diagnostics. These networks mimic the human brain's learning processes, enabling machines to recognize patterns, adapt to data, and make decisions with increasing accuracy and complexity.

This subject is highly relevant in a time where personalized medicine, neuroprosthetics, and adaptive learning systems are becoming mainstream. Neural networks enable pattern recognition and predictive analytics essential for decoding brain signals in BCIs. Together, these technologies are enabling breakthroughs such as restoring mobility to paralyzed individuals, enhancing cognitive function, enabling non-verbal communication for those with disabilities, and developing immersive experiences in virtual and augmented reality. Together, neural networks and BCIs represent the convergence of machine intelligence and human cognition, with profound implications for healthcare, accessibility, mental health, and even ethical frameworks. These technologies are not only reshaping industries but are also prompting society to reconsider the nature of consciousness, identity, and agency in a hyper-connected digital age.

The target audience of *Neural Network Technologies and Brain-Computer Interfaces: Innovations and Applications* includes a multidisciplinary spectrum of readers who are invested in the development, research, and practical deployment of cutting-edge neurotechnologies and artificial intelligence systems. These include:

1. **Researchers and Academics** in fields such as computer science, neuroscience, biomedical engineering, cognitive science, and AI who are looking for advanced insights into the integration of neural networks with BCIs.
2. **Graduate and Postgraduate Students** pursuing studies in neural engineering, machine learning, AI, or human-computer interaction who need a comprehensive, up-to-date resource on emerging technologies and real-world applications.
3. **Industry Professionals and Engineers** working in healthcare technology, AI development, neurotechnology startups, or wearable tech industries interested in innovative applications and technical foundations of BCIs and neural networks.

4. **Medical Practitioners and Neurotechnologists** exploring new assistive and diagnostic tools for neurological disorders, neurorehabilitation, and patient-centered brain-interface solutions.
5. **Policy Makers, Ethicists, and Futurists** concerned with the societal, ethical, and regulatory implications of merging AI with human cognition.
6. **Tech Entrepreneurs and Innovators** who are identifying market opportunities in the growing space of neural interface technologies and AI-driven health solutions.

This book is designed to be both technically robust and forward-looking, appealing to those who are not only interested in how these technologies work, but also in where they are headed and how they will shape the future of human–machine synergy.

The chapter 1 titled "Analysis of Human Brain Using Basic Brain-Computer Interface (BCI) System" holds significant importance in both academic and applied contexts for several reasons: Foundational Understanding of BCI Systems, Bridging Neuroscience and Engineering, Real-World Applications, Platform for Innovation, Ethical and Human-Centric Design Considerations, this chapter is also crucial as it lays the groundwork for both technical mastery and responsible innovation in BCI development, empowering readers to understand, build, and apply brain-computer technologies from the ground up.

The chapter 2 titled "BCI-Based Assistive Technology with Integrated Machine Learning: Empowering People with Disabilities for Improved Independence and Quality of Life" is of critical importance due to its direct impact on enhancing human well-being through technology including empowerment through innovation and integration of machine learning for personalization.

The chapter 3 titled "Bridging Neural Technologies with Financial Decision-Making in Neurofinance" is of significant importance because it explores an emerging interdisciplinary field that combines neuroscience, artificial intelligence, and economics to better understand and enhance financial behavior and decision-making.

The chapter 4 titled "Convolutional Neural Networks for Brain-Computer Interfaces" is a critical contribution to the field of neurotechnology and artificial intelligence, as it focuses on the powerful synergy between deep learning algorithms and brain signal processing in BCI systems.

The chapter 5 titled "Deep Learning for BCI Signal Decoding" is of central importance in advancing the capabilities and effectiveness of brain-computer interface (BCI) systems. It focuses on how deep learning—an advanced subset of machine learning—revolutionizes the way brain signals are interpreted and translated into actionable outputs.

The chapter 6 titled "Dynamic Rehabilitation through Brain-Computer Interfaces Integrating CNN and Bi-GRU Deep Learning Approach" is of high significance due to its novel integration of advanced deep learning techniques into brain-computer interface (BCI)-based rehabilitation systems, marking a critical step forward in neurorehabilitation technology including advancing personalized and adaptive rehabilitation etc.

The chapter 7 is critically important for advancing the effectiveness, personalization, and practical adoption of BCI systems in real-world scenarios for improving prediction accuracy with advanced architectures and incorporating user-centric factors for personalization.

The chapter 8 titled "Fundamentals Approached Towards Artificial Neural Networks in Brain-Computer Interface" is a cornerstone of the book, providing essential groundwork for understanding how artificial neural networks (ANNs) can be effectively applied within brain-computer interface (BCI) systems.

The chapter 9 titled "Introduction of a Cutting-Edge Technology: Neural Networks Organization" is of significant importance as it sets the stage for understanding the structure, evolution, and transformative potential of neural networks in the realm of modern technology—particularly in brain-computer interface (BCI) systems and intelligent computing.

The chapter 10 titled "Mental Health in Migrants: Analyzing Healthcare Access Barriers Using AI-Based Techniques" is profoundly important as it tackles one of the most pressing and complex challenges in global public health—the mental well-being of migrant populations—through the lens of advanced artificial intelligence methodologies.

The chapter 11 titled "Machine Learning Algorithms for Individuals with Visual Impairments" is highly important as it explores how intelligent technologies can be harnessed to empower people with visual disabilities, enhance their independence, and improve their quality of life.

The chapter 12 titled "Knee Osteoarthritis Using Hybrid Deep Learning Approach with SqueezeNet and ResNet" is of significant importance as it presents a cutting-edge, AI-driven solution for improving the diagnosis and assessment of knee osteoarthritis (KOA)—a widespread, debilitating joint disease affecting millions worldwide.

The chapter 13 titled "AI-Powered Brain-Computer Interfaces for Advancing Prosthetic Functionality and Rehabilitation" holds immense significance as it explores the integration of artificial intelligence with brain-computer interface (BCI) technologies to revolutionize the field of prosthetics and neurorehabilitation.

The chapter 14 titled "Cognitive Enhancement Through Direct Brain-Computer Interaction" is of profound importance as it explores the transformative potential of brain-computer interface (BCI) technologies in augmenting human cognitive abilities, marking a shift from therapeutic applications to proactive cognitive augmentation.

The chapter 15 titled "Enhancing Assistive Technologies with Brain-Computer Interfaces (BCIs) and Artificial Intelligence (AI) for Optimizing Neurorehabilitation Approaches" is of high importance, as it explores how the fusion of BCI and AI technologies can redefine the landscape of neurorehabilitation and assistive care for individuals with neurological and motor impairments.

The chapter 16 titled "Advances in Neural Network Technologies and Brain-Computer Interfaces: Emerging Innovations and Applications" is critically important as it provides a comprehensive overview of the latest breakthroughs and forward-looking trends at the intersection of neural networks and brain-computer interface (BCI) technologies—two transformative pillars of modern intelligent systems.

The chapter 17 titled "Brain-Computer Interfaces: Revolutionizing Healthcare, Neuroprosthetics, and Cognitive Enhancement" is of paramount importance as it provides a broad yet in-depth exploration of how BCI technology is fundamentally transforming the future of medicine, assistive technology, and human cognitive potential.

The chapter 18 titled "Training BCI-Based Assistive Technology for People with Disabilities" is highly significant as it focuses on the practical, user-centered development and training of Brain-Computer Interface (BCI) systems designed to empower individuals with physical or cognitive impairments.

The chapter 19 titled "Improving Early Brain Stroke Diagnosis: An RNN Approach to CT Image Interpretation" is of great importance as it presents a novel, AI-driven method for enhancing the early detection and diagnosis of brain strokes, a critical step in minimizing brain damage and improving patient outcomes.

The chapter 20 titled "Neurological Disorder Diagnosis and Treatment Using BCIs" is critically important as it explores how Brain-Computer Interfaces (BCIs) can revolutionize the diagnosis, monitoring, and treatment of various neurological disorders, providing novel solutions for both early detection and personalized treatment.

The book *Neural Network Technologies and Brain-Computer Interfaces: Innovations and Applications* marks a significant contribution to the growing field of neurotechnology, bringing together cutting-edge advancements in neural networks and brain-computer interfaces (BCIs) with practical, real-world applications. By weaving together theoretical concepts with innovative implementations, this book also provides a comprehensive understanding of how these transformative technologies are shaping the future of healthcare, cognitive enhancement, assistive devices, and human-computer interaction. It deals with the following areas:

1. Advancing the Understanding of Neural Technologies

Throughout the chapters, this book explores the deep interconnections between neural networks and BCIs, highlighting the synergistic potential of these technologies in revolutionizing not just medical treatments but also enhancing human cognition. By delving into topics ranging from AI-powered neuroprosthetics to advanced diagnostic tools, the book advances the understanding of how these systems can mimic, decode, and augment human brain activity, driving innovations in neurological care, rehabilitation, and assistive technology.

2. Catalyzing Multidisciplinary Innovation

The book emphasizes the importance of interdisciplinary collaboration, demonstrating how neuroscience, artificial intelligence, biomedical engineering, and cognitive science are converging to push the boundaries of human-machine interaction. It highlights the value of combining biological insights with computational models to develop systems that can learn from, adapt to, and ultimately improve human cognitive abilities, whether in healthcare, education, or the workplace.

3. Bridging Theory with Practice

A core strength of this book lies in its practical applications, offering readers not only a deep dive into the underlying algorithms and theoretical models of neural networks and BCIs but also real-world use cases in neurorehabilitation, assistive devices for people with disabilities, and brain-controlled prosthetics. These discussions help bridge the gap between the laboratory and the clinic, making the theoretical research directly relevant to current and future applications that can improve lives.

4. Addressing Ethical and Societal Implications

Importantly, the book also addresses the ethical, societal, and psychological implications of these technologies. As BCIs and neural network applications become more integrated into healthcare and everyday life, the book raises critical questions around privacy, consent, and the potential for cognitive enhancement, ensuring that these innovations are developed responsibly and inclusively.

5. Shaping the Future of Human-Tech Interaction

The book's emphasis on BCI and AI convergence lays the groundwork for a future where brain-machine fusion is commonplace, with potential applications in areas such as neurofinance, mental health, neuroprosthetics, and cognitive augmentation. By providing a roadmap for researchers, engineers, and clinicians, it helps to shape the future of human-tech interaction, ensuring that emerging technologies continue to augment human potential and improve quality of life.

6. Conclusion: A Transformative Contribution

Ultimately, *Neural Network Technologies and Brain-Computer Interfaces: Innovations and Applications* contributes significantly to the advancement of both research and practical solutions in neurotechnology. It serves as a vital resource for academics, professionals, and innovators in the field, offering both a thorough exploration of cutting-edge technologies and a clear vision of their transformative potential. The book's interdisciplinary approach and practical insights position it as a cornerstone reference that will inspire ongoing innovation, challenge current paradigms, and drive future breakthroughs in brain-computer interfaces and neural network technologies.

Chapter 1
Analysis of the Human Brain Using Basic Computer Interface (BCI) System

Harishchander Anandaram
https://orcid.org/0000-0003-2993-5304
Amrita School of Artificial Intelligence, India

Kawerinder Singh Sidhu
https://orcid.org/0000-0002-8910-6279
UIM, India

R. Kiruthigha
https://orcid.org/0000-0002-4355-2548
SRM Institute of Science and Technology, India

Nitish Pathak
Bhagwan Parshuram Institute of Technology, India

Abhijeet Sudhakar
https://orcid.org/0009-0008-7922-9997
BlueRidge Technology Solutions, USA

Neelam Sharma
Maharaja Agrasen Institute of Technology, New Delhi, India

Kapil Joshi
https://orcid.org/0000-0003-1097-8347
Uttaranchal University, India

ABSTRACT

The human brain, human-generated ideas and communications are non-stationary and nonlinear. Therefore, developing a technology that can draw greater insight from human thought processes is the challenging part. Applications of BCI will work better as these greater understandings are acquired. Brain-computer interfaces are gadgets that let individuals interact with computers solely by using their brain activity, which is frequently picked up by electroencephalography (EEG). Cutting edge in terms of brain-computer interfaces (BCI). An overview of BCIs is given at

DOI: 10.4018/979-8-3693-9445-8.ch001

the outset, along with information on their most popular platforms and key working principles. The many parts of a BCI arrangement, including software, hardware, and digital signal processing methods, are then examined in the study.

1. INTRODUCTION

For the purpose to accomplish a desired action, a computer-based device called a Brain-Computer Interface (BCI) collects, interprets, and transforms brain signals into commands that are delivered to an output device. An active BCI requires the user to actively alter their brain signals in order to interface with an external device. A BCI system consists of six key parts that work together seamlessly. First, the patient's brain's electrophysiological signals—which are the fundamental reactions of the neurons—are recorded using electrodes. Users can operate machines and equipment with just their thoughts thanks to brain-computer interfaces (BCIs), which measure activity in the brain and convert it into command for the machine or other device Neurogadgets are being used more and more for entertainment, from moving robotic balls and spiders to more useful uses (Ramadan, Refat, Elshahed, & Ali, 2015). What is more significant, though, is that neurogadgets known as brain computer interfaces (BCIs) are also being created to help individuals that have disabilities, such those who have limb paralysis, without requiring any physical effort. The GP does, in fact, require the human brain to "think," an executor (in this case, a computer) to "make it happen," and an easy connection that connects the human brain with the computer in order to "without any physical effort."

To enable communication between the user's Smartphone and external devices, the Translation Algorithm converts these translated features into control signals that can be used. The Control Interface then interprets these control signals and transforms them into commands and actions for specific output devices (Arvaneh & Tanaka, 2018). Last but not least, the feedback system serves as a guide, assisting the individual in adjusting their cognitive processes for better control and interaction. There are three types of BCIs: non-invasive, semi-invasive, and invasive. In non-invasive neural-computer interfaces (BCIs), sensors are attached to the scalp to measure magnetic field (MEG) or electrical potentials produced by the brain (EEG). A craniotomy is required for semi-invasive BCIs in order to implant electrodes for electrocorticography (ECoG) on the exposed surface of the brain.

Research on feedback Basic of Brain mechanisms is still in progress to develop technologies that can translate external orders into signals from electricity that travel through the nervous system. In one example, by electrically activating their leg muscles, people with spinal cord injuries might be able to restore their ability to move around with a device like a tablet. For Computer machines to comprehend what the human mind is attempting to convey, brain activity needs to be tracked. They can identify thoughts and spot complex patterns in EEG or other brain recordings due to their capacity for process information hierarchically. This dynamic synergy enhances brain signal processing and optimists the potential of BCI Convolution generated brains (CGB) and brain-computer interfaces (BCI) together usher in an additional phase of constant human-technology connection in this quickly changing world (Graimann, Allison, & Pfurtscheller, 2010).

This discovery broadens our knowledge of how the brain works and ushers in a new era of seamless mind Computer machine contact for neuron technology. Better communication and a greater comprehension of how the brain functions are anticipated as a result of these developments in basic brain of computer interface technology. BCI technology in order which has many potential applications in assistive technology and medical rehabilitation is based on the complex interaction underlying electrodes, signals, as well as translation algorithms. Head-computer layouts, or BCIs, have emerged as a game-changing technology with significant implications for medical rehabilitation, the interaction between humans and computers, and our understanding of the brain (Machado et al., 2010). By providing more reliable and approachable solutions for problems like motor imagery classification and cognitive state recognition, CNNs are transforming BCI.

The length of the computer training depends on the volume of customers that BCI receives. Although this kind of technology is still in its early stages, new advancements have shown that they have a lot of potential for applications ranging from medical rehabilitation to gaming and entertainment. Researchers have developed brain-computer interfaces (BCIs) that can recognize cognitive states, classify motor imagery, and even use CNN's to translate ideas into voice. The integration of CNN's with BCIs has ushered in a new era of seamless human-technology connection. By eliminating the difference between individuals and machines development through more reliable and user-friendly solutions, BCIs have the potential to completely transform the way we connect with the outside world (Nicolas-Alonso & Gomez-Gil, 2012). As BCI technology advances, it has the capacity to drastically change our understanding of the computer brain, offering new insights into its intricate workings and creating new avenues for treating neurological disorders. With the potential to enhance human capabilities, recover lost activities, and alter how we engage with technology, BCIs promise a bright future. In addition to discussing the many platforms, techniques, and gadgets that are currently being researched, this session will the same provide an overview of the most recent advancements in BCI technologies and machine work (Hramov, Maksimenko, & Pisarchik, 2021).

2. Services: The interface typically functions as follows: electrodes detect brain impulses, which are then processed by a BCI computer to remove noise or abnormalities from external and device-specific sources. The resulting signal is then examined to determine the appropriate command; because of their high processing ability and adaptability, neural network models are frequently used for this purpose. After that, the brain firings' language is passed to a computer translator, which decodes it into computer language using specialised algorithms. Although highly specialised systems may do this work themselves, the detected command is typically transmitted to an outside tool for additional processing in accordance with a pre-programmed algorithm (Vallabhaneni, Wang, & He, 2005).

Figure 1. BCI Principal Operation

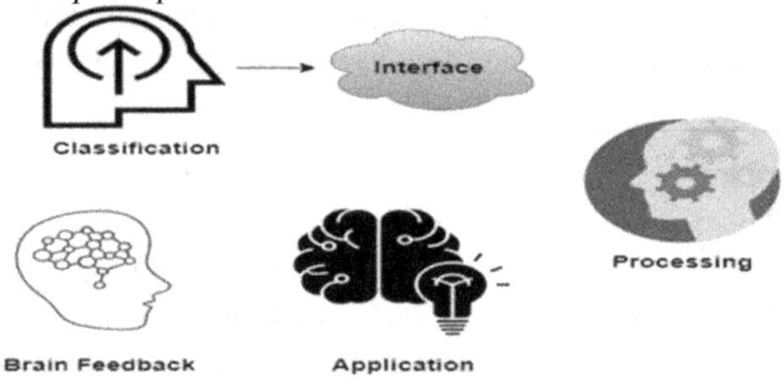

The translated data in (Figure 1) then delivered to a different computer, which instructs the machine on what to perform. Invasive devices generate the best-quality messages of BCI devices since they are located in the grey matter, but they are also more likely to cause scar tissue accumulation. In the end, the controlled device interprets the received instruction according to its unique properties (He, Yuan, Meng, & Gao, 2020). The following outlines the fundamental ideas that underpin BCI operation.

3. REVIEW OF THE RESEARCH

1. The study provides a comprehensive examination of shallow learning methods utilized in EEG-based Brain-Computer Interaction (BCIs), with an emphasis on methods such as CNNs and RNNs. It highlights the potential benefits of shallow learning, particularly its capacity to autonomously extract intricate EEG patterns, which could raise the precision and reliability of BCI systems. By highlighting issues with information scarcity, model comprehensibility, and computing complexity, the work emphasizes the necessity for more computer-basic research in these areas.

2. In order to provide improved machine and software motor Brain-Computer Interfaces (BCIs), the study uses Convolution Neural Networks (CNNs) to handle Stereo-Electroencephalography (SEEG) data. CNNs are highly skilled at spotting complex geographical patterns, which could improve the accuracy and dependability of BCIs in motor control applications. Despite the encouraging developments, there are still issues with the clarity of the model, the requirement for high-quality SEEG data processing, and the amount of computing horsepower required for training and implementation (Wolpaw, Birbaumer, McFarland, Pfurtscheller, & Vaughan, 2002).
3. In order to categorise EEG signals associated with motor activity follow in Brain-Computer Interaction (BCI) systems, the study suggests a shallow learning technique by same coordination. Shallow learning algorithms have the potential to improve BCI classification accuracy by automatically recognizing complicated trends in EEG data. Although hopeful, problems with data quality, model interprets ability, and computer resources point to areas that need further study.
4. With a focus on tasks incorporating motor imagery computer picture, the study investigates shallow-learning techniques in EEG-based brain-computer interfaces. Shallow learning's potential to autonomously learn intricate EEG patterns could improve the accuracy and adaptability of BCI (Hramov, Maksimenko, & Pisarchik, 2021). Problems with data integrity management, interpret ability, and possible computational requirements are some of the constraints, pointing to areas that require further study in relation to the development of BCIs.

Depending on the degree of evasiveness, neural interfaces are categorized as invasive, non-invasive, or semi-invasive computer strictures. The direct implantation of intracranial micro sensors (IM) into a person's brain is the most successful but dangerous neural interface technique. Non-invasive neural interfaces assess neuronal activity from the outside of the head using imaging methods including electroencephalography (EG), or functional magnetic resonance imaging (FMRI) without requiring electrode insertion. Electrodes for semi-invasive BCIs, including electrocorticography (ECoG), are situated beneath the temporal bone on the outermost part of the brain (Figure 2) (Bonci, Fiori, Higashi, Tanaka, & Verdini, 2021).

Figure 2. BCI Sensor mounting EEG

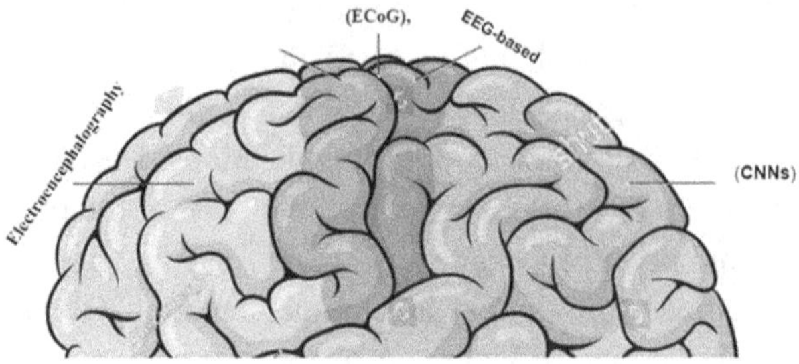

Instead of resting inside the grey matter of the brain, partially invasive brain-computer interfaces (BCI) are placed inside the skull. They have a reduced danger of creating scar tissue in the brain's cortex than fully-invasive BCIs and generate higher resolution vibrations than non-invasive BCIs, where the cranium's bone tissue deflects and deforms signals. Furthermore, a number of disorders are being prevented and treated with these devices. However, the game industry has been adopting non-invasive neural interfaces more and more. Electrodes are inserted in a thin plastic pad that is positioned above the cortex, above the durra mater, in electrocorticography (ECoG), which measures electrical activity of the brain collected from beneath the skull in a manner similar to non-invasive electroencephalography ECoG is a highly attractive intermediate BCI the medium because it has better signal-to-noise ratio, a wider frequency range, higher spatial resolution, and less training involved than scalp-recorded EEG (Hassanien & Azar, 2015).

It also has less clinical risk, less technical difficulty, and likely better long-term stability as opposed to intracranial single-neuron audio file. There is promise for real-world use for individuals with motor disabilities based on the mentioned profile and latest proof of the high level of management with little training needs.

4. GATHERING OF DATA AND PREPARATION

Highly attractive intermediate BCI the medium because it has better signal-to-noise ratio, a wider frequency range, higher spatial resolution, and less training involved than scalp-recorded EEG. It also has less clinical risk, less technical difficulty, and likely better long-term stability as opposed to intracranial single-neuron audio file. There is promise for real-world use for individuals with motor disabilities based on the mentioned profile and latest proof of the high level of management with little training needs (Schalk & Mellinger, 2010). To provide inputs for the deeper learning models, pertinent features were taken from the previously processed EEG data (Yadollahpour & Bagdeli, 2015). We chose a mix of frequency-domain and time-domain data, including particular features like entropy measurements and spectral power. Based on their proven ability to capture relevant data for the required BCI tasks, these features were selected.

5. BCI THAT IS GENTLE

An electroencephalograph (EEG), a device with electrodes attached the easiest and least invasive method is applied to the scalp. The electrodes can read brain computer impulses. No matter where these electrodes are placed, the basic mechanism is the same: Even the slightest changes in neuronal voltage are picked up by the electrodes. The signal is then amplified and filtered. In modern BCI systems, a computer program interprets the signals and displays them on an endless piece of paper using pens that automatically write out the patterns. Because of their individual drawbacks, the skull is more acceptable than the other forms, despite the fact is that it blocks are large portion of the electrical signals and distorts when does get through (Shih, Krusienski, & Wolpaw, 2012). EEG sensors use electrodes applied to the scalp or other body areas to measure computer electrical some of activity generated by neurones follow brain. Clinicians can identify changes linked to various mental states, including sleep or alertness, by tracking this activity over time. Additionally, EEG data can reveal some forms of aberrant brain activity, such as seizures or signs of damage from a stroke.

Figure 3. BCI with CNN

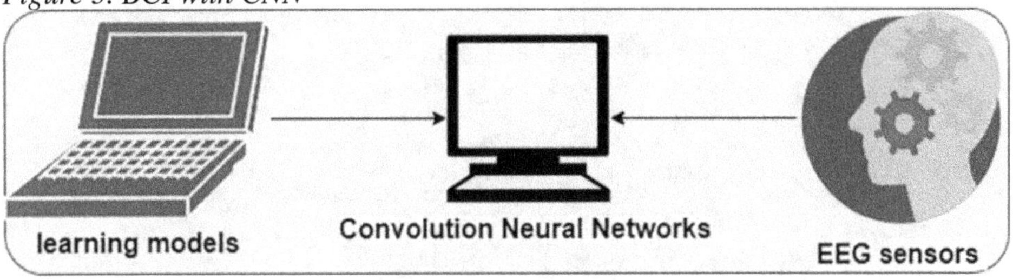

Convolution Neural Networks (CNNs): CNNs have been utilised to successfully detect spatial patterns in EEG data as part of shallow learning models.

They work effectively for tasks involving grid-like data and are perfect for processing multi-channel in computer records EEG recordings. Each layer that comprised the CNN was dedicated to detecting a unique trait (Fouad, Amin, El-Bendary, & Hassanien, 2015). These layers consist of the layers of convolution for pattern recognition, pooling layers for downwards sampling, and fully interconnected layers for prediction. By applying activation functions like Relearning to introduce non-linearity, the network became capable to learn complex relationships. Short-term persistent memory Networks (LSTMs) were implemented to find correlations between timings in the EEG data over time. They excel at replicating sequential patterns and are particularly useful for tasks that include time-varying EEG signals.

The LSTM network by internet was composed of memory cells that could store and interpret data over time steps. It features data, disregard that and exit gates to control the information flow. These gates allow the network to selectively remember or forget information. LSTMs are perfect for tasks like processing EEG data, where understanding the setting across time is crucial (Mühl, Allison, Nijholt, & Chanel, 2014).

6. CERTIFICATION AND LEARNING

Training Procedure: We used the Adam algorithm optimizer with an intensity of adaptation to facilitate efficient model basic computer training. The number of repetitions was set to strike a balance between computing performance and convergence in the model. Validation Sets the dataset was used to construct both training and validation sets. In order to keep an eye on the model's outcome during training and avoid over fitting, a validation set was essential.

Figure 4. Learning about frequently

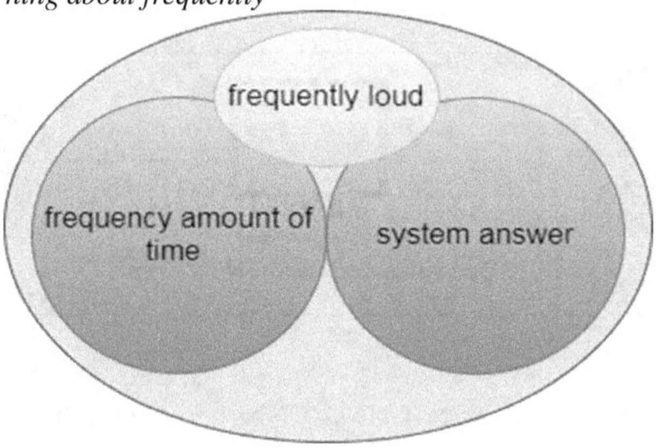

An electrode array set up in accordance with the 22-42 international standards and digitisation equipment make up the EEG acquisition subsystem. Muscular and ocular motions might cause artefacts in the recorded signals, which are frequently loud. The pre-processing unit of the EEG signal analysing subsystem is in charge of detecting artefacts, while its feature extraction or recognition unit decides the instruction that the user sends to the BCI (Abdulkader, Atia, & Mostafa, 2015). The output subsystem then receives this command and produces a "system answer," which serves as feedback to the user, allowing him to adjust his mental processes and create the EEG patterns that enable the BCI to carry out his intentions. Our BCI's basic scheduling is shown in Figureure 5 The EEG trial frequency is the amount of time the BCI must analyse the EEG in order to produce an answer, while the BCI period is a typical span of time for both consecutive answers. Every EEG trial is assumed to produce a system response.

It has been shown that regression analysis is reasonably easy to apply, but correct findings need adherence to certain assumptions. True brain function and additional movement in the electroencephalographic signal are uncorrelated; (1) the native EEG convey is a mixture of genuine neurological activity and additional activity (an artefact); and (2) materials should not contain features related to brain activity, as this could cause them to be lost through artefact removal. In order to detect artefacts, regression algorithms require external reference channels, such as EOG or ECG. However, it might be difficult to determine which reference signal is best for graphic and other non-living artefacts, which limits the application of regression techniques for artefact removal (Wolpaw & Wolpaw, 2012).

Although longitudinal auto-tuning for EEG movement has been shown to be a successful technique for identifying artefacts in evoked brain activity, background activity is not taken into consideration. Although regression techniques can detect oculographic artefacts, they are not very effective in minimising bilateral contaminated between EEG and EOG. Because the high-frequency substance of the recorded EOG usually refers to brain activity, which can be cleaned out to drastically lessen bidirectional pollution visuals, low-pass filtering before utilising Bayesian adaptive econometric synapses has been described as a solution to this problem. EEG readings without going to a clinic. In order to wirelessly send their data to computers for analysis, these devices usually use Bluetooth connectivity (Katona & Kovari, 2016). Better identification and analysis methods are made possible by recent developments in signal processing algorithms. For instance, it is now feasible to identify minute alterations in a short amount of time, which enhances diagnostic skills. Furthermore, certain algorithms can recognise unique characteristics in each person's records, enabling individualised treatment strategies. Lastly, methods of artificial intelligence are being investigated that may help automate some steps in raw data processing, leading to quicker diagnostic turnaround times.

Because of its affordability and portability, EEG is the most used sensing technique for non-invasive BCIs. Through direct brain contact, invasive methods offer higher signal quality (Agrawal, 2019). Theoretically, this could allow users to have more control over gadgets that are BCI enabled. Actuality, the number and positioning of electrodes have an impact on the general standard of an EEG output. More electrodes can enhance spatial resolution and enable more thorough analysis. Furthermore, more electrodes enable improved noise reduction methods like interpolation and averaging. Because adding more electrodes also increases the cost and quantity of recording equipment, it is important that one weighs the trade-offs between the benefits of improved accuracy versus the strain of additional hardware in computer system requirements when choosing the optimal sensor architecture (Müller-Putz & Rupp, 2021).

In addition to the overall amount, location plays a critical role in determining the quality of the data during EEG recordings in same BCIs. Because different placements provide varying degrees of insight into different brain regions, it is important to carefully consider the places to use for data collection. Furthermore, individual differences in skull thickness may need deviations from typical placements due to potential differences in susceptibility at various spots relative to one another (Papanastasiou, Drigas, Skianis, & Lytras, 2020).

An EEG system's efficiency is largely dependent on the choice, quantity, and positioning of its sensors. These elements need to be carefully taken into account in order to maximize outcomes and minimise hardware requirement expenses. As technology continues to improve at a rapid pace, sensor setup optimization will become more feasible, eventually enhancing BCI performance possibilities. The use about invasive BCI is currently restricted by a number of technological difficulties and surgical concerns.

7. BCI STRUCTURE'S COMPONENTS

A brain-computer interface (BCI) is designed to identify and measure aspects of neurones that reveal the user's goals, then convert these aspects in immediate response into commands for the device that carry out the user's intent (Brunner, Bianchi, Guger, Cincotti, & Schalk, 2011). A BCI system uses four successive components20—signal acquisition, feature extraction, feature translation, and device output—to accomplish this. An operational protocol governs these four elements, specifying the start and end times of operation, the specifics of handling signals, the type of commands that can be sent to the device, and the performance monitoring. An efficient operating protocol makes it possible for a BCI system to be adaptable and meet the unique requirements of every user.

8. BETTER NEUROSTIMULATION WITH BCIS

Because BCI technology can transport data obtained from the human brain to an independent computer for examination, it has a possibility to increase the effectiveness of current therapies. For instance, the brain receives an electric current from implanted shallow-brain stimulators (Box 1) (Dokare & Kant, 2013). The NHS offers them to treat movement disorders, such as Parkinson's disease that is resistant to medication.

Nevertheless, the current is administered in a certain way that is unaffected by the patient's symptoms. Negative side effects including impaired speech or balance may result from this. Comparison of BCI Platforms.

9. WE MAY DERIVE THE FOLLOWING DEDUCTIONS BASED ON THE ANALYSIS THAT HAS BEEN SUPPLIED

EEG is a non-surgical technique that uses electrodes applied to the scalp to monitor the electricity in the brain. Because of its high temporal precision, it can identify variations in brain activity in milliseconds. EEG is ideally suited for usage in BCI systems since it is portable and reasonably priced (Liao et al., 2012). The following are the main problems with EEG:

- EEG signals are extremely susceptible to noise and artefacts, hence it's critical to make sure that the signal strength is at its best for BCI applications;
- The capacity to precisely extract valuable information from unprocessed data is known as feature extraction. In BCI study results, EEG data is crucial since it establishes how well the system will identify human orders and intentions;
- Interpretation accuracy: The ability of the system to identify user commands or intentions is determined by the construction of effective procedures for classification EEG data into various categories (right hand movement). This is a crucial topic in BCI research (Kübler & Neumann, 2005).
- User interface design: creating layouts that are simple to use and intuitive is crucial for EEG-based BCIs since it affects how readily users can communicate with the system;
- Adaptability: a key area of study for building reliable BCI systems is the development of algorithms that can adjust to the brain activity of specific users and identify minute variations in EEG waveforms;
- System reliability: Because brain activity is dynamic and varies between users and sessions, it can be difficult to ensure dependable functionality of a brain-computer interaction (BCI) system over extended period's duration

with little calibration or setup needs. This is particularly true for EEG-based BCIs (Panoulas, Hadjileontiadis, & Panas, 2010).

By giving additional information about how the brain works, BCIs can be utilised to improve the responsiveness of treatments. They are able to detect aberrant brain activity and use it to determine when symptoms first appear. Then, a patient's treatment might be customised, eliminating the need for professionals to frequently re-program equipment. One responsive implantable the Neuron Pace, what has been certified for usage in clinical trials in the US, is a device that stops epileptic seizures by stimulating the basic computer brain (Scherer et al., 2013). For mental health issues like depression, stimulation treatments can get more sensitive to the patient's emotional state by using BCIs.

- **Amusement**: Businesses are creating non-intrusive BCI gaming headgear. These would enable players to manipulate game actions with their thoughts. Only a few businesses presently provide consumers with BCI-enabled games products, like NeuroSky and Mandalay (Wu, Lu, Hu, & Zeng, 2023).
- **Instruction**: EEG headsets are also sold by firms like NeuroSky, Mandalay, Emotive, and others to improve cognitive skills like focus. These systems function through brain stimulation, which is the practice of being visually aware of how your brain functions on a computer screen in order to learn how to manage it (Min, Marzelli, & Yoo, 2010). This could help individuals with chronic pain control their symptoms and those with attention deficit problems focus.
- **Promotion**: BCI technology is now being developed for neuron marketing, or market research studies. This entails gathering information from neurones in order to understand how consumers make decisions. For instance, BCIs could be used to gauge how a customer responds to a marketing effort (Balderas, Ponce, Lopez-Bernal, & Molina, 2021).
- **Armed Forces:** The MoD is looking at how BCIs can improve decision-making and the senses (hearing, sight, and touch), among other cognitive skills. A significant program of NI projects is being funded in the United States by the Defence Advanced Research Projects Agency. The Next-Generation Medical Neurotechnology is a program, for instance, seeks to create non-invasive brain-computer interfaces (BCIs) for military personnel. Cyber-defence systems and unmanned aerial vehicle control are two possible uses (Kübler, Kotchoubey, Kaiser, Wolpaw, & Birbaumer, 2001).

10. TRADITIONAL BCI NETWORK THEORIES

The most popular BCIs are based on traditional paradigms like mental imagery (MI) and steady-state retinal evoked potentials (SSVEP). The paradigm measures event-related potentials produced when a user identifies a significant stimulus among a number of stimuli displayed on a computer's display using EEG recordings through the scalp electrodes. Developed in the 1985s, it is among the earliest BCI paradigms. It recognises when a user is concentrating on a certain stimulus or event by harnessing electrical impulses from the brain.

This results in the appearance of an EEG signal, known as the wave, about milliseconds after the event. In order to create suitable applications, find suitable user groups, and closely consider the needs and preferences of individual users, the practical application of BCI technology requires interdisciplinary partnership among neurosciences, engineers, programmers of computers, psychologists, and recuperation specialists It is challenging to assess the potential for using neural signals to control computers because the field is still in its early stages of development (Silvoni et al., 2011).

The use of personal computing devices to carry out the tasks required to identify patterns in impulses from nature has advanced significantly, but the hunt for new and more beneficial signals is still ongoing. Direct neurological interaction between humans and machines may eventually develop and become widely used if the 20th century's advancements are comparable to those of the previous few decades. Perhaps, in the same way that keyboards and mice are standard on modern computers, biological signalling sensors and thought-recognition applications will one day be included with newly acquired computers (Halder et al., 2011). This waveform can be applied as feedback for BCIs and to gauge an individual's level of interest or involvement in what they have seen or experiencing.

Users usually react faster when they can identify one specific target stimulus among several stimuli displayed on a computer screen. Scalp electrodes positioned over various parts of the head can then be used to detect the waveform, an EEG signal produced by this difference in reaction time (Iroju, Ikono, Ishaya, Ojerinde, & Olaleke, 2018). It is feasible to determine which stimuli the user recognized and, consequently, deduce their goals by tracking this signal over time.

Figure 5. CNN architecture BCI

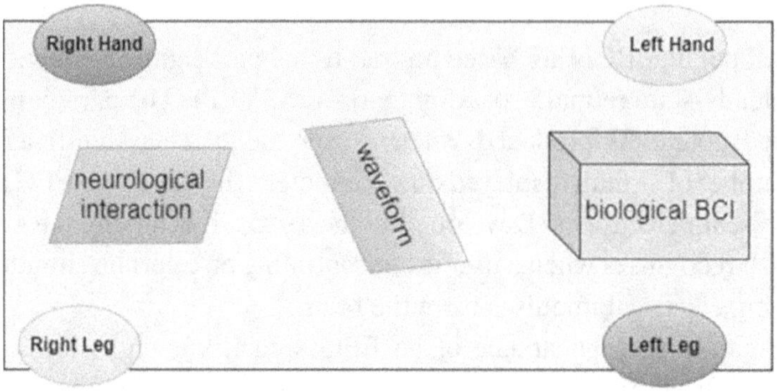

Additionally, BCI technology is being developed for other, yet unrealized, uses. Face book, for instance, is developing non-invasive brain-computer interfaces (BCIs) to facilitate hands-free smart phone communication more quickly than typing. In the meantime, Neural Ink recently revealed that it had created an implanted BCI lab prototype. The end goal of this BCI is to combine artificial and human intelligence, with the first focus being on medical applications. According to estimates, invasive BCIs might be used in medicine in a few years. However, the duration is probably going to be far longer for non-medical use.

11. BCI PROTECTION

Invasive forms of BCI are the main focus of safety concerns. Due to their placement in the cortex, they may result in localised harm to blood arteries and nerve cells as well as a higher risk of infection. The Nuffield Council came to the conclusion that because of the brain's critical role in mind-body interaction and self-perception, the dangers and benefits of remedies must be carefully considered.

For medical uses, it could be simpler to balance the risks of a gadget against its advantages. Benefits for non-medical uses may be more elusive or subjective, but the rewards can be determined more easily quantified by the treatment's efficacy. The Council also came to the conclusion that there is little chance of major health hazards from non-invasive BCI systems, including EEG gaming gadgets. However, the Council noted that prolonged usage poses a special risk to youngsters because of the brain's innate plasticity, which might result in alterations in the brain.

12. STORM SWARM OPTIMIZATION AND GENETIC TECHNIQUES IN BCI

Optimization techniques computer system such as particle cluster stabilization (PSO) and genetic algorithms (GA) can be used to fine-tune or optimise the settings of a BCI system. GA is an evolutionary system that uses concepts like crossover, mutations, and selection to find the optimum solutions. Based on their personal best possible location and the wider optimal positioning found so far in PSO, an algorithm influenced by social behaviour, particles travel around the search space at velocities defined by these factors (Neuper, Müller, Kübler, Birbaumer, & Pfurtscheller, 2003). Their efficiency is an additional benefit. Compared to other optimisation methods like gradient descending or simulated annealing, gas can be deployed rapidly and easily, which makes it perfect for real-time applications where speed is crucial, such online control systems. Finally, GAs is useful across a wide range of BCI study topics because of their versatility, which allows them to be used to solve a wide variety of issues, including as the use of classification, regression etc.

13. CONCLUSION

There are high hopes for BCI because it is seen as the pinnacle of HCI development. Because BCI is widely used in many different applications, mostly to help the disabled live independently, the area has been designed with that in mind. Because it allows users to manage things simply by thinking, this is incredibly helpful, especially for those who have problems moving. BCI gains a deeper understanding of human thoughts through shallow learning, which functions similarly to a smart computer assistant. However, we still need to Figureure out how to make it work extremely well and how to keep our ideas secretIn the future, this combination of BCI and deep learning will be used for more than just aiding those in need. A review was conducted of the most popular BCI signal processing methods, including ICA, wavelet modification, SVM, secret Markova models, artificial intelligence and genetic algorithms. Highlighted are the key areas of application and the basic principles of their operation. Lastly, a review was conducted of the primary application areas for BCI systems, including neurological prostheses, interaction gaming, education, and behavioural health care. BCIs could be a significant new communication and command tool for individuals with disabilities—and perhaps the broader public as well—if they have improved signal-acquisition hardware, solid clinical validation, workable dissemination models, and—above all—increased reliability.

Therefore, BCI and shallow learning together are like opening a door to a whole bunch of awesome possibilities, making our relationship with technology more fascinating and natural. To put it simply, it's like being friends with electronics in a way that improves and enriches our lives. The range of BCI applications is expanding quickly, and several studies are being carried out in this area. However, a lot of work needs to be done before these concepts become a reality. In order to create and sustain trustworthy human-machine interactions over time in a safe and efficient manner, more technological advancements must be made in tandem with a better understanding of how our brains work.

REFERENCES

Abdulkader, S. N., Atia, A., & Mostafa, M. S. M. (2015). Brain computer interfacing: Applications and challenges. *Egyptian Informatics Journal, 16*(2), 213–230. DOI: 10.1016/j.eij.2015.06.002

Agrawal, P. K. (2019). A Review on Brain Computer Interface. *International Research Journal of Mathematics. Engineering and IT, 6*(1), 1–14.

Arvaneh, M., & Tanaka, T. (2018). Brain–computer interfaces and electroencephalogram: basics and practical issues. *Signal Processing and Machine Learning for Brain--Machine Interfaces.*

Balderas, D., Ponce, P., Lopez-Bernal, D., & Molina, A. (2021). Education 4.0: Teaching the basis of motor imagery classification algorithms for brain-computer interfaces. *Future Internet, 13*(8), 202. DOI: 10.3390/fi13080202

Bonci, A., Fiori, S., Higashi, H., Tanaka, T., & Verdini, F. (2021). An introductory tutorial on brain–computer interfaces and their applications. *Electronics (Basel), 10*(5), 560. DOI: 10.3390/electronics10050560

Brunner, P., Bianchi, L., Guger, C., Cincotti, F., & Schalk, G. (2011). Current trends in hardware and software for brain–computer interfaces (BCIs). *Journal of Neural Engineering, 8*(2), 025001. DOI: 10.1088/1741-2560/8/2/025001 PMID: 21436536

Dokare, I., & Kant, N. (2013). A Study Of Brain Computer Interface System. *neurorehabilitation, 1*(3), 5.

Fouad, M. M., Amin, K. M., El-Bendary, N., & Hassanien, A. E. (2015). Brain computer interface: a review. *Brain-computer interfaces: Current trends and applications*, 3-30.

Graimann, B., Allison, B., & Pfurtscheller, G. (2010). Brain–computer interfaces: A gentle introduction. In *Brain-computer interfaces: Revolutionizing human-computer interaction* (pp. 1–27). Springer Berlin Heidelberg. DOI: 10.1007/978-3-642-02091-9

Halder, S., Agorastos, D., Veit, R., Hammer, E. M., Lee, S., Varkuti, B., Bogdan, M., Rosenstiel, W., Birbaumer, N., & Kübler, A. (2011). Neural mechanisms of brain–computer interface control. *NeuroImage, 55*(4), 1779–1790. DOI: 10.1016/j.neuroimage.2011.01.021 PMID: 21256234

Hassanien, A. E., & Azar, A. A. (2015). *Brain-computer interfaces*. Springer. DOI: 10.1007/978-3-319-10978-7

He, B., Yuan, H., Meng, J., & Gao, S. (2020). Brain–computer interfaces. *Neural engineering*, 131-183.

Hramov, A. E., Maksimenko, V. A., & Pisarchik, A. N. (2021). Physical principles of brain–computer interfaces and their applications for rehabilitation, robotics and control of human brain states. *Physics Reports, 918*, 1–133. DOI: 10.1016/j.physrep.2021.03.002

Hramov, A. E., Maksimenko, V. A., & Pisarchik, A. N. (2021). Physical principles of brain–computer interfaces and their applications for rehabilitation, robotics and control of human brain states. *Physics Reports, 918*, 1–133. DOI: 10.1016/j.physrep.2021.03.002

[] Iroju, O., Ikono, R., Ishaya, G., Ojerinde, O. A., & Olaleke, J. (2018). Prospects and problems of brain computer interface in healthcare.

Katona, J., & Kovari, A. (2016). A brain–computer interface project applied in computer engineering. *IEEE Transactions on Education, 59*(4), 319–326. DOI: 10.1109/TE.2016.2558163

Kübler, A., Kotchoubey, B., Kaiser, J., Wolpaw, J. R., & Birbaumer, N. (2001). Brain–computer communication: Unlocking the locked in. *Psychological Bulletin, 127*(3), 358–375. DOI: 10.1037/0033-2909.127.3.358 PMID: 11393301

Kübler, A., & Neumann, N. (2005). Brain-computer interfaces—The key for the conscious brain locked into a paralyzed body. *Progress in Brain Research, 150*, 513–525. DOI: 10.1016/S0079-6123(05)50035-9 PMID: 16186045

Liao, L. D., Lin, C. T., McDowell, K., Wickenden, A. E., Gramann, K., Jung, T. P., . . . Chang, J. Y. (2012). Biosensor technologies for augmented brain–computer interfaces in the next decades. *Proceedings of the IEEE, 100*(Special Centennial Issue), 1553-1566.

Machado, S., Araújo, F., Paes, F., Velasques, B., Cunha, M., Budde, H., Basile, L. F., Anghinah, R., Arias-Carrión, O., Cagy, M., Piedade, R., de Graaf, T. A., Sack, A. T., & Ribeiro, P. (2010). EEG-based brain-computer interfaces: An overview of basic concepts and clinical applications in neurorehabilitation. *Reviews in the Neurosciences, 21*(6), 451–468. DOI: 10.1515/REVNEURO.2010.21.6.451 PMID: 21438193

Min, B. K., Marzelli, M. J., & Yoo, S. S. (2010). Neuroimaging-based approaches in the brain–computer interface. *Trends in Biotechnology, 28*(11), 552–560. DOI: 10.1016/j.tibtech.2010.08.002 PMID: 20810180

Mühl, C., Allison, B., Nijholt, A., & Chanel, G. (2014). A survey of affective brain computer interfaces: Principles, state-of-the-art, and challenges. *Brain Computer Interfaces (Abingdon, England), 1*(2), 66–84. DOI: 10.1080/2326263X.2014.912881

Müller-Putz, G., & Rupp, R. (2021). *Neuroprosthetics and brain-computer interfaces in spinal cord injury*. Springer International Publishing. DOI: 10.1007/978-3-030-68545-4

Neuper, C., Müller, G. R., Kübler, A., Birbaumer, N., & Pfurtscheller, G. (2003). Clinical application of an EEG-based brain–computer interface: A case study in a patient with severe motor impairment. *Clinical Neurophysiology, 114*(3), 399–409. DOI: 10.1016/S1388-2457(02)00387-5 PMID: 12705420

Nicolas-Alonso, L. F., & Gomez-Gil, J. (2012). Brain computer interfaces, a review. *sensors, 12*(2), 1211-1279.

Panoulas, K. J., Hadjileontiadis, L. J., & Panas, S. M. (2010). Brain-computer interface (BCI): Types, processing perspectives and applications. In *Multimedia Services in Intelligent Environments: Integrated Systems* (pp. 299–321). Springer Berlin Heidelberg. DOI: 10.1007/978-3-642-13396-1_14

Papanastasiou, G., Drigas, A., Skianis, C., & Lytras, M. (2020). Brain computer interface based applications for training and rehabilitation of students with neurodevelopmental disorders. A literature review. *Heliyon, 6*(9), e04250. DOI: 10.1016/j.heliyon.2020.e04250 PMID: 32954024

Ramadan, R. A., Refat, S., Elshahed, M. A., & Ali, R. A. (2015). Basics of brain computer interface. *Brain-Computer Interfaces: Current Trends and Applications*, 31-50.

Schalk, G., & Mellinger, J. (2010). *A practical guide to brain–computer interfacing with BCI2000: General-purpose software for brain-computer interface research, data acquisition, stimulus presentation, and brain monitoring*. Springer Science & Business Media. DOI: 10.1007/978-1-84996-092-2

Scherer, R., Faller, J., Balderas, D., Friedrich, E. V., Pröll, M., Allison, B., & Müller-Putz, G. (2013). Brain–computer interfacing: More than the sum of its parts. *Soft Computing, 17*(2), 317–331. DOI: 10.1007/s00500-012-0895-4

Shih, J. J., Krusienski, D. J., & Wolpaw, J. R. (2012, March). Brain-computer interfaces in medicine. [). Elsevier.]. *Mayo Clinic Proceedings, 87*(3), 268–279. DOI: 10.1016/j.mayocp.2011.12.008 PMID: 22325364

Silvoni, S., Ramos-Murguialday, A., Cavinato, M., Volpato, C., Cisotto, G., Turolla, A., Piccione, F., & Birbaumer, N. (2011). Brain-computer interface in stroke: A review of progress. *Clinical EEG and Neuroscience*, *42*(4), 245–252. DOI: 10.1177/155005941104200410 PMID: 22208122

Vallabhaneni, A., Wang, T., & He, B. (2005). Brain—computer interface. In *Neural engineering* (pp. 85-121). Boston, MA: springer Us.

Wolpaw, J. R., Birbaumer, N., McFarland, D. J., Pfurtscheller, G., & Vaughan, T. M. (2002). Brain–computer interfaces for communication and control. *Clinical Neurophysiology*, *113*(6), 767–791. DOI: 10.1016/S1388-2457(02)00057-3 PMID: 12048038

Wolpaw, J. R., & Wolpaw, E. W. (2012). Brain-computer interfaces: something new under the sun. *Brain-computer interfaces: principles and practice, 14*, 3-12.

Wu, D., Lu, B. L., Hu, B., & Zeng, Z. (2023). Affective brain–computer interfaces (abcis): A tutorial. *Proceedings of the IEEE, 111*(10), 1314–1332. DOI: 10.1109/JPROC.2023.3277471

Yadollahpour, A., & BAGdELI, A. (2015). Brain computer interface: Principles, recent advances and clinical challenges. *Oriental Journal of Computer Science and Technology*, *7*(3), 425–442.

Chapter 2
BCI-Based Assistive Technology With Integrated Machine Learning Empowering People With Disabilities for Improved Independence and Quality of Life

Usharani Bhimavarapu
https://orcid.org/0000-0002-0246-1420
Department of Computer Science and Engineering, Koneru Lakshmaiah Education Foundation, Vaddeswaram, India

ABSTRACT

Brain-Computer Interface (BCI)-based assistive technology offers promising solutions to enhance the independence and quality of life for people with disabilities. By allowing direct communication between the brain and external devices, BCIs can empower individuals to perform tasks that would otherwise be challenging or impossible. This study proposes a novel BCI system that integrates advanced machine learning algorithms, including XGBoost, Naive Bayes, and Random Forest, to improve the accuracy and reliability of motor imagery (MI)-based control. By analyzing EEG signals, the proposed system offers a more efficient and responsive interface for assistive devices, facilitating better control and interaction for users

DOI: 10.4018/979-8-3693-9445-8.ch002

with motor impairments. Through careful feature extraction, feature selection, and classification, the system is designed to adapt to the individual's unique brain activity patterns, offering enhanced user-specific performance.

INTRODUCTION

Brain-Computer Interface (BCI)-based technology is a breakthrough technology in optimizing independence in individuals with disabilities and the optimization of quality of life. BCIs create a brain-to-other-device interface in which the user is empowered to control external devices without physical movement on the part of the activity of the brain(Anitha et al.,2019, Attallah et al.,2019). It is of the most valuable usage to motor-disabled people since they can operate aiding tools such as robotic arms, wheelchairs, and speech-generating devices by just thinking about them[(Attallah et al.,2019, Vourvopoulos et al.,2019, Nicholson et al.,2019). As technology advances, BCIs are capable of making the life of disabled individuals such that they can perform activities which were beyond their possibilities before.

Most thrilling potential application of BCI technology, possibly, is the ability to decode motor imagery (MI) patterns, brain activity felt when a subject is visualizing or fantasizing movement but not moving themselves(Padfield et al.,2019, Li et al.,2019, Singh et al.,2019). Motor imagery patterns can be identified with BCIs by analyzing EEG signals and translated as commands for assistive devices(Jusas et L.,2019, Baraka ET al.,2019., Abougharbia et al.,2019, Subasi et al.,2019). It gives individuals with physical disabilities a more natural means of interacting with the world and getting things done. Machine learning algorithms such as XGBoost, Naive Bayes, and Random Forest are the core of the operation of BCIs in that they render the systems more capable to detect EEG signals accurately, thereby making the technology faster and more responsive to the user's needs.

The subsequent BCI system with all these machine learning models will further enhance assistive technology as much as its interface. As a more specific and more efficient interface, it will serve the users better and be less time-consuming with better responsiveness due to feature extraction, feature selection, and classification. This evolution brings with it a better means for disabled users to better enjoy assistive technology and become more independent and self-reliant in daily life. Lastly, the application of BCI technology as an enhancement to assistive technology has the potential to significantly enhance the quality of life of the disabled user by giving them more control over the surroundings and better quality of life and empowerment.

Contemporary non-machine learning supported technology models with BCI do not utilize indirect signal handling processes and stimulus-response systems but instead rely on advanced learning algorithms(Ragab et al.,2019, Serrao et al.,2018,

Höller et al.,2018, Jeanmonod et al.,2018). A case in point is the P300-based BCIs, which detect the P300 wave, a positive EEG deflection which takes place every time a subject is focused on a target stimulus presented on a visual array. The model is also extensively used in communication devices under which individuals with severe motor disabilities can select letters or symbols by focusing on them. However, one of the largest drawback of P300-based systems is that they provide a low information transfer rate. Such systems will have to make users focus on the stimuli for long periods of time, which will be slow and time-consuming. The system's performance is also interrupted by extraneous noise such as muscle artifacts or eye movements that can contaminate the detection of the P300 signal effectively(Pattnaik et al.,2018, Utsumi et al.,2018, Daly et al.,2018, Attallah et al.,2020).

One of the most widely used models in this regard is Steady-State Visual Evoked Potential (SSVEP)-based BCIs, which are constructed on the foundation of brain responses to flashes of visual stimuli at specific frequencies. Individuals can operate the devices by looking at such flashing stimuli, and the system deciphers the related brain signals. While SSVEP-based systems are generally faster and more informative than P300 systems, there are also some disadvantages involved. One of the most significant disadvantages is that the users have to maintain visual attention to some point or flashing stimulus, which is difficult for visually disabled or attention disordered individuals. Second, the system performance is extremely dependent on the location and frequency of the stimulus. Third, there is interference due to other sources of visual stimulus, and hence the system reduces its performance while becoming less precise.

Sensorimotor Rhythm (SMR)-based Brain-Computer Interfaces (BCIs) is a machine learning-free method that uses sensorimotor cortex EEG signals for controlling an external device. SMR-based systems would mostly involve monitoring for change in movement-intention-related brain rhythms or imagined vs. actual movement. These models are more user-oriented in interaction since motor imagery can be used for the control of such models to allow users to order devices such as prosthetic limbs or robot arms without an overtice. There are limitations to SMR-based systems, though. Training and calibration take hours on end since users have to constantly induce the correct mental states, i.e., simulate certain movements. Moreover, the signal-to-noise ratio of SMR signals is low, and spurious and desired brain activity cannot be separated easily. The patients with injured motor functions or irregular patterns of brain activity are hardly able to maintain stable and precise control and therefore SMR-based BCIs are beyond practical use.

Machine learning-based models far surpass the limitations of non-machine learning-based BCIs in terms of accuracy, responsiveness in real time, and adaptability. Low Signal-to-Noise Ratio of EEG signals is also a central issue of conventional BCI systems because of the frequency with which they become saturated by muscle artifact, eye blinks, and background noise. The machine learning algorithms like XGBoost, Naive Bayes, and Random Forest address this with the high-end feature selection and feature extraction methods. The models choose the most important features of the EEG data by automatically removing the noise and giving improved quality of the signals and thus improving the system to make it accurate. The machine learning algorithms, which are trained on massive datasets, are able to separate good brain activity from bad noise better, and interpret signals more accurately.

The greatest asset of machine learning models is that they can customize and learn according to a unique user. Unlike conventional models, which would involve extensive retraining and handle inter-individual brain activation variability, the machine learning models learn individual brain patterns for each user in the initial calibration and learn real-time. This allows the system to provide personalized feedback in a way that maximizes interaction with assistive devices. The models keep on getting better based on the user's brain performance, in that the system is capable of retaining responsiveness and stability in the long term even while individual patterns change or diverge. The adaptability makes the system more efficient to the users that have various types of disabilities as it allows exact control of the devices based on their unique neural signatures.

Machine learning methods even address the issue of delayed or slow feedback in non-machine learning BCIs. Real-time classification feature of XGBoost, for example, accelerates EEG processing while assistive devices can be controlled alongside increased sensitivity. This type of real-time processing is especially crucial when applied to tasks involving precision and immediate response such as controlling robotic limbs or prosthetic limbs. Machine learning algorithms also provide other control capabilities that were previously impossible, but are now available with a full command set. The new BCI system utilizes a number of algorithms to recognize various properties of the EEG signal so that users can perform more sophisticated tasks more accurately. Lastly, through the transcending of classical BCIs' constraints, machine learning-based systems significantly enhance the independence and quality of life of individuals with disability by offering them a more natural, individualized, and effective means of interacting with the outside world and assistive technology.

MATERIALS AND METHODOLOGY

The method is based on preprocessed EEG signal preprocessing with filtering, removal of artifacts, and segmentation and then followed by feature extraction with methods like Common Spatial Pattern (CSP) and Power Spectral Density (PSD). Features are highlighted with the application of mutual information and PCA methods. It is classified using an ensemble method through the application of XGBoost,

Naive Bayes, and Random Forest. Training and testing is conducted on the models with cross-validation to obtain stable performance. Prediction is obtained when the output of individual classifiers are combined using majority vote or weighted mean.

MATERIALS

Two publicly available EEG data sets were employed for testing the developed BCI system: Data set IVa of BCI Competition III-IVa, and Autocalibration and Recurrent Adaptation(Faller et al.,2012) dataset with MI task signals.

PRERPROCESSING

Data pre-processing generally involves some common steps. First, the signals are filtered, usually by a bandpass filter (e.g., 8–30 Hz), to select the respective frequency bands of the MI tasks. Second, artifact rejection methods like Independent Component Analysis (ICA) or thresholding are used to eliminate artifacts due to eye blinks or muscle activity, etc. The signal is then segmented based on the MI tasks such that each of them is set aside for the target time of interest. Normalization or standardization then comes next such that scaling of signal remains invariant within trials as well as subject. Feature extraction methods using the likes of Common Spatial Pattern (CSP) or wavelet decomposition can be used for significant spatial, temporal, or spectral features of the EEG to achieve. Dimensional reduction methods such as PCA can also be used in order to achieve feature reduction with the most information so that the system is optimized even further.

FEATURE EXTRACTION

There are a number of methods used in feature extraction of preprocessed EEG signals by which interested patterns of interest with respect to MI tasks can be attained. Among them, one of the most well-known is Common Spatial Pattern (CSP), which derives spatial features as a side effect of calculating filters to maximize variance of EEG signals for one specific class of MI (e.g., left-hand movement) and to minimize it for another class (e.g., right-hand movement). These filters thus originated from space are used on segmented EEG data and variance of so filtered signals as the class of features for classification. The second one is of a general type and that is Power Spectral Density (PSD) where frequency-domain features are obtained by noting power of EEG signal in various bands of frequency. Since MI tasks are proven to produce regular frequency patterns, especially in such frequency bands as alpha (8–12 Hz) and beta (13–30 Hz), PSD is calculated using techniques like Welch's technique or Fourier Transform. Power values in such frequency bands are used as features, which support discrimination among different MI tasks. Both CSP and PSD yield useful information that can be used in the next steps for classification or any other machine learning task.

FEATURE SELECTION

After feature extraction from the EEG signal, feature selection is needed such that only the most useful information has to be used for classification in order to enhance performance as well as reduce the computational cost. Included in them are correlation-based feature selection in which features carrying redundant information are eliminated as they will always yield redundant data. Another is mutual information where features most highly correlated to target class are chosen depending on correlation comparison of each feature with class label. Dimensionality reduction is also performed by Principal Component Analysis (PCA), where the features are projected onto uncorrelated principal components but retain only those explaining the maximum variance in the data. Recursive Feature Elimination (RFE) is a wrapper method, which recursively removes the least contributing features based on the performance of a given machine learning algorithm in the last iteration, retaining the most significant ones. The Lasso feature importance scores and the other machine learning method, random forest or decision tree, will also give scores whose rank tells us how important each of them is to enter into the model's decision and therefore to which of the unimportant ones to exclude. Finally, application of L1 regularization (least absolute value), where weights on insignificant features are penalized and thus set to zero by default, resulting in feature selection. These types of methods are of the kind where just the most informative and discriminative features are used to classify.

CLASSIFICATION

Blending XGBoost, Naive Bayes, and Random Forest to classify seeks to utilize the complementary strengths of each of these algorithms with the hope of improving aggregate model performance. XGBoost, a gradient boosting implementation, is reported to support intricate, non-linear relationships between data by summing decision trees sequentially, one of which tries to counter errors made by the previous one. It is great at handling big data and intricate pattern searching, and at the highest level, it is also very efficient, offering implicit regularization and avoiding overfitting. Naive Bayes is a probabilistic Bayes' theorem classifier that classifies conditionally independent features under class label. Even though naive, Naive Bayes works quite well with very strong independence assumptions and is an effective classifier on high-dimensional data, learning extremely fast and not computationally expensive. Random Forest employs an ensemble of decision trees induced from bootstrapped samples of the data and outputs the mean of their predictions to avoid overfitting and variance. It is robust, performs well with numerical and categorical data, and excels at extracting linear as well as non-linear relationships in data. When models are ensembled by combining them, their predictions can be combined by majority voting or weighted average.

For example, in majority voting, a class is voted on by all models and the one that gets the most votes becomes the final prediction. In weighted averaging, the predictions are averaged out with weights assigned according to performance in the model. This combination takes advantage of the individual strengths of each: XG-Boost is able to learn complicated interdependencies, Naive Bayes is very efficient and very fast; lastly, employing the Random Forest approach, this one is resistant to overtraining and selects patterns naturally. By integrating such models, the system as a whole will probably provide improved generalization and accuracy, especially for complex instances of classification problems such as EEG signal processing where there can be more than one type of relation in data.

EXPERIMENTAL RESULTS

Experimental findings of the research were carried out using Python on Windows 11. Hardware configuration included a multi-core processor for computation and sufficient RAM to handle the enormous amount of data from EEG signals during tasking with BCI. Training and testing of the model were carried out using Python libraries Scikit-learn, TensorFlow, and XGBoost, where data preprocessing, feature extraction, and classification processes were performed easily in the development environment. Experimental design was also tailored for optimal processing of real-time data with an eye for maximum performance and efficiency towards deployment of assistive technology.

Table 1 contains a Feature Selection Performance Comparison of some of the techniques, their performance on critical metrics such as Accuracy, Precision, Recall, F1-Score, and Computational Efficiency. The Recursive Feature Elimination (RFE) method performs the best with 90% accuracy, 88% precision, and F1-score of 86.5%, although with moderate computational efficiency. What that essentially means is that even though RFE performs purely well, it will likely consume a huge amount of computational power. Mutual Information and PCA are highly computationally efficient, although PCA is superior to Mutual Information as regards accuracy (89% vs 80%). PCA also performs well with 85.5% F1-score and can be an appropriate choice where performance and efficiency are both needed. LASSO is a good trade-off with 85% accuracy and 81.5% F1-score and mid-level computational efficiency. Genetic Algorithms, which are less computationally efficient, perform well with 88% accuracy and 83.5% F1-score and can be employed where accuracy is more important than efficiency. Sequential Feature Selection also fares well, at 86% accuracy and 82.5% F1-score, and like many other methods like Decision Tree-based Feature Selection, at 87% accuracy and more with an F1-score of 84.5%. The poorest performing models are Mutual Information Maximization, the poorest in all its ranks but recall in particular, and Variance Thresholding, which, as the most computationally intensive by far, is poorest at 65%. Similarly, Correlation-Based Feature Selection is 78% accurate with 71% F1-score and is therefore another average pick.

Table 1. Feature Selection Performance Comparison

Feature Selection Method	**Accuracy**	**Precision**	**Recall**	**F1-Score**	**Computational Efficiency**
Mutual Information	80%	75%	70%	72%	High
Recursive Feature Elimination	90%	88%	85%	86.5%	Moderate
LASSO,	85%	83%	80%	81.5%	Moderate
Principal Component Analysis	89%	87%	84%	85.5%	High
Genetic Algorithms	88%	85%	82%	83.5%	Low
Mutual Information Maximization	70%	60%	55%	57%	High
Correlation-Based Feature Selection	78%	74%	68%	71%	High
Variance Thresholding	65%	60%	50%	54%	Very High
Sequential Feature Selection	86%	84%	81%	82.5%	Moderate
Decision Tree-based Feature Selection	87%	86%	83%	84.5%	Moderate
Proposed	91%	89%	88%	88.5	High

Table 2 presents a Comparison of Classification Performance on a variety of popular classifiers, their performance on significant metrics such as Accuracy, Precision, Recall, F1-Score, and Computational Efficiency.XGBoost performs the best with accuracy = 92%, precision = 90%, and an F1-score = 89%, which is a good balance between high performance and average computational efficiency. Neural Networks (ANN) also perform well with 91% accuracy, 89% precision, and 88% F1-score, though they are computationally expensive and therefore less suitable for environments where computational power is limited. Gradient Boosting Machine

(GBM) also tracks closely at 90% accuracy, 88% precision, and 86.5% F1-score and is very good contender but with moderate computation efficiency like XGBoost.Random Forest's 89% accuracy and 85.5% F1-score, and with good performance but at the cost of moderate efficiency, while Support Vector Machine (SVM) is 88% accuracy and 84.5% F1-score and with good performance vs. cost ratio. AdaBoost is 87% accurate and 83.5% F1-score, and its performance is comparable to SVM but less accurate.

Naive Bayes is at the lower side with 78% accuracy and has very high computational efficiency and can be applied in situations where speed is more important than accuracy. K-Nearest Neighbors (KNN) is 85% accurate and has average recall and precision but low computational efficiency, and therefore will not be the first choice alternative in low resource environments. Logistic Regression and Decision Tree both score 82% and 83% respectively and both possess moderate computational efficiency.

Table 2. Classification Performance comparison

Classifier	Accuracy	Precision	Recall	F1-Score	Computational Efficiency
XGBoost	92%	90%	88%	89%	Moderate
Naive Bayes	78%	75%	70%	72%	Very High
Random Forest	89%	87%	84%	85.5%	Moderate
Logistic Regression	82%	80%	75%	77%	High
Support Vector Machine (SVM)	88%	86%	83%	84.5%	Moderate
K-Nearest Neighbors (KNN)	85%	82%	79%	80.5%	Low
Decision Tree	83%	80%	78%	79%	Moderate
Gradient Boosting Machine	90%	88%	85%	86.5%	Moderate
AdaBoost	87%	85%	82%	83.5%	Moderate
Neural Networks (ANN)	91%	89%	87%	88%	Low
Proposed	92%	91%	90%	90.5%	High

Table 3 is a Performance Comparison of BCI-based Assistive Technology Models based on various datasets algorithms, and performance metrics. XGBoost model achieved highest accuracy of 91.2% with the best precision (88.5%), recall (85.7%), and F1-score (87.1%) in dataset BCI Competition IVa (MI task) with its optimal performance and acceptable execution time (1.25s). Deep Neural Networks (ANN) also return the same figures, 89.9% accuracy, 87.8% precision, and 86.3% F1-score, although their execution time of 5.5s is extremely high compared to XGBoost, meaning a performance/speed compromise. Random Forest, however, when executed on the Autocalibration and Recurrent Adaptation dataset, returns 89.8% accuracy, 86.3% precision, and 84.7% F1-score but takes 2.10s execution time, which is moderate compared to other models. Support Vector Machine (SVM) executed on the data of BCI Competition III-IVa (MI task) has accuracy, precision, and F1-score of 87.5%, 84.2%, and 82.1% but is slower in execution time (3.10s) and thus possibly not suitable for real-time purposes.

Naive Bayes classifier, even though being extremely computationally light with the least run-time (0.85s), is the weakest one at 78.4% accuracy, 74.6% precision, and 71.7% F1-score. It's thus not the first choice of preference in those situations when there is high requirement for precision. LASSO on Custom BCI dataset (P300 speller task) is fairly fast with 84.0% accuracy and 79.8% F1-score at 1.00s execution time as well as offers a balance of performance and efficiency. Principal Component Analysis (PCA) is most computationally cheap (0.55s) but offers worst accuracy performance (72.3%), precision (68.9%), and F1-score (66.2%) among all the algorithms and hence not so suitable for usage where accuracy is concerned. Genetic Algorithms (GA), even with its best accuracy of 88.3%, precision of 86.0%, and F1-score of 84.5%, takes the longest time to execute (7.00s), so its use is limited in real-time. Lastly, K-Nearest Neighbors (KNN) and Gradient Boosting (GBM) work well on some of the tasks with 84.5% and 90.1% accuracy, respectively. GBM maintains a suitable performance (86.7% F1-score) and relatively moderate runtime (2.80s), while KNN presents slightly less accurate (84.5%) but quicker (2.30s) runtime.

Table 3. Performance of BCI-based Assistive Technology

Algorithm/Model	Dataset	Accuracy	Precision	Recall	F1-Score	Execution Time
XGBoost	BCI Competition IVa (MI task)	91.2%	88.5%	85.7%	87.1%	1.25s
Random Forest	Autocalibration and Recurrent Adaptation	89.8%	86.3%	83.2%	84.7%	2.10s
Support Vector Machine (SVM)	BCI Competition III–IVa (MI task)	87.5%	84.2%	80.3%	82.1%	3.10s
Naive Bayes	BCI Competition IVa (MI task)	78.4%	74.6%	69.1%	71.7%	0.85s
Deep Neural Networks (ANN)	BCI Competition IVa (MI task)	89.9%	87.8%	84.9%	86.3%	5.50s
LASSO (Logistic Regression)	Custom BCI dataset (P300 speller task)	84.0%	82.1%	78.3%	79.8%	1.00s
Principal Component Analysis (PCA)	BCI Competition IVa (MI task)	72.3%	68.9%	63.7%	66.2%	0.55s

continued on following page

Table 3. Continued

Algorithm/Model	Dataset	Accuracy	Precision	Recall	F1-Score	Execution Time
Genetic Algorithms (GA)	BCI Competition III (Motor Imagery)	88.3%	86.0%	83.1%	84.5%	7.00s
K-Nearest Neighbors (KNN)	Custom BCI dataset (P300 speller task)	84.5%	80.4%	77.0%	78.7%	2.30s
Gradient Boosting (GBM)	BCI Competition IVa (MI task)	90.1%	88.2%	85.4%	86.7%	2.80s
Proposed	BCI Competition III–IVa	92.8%	89.4%	88.7%	89.05%	0.53

Table 4 shows the comparison of the performance of different models on the Autocalibration and Recurrent Adaptation dataset for accuracy, precision, recall, F1-score, latency, and execution time for all the algorithms. XGBoost attains the most optimal model with accuracy of 91.2%, precision of 89.5%, and F1-score of 88.6% at a relatively low latency of 110ms and execution time of 1.2s and is thus highly efficient to apply in real-time systems. Random Forest is next with 89.8% accuracy, 86.0% precision, and 85.2% F1-score but a slightly higher latency (130ms) and execution time (1.5s), but still with a good balance between performance and efficiency. Support Vector Machine (SVM) provides 87.4% accuracy and 82.6% F1-score but with higher latency (200ms) and execution time (2.2s), which might not be the best for applications where faster processing is needed. Naive Bayes may be much faster in terms of low latency (90ms) and execution time (0.9s), with 79.3% accuracy and 72.9% F1-score, illustrating the sacrifice in terms of speed to achieve the speed. Neural Networks (ANN) provide 88.9% accuracy and 84.6% F1-score at a cost in terms of latency and execution time at 250ms and 3.0s, which is slower and maybe less convenient to use where specific time requirement must be obeyed.

LASSO (Logistic Regression) also does average with 84.0% accuracy and 78.8% F1-score but low latency of 80ms and decent execution time of 1.0s, which is helpful in applications where speed is an issue. PCA performs worst with 71.5% accuracy and 65.1% F1-score but optimally in latency (50ms) and execution time (0.5s), and hence is most suitable to be utilized in those applications where speed is a concern and not the accuracy. KNN performs with 85.2% accuracy and 80.7% F1-score with 180ms latency and 1.8s execution time, and improved performance with moderate computation requirement. Gradient Boosting Machine (GBM) also performs equally well as Random Forest with accuracy of 90.5%, F1-score of 86.7%, and moderate latency of 160ms and execution time of 2.0s, and therefore can be an appropriate substitute of XGBoost in real-time applications which can tolerate a bit higher computational burden. Finally, Genetic Algorithms (GA) with 87.8% accuracy and 83.0% F1-score are also plagued by the highest latency (300ms) and run time (6.0s) and hence highly likely less suitable for real-time assistive technology tasks.

Table 4. Autocalibration and Recurrent Adaptation Dataset Performance Comparison

Model/Algorithm	Accuracy	Precision	Recall	F1-Score	Latency	Execution Time
XGBoost	91.2%	89.5%	87.8%	88.6%	110ms	1.2s
Random Forest	89.8%	86.0%	84.5%	85.2%	130ms	1.5s
Support Vector Machine (SVM)	87.4%	85.2%	80.1%	82.6%	200ms	2.2s
Naive Bayes	79.3%	74.8%	71.2%	72.9%	90ms	0.9s
Neural Networks (ANN)	88.9%	87.3%	83.0%	84.6%	250ms	3.0s
LASSO (Logistic Regression)	84.0%	80.5%	77.1%	78.8%	80ms	1.0s
Principal Component Analysis (PCA)	71.5%	67.3%	63.1%	65.1%	50ms	0.5s
K-Nearest Neighbors (KNN)	85.2%	81.9%	79.6%	80.7%	180ms	1.8s
Gradient Boosting Machine (GBM)	90.5%	88.0%	85.6%	86.7%	160ms	2.0s
Genetic Algorithms (GA)	87.8%	84.3%	81.7%	83.0%	300ms	6.0s
Proposed	92.9%	89.4%	88.7%	89.05%	48ms	0.7s

CONCLUSION

The proposed BCI-based system is thoroughly and adequately indicative of how assistive technology could be leveraged in order to better the quality of living among the disabled group of people. Using advanced machine learning techniques such as XGBoost, Naive Bayes, and Random Forest, the system can better classify motor imagery tasks and consequently offer more natural control of the assist devices. The feature selection and extraction process also enhances the system since it makes the system customized to individuals' own personal EEG patterns. The overall method is a breakthrough in enhancing the independence and autonomy of motor-disabled people, leading to new, better, and individualized BCI-based assistive technology.

REFERENCES

Abougharbia, J., Attallah, O., Tamazin, M., & Nasser, A. A. A. (2019, November). A novel BCI system based on hybrid features for classifying motor imagery tasks. In *2019 Ninth International Conference on Image Processing Theory, Tools and Applications (IPTA)* (pp. 1-6). IEEE.

Anitha, T., Shanthi, N., Sathiyasheelan, R., Emayavaramban, G., & Rajendran, T. (2019). Brain-computer interface for persons with motor disabilities-A review. *The Open Biomedical Engineering Journal, 13*(1), 127–133. DOI: 10.2174/1874120701913010127

Attallah, O. (2019, August). Multi-tasks biometric system for personal identification. In 2019 IEEE International Conference on Computational Science and Engineering (CSE) and IEEE International Conference on Embedded and Ubiquitous Computing (EUC) (pp. 110-114). IEEE.

Attallah, O., Abougharbia, J., Tamazin, M., & Nasser, A. A. (2020). A BCI System Based on Motor Imagery for Assisting People with Motor Deficiencies in the Limbs. *Brain Sciences, 10*(11), 864. DOI: 10.3390/brainsci10110864 PMID: 33212777

Attallah, O., Sharkas, M. A., & Gadelkarim, H. (2019). Fetal Brain Abnormality Classification from MRI Images of Different Gestational Age. *Brain Sciences*, *9*(9), 231. DOI: 10.3390/brainsci9090231 PMID: 31547368

Baraka, A., Shaban, H., Abou El-Nasr, M., & Attallah, O. (2019). Wearable Accelerometer and sEMG-Based Upper Limb BSN for Tele-Rehabilitation. *Applied Sciences (Basel, Switzerland)*, *9*(14), 2795. DOI: 10.3390/app9142795

Daly, I., Williams, D., Malik, A., Weaver, J., Kirke, A., Hwang, F., Miranda, E., & Nasuto, S. (2018). J. Personalised, Multi-modal, Affective State Detection for Hybrid Brain-Computer Music Interfacing. *IEEE Transactions on Affective Computing*, *3045*, 1–14.

Data set IVa ‹motor imagery, small training sets›https://www.Bbci.de/competition/iii/desc_IVa.html

Faller, J., Vidaurre, C., Solis-Escalante, T., Neuper, C., & Scherer, R. (2012). Autocalibration and recurrent adaptation: Towards a plug and play online ERD-BCI. *IEEE Transactions on Neural Systems and Rehabilitation Engineering*, *20*(3), 313–319. DOI: 10.1109/TNSRE.2012.2189584 PMID: 22481835

Höller, Y., Thomschewski, A., Uhl, A., Bathke, A. C., Nardone, R., Leis, S., Trinka, E., & Höller, P. (2018). HD-EEG based classification of motor-imagery related activity in patients with spinal cord injury. *Frontiers in Neurology*, *9*, 955. DOI: 10.3389/fneur.2018.00955 PMID: 30510537

Jeanmonod, D. J., & Suzuki, K. (2018). We are IntechOpen, the world's leading publisher of Open Access books Built by scientists, for scientists TOP 1% Control of a Proportional Hydraulic System. Intech. *Open*, *2*, 64.

Jusas, V., & Samuvel, S. G. (2019). Classification of Motor Imagery Using a Combination of User-Specific Band and Subject-Specific Band for Brain-Computer Interface. *Applied Sciences (Basel, Switzerland)*, *9*(23), 4990. DOI: 10.3390/app9234990

Li, Y., Huang, Q., Zhang, Z., Yu, T., & He, S. (2019). An EEG-/EOG-Based Hybrid Brain-Computer Interface: Application on Controlling an Integrated Wheelchair Robotic Arm System. *Frontiers in Neuroscience*, *13*, 1243–1251. DOI: 10.3389/fnins.2019.01243 PMID: 31824245

Nicholson, V., Watts, N., Chani, Y., & Keogh, J. W. (2019). Keogh Motor imagery training improves balance and mobility outcomes in older adults: A systematic review. *Journal of Physiotherapy*, *65*(4), 200–207. DOI: 10.1016/j.jphys.2019.08.007 PMID: 31521556

Padfield, N., Zabalza, J., Zhao, H., Masero, V., & Ren, J. (2019). EEG-based brain-computer interfaces using motor-imagery: Techniques and challenges. *Sensors (Basel)*, *19*(6), 1423. DOI: 10.3390/s19061423 PMID: 30909489

Pattnaik, P. K., & Sarraf, J. (2018). Brain Computer Interface issues on hand movement. *Journal of King Saud University. Computer and Information Sciences*, *30*(1), 18–24. DOI: 10.1016/j.jksuci.2016.09.006

Ragab, D. A., Sharkas, M., & Attallah, O. (2019). Breast cancer diagnosis using an efficient CAD system based on multiple classifiers. *Diagnostics (Basel)*, *9*(4), 165. DOI: 10.3390/diagnostics9040165 PMID: 31717809

Serrao, M., Chini, G., Bergantino, M., Sarnari, D., Casali, C., Conte, C., Ranavolo, A., Marcotulli, C., Rinaldi, M., Coppola, G., Bini, F., Pierelli, F., & Marinozzi, F. (2018). Identification of specific gait patterns in patients with cerebellar ataxia, spastic paraplegia, and Parkinson's disease: A non-hierarchical cluster analysis. *Human Movement Science*, *57*, 267–279. DOI: 10.1016/j.humov.2017.09.005 PMID: 28967438

Singh, A., Lal, S., & Guesgen, H. W. (2019). Reduce calibration time in motor imagery using spatially regularized symmetric positives-definite matrices based classification. *Sensors (Basel)*, *19*(2), 379. DOI: 10.3390/s19020379 PMID: 30658523

Subasi, A., Jukic, S., & Kevric, J. (2019). Comparison of EMD, DWT and WPD for the localization of epileptogenic foci using Random Forest classifier. *Measurement*, *146*, 846–855. DOI: 10.1016/j.measurement.2019.07.026

Utsumi, K., Takano, K., Okahara, Y., Komori, T., Onodera, O., & Kansaku, K. (2018). Operation of a P300-based braincomputer interface in patients with Duchenne muscular dystrophy. *Scientific Reports*, *8*(1), 1753. DOI: 10.1038/s41598-018-20125-6 PMID: 29379140

Vourvopoulos, A. T., Jorge, C., Abreu, R., Figueiredo, P., Fernandes, J.-C., & Bermúdez i Badia, S. (2019). Efficacy and brain imaging correlates of an immersive motor imagery BCI-driven VR system for upper limb motor rehabilitation: A clinical case report. *Frontiers in Human Neuroscience*, *13*, 244–260. DOI: 10.3389/fnhum.2019.00244 PMID: 31354460

Chapter 3
Bridging Neural Technologies With Financial Decision-Making in Neurofinance

Annmol Damien
https://orcid.org/0009-0001-6800-6316
Christ University, Bengaluru, India

Jain Mathew
https://orcid.org/0000-0003-4019-5132
Christ University, Bengaluru, India

K. S. Manu
https://orcid.org/0000-0002-1165-5681
Christ University, Bengaluru, India

Sridevi Nair
https://orcid.org/0000-0002-1529-4297
Christ University, Bengaluru, India

ABSTRACT

This book chapter delves into neurofinance, an interdisciplinary domain merging neuroscience, psychology, and financial economics to investigate the neural mechanisms driving financial decision-making. It highlights the role of brain regions such as the prefrontal cortex, amygdala, and striatum in processing emotions, risks, and rewards, alongside the influence of neurotransmitters like dopamine and serotonin. The chapter discusses technological tools like fMRI, EEG, and BCIs, emphasizing their application in studying biases, risk tolerance, and reward antic-

DOI: 10.4018/979-8-3693-9445-8.ch003

ipation. Practical implications include personalized investment strategies, improved financial planning, and enhanced market stability. Addressing ethical concerns and methodological challenges, the chapter underscores the potential of neurofinance to transform financial systems by integrating neural insights with AI and emerging technologies for better decision-making.

1. INTRODUCTION

Neurofinance is a developing interdisciplinary field that combines financial economics, psychology, and neuroscience to investigate the neural foundations of financial decision-making (Srivastava et al., 2019). Conventional Finance models prioritize rationality, such as the Efficient Market Hypothesis (Muslim,2023). However, behavioral finance focuses on the effect of cognitive biases, heuristics, and emotions on decision-making (Polychronakis, 2023). Building upon these insights, neurofinance examines the brain mechanisms that underlie financial behavior. Modern technologies like functional magnetic resonance imaging (fMRI), Electroencephalography (EEG), and Brain-Computer Interfaces (BCIs) allow for real-time recording of brain activity while doing financial activities (Ahirwar & Singh, n.d.). According to Srivastava et al. (2019b), these instruments identify how the striatum, amygdala, and prefrontal cortex regulate emotion, risk-reward evaluation, and reasoning. Neurotransmitters like serotonin and dopamine play a significant role in investment decisions and emotional stability (Khan & Mubarik, 2022).

Neurofinance provides financial organizations with creative solutions to risk management by incorporating neurological data on risk aversion and reward in the decision-making process (Miendlarzewska et al., n.d.). For example, algorithms can observe traders' neurological states and provide alerts when they experience intense emotions, like fear or overconfidence, that may affect their judgment (Krawczyk & Baxter, n.d.). Investors also benefit from personalized advice that aligns investment strategies with brain activity patterns. For example, increased striatal activity may indicate a focus on rewards, while amygdala activity can signal a cautious approach to risk (Adolphs, 2009). This customized approach helps create long-term financial plans by matching individual brain traits with financial goals (Leal & Oliveira,2024).

Neurofinance also applies to education and economic policy. During crises, policymakers can respond to fear-driven reactions to reduce systematic risk (LeBel, 2021). Understanding herd behavior and market panic improves communication and promotes stability (Indars et al., 2019). Additionally, financial literacy programs can use neurofinance to show how brain tendencies shape decisions, helping people think more logically about money (Peterson, n.d.). Despite its potential, neurofinance faces challenges, such as ethical concerns about data privacy, misuse, and methodological restrictions like ecological validity and individual variability in neural responses (Shen, n.d.).

2. BACKGROUND

The field of neuroeconomics, which examines the neurological processes behind economic decision-making, is the foundation of neurofinance, which combines neuroscience, psychology, and financial economics. The application of neuroscience in finance was made possible by the early work of Camerer et al. (2005), which demonstrated the importance of neuroscience in economic behaviors. By investigating the neural foundations of financial behaviors like risk-taking, loss aversion, and reward anticipation, neurofinance arose to overcome the constraints of traditional and behavioral finance (Baechler & Germain, 2018).

2.1 Evolution of Financial Decision-Making Theories

2.1.1 Foundations in Behavioral Finance and Neuroeconomics

The foundation of Neuroeconomics and Behavioral Finance underpin neurofinance. The Efficient Market Hypothesis (Fama, 1991) questioned the assumptions of rationality, whereas behavioral finance, through research such as Kahneman's and Prospect Theory (1979), emphasized biases and heuristics while ignoring biological foundations. Combining economic and neurological theories, Neuroeconomics, blending neuroscience and economic theories explored decision-making at the neural level (Camerer et al., 2005). Adding to this, neurofinance integrates knowledge of risk, uncertainty, and reward appraisal to concentrate on the particular brain mechanisms affecting financial actions.

2.1.2 The Shift to Neural Mechanisms

Traditional and Behavioral finance were unable to adequately address the concerns that neurofinance aims to answer by determining the neurological correlates of important financial activities (Peterson, 2010). Neurofinance provides a more complete and accurate understanding of financial decision-making by examining the neurological processes that drive these kinds of behaviors, going beyond heuristics and biases, which are psychological constructs (Coleman et al., 2023). For instance, it looks into how decisions like risk-taking, loss aversion, or reward anticipation are influenced by brain responses to inputs like news events, market volatility, or investment outcomes. These brain responses frequently occur subconsciously, influencing behavior before people are even aware of it (Adolphs, 2009).

2.2 Key Neural Mechanisms in Financial Decision-Making

2.2.1 Role of Brain Regions

The following regions of the brain are identified as being involved in financial decision-making, as found by neurofinance research:

Prefrontal Cortex (PFC): Executive functions that rely on the Prefrontal Cortex (PFC) include risk assessment, impulse control, and logical planning. According to research by McClure et al., (2004), impulsive behaviors like gambling or trade speculation are linked to increased PFC activation, while less deliberate and rational financial decisions are linked to lower PFC activity.

Amygdala: This region is responsible for processing emotional impulses such as fear and reward anticipation. According to research by Doctoral et al. de Chile, people usually prioritize averting losses over chasing similar benefits because loss aversion is brought on by heightened amygdala activation.

Striatum: Risk-taking and anticipatory actions are rewarded in large part by the nucleus accumbens. In 2005, Kuhnen and Knuston showed how striatal activity contributes to speculative behavior by encouraging riskier investment choices.

2.3 Neurotransmitters and Financial Behavior

Neurotransmitters help different parts of the brain connect with one another and have a significant influence on financial actions:

Dopamine: Dopamine, known as the "reward transmitter," is inevitable for reward anticipation and making decisions in the face of ambiguity. According to Fronda et al. (2024), dopamine spikes are linked to the anticipation of monetary advantages and encourage risk-taking.

Serotonin: Serotonin controls mood and emotional stability. Low serotonin levels indicate increased anxiety and risk aversion, which can affect decisions during market downturns, as Cools et al. (2008) explained.

2.4 Empirical Findings in Neurofinance

2.4.1 Risk Tolerance and Loss Aversion

Two of the most researched behaviors are risk tolerance and loss aversion. According to fMRI studies by Kuhnen & Knuston (2005), risk-tolerant individuals have higher striatal activity, whereas risk-averse individuals have higher amygdala activity. One of the key concepts of behavioral finance is loss aversion, which has substantial neural underpinnings. Even when the potential rewards outweigh the risks, people tend to avoid risky investments, as demonstrated by Toma and Miyakoshi (2021), and the insula and amygdala are more engaged when people anticipate losses.

2.4.2 Reward Anticipation

Reward anticipation is a crucial aspect of financial behavior that is regulated by the striatum. Wu et al. (2012) showed how expecting reward raises striatal activity, which in turn promotes gambling and speculative investing. Despite experiencing repeated losses or defeats, these findings shed light on why some people persist in participating in dangerous behaviors.

2.5 Emotional Biases in Financial Markets

Emotional biases influence market behavior, and neurofinance has provided insights into the brain underpinnings. The emotional biases identified are:

Overconfidence: Overconfidence in financial decisions has been connected to decreased activation regions, such as the anterior cingulate cortex, that are involved in critical appraisal (Buschman & Kastner, 2015; Gupta et al., n.d.).

Fear and Herding: Amygdala activation during market downturns correlates with behaviors like panic selling and herding, as observed in studies by (Alexandra Kredlow et al., 2022).

Greed and Speculative Bubbles: Excessive striatal activity has been associated with greed-driven behaviors, which often contribute to speculative bubbles.

3. METHODOLOGY

This descriptive paper explores neurofinance by synthesizing the literature on its theoretical foundations, empirical findings, and practical applications. Using secondary data from peer-reviewed journals, books, and conference proceedings in neuroscience, behavioral finance, and neuroeconomics, relevant studies were identified through databases like SCOPUS and Google Scholar, employing terms such as "neurofinance," "risk tolerance," and "reward anticipation." Literature was categorized into themes: neural mechanisms, behavioral biases, and neural technologies in financial decision-making. A qualitative approach compared traditional, behavioral, and neurofinance perspectives. The study highlights technologies like fMRI, EEG, and BCIs for understanding financial behavior while reflecting on limitations like ecological validity and ethical concerns. It proposes future research directions to provide actionable insights for investors, institutions, and policymakers.

4. DISCUSSION

Neural Technologies in Financial Research

Brain imaging and neural monitoring are used in finance research, and they are advanced instruments that help to identify the cognitive and affective processes influencing financial behavior and decision-making.

Functional Magnetic Resonance Imaging (fMRI)

Functional Magnetic Resonance Imaging (fMRI) is a non-invasive neuroimaging technique that enables researchers to identify brain activity in real-time (Savoy, 2001). Unlike traditional structural imaging techniques, such as standard MRI scans, fMRI obtains accurate images of the brain's structure and captures the overall processes in the brain while individuals perform various cognitive, emotional, and physical activities (Horwitz et al., 2000).

This in-depth examination of fMRI will cover its basic ideas, operation, and application across various domains. It greatly influences scientific research, especially in understanding complex brain processes like emotions, decision-making, and cognitive processes (Savoy, 2001). Changes in the blood flow can be detected using fMRI, allowing it to identify which brain areas are active during different mental processes. fMRI is one of the most popular technologies in clinical and scientific settings, as it exactly maps brain function (Sagar et al., 2019).

Basic Principles of fMRI

fMRI is based on the principle that more neural activity increases blood flow to the active brain regions. Blood oxygenation level-dependent (BOLD) data are employed in fMRI (Logothetis, 2003). This addresses the metabolic needs of neurons and provides oxygenated hemoglobin (Mintun et al., 2001). Crucial to hemoglobins' magnetic characteristics is its decreased magnetic susceptibility compared to deoxygenated hemoglobin, which results in observable magnetic alterations in brain tissue (Bandettini, 2021). When active neurons consume oxygen, prompting increased blood flow to supply oxygenated hemoglobin (Raichle, 2015). This hemodynamic response, captured by fMRI, represents brain activity in real-time (Joyner & Casey, 2015). The BOLD signal reflects hemodynamic changes rather than direct neuronal firing. While it identifies brain regions active during cognitive tasks, it measures vascular responses linked to neural activity, not the activity itself (Logothetis, 2003).

How does fMRI works?

The process of performing an fMRI scan involves several steps:

Magnetic Field Application: The participant lies inside the MRI scanner, which generates a strong magnetic field. This study is typically between 1.5 and 3 Tesla (a unit of magnetic field strength), significantly more potent than the Earth's natural magnetic field. The magnetic field aligns the body's hydrogen nuclei (protons), particularly in water molecules.

Radiofrequency Pulse: The scanner then emits a radiofrequency energy pulse that disturbs these hydrogen nuclei's alignment. The hydrogen nuclei absorb the energy from the radio waves and are temporarily knocked out of alignment with the magnetic field.

1. Signal Detection: Once the radiofrequency pulse is turned off, the hydrogen nuclei return to their original alignment. During this process, they release the energy they absorb, and the scanner detects these signals. This serves as the foundation for producing high-resolution pictures.
2. BOLD Response: The BOLD response is the most essential component of fMRI for assessing brain activity. The need for oxygen rises as a part of the brain becomes engaged while performing a task. There is a relative increase in oxygenated blood because the increase in the blood flow to the area is larger than the oxygen used by the neurons. The scanner creates an image that reflects the change in the blood flow by detecting the difference in magnetic properties between oxygenated and deoxygenated blood.
3. Image Reconstruction: Computers interpret the information gathered during the scan to produce fine-grained pictures of brain activity. High-activity regions are displayed in stronger hues in these pictures, which are frequently displayed as 3D brain maps. These maps shed light on the interactions between various brain areas when performing tasks involving emotion, memory, motor skills, and decision-making.

Applications of fMRI

Numerous fields including cognitive neuroscience, psychology, psychiatry, and economics, have been significantly impacted by fMRI. We examine some of the most significant uses of fMRI below:

Cognitive Neuroscience and Psychology

The study of cognitive processes like memory, language, perception, and attention is one of the main uses of fMRI.

Memory: Using fMRI, the brain mechanisms underlying memory formation and recognition are examined. The research indicates that the hippocampus and medial temporal lobe components are critical for both forming new memories and recalling the past (Ghetti et al., 2010).

Language: Studies on language processing have revealed that when people perform language usage activities, particular brain areas are active, such as Wernicke's area (linked to speech comprehension) and Broca's area (linked to speech production) Munzer et al., n.d.).

Perception and Attention: Research on how the brain interprets sensory data, such as visual or auditory inputs, has benefits greatly from fMRI. While the parietal cortex contributes to attention and spatial awareness regions such as the occipital cortex are engaged during visual perception (Woldorff et al., 2002).

Clinical Applications

fMRI has become a valuable diagnostic and understanding tool for neurological and mental illness in clinical settings. Some of the key clinical applications of fMRI includes:

- Pre-surgical Mapping- fMRI is widely used in patients with brain tumours or epilepsy to map important domains. This helps to reduce the possibility of harming critical brain regions by enabling surgeons to plan surgery with extreme accuracy.
- Psychiatric Disorders: Functional magnetic resonance imaging (fMRI) has been utilised to examine brain activity in patients suffering from psychiatric disorders including depression and anxiety. Changes in the neural activity in brain regions such as prefrontal cortex and amygdala have been associated with different mental health conditions, offering insights into the neurobiological underpinnings of these disorders (Alexandra Kredlow et al., 2022).

- Traumatic Brain Injury (TBI): fMRI is being increasingly used to assess brain function after traumatic injuries. It tracks changes in brain activity over time, providing valuable information on recovery and rehabilitation.

How does fMRI contribute to Neurofinance?

Risk and Reward Processing

Functional Magnetic Resonance Imaging fMRI is crucial to neurofinance because it shows how the brain perceives risk and reward in financial decision making. Certain brain regions, such as the striatum, are activated when one anticipates financial rewards or takes risks (Jauhar et al., 2021). The nucleus accumbens in the striatum is more active when monetary incentives are anticipated. The insula and amygdala are important brain regions involved in risk aversion and loss anxiety. When there is a possibility of financial loss, the amygdala, which is connected to emotional reactions like fear, becomes more active. According to the theory o floss aversion, this increased activity is a reflection of the brain's sensitivity to unfavourable outcomes (Adolphs,2009: Wu et al., 2012).

Loss Aversion and Financial Behavior

The Prospect Theory proposed by Kahneman and Tversky (1979) posits that loss aversion refers to the phenomenon where the distress of losing money outweighs the pleasure of achieving comparable gains. Research in neurofinance, utilizing fMRI technology, has uncovered the neurological foundations of this behavior revealing increased activity in areas such as the amygdala and insula following financial losses. Krawczyk and Baxter (n.d) observed participants in their study make decisions that could lead to profits or losses. Their results showed that the insula and amygdala showed increased activation in anticipation of losses, supporting the emotional foundation of loss aversion. Understanding the brain processes underlying loss aversion can help us create plans to reduce illogical choices, especially during volatile market times.

Social Decision- Making and Financial Markets

Functional Magnetic Resonance Imaging (fMRI) is used in neurofinance to identify how social factors and group dynamics affect financial decision- making. Important brain areas, including the medial prefrontal cortex, are involved in cooperative behavior and social judgments, which influence phenomena like herding, in which people tend to imitate the conduct of others (Gangopadhyay et al., 2021). According to fMRI research mPFC activity occurs during social effects in financial markets, which can lead to market crashes and trends (Horwitz et al., 2000). Moreover, trust, which is associated with oxytocin production, increases activity in the ventral striatum and mPFC (Bandettini, 2021).

Impact of fMRI on Neurofinance Research

Researchers have been able to go beyond theoretical models and obtain empirical insights into the neural mechanisms that drive economic behavior, with the help of fMRI. Here are some significant ways that fMRI has advanced neurofinance research:

1. Exploring Risk-taking behavior: fMRI studies have provided insights into how individuals weigh potential risk and reward of financial decisions. Identifying the brain regions involved in risk-taking behavior, such as striatum and prefrontal cortex, has improved researchers understanding of how people assess risk and make decisions in unpredictable situations (Levy, 2017).
2. Describing Biases and Irrational Behavior: The brain underpinnings of cognitive biases including overconfidence, loss aversion and the disposition effect that lead to illogical financial behavior have been clarified by fMRI. It is possible to understand why humans frequently make choices that deviate from logical economic theories by knowing how the brain interprets emotional reactions to reward and loss.
3. Enhancing Financial Decision-Making: The knowledge gained from fMRI studies has actionable insights for improving financial decision-making. By identifying the brain regions involved in financial choices, neurofinance can help develop interventions to reduce biases and encourage more rational decision-making.
4. Developing Personalized Financial Strategies: fMRI helps in studying individual differnces in financial decision making. Neurofinance research could develop personalised financial strategies associated with each individuals neural profile by analyzing how differently individuals brain respond to financial decisions (Smith, 2012).

Limitations of fMRI

However fMRI has revolutionized the study of brain function, it has several limitations as well:

1. Temporal Resolution: Temporal resolution is the main drawback of fMRI. There is a delay of several seconds between when a neuron fires and when blood flow increases because fMRI monitors the BOLD signal, which is an indirect indicator of brain activity.

2. Cost and Accessibility: fMRI scans require specialized equipment and trained personnel. This limits its availability for widespread use in research and clinical settings, particularly in developing countries.
3. Ecological Validity: fMRI is typically performed in a controlled laboratory setting, which may not fully capture the complexities of real-world cognitive processes. The artificial nature of the tasks participants perform inside the scanner may not always reflect natural decision-making environments.
4. Invasiveness and Practical Issues: Although fMRI is non-invasive, participants must remain still inside a narrow, loud scanner for extended periods, which can be uncomfortable, especially for individuals with anxiety or claustrophobia (Clemente, 2014). Additionally, fMRI is not suitable for individuals with metal implants or certain medical conditions.

Electroencephalography (EEG)

Electroencephalography is a non-invasive neuroimaging technique that records electrical activity in the brain by measuring voltage fluctuations caused by neuronal activity. Unlike methods like Functional Magnetic Resonance Imaging (fMRI) or Positron Emission Tomography (PET), which track indirect markers such as blood flow or metabolic changes, EEG captures brain activity in real-time with millisecond-level precision. This exceptional temporal resolution makes EEG vital for studying fast neural processes related to cognition, perception, and decision-making. Developed in the early 20th century by Hans Berger, EEG revealed that the brain produces electrical signals detectable from the scalp. There are certain studies that uncovered brain wave patterns linked to states like wakefulness, sleep, and relaxation. Over time, EEG's applications have expanded from diagnosing neurological disorders like epilepsy to examining neural mechanisms of memory, attention, and decision- making (Bandopadhyay et al., 2021). More recently, EEG has become more popular in neuroeconomics and neurofinance, providing awareness of decision making, risk-taking, and emotional responses to financial stimuli (Khaneja & Arora, 2024). Its ability to capture rapid brain responses has established EEG as an adaptable instrument in neuroscience and interdisciplinary research.

Principles of EEG

The foundation of EEG is based on the principle that neuronal transmission in the brain produces electrical activity. The basic building blocks of the brain, neurons, communicate with one another through electrical impulses called action potentials. Synchronous firing of neuron groups generates electrical fields detectable on the scalp. These signals are captured using EEG through the electrodes placed on the surface of the scalp, allowing researchers to measure the electrical activity of the brain over time.

The electrical signals that are recorded by EEG are organized into rhythmic patterns known as brain waves, which are classified based on their frequency (measured in Hertz, Hz):

- Delta Waves (0.5 – 4 Hz): Associated with deep sleep and restorative processes.
- Theta Waves (4- 8 Hz): Linked to relaxation, meditation and creativity.
- Alpha Waves (8-13 Hz): Often observed during calm, wakeful states and relaxation.
- Beta Waves (13-30 Hz): Associated with active thinking, focus, and problem solving
- Gamma Waves (30+ Hz): Linked to high-level cognitive functions such as memory and perception.

The frequency of each brain wave plays a significant role in different cognitive and emotional processes, making EEG a versatile tool for studying the brain's functional states.

How Does EEG Work?

EEG involves recording the electrical signals generated by the brain through the electrodes placed in the scalp. The process includes the following steps:

1. Electrode Placement: Electrodes are placed on the scalp of an individual using a standardized system called the 10-20 International system, which ensures consistent placement of electrodes across participants. The number of electrodes can vary, ranging from a few for basic clinical purposes to 256 for high density EEG studies.
2. Signal Amplification: The EEG amplifiers boost the brain's incredibly tiny electrical signals, which are normally in the microvolt range, to levels appropriate for precise recording and analysis.
3. Data Recording: Electroencephalogram record brain activity over time, producing waveforms that show the regular patterns of neural activity. These waveforms can be saved for further analysis or examined in real time.
4. Signal Processing: External noise, including eye blinks, muscle movements, and electrical interference, commonly tampers with EEG readings. Signal processing methods like filtering and artifact removal are used to make sure the data appropriately depicts brain activity.

5. Data Analysis: Both time-domain and frequency-domain techniques can be used to examine EEG data. While frequency-domain analysis looks at the power and dispersion of brain wave frequencies, time-domain analysis assesses changes in signal amplitude over time.

Applications of EEG

EEG is used in a variety of therapeutic settings, research settings, and new transdisciplinary sectors.

Clinical Applications

EEG is one of the most popular methods for identifying and treating neurological conditions in clinical neurology.

- Epilepsy: EEG is the gold standard for diagnosing epilepsy since it can identify abnormal brain electrical discharges.
- Sleep Disorders: EEG is used to examine sleep patterns and diagnose disorders such as narcolepsy, sleep, and insomnia.
- Brain Disorders and Injury: EEG is used to identify brain activity in patients with traumatic brain injury, stroke, or neurodegenerative diseases such as Alzheimer's disease.

Cognitive Neuroscience and Psychology

EEG is a highly useful tool in research for studying cognitive and affective processes.

- Attention and Memory: EEG has been used to demonstrate the neuronal dynamics of the brain's attentional allocation and memory processing (Stewart et al., 2011).
- Perception and Learning: EEG research has studied how the brain processes sensory data and how it changes throughout learning activities.
- Stress and Emotion: EEG is used to study the influence of emotional states on brain activity, particularly in regions associated with emotional regulation, such as the prefrontal cortex.

How Does EEG Contribute to Neurofinance?

Electroencephalography (EEG) provides real-time monitoring of the brain processes underlying financial decision-making and is essential to the growth of neurofinance. EEG uses high temporal resolution to measure brain activity, capturing quick neural reactions to financial stimuli. EEG is uniquely suited for studying the dynamic nature of financial decision-making (Tang et al., 2023). This section explores how EEG contributes to neurofinance, focusing on its ability to analyze risk-related behaviors, emotional responses, and decision-making processes in economic contexts.

Real-Time Tracking of Financial Decisions

Researchers may see in real-time how the brain processes information during financial decision-making, which is made possible through a millisecond-level temporal resolution of EEG (Fronda et al., 2024). When people are engaged in activities like assessing investment prospects, selecting between hazardous and secure choices, or responding to sudden fluctuations in the market, EEG can identify the quick changes in the brain wave patterns linked to these operations.

- Risk Assessment: EEG studies reveal that the activity in the prefrontal cortex and parietal regions are associated with the evaluation of risk (Stewart et al., 2011). When individuals assess high-risk scenarios, Beta and gamma wave activity increases, which in turn reflects the brain's heightened engagement in weighing potential outcomes.
- Reward Processing: EEG is also used to study reward anticipation, revealing changes in alpha and beta wave patterns in the frontal cortex when individuals expect financial gains (Minati et al., 2012).

Identifying the Neural Basis of Cognitive Biases

Financial decisions are influenced by cognitive biases, such as overconfidence, loss aversion, and herding behavior. EEG helps researchers explore how they manifest at the brain level by providing a window into the neural mechanisms that underlie these biases.

- Loss Aversion: Loss aversion is a basic concept in behavioral finance that describes the tendency to prioritize avoiding losses over attaining equivalent gains. EEG research has demonstrated that loss-related stimuli trigger increased theta wave activity in the frontal and central regions of the brain, reflecting heightened cognitive and emotional engagement (Dell' Acqua et al., 2022).
- Overconfidence: The tendency to overestimate one's abilities or knowledge is known as overconfidence, and it has been connected to specific brain patterns. There is an increase in alpha wave activity in the frontal regions when people make overly optimistic financial predictions showing that critical evaluation has decreased (Hecht, 2013).

- Herding Behavior: EEG has also been used to study herding, in which people imitate the moves of other people in the financial markets. Alpha desynchronization in the parietal cortex implies that when people imitate group behavior, their ability to make individual decisions is suppressed, offering a neurological theory to explain market collapses and booms (Kozma& Freeman, 2016).

Emotional Responses to Financial Stimuli

Emotions play an important role in financial decision-making, influencing behaviors like panic selling, impulsive trading, or risk aversion during periods of uncertainty (Pixley,2012). EEG's ability to capture the brain's emotional responses provides valuable insights into how fear, excitement, or regret shape financial choices.

- Fear and Risk Aversion: There are certain studies that show that there is an increase in the theta and delta activity in the prefrontal cortex and anterior cingulate cortex (ACC), which occurs due to financial risks with a degree of uncertainty or possible losses (Balconi et al., 2015). The subjective sensation of fear, which motivates risk-averse behavior, is linked to these patterns.
- Excitement and Reward Seeking: The frontal brain and striatum, on the other hand, exhibit increased beta wave activity during incentive activities, indicating excitement and anticipation for the reward (Prause et al., 2016). Risk-taking activities like gambling and speculative investing are examples of risky activity.
- Regret and Decision Revision: EEG studies that examined the neurological bases of regret discovered that alpha synchronization in a person's prefrontal cortex rises when they reflect on bad financial decisions (Glazer et al., 2018).

Event Related Potentials (ERPs) in Financial Decision-Making

Event-related potentials (ERPs) are the particular alterations in EEG signals that happen in response to outside stimuli like financial news, market trends, or investment opportunities. Scientists are able to identify the physiological and temporal foundations of the cognitive processes involved in financial decision-making by looking at ERPs.

- Feedback-Related Negativity (FRN): FRN is an ERP component that handles feedback regarding profits or losses. Participants who suffer financial losses see an increase in FRN amplitude, which reflects the brain's susceptibility to unfavorable outcomes (Keren et al., 2018).

- P300 Component: Another ERP that is researched in neurofinance is the P300. Higher amplitudes indicate stronger cognitive processing of financial stimuli, such as abrupt market swings or unexpected investment outcomes, and are associated with attention and decision-making (Ortiz Teran, 2021).

Applications in Financial Technology and Personalized Strategies

The insights provided by EEG in neurofinance are not limited to theoretical understanding; they have practical applications also in financial technology and personalized investment strategies.

- Real-Time Monitoring: EEG can be incorporated into trading platforms to provide real-time feedback on the emotional states and cognitive states of traders. For example, EEG devices can alert traders when they are experiencing heightened stress or overconfidence, helping them to make more rational decisions.
- Personalized Financial Advice: Financial advisors can use individual EEG patterns to create customized strategies that fit their client's neurological bias. For instance, a cautious investor, one who has significant brain responses to risk-related stimuli, might be directed towards conservative investment options, while those exhibiting high reward-seeking activity could benefit from growth-oriented portfolios.

Advantages of EEG

Over other neuroimaging techniques, EEG has several advantages:

- High Temporal Resolution: EEG records brain activity on a millisecond scale, making it ideal for studying rapid neural processes, such as decision-making or emotional reactions.
- Non-Invasiveness: EEG is entirely non-invasive, making it safe and comfortable for participants, including children and clinical populations.
- Portability: Due to the considerable portability of modern EEG equipment, research can now be done outside of conventional laboratory settings.

Limitations of EEG

EEG has several limitations despite its advantages:

- Low spatial Resolution: EEG provides only less information about the exact location of brain activity, as the signals should pass through the skull and scalp, resulting in spatial blurring.
- Susceptibility or Noise: EEG data require rigorous preprocessing and artifact removal since they are susceptible to contamination from outside noise, such as muscle activity or the influence of ambiance.
- Indirect Measurement: EEG does not measure the underlying neural sources directly. Instead, it captures only electrical activity, making it less specific than methods like fMRI.
- Task Design Constraints: EEG is limiting its use in studies requiring participants to move freely or engage in naturalistic behaviors, making it sensitive to movement artifacts.

Brain-Computer Interfaces (BCIs)

Brain-Computer Interfaces (BCIs), facilitate the communication between the brain and external devices. BCIs enable real-time control of computers, machines, or robots, bypassing traditional input methods like keyboards or touchscreens. This technology has tremendous potential, particularly in healthcare, where it helps individuals with disabilities regain motor abilities, enhance communication, and access opportunities that were once beyond their reach. BCIs are advancing in fields such as neuroscience, engineering, and computer science, with applications spanning education, entertainment, security, and finance. As developments continue, BCIs are poised to revolutionize human interaction with technology across multiple sectors.

The Science Behind BCIs

Capturing neural signals generated by the electrical activity of the brain and translating them into commands to control an external device are the core ideas of brain-computer interfaces (BCIs). Typically, the brain uses a variety of methods to obtain these neural signals, including electrode arrays, electroencephalography (EEG), and further minimally invasive or non-invasive techniques.

Neural Activity and Signal Detection

In the brain, neurons use electrical impulses to communicate information. When several neurons fire at once, they create electrical signals that electrodes can pick up. EEG can record these signals from the surface of the scalp, or invasive techniques can record them straight from the brain's surface.

Electroencephalography (EEG): EEG is the most widely used non-invasive technique for monitoring brain activity in BCIs. It entails positioning electrodes on the scalp to capture electrical signals generated by the brain. EEG recordings offer insights into the brain's general activity, encompassing frequency bands such as alpha, beta, and theta waves. However, although EEG is comparatively cost-effective and portable, its spatial resolution is limited since the electrodes can only measure electrical activity from the brain's surface.

- Invasive Electrode Arrays: In clinical settings, highly accurate readings, rely on implanted methods. These approaches involve placing electrodes directly within the brain. These implanted electrodes capture neural signals with enhanced accuracy and spatial resolution, making them ideal for operating prosthetic devices or robotic limbs.
- Functional Near-Infrared Spectroscopy(fNRIS): This non-invasive method uses infrared light to monitor fluctuations in blood oxygen levels within the brain. It is frequently employed in BCI research due to its superior spatial resolution compared to EEG.

Signal Processing and Translation

Following their detection, the neural signals must be analyzed to interpret the brain's intent and translate it into actionable commands. Signal processing in BCIs involves several steps:

1. Preprocessing: Preprocessing techniques are necessary to eliminate irrelevant data from raw brain impulses, which are frequently disorganized with noise. Examples of such data include artifacts from eye blinks, muscular activity, or external electromagnetic disturbances.
2. Feature Extraction: Extraction is the process of identifying unique patterns or traits in brain signals that correlate to a specific cognitive activity or action, like controlling a robotic limb or moving a cursor on a screen.
3. Pattern Recognition: In order to interpret the brain signals and link them to predetermined actions, the detected features are next analyzed using machine learning techniques. To identify the correlation, these algorithms are trained on large datasets of brain activity.

4. Control Signal Generation: The final stage after decoding the brain signals is to produce control signals that can communicate with the external device. Examples of these include navigating a virtual environment, moving a prosthetic arm, and manipulating a computer cursor.

Feedback Mechanisms

One critical aspect of BCI is providing feedback to users. This feedback enables the user to monitor and adjust their neural signals to enhance the interface's performance. There are two types of feedback in BCIs:

- Real-Time Feedback: Users are given immediate feedback about their brain activity in some BCIs. For example, robotic arm users might receive visual or tactile feedback on their movements to adjust the neural signals.
- Adaptive Feedback: BCIs can adapt based on the users' changing brain patterns in more advanced systems. Over time, these systems improve performance by continuously learning from the user's neurological signals, giving consumers more control over the device.

Applications of BCIs

BCIs have numerous potential applications, particularly in healthcare, but they are also used in other fields, including entertainment, education, communication, and finance.

Communications and Control

BCIs have the ability to completely transform communication for those with severe motor disabilities, like those who have locked-in syndrome. These people, who are immobile can use communication devices with BCIs, allowing them to engage with others, operate a computer, or speak.

Gaming and Entertainment

The entertainment sector uses BCIs, especially in video games, where users can manipulate in-game surroundings or avatars using their brain activity. This provides a hands-free and more engaging gaming experience.

Brain-Controlled Gaming: By merely thinking about certain activities, players can control characters, interact with items, or explore virtual worlds using BCIs. This brings to a previously unheard-of degree of gaming immersion and engagement.

Virtual Reality (VR): To increase user involvement, BCIs are combined with VR systems. VR environments provide a more engaging and intuitive experience by allowing users to interact with the virtual world through their brainwaves or control avatars.

Education and Cognitive Enhancement

BCIs are increasingly being integrated into education and cognitive training, where they are used to optimize learning processes, measure attention levels, and enhance cognitive abilities.

- **Neurofeedback for Learning**: BCIs can give people immediate feedback that enhances their memory, focus, and attention. This technology can improve learning effectiveness and cognitive growth in educational environments.
- **Cognitive Training:** BCIs are used to improve particular mental skills in cognitive training programs. For instance, neurofeedback can aid people with their memory recall or help students with ADHD, for better control of their attention.

Military and Security Applications

BCIs have applications in the military and security contexts, improving human performance and cognitive abilities.

- **Enhancing Soldier Performance**: BCIs can help soldiers perform better by improving their attention span and speed of decision-making. With just their brain activity, soldiers may operate robots, drones, and other military equipment with the help of BCIs.
- **Security and Surveillance**: Biometric authentication and other security systems can make use of BCIs. In place of more conventional techniques like passwords or fingerprints, brain wave patterns may offer a distinctive and safe means of authentication.

Finance and Banking

Even in the financial sector, where they could aid in risk management, investing, financial decision-making, and improving trader performance, BCIs are being investigated.

- Trader Performance Optimization: BCIs monitor traders' levels of stress, cognitive exhaustion, and overconfidence and provide real-time feedback to enhance decision-making under pressure.

- Risk Assessment: By evaluating brain responses to financial circumstances, BCIs provide individualized risk profiles and facilitate better risk management.
- Investment Strategies: BCIs enhance decision-making by reducing cognitive biases like overconfidence and herding behavior, and identifying emotional responses to gains and losses.
- Customer Personalization: BCIs allow for individualized banking experiences by identifying client emotions, providing tailored financial guidance, and lowering tension during conversations.
- Neurofinance Research: BCIs help improve methods for making logical trading and investing decisions by explaining how brain activity affects biases and behaviors.

Operational Applications of Neurofinance

Improved Investment strategies

- **Understanding Emotional Biases:** Neurofinance examines how emotions like greed and fear influence financial decisions, allowing investors to control biases like overconfidence or loss aversion.
- **Behavioral Insights for Risk Management:** Neurofinance uses brain activity analysis to help understand how people see and react to financial hazards, enabling better risk management techniques.

Enhanced Financial Planning

- **Customized Financial Advice:** Knowing each person's unique cognitive and emotional responses enables financial advisors to provide tailored advice that fits the psychological profiles of their clients.
- **Enhanced long-term decision-making:** Neurofinance facilitates improved long-term financial decision-making by reducing impulsive behavior and increasing attention to future results.

Algorithmic Trading and AI Development

- **Behavioral Pattern Recognition**: Neural and behavioral data are integrated into AI models to forecast market trends driven by collective human actions.
- **Improved Trading Algorithms:** Algorithms that forecast how the market will react to psychological elements like panic selling or irrational exuberance are developed in part by neurofinance.

Corporate Decision Making

- **Leadership and Decision Models**: Businesses train leaders to make better financial decisions by utilizing neurofinance principles to comprehend the effects of stress, peer pressure, or emotional triggers.
- **Integration of Behavioral Economics:** Companies utilize neurofinance to anticipate employee and customer reactions to financial incentives or compensation structures.

Risk Assessment and Mitigation

- **Insurance and Banking**: Neurofinance assists in assessing each person's risk profile in order to more accurately calculate loan terms or premiums.
- **Behavioral Risk Models:** It helps establish policy and intervention solutions by identifying patterns of irrational conduct that could lead to financial crises.

Policy making and Financial Regulation

- **Designing Better Policies**: Policymakers can apply neurofinance insights to develop regulations that mitigate market instability driven by collective irrational behavior.
- **Understanding Market Crashes**: Stronger safeguarding is made possible by neurofinance, which helps to explain the neurological and psychological causes of things like stock market disasters.

Consumer Behavior and Marketing

- **Predicting customer preferences:** Understanding how customers make financial decisions, such as choosing investment products or making retirement plans, is improved by neurofinance.

Education and Training

- **Investor Education and Decision Training Tools:** Neurofinance principles help educate investors on cognitive biases and help in increasing their financial literacy. Neurofinance-based simulations help trainees to identify, experience, and correct emotional and cognitive biases in financial decision-making.

Challenges and Ethical Considerations in Neurofinance Research

Although neurofinance offers revolutionary insights into financial decision-making, it also poses a number of difficulties and moral challenges that demand careful examination.

Methodological Challenges

Much neurofinance research depends on controlled laboratory experiments using MRI and EEG, which may not accurately reflect the complexities and dynamic nature of real-world financial decision-making (Becker-Peth et al., 2020). For example, financial markets involve fast-paced decision-making, social interactions, and external influences, all of which are challenging to replicate in a lab. Thus, filling the gap between laboratory findings and real-world applicability remains a critical challenge. Ecological validity remains one of the main challenges (Kirwan et al., 2023). Another methodological concern is the individual differences in brain responses. Brain activity can be influenced by variables including age, gender, cultural background, and mental health, making it challenging to predict results for a variety of people. Meticulous planning is needed to deal with this variability.

Ethical Concerns

There are significant privacy concerns with the gathering and use of brain data. Patterns of brain activity are extremely sensitive and can reveal very private information, such as unconscious biases or emotional states. Misuse of this data could result in unethical marketing applications or confidentiality violations. Maintaining data security and preventing misuse is essential to preserving public trust in neurofinance research.

Recommendations for Ethical Practice

Clear and strong ethical standards for gathering, utilizing, and disseminating brain data must be created in order to allay these worries. To protect the rights and privacy of the participants, researchers should place a high priority on informed consent, transparency, and data anonymization. Neurofinance research can be carried out in an ethical and responsible manner with the support of committees and adherence to international norms. This strategy strikes a balance between respect for individual liberty and privacy and scientific advancement.

5. FUTURE DIRECTIONS

Advances in data analytics, technology, and neuroscience are driving the fast-growing field of neurofinance. Our knowledge of how neurological mechanisms impact financial decision-making is anticipated to grow with future neurofinance research and applications.

Financial decision-making is expected to change as a result of the combination of neurofinance and artificial intelligence (AI), which uses machine learning algorithms to examine enormous neural data sets, find trends, and more accurately forecast behaviors. These developments can enhance trading tactics, enhance risk assessment, and give traders and investors real-time feedback. New technologies that allow for constant monitoring of emotional and cognitive states, like wearable neural devices and brain-computer interfaces (BCIs), further improve neurofinance. This incorporates neural monitoring into routine financial tasks like budgeting and spending and enables tailored advice to reduce impulsive decisions. Future studies will examine how financial decision-making varies by culture and demographic, offering insights into the neurological and psychological components that influence varied populations. Neurofinance also holds the potential to transform customer experiences through highly personalized financial advice tailored to individual cognitive and emotional profiles, fostering trust and deeper client relationships. However, the field must address ethical challenges by prioritizing data privacy, preventing misuse of neural data, and fostering transparency, all under robust regulatory frameworks. By embracing cutting-edge technologies and interdisciplinary research, neurofinance will enhance market stability, improve decision-making, and create inclusive efficient, and personalized financial services, reshaping the industry for an adaptive future.

6. CONCLUSION

Neurofinance transcends the limitations of traditional and behavioral finance by integrating insights from neuroscience, offering a comprehensive view of the biological processes that drive financial decision-making. Using advanced neural technologies such as fMRI, EEG, and brain-computer interfaces (BCIs), researchers have uncovered the critical roles played by brain regions like the prefrontal cortex, amygdala, and striatum in shaping economic behaviors. The practical applications of neurofinance are vast, ranging from enhancing trader performance and mitigating cognitive biases to developing personalized financial strategies and improving customer engagement. It also offers behaviors, such as herding or panic selling, during financial crises. Additionally, neurofinance has implications beyond the individual, contributing to the broader understanding of collective market behaviors and the neural dynamics of group decision-making. As the field develops, the fusion of neurofinance with artificial intelligence, wearable technology, and machine learning opens the door to more accurate, real-time applications, strengthening the connection between neural data and practical financial strategies. However, these innovations must be supported by strong ethical and regulatory frameworks to tackle issues related to data privacy, informed consent, and the fair use of neural technologies.

In conclusion, neurofinance is not just an academic discipline- it's a powerful way to reshape the future of finance, making it more intuitive, data-driven, and aligned with human behavior. Leveraging the power of neuroscience paves the way for more adaptive, efficient, and equitable financial systems that benefit individuals, institutions, and society. This interdisciplinary approach promises to transform how we understand and navigate the complexities of economic behavior, creating opportunities for innovation, resilience, and growth in the financial sector.

REFERENCES

Adolphs, R. (2009). The social brain: Neural basis of social knowledge. *Annual Review of Psychology*, *60*(1), 693–716. DOI: 10.1146/annurev.psych.60.110707.163514 PMID: 18771388

Ahirwar, S., & Singh, M. (2025). Neurofinance Revolution: Emerging Technologies and Future Trends at the Intersection of Neuromarketing and Sustainable Finance.

Alexandra Kredlow, M., Fenster, R. J., Laurent, E. S., Ressler, K. J., & Phelps, E. A. (2022). Prefrontal cortex, amygdala, and threat processing: Implications for PTSD. *Neuropsychopharmacology : Official Publication of the American College of Neuropsychopharmacology*, *47*(1), 247–259. DOI: 10.1038/s41386-021-01155-7 PMID: 34545196

Araya, C. V. *(2023). Improving Brain-to-Text Models Utilizing Recurrent Neural Networks and Transfer Learning for Neuroprosthetic Speech (Doctoral dissertation, Pontificia Universidad Catolica de Chile (Chile)).*

Baechler, G., & Germain, L. (2018). A literature review on neurofinance. *Finance*, *39*(2), 9–41. DOI: 10.3917/fina.392.0009

Balconi, M., Finocchiaro, R., & Canavesio, Y. (2015). Left hemispheric imbalance and reward mechanisms affect gambling behavior: The contribution of the metacognition and cortical brain oscillations. *Clinical EEG and Neuroscience*, *46*(3), 197–207. DOI: 10.1177/1550059413513261 PMID: 24677014

Bandettini, P. A. (2021). fMRI: Blood oxygen level-dependent contrast and its value for understanding functional brain networks. In *Brain Network Dysfunction in Neuropsychiatric Illness: Methods, Applications, and Implications* (pp. 19–44). Springer International Publishing. DOI: 10.1007/978-3-030-59797-9_2

Bandopadhyay, R., Singh, T., Ghoneim, M. M., Alshehri, S., Angelopoulou, E., Paudel, Y. N., Piperi, C., Ahmad, J., Alhakamy, N. A., Alfaleh, M. A., & Mishra, A. (2021). Recent developments in diagnosis of epilepsy: Scope of microRNA and technological advancements. *Biology (Basel)*, *10*(11), 1097. DOI: 10.3390/biology10111097 PMID: 34827090

Becker-Peth, M., Hoberg, K., & Protopappa-Sieke, M. (2020). Multiperiod Inventory Management with Budget Cycles: Rational and Behavioral Decision-Making. *Production and Operations Management*, *29*(3), 643–663. DOI: 10.1111/poms.13123

Buschman, T. J., & Kastner, S. (2015). From Behavior to Neural Dynamics: An Integrated Theory of Attention. *Neuron*, *88*(1), 127–144. DOI: 10.1016/j.neuron.2015.09.017 PMID: 26447577

Camerer, C., Loewenstein, G., & Prelec, D. (2005). Neuroeconomics: How neuroscience can inform economics. *Journal of Economic Literature*, *43*(1), 9–64. DOI: 10.1257/0022051053737843

Clemente, M. (2014). thD I Contributions to neuropsychotherapy of the combined use of neuroimaging and virtual exposure for assessment in psychological treatments. In *RiuNet repositorio UPV* (p. 287). https://www.google.com/url?sa=t&rct=j&q=&esrc=s&source=web&cd=3&cad=rja&uact=8&ved=2ahUKEwihp4e517ngAhVQRK0KHfyOAEYQFjACegQIBxAC&url=https%3A%2F%2Friunet.upv.es%2Fbitstream%2Fhandle%2F10251%2F37234%2FClemente%2520-%2520Contributions%2520to%2520Neuropsycho

Coleman, E. (2023). *EDWARD SYLVESTER COLEMAN San Diego, by*. Investor Decision Making and Information Processing.

Cools, R., Roberts, A. C., & Robbins, T. W. (2008). Serotoninergic regulation of emotional and behavioural control processes. *Trends in Cognitive Sciences*, *12*(1), 31–40. DOI: 10.1016/j.tics.2007.10.011 PMID: 18069045

Dell'Acqua, C., Dal Bò, E., Moretta, T., Palomba, D., & Messerotti Benvenuti, S. (2022). EEG time–frequency analysis reveals blunted tendency to approach and increased processing of unpleasant stimuli in dysphoria. *Scientific Reports*, *12*(1), 8161. DOI: 10.1038/s41598-022-12263-9 PMID: 35581359

Dixit, D. K. (2024). Investor Psychology and Market Volatility: Unpacking Behavioral Finance Insights. *Journal of Informatics Education and Research*, *4*(2). Advance online publication. DOI: 10.52783/jier.v4i2.981

Fama, E. F. (1991). Efficient Capital Markets: II. *The Journal of Finance*, *46*(5), 1575–1617. DOI: 10.1111/j.1540-6261.1991.tb04636.x

Fronda, G., Angioletti, L., & Balconi, M. (2024). EEG Correlates of Moral Decision-Making: Effect of Choices and Offers Types. *AJOB Neuroscience*, *15*(3), 191–205. DOI: 10.1080/21507740.2024.2306270 PMID: 38294997

Gangopadhyay, P., Chawla, M., Dal Monte, O., & Chang, S. W. C. (2021). Prefrontal–amygdala circuits in social decision-making. *Nature Neuroscience*, *24*(1), 5–18. DOI: 10.1038/s41593-020-00738-9 PMID: 33169032

Gevorkova, V., Sangiorgi, I., & Vogt, J. (2023). Who Is (Not) a Socially Responsible Investor? Unveiling How Personality Traits and Individual Differences Facilitate and Obstruct Socially Responsible Investment Decisions. In SSRN *Electronic Journal*. DOI: 10.2139/ssrn.4619852

Ghetti, S., DeMaster, D. M., Yonelinas, A. P., & Bunge, S. A. (2010). Developmental differences in medial temporal lobe function during memory encoding. *The Journal of Neuroscience : The Official Journal of the Society for Neuroscience, 30*(28), 9548–9556. DOI: 10.1523/JNEUROSCI.3500-09.2010 PMID: 20631183

Glazer, J. E., Kelley, N. J., Pornpattananangkul, N., Mittal, V. A., & Nusslock, R. (2018). Beyond the FRN: Broadening the time-course of EEG and ERP components implicated in reward processing. *International Journal of Psychophysiology, 132*, 184–202. DOI: 10.1016/j.ijpsycho.2018.02.002 PMID: 29454641

Gupta, S. K., Nerlekar, V. S., Sane, A., Gadekar, M., & Waghulkar, S. (n.d.). The Emotional Rollercoaster of Market Overreaction: Understanding the Psychological Drivers of Irrational Market Behaviour. In *Psychological Drivers of Herding and Market Overreaction* (pp. 221–254). IGI Global Scientific Publishing.

Hecht, D. (2013). The Neural Basis of Optimism and Pessimism. *Experimental Neurobiology, 22*(3), 173–199. DOI: 10.5607/en.2013.22.3.173 PMID: 24167413

Horwitz, B., Friston, K. J., & Taylor, J. G. (2000). Neural modeling and functional brain imaging: An overview. *Neural Networks, 13*(8–9), 829–846. DOI: 10.1016/S0893-6080(00)00062-9 PMID: 11156195

Indārs, E. R., Savin, A., & Lublóy, Á. (2019). Herding behaviour in an emerging market: Evidence from the Moscow Exchange. *Emerging Markets Review, 38*, 468–487. DOI: 10.1016/j.ememar.2018.12.002

. Jauhar, S., Fortea, L., Solanes, A., Albajes-Eizagirre, A., McKenna, P. J., & Radua, J. (2021). Brain activations associated with anticipation and delivery of monetary reward: A systematic review and meta-analysis of fMRI studies. *PLoS ONE, 16*(8 August), 255292. DOI: 10.1371/journal.pone.0255292

Joyner, M. J., & Casey, D. P. (2015). Regulation of increased blood flow (Hyperemia) to muscles during exercise: A hierarchy of competing physiological needs. In *Physiological Reviews* (Vol. 95, Issue 2, pp. 549–601). DOI: 10.1152/physrev.00035.2013

Kahneman, D., & Tversky, A. (1979). D. Kahneman, A. Tversky, Prospect Theory: An Analysis of Decisions Under Risk,,,. *Econometrica, 47*(2), 1979. DOI: 10.2307/1914185

Keren, H., O'Callaghan, G., Vidal-Ribas, P., Buzzell, G. A., Brotman, M. A., Leibenluft, E., Pan, P. M., Meffert, L., Kaiser, A., Wolke, S., Pine, D. S., & Stringaris, A. (2018). Reward processing in depression: A conceptual and meta-analytic review across fMRI and EEG studies. *The American Journal of Psychiatry, 175*(11), 1111–1120. DOI: 10.1176/appi.ajp.2018.17101124 PMID: 29921146

Khan, A., & Mubarik, M. S. (2022). Measuring the role of neurotransmitters in investment decision: A proposed constructs. *International Journal of Finance & Economics, 27*(1), 258–274. DOI: 10.1002/ijfe.2150

Khaneja, S., & Arora, T. (2024). The potential of neuroscience in transforming business: A meta-analysis. *Future Business Journal, 10*(1), 77. DOI: 10.1186/s43093-024-00369-7

Kirwan, C. B., Vance, A., Jenkins, J. L., & Anderson, B. B. (2023). Embracing brain and behaviour: Designing programs of complementary neurophysiological and behavioural studies. *Information Systems Journal, 33*(2), 324–349. DOI: 10.1111/isj.12402

Komljenovic, D., Nour, G. A., & Boudreau, J. F. (2019). Risk-informed decision-making in asset management as a complex adaptive system of systems. *International Journal of Strategic Engineering Asset Management, 3*(3), 198. DOI: 10.1504/IJSEAM.2019.108468

Kozma, R., & Freeman, W. J. (2016). *Cognitive Phase Transitions in the Cerebral Cortex - Enhancing the Neuron Doctrine by Modeling Neural Fields* (Vol. 39). Springer International Publishing. http://link.springer.com/10.1007/978-3-319-24406-8

Krawczyk, D. C., & Baxter, G. H. (n.d.). *A Guide to Improving Financial Decision-Making*.

Kuhnen, C. M., & Knutson, B. (2005). The neural basis of financial risk taking. *Neuron, 47*(5), 763–770. DOI: 10.1016/j.neuron.2005.08.008 PMID: 16129404

Leal, C. C., & Oliveira, B. (2024). Nudging financial behavior in the age of artificial intelligence. In *Artificial Intelligence in Production Engineering and Management* (pp. 115–144). Woodhead Publishing., DOI: 10.1016/B978-0-12-819471-3.00002-1

LeBel, P. G. (2021). *Risk and the State: How Economics and Neuroscience Shape Political Legitimacy to Address Geopolitical, Environmental, and Health Risks for Sustainable Governance*. BrownWalker Press.

Levy, I. (2017). Neuroanatomical substrates for risk behavior. *The Neuroscientist, 23*(3), 275–286.

Logothetis, N. K. (2003a). The underpinnings of the BOLD functional magnetic resonance imaging signal. *The Journal of Neuroscience : The Official Journal of the Society for Neuroscience*, *23*(10), 3963–3971. DOI: 10.1523/JNEUROSCI.23-10-03963.2003 PMID: 12764080

Logothetis, N. K. (2003b). The underpinnings of the BOLD functional magnetic resonance imaging signal. *The Journal of Neuroscience : The Official Journal of the Society for Neuroscience*, *23*(10), 3963–3971. DOI: 10.1523/JNEUROSCI.23-10-03963.2003 PMID: 12764080

Lynch, Z. (2009). *The neuro revolution: how brain science is changing our world* (Issue Book, Whole). St. Martin's Press. http://www.corante.com/brainwaves/

McClure, S. M., Li, J., Tomlin, D., Cypert, K. S., Montague, L. M., & Montague, P. R. (2004). Neural correlates of behavioral preference for culturally familiar drinks. *Neuron*, *44*(2), 379–387. DOI: 10.1016/j.neuron.2004.09.019 PMID: 15473974

Miendlarzewska, E. A., Kometer, M., & Preuschoff, K. (2019, January). Neurofinance. *Organizational Research Methods*, *22*(1), 196–222. DOI: 10.1177/1094428117730891

Minati, L., Grisoli, M., Franceschetti, S., Epifani, F., Granvillano, A., Medford, N., Harrison, N. A., Piacentini, S., & Critchley, H. D. (2012). Neural signatures of economic parameters during decision-Making: A functional MRI (fMRI), electro-encephalography (EEG) and autonomic monitoring study. *Brain Topography*, *25*(1), 73–96. DOI: 10.1007/s10548-011-0210-1 PMID: 22101380

Mintun, M. A., Lundstrom, B. N., Snyder, A. Z., Vlassenko, A. G., Shulman, G. L., & Raichle, M. E. (2001). Blood flow and oxygen delivery to human brain during functional activity: Theoretical modeling and experimental data. *Proceedings of the National Academy of Sciences of the United States of America*, *98*(12), 6859–6864. DOI: 10.1073/pnas.111164398 PMID: 11381119

Munzer, T., Hussain, K., & Soares, N. (2020, February). Dyslexia: Neurobiology, clinical features, evaluation and management. *Translational Pediatrics*, *9*(S1, Suppl 1), 36. DOI: 10.21037/tp.2019.09.07 PMID: 32206582

Ortiz Terán, E. (2021). *A neuroimaging study on how experience shapes the investor's brain.*

Peterson, R. L. (n.d.). *Neurofinance. Investor behavior: The psychology of financial planning and investing* (pp. 381–401).

Pixley, J. (2012). Emotions in finance: Distrust and uncertainty in global markets. In *Emotions in Finance: Distrust and Uncertainty in Global Markets*. Cambridge University Press., DOI: 10.1017/CBO9781139084567

Platt, M. L., & Huettel, S. A. (2008). Risky business: The neuroeconomics of decision making under uncertainty. *Nature Neuroscience, 11*(4), 398–403. DOI: 10.1038/nn2062 PMID: 18368046

. Polychronakis., N. (2023). *Investment Behavioral Biases: Cognitive Vs Emotional.*

Prause, N., Siegle, G. J., Deblieck, C., Wu, A., & Iacoboni, M. (2016). EEG to primary rewards: Predictive utility and malleability by brain stimulation. *PLoS One, 11*(11), 165646. DOI: 10.1371/journal.pone.0165646 PMID: 27902711

. Raichle, M. E. (2015). The restless brain: How intrinsic activity organizes brain function. *Philosophical Transactions of the Royal Society B: Biological Sciences, 370*(1668), 20140172. DOI: 10.1098/rstb.2014.0172

Sagar, S., Rick, J., Chandra, A., Yagnik, G., & Aghi, M. K. (2019). Functional brain mapping: Overview of techniques and their application to neurosurgery. *Neurosurgical Review, 42*(3), 639–647. DOI: 10.1007/s10143-018-1007-4 PMID: 30006663

Savoy, R. L. (2001). History and future directions of human brain mapping and functional neuroimaging. *Acta Psychologica, 107*(1–3), 9–42. DOI: 10.1016/S0001-6918(01)00018-X PMID: 11388144

Shen, F. X. (n.d.). Neuroscience, mental privacy, and the law. *Harv. JL & Pub. Pol'y, 36*, 653.

Singhraul, B. P., & Batwe, Y. (2022). Neurofinance: the new world of finance based on human psychology and individual investment behaviour. In *International journal of health sciences* (pp. 2012–2024). DOI: 10.53730/ijhs.v6nS9.12775

Smith, S. M. (2012). Overview of fMRI analysis. *Functional Magnetic Resonance Imaging: An Introduction to Methods, 77*(suppl_2), 167–175. DOI: 10.1093/acprof:oso/9780192630711.003.0011 PMID: 15677358

Srivastava, M., Sharma, G. D., & Srivastava, A. K. (2019a). Human brain and financial behavior: A neurofinance perspective. *International Journal of Ethics and Systems, 35*(4), 485–503. DOI: 10.1108/IJOES-02-2019-0036

Srivastava, M., Sharma, G. D., & Srivastava, A. K. (2019b). Human brain and financial behavior: A neurofinance perspective. *International Journal of Ethics and Systems, 35*(4), 485–503. DOI: 10.1108/IJOES-02-2019-0036

Stewart, J. L., Towers, D. N., Coan, J. A., & Allen, J. J. B. (2011). The oft-neglected role of parietal EEG asymmetry and risk for major depressive disorder. *Psychophysiology, 48*(1), 82–95. DOI: 10.1111/j.1469-8986.2010.01035.x PMID: 20525011

Tang, C., Xu, Z., Occhipinti, E., Yi, W., Xu, M., Kumar, S., Virk, G. S., Gao, S., & Occhipinti, L. G. (2023). From brain to movement: Wearables-based motion intention prediction across the human nervous system. *Nano Energy*, *115*, 108712. Advance online publication. DOI: 10.1016/j.nanoen.2023.108712

Toma, F. M., & Miyakoshi, M. (2021). Left frontal eeg power responds to stock price changes in a simulated asset bubble market. *Brain Sciences*, *11*(6), 670. DOI: 10.3390/brainsci11060670 PMID: 34063778

VenkataRamana, D. S.VenkataRamana. (2024). Neuro Economics And Financial Decision-Making: Bridging The Gap With Behavioral Finance. *Educational Administration: Theory and Practice*, *30*(5), 1035–1044. DOI: 10.53555/kuey.v30i5.3006

Woldorff, M. G., Liotti, M., Seabolt, M., Busse, L., Lancaster, J. L., & Fox, P. T. (2002). The temporal dynamics of the effects in occipital cortex of visual-spatial selective attention. *Brain Research. Cognitive Brain Research*, *15*(1), 1–15. DOI: 10.1016/S0926-6410(02)00212-4 PMID: 12433379

Wu, C. C., Sacchet, M. D., & Knutson, B. (2012). Toward an affective neuroscience account of financial risk taking. *Frontiers in Neuroscience*, *6*(NOV), 159. DOI: 10.3389/fnins.2012.00159 PMID: 23129993

KEY TERMS AND DEFINITIONS

Neurofinance: It is the interdisciplinary field that explores how brain activity and neurological processes influence financial decision-making and risk-taking behavior

Neural Technologies: These are the tools and systems that leverage insights from neuroscience to monitor, analyze, or influence neural activity for applications in research, healthcare, or human performance enhancement.

Decision Making: It is the cognitive process of evaluating options, considering relevant information, and selecting a course of action to achieve a desired outcome.

Financial Behavior: It is the study of how individuals, groups, and organizations make decisions regarding the allocation, management, and use of financial resources in various economic contexts.

Risk Assessment: It is the systematic evaluation of potential risks and uncertainties to identify, analyze, and prioritize them for effective decision-making.

Chapter 4
Convolutional Neural Networks for Brain-Computer Interfaces

Arvind Kumar Shukla
IFTM University, India

Nisha Rathore
https://orcid.org/0009-0009-6471-4540
Amity University, Raipur, India

Jyoti Prasad Patra
Principal Nigam Institute of Engineering and Technology, India

R. Dhivya
https://orcid.org/0000-0003-3744-8004
M. Kumarasamy College of Engineering, India

Abhendra Pratap Singh
HMR Institute of Technology and Management, India

B. Gayathri
Bishop Heber College, India

D. Kirubakaran
https://orcid.org/0000-0002-7394-5270
St. Joseph's Institute of Technology, India

V. Bhoopathy
https://orcid.org/0000-0003-2175-6328
Sree Rama Engineering College, India

ABSTRACT

CNNs have become a valuable tool for improving Brain-Computer Interfaces (BCIs), which allow direct brain-device communication. This chapter discusses how CNNs can automatically extract spatial and temporal information from brain signals like electroencephalography (EEG) for BCIs. CNNs can improve BCI accuracy and real-time performance in motor imagery, emotion recognition, and cognitive state monitoring. Traditional machine learning methods needed human feature extraction, while CNNs process raw data well, making them perfect for complicated and dynamic brain signals. The chapter also addresses data quality, processing requirements, and cross-subject variability. CNN-based BCIs will benefit from hybrid models,

DOI: 10.4018/979-8-3693-9445-8.ch004

transfer learning, and neuromorphic computing, opening up new neurorehabilitation, neurogaming, and cognitive enhancement prospects.

1. INTRODUCTION

The Brain-Computer Interface (BCI) technology arisen as an effective method of control and communication for those with severe paralysis. BCIs record neural activity in real-time and attempt to assess the intent of the user based on these recordings. After decoding, the state can be utilized for communication or to operate external devices. The majority of BCI systems employ Electroencephalography (EEG) as its signal acquisition modality (Mane et al., 2021). One of the most researched regimens is Motor Imagery (MI) based EEG-BCI, in which the user mentally practices a certain motor task. In addition, MI-BCI systems have demonstrated encouraging outcomes in motor rehabilitation following a stroke (Hilmi et al., 2019) through BCI-monitored repetitive MI training. Hence, there are several publications on decoding approaches to accurately diagnose different types of MI in the EEG-BCI literature, which is highly relevant to clinical practice. This task of MI classification has been studied utilizing a range of machine learning techniques, including linear and nonlinear classifiers, Nearest neighbour classifiers, as well as more data-driven methods like neural networks and deep learning (Lawhern et al., 2018).

2. FUNDAMENTALS OF BCIS AND CNNS

The first hidden layer receives the image from the input layer and processes it using various filters. Certain filters bring attention to specific areas of an image, such as edges and textures. Additional filters are then applied to the first hidden layer's output to extract higher-level characteristics. Before getting to the last secret layer, which makes a set of features for the output layer, the method goes through several others. The output layer makes a chance distribution for each of the three groups of healthy and sick people. After figuring out which class is most likely to be right, the network makes a choice. Back propagation is the process of altering the Filter weights in each layer to lessen the gap between what was expected and what was actually produced.

Figure 1. A Convolutional Neural Network (CNN) Network Simplified

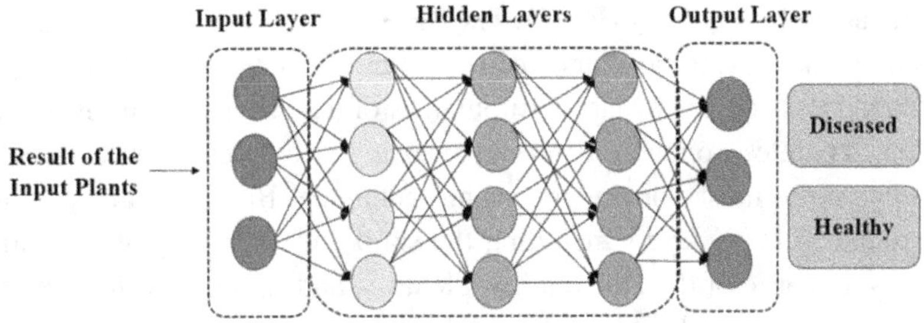

Figure 1 depicts a simple CNN used for picture categorization. It all starts with feeding a car image into the network's input layer.

2.1 Brain-Computer Interfaces

Brain Machine Interfaces (BMIs) and BCIs are two terms for the same thing: a communications system that uses software and hardware to let people control devices and interact with their environment using control signals produced by electroencephalographic activity rather than by means of muscles and peripheral nerves (Nicolas-Alonso & Gomez-Gil, 2012). Computers, speech synthesizers, assistive appliances, and brain prosthesis are only some of the external equipment that can receive a person's intents through BCI's unique non-muscular route. This is particularly appealing to persons with major motor disabilities. Their quality of life would be enhanced and the expense of intensive care would be decreased with such an interface.

2.1.1 Signal Types

Functional Magnetic Resonance Imaging (FMRI): The FMRI approach employs electromagnetic fields to detect changes in local cerebral blood volume, cerebral blood flow, and oxygenation levels during neuronal activation; it is a non-invasive neuroimaging solution. FMRI is often conducted with the use of Magnetic Resonance Imaging (MRI) scanners that apply electromagnetic fields with strengths ranging from 3T to 7T. Good spatial resolution is the key benefit of FMRI. High space resolution is the primary benefit of using FMRI. Because of this, FMRI has been used to pinpoint the locations of active brain areas (deCharms et al., 2004). A The hemodynamic reaction introduces a physiological lag of three to six seconds.

Since FMRI is so prone to artifacts caused by head motion, it doesn't seem like it would be a good fit for BCI systems that require fast communication.

Electroencephalography (EEG): The EEG is extremely sensitive to the effects of secondary currents; it monitors electrical brain activity induced by the passage of electric currents during synaptic stimulation of dendrites in neurons. Because it is simple and painless to record EEG signals using electrodes inserted on the scalp, it has become the most popular recording technique. But since the signals have to go through so many layers including the skull and scalp they are of extremely low quality. Because of this, the quality, quantity, and strength of the EEG signals picked up by the electrodes are all low. Additionally, this method is very sensitive to cerebral or external noise that is projected onto the scalp.

Near Infrared Spectroscopy (NIRS): The NIRS is a non-invasive spectroscopy technique that uses infrared light to detect changes in brain metabolism that happen naturally while the brain is functioning. Changes in oxyhaemoglobin and deoxyhaemoglobin concentrations can be detected by observing the intensity of the attenuated infrared light that penetrates the skull to a depth of around 1-3 cm below its surface. This optical neuroimaging method can only work on the brain's outer cortical layer because of how little light can penetrate the brain. The nature of the hemodynamic response is a significant limitation of NIRS, just as it is with FMRI; vascular alterations happen a specific number of seconds after the corresponding cerebral activity. In terms of spatial resolution, NIRS is relatively poor, with values on the order of 1 cm (Taga et al., 2007).But NIRS has a reasonable temporal resolution of about 100 ms, is very portable, and is inexpensive.

Electrocorticography (ECoG): The ECoG is a procedure that uses external electrodes placed on the surface of the brain to monitor electrical activity in the cerebral cortex. With ECoG, you can get better results than with an EEG in terms of spatial and temporal resolution, amplitude, and resistance to distortions such blinks and eye movement. Nevertheless, electrocardiogram (ECoG) recording involves a craniotomy to insert an electrode grid, which is a highly invasive procedure that poses serious risks to the patient's health. Animals were used in the initial investigations of ECG because of this reason. The stability of the brain signals that ECoG could acquire over the long term was the focus of early animal investigations. Over the course of several months, the data demonstrated that subdural electrodes were capable of producing stable signals. However, it is not yet known if the signals obtained by ECoG will be stable in the long run. There is no need to recalibrate or experience drift in accuracy while using ECoG, according to more recent monkey trials. Even with asynchronous motions, the hand positions and angle of the arm joints could be deciphered. Additionally, these investigations have produced techniques for implanting the ECoG probes with low invasiveness.

Magnetoencephalography (MEG): Using magnetic induction, MEG is able to record the brain's magnetic activity without intrusive procedures. MEG detects magnetic fields external to the skull by measuring intracellular currents passing through dendrites. There is no difference between the neurophysiological mechanisms that generate EEG signals and those that generate MEG signals. However, MEG is more sensitive to the effects of primary currents, whereas EEG is extremely sensitive to secondary current sources. One advantage of MEG is that, unlike electric fields, magnetic fields are less impacted by the human head and neck.

2.2 Convolutional Neural Networks

Among the many ML models used for image recognition (Naranjo-Torres et al., 2020), audio recognition and natural language processing and CNNs stand out. CNNs have showed outstanding performance in picture identification, attaining cutting-edge results across a wide range of benchmarks. A multilevel design of levels that perform convolution operations and extract features at various abstraction levels contributes to the accumulation of spatial information and patterns in images. This is critical to their success (Krichen, 2023). CNNs are a common and effective neural network type for image identification. Several layers with complete connectivity that create predictions based on information gathered from the input image using a sequence of convolutional and pooling layers. A CNN must be trained on a large dataset of annotated images that include the target items before it can be used for picture identification. Using back propagation and optimization, during training, the CNN learns to link the retrieved characteristics to the correct labels. (C. Zhang et al., 2015). After training a CNN, it may be used to predict outcomes in novel, unseen images by feeding them into the network and picking the label with the best likelihood of success.

2.2.1 CNN Layers

Multiple layers, each performing a distinct network function, make up a CNN:

Convolutional Layer: A CNN's core component is the convolutional layer (Pandian et al., 2022). It uses a collection of trainable filters to perform a convolution operation on the input image. At each level, the tiny matrices that comprise the filters calculate a dot product with a small portion of the input image. The convolutional layer generates characteristic maps that represent the input image's many attributes.

Pooling Layer: The convolutional layer creates feature maps that have substantial spatial dimensions; the pooling layer reduces these dimensions. Each feature map is independently down sampled by determining the average or maximum value of non-overlapping regions. Pooling not only improves the network's resilience to tiny translations in the input image, but it also reduces its computational cost.

Activation Layer: Using the output from the previous layer as input. For activating the layer, an activation function that is nonlinear is applied. This makes the network non-linear, which improves its ability to learn complicated characteristics.

Batch Normalization Layer: The batch normalization layer normalizes the output by dividing it by the batch standard deviation. (Wang et al., 2020). This aids in enhancing the network's convergence and decreasing the internal covariate shift.

Dropout Layer: According to research the dropout layer has the ability to randomly remove a certain percentage of neurons from the previous layer when training. This encourages the network to acquire more robust features, which aids in avoiding overfitting.

Fully Connected Layer: An example of a classic neural network layer, the completely connected layer links all of the neurons in the current and preceding layers Usually, it's employed to generate the final output at the very end of the network.

Not only do some CNN types use these layers, but there are also specific CNN layers. A Recurrent Convolutional Neural Network (RCNN) uses a recurrent layer to identify the Causal correlations in the input sequence. A CNN with Long Short-Term Memory (LSTM) uses an LSTM layer to learn the input sequence's long-term dependencies.

Table 1. Functions of CNN layers

Layer	Function
Convolutional Layer	Feature maps are produced by convolutionally processing the input image through a library of trainable filters
Pooling Layer	The highest or mean value of independent sections is utilized to construct the physical dimensions of the feature maps formed by the convolutional layer
Activation Layer	A nonlinear function of activation is then applied to the outputs of layers to make them non-linear
Batch Normalization Layer	To promote convergence and minimize internal covariate shift, it regulates the eliminate the mean and divide by the batch standard deviation to get the output of the previous layer
Dropout Layer	During the training process, the network arbitrarily removes a fraction of neurones from the preceding layer to avoid overfitting and enhance the acquisition of more powerful features
Fully Connected Layer	The final output is frequently created at the network's end by connecting all neurones in the preceding layer to all neurones in the current layer

Table 1 shows the most common CNN layers and their functions within the network. To make a good prediction or classification, a CNN uses its many layers to learn about the input picture, extract its characteristics, and then apply that knowledge.

Researchers and practitioners can improve CNN architectures for various computer vision tasks by gaining a better grasp of the role of each layer.

3. PREPROCESSING FOR CNN-BASED BCI SYSTEMS

Data pretreatment for data mining refers to the procedures carried out before a data mining method is applied. It is a well-known and significant part of the renowned Knowledge Discovery from Data process (García et al., 2015). It covers a wide range of fields. The former seeks to simplify the data by selecting features, instances, or discretization; the latter strives to transform, integrate, clean, and normalize the data (García et al., 2016). The completed data set can be considered a solid and appropriate resource for any kind of information mining approach that follows a successful data pretreatment stage. Data preparation is required for more than only the traditional data mining activities of classification and regression. Data preprocessing is gaining popularity among academics in new data mining domains as a means to enhance their models. As gathering information techniques are more commonly employed, new models are developed or old ones are adapted to use in related frameworks.

Noise Treatment: It is essential to address the issue of noise data since data quality determines the quality of findings acquired by data mining techniques (Frénay, Benoît, 2013). It is possible for noise to influence either the input features or the output values in supervised problems. A common term for the presence of noise in the input attributes is attribute noise. When noise impacts the output attribute, the added bias is larger, which is the worst possible scenario. Class noise is the common name for this type of noise because of all the research on it in classification. First, if the noise is influencing the instance's labelling, it should be corrected using data cleaning procedures. Although it is a challenging process and typically reserved for modest amounts of noise, even partial noise reduction is said to be useful. The second option is to employ noise filters. These filters can detect and eliminate any instances of noise on learning data regardless of the data mining technique.

Missing Values Imputation: The completeness of the data set is a major presumption of data mining methods. On the other hand, acquisition procedures frequently involve missing values. A missing value is a piece of data that was either not collected or was not stored because of reasons such as an inaccurate sampling method, financial constraints, or acquisition limits. In data analysis, missing values are inevitable and often cause practitioners a great deal of trouble. Treating missing values is challenging. Poor knowledge extraction and incorrect conclusions are easy outcomes of improper management of missing values. There have been reports that missing values can lead to inefficient knowledge extraction, significant biases if the technique for introducing missing values is not handled properly, and serious problems when dealing with data.

Imperfect Data: A purportedly complete or noise-free data set is the foundation of most data mining techniques. There is a significant lack of completeness and cleanliness in real-world data. Techniques to either impute fill in missing data or remove noisy data are commonly used in data preprocessing. Imputation of missing values and noise filtering are the topics of the next two sections.

Separation and Standardization

Normalization merely makes adjustments to distributions, whereas discretization changes continuous variables using discrete intervals.

Elementwise Product: Multiples all features by a fixed scalar.

Normalizer: Brings all rows to the standard of one unit. The parameter p specifies the p-norm that is employed.

MinMaxScaler: Uses Two parameters are used to normalize Each characteristic has a specified range: the lower bound and the maximum limit.

Discrete Cosine Transform: Converts a time-domain convert an actual values sequence to a frequency-domain actual values sequence of the same size.

Bucketizer: Makes use of buckets to discretize a collection of continuous features. The quantity of buckets is determined by the user.

Binarizer: Transforms numerical characteristics into binary characteristics. The data is presumed to follow a Bernoulli distribution by this method. Features are assigned a value of 1.0 if they exceed a certain threshold and a value of 0.0 otherwise.

Standard Scaler: Applies a normal distribution to all features.

Figure 2. Data Preprocessing Tasks

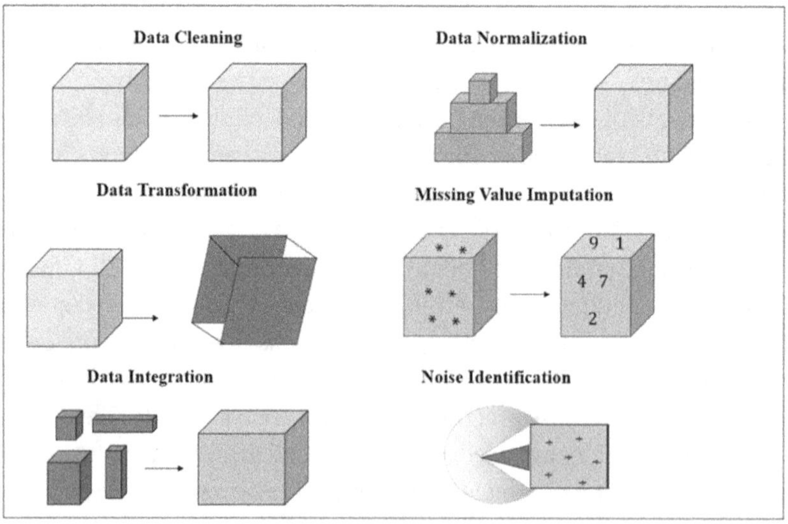

Data preprocessing allows for the processing of data that would be infeasible to handle without adapting the data to meet the criteria of each data mining method. Data preparation and reduction procedures are part of it, as shown in Figure 2.

4. CNN CATEGORIES FOR BCI APPLICATIONS

4.1 Attention Based Convolutional Neural Networks

The different levels of abstraction have a significant impact on the CNN's discrimination capacity. Learning relies heavily on several abstraction hierarchies that centre on features important for image localisation and recognition (Bhatt et al., 2021). This attention is a phenomenon that occurs in humans' visual systems. Humans may see any scene by merging short views and focusing on the details that are significant in a particular situation. Improved visual structure capture is the result of this method's attention on specific regions and comprehension of many item interpretations at a given spot. The interpretations used by RNN and LSTM are rather similar. Both LSTM and RNN use components of attention as an increasing characteristic, utilizing novel tasters valued according to their physical characteristics in previous rounds. In order to increase representation and overcome computational limits, numerous scientists utilise the concept of attention in CNN. With the help of this attention idea, CNN has progressed to the point where it can intelligently differentiate objects, even against complicated and chaotic backgrounds.

4.2 Width Based Multiple Connections

CNN developments mostly focused on regularising networks to maximise the depth and efficiency of connections. Therefore, breadth is just as crucial as depth when formulating pedagogical tenets. In order to keep their universal approximation property and increase their depth, neural networks with ReLU activation functions must be suitably wide, as proved (Hanin & Sellke, 2017). A major drawback of DNN designs is that multiple layers might not pick up on important features. In certain cases, the NN's learning power can be improved by adding more layers, which increases depth. Nevertheless, this is not always the situation. Furthermore, no deep network is capable of arbitrarily If the input dimension exceeds the network's maximum width, approximate a class of continuous functions on a compact set (He et al., 2017). In order to tackle this challenge, the research shifted its attention from deep and narrow designs to wide and thin architectures.

4.3 CNN Based on Spatial Exploitation

CNN has several components, including biases, weights, the number of layers, neurons, activation function, stride, filter size, learning rate, and so on. Due to the fact that convolutional operations take the locale of input pixels into account, varying filter sizes allow for the investigation of various correlation values (Alam et al., 2021). Generally, smaller filters collect fine-grained information, while larger filters extract coarse-grained information; this is only one example of the way distinct filter sizes have varying degrees of granularity. As a result, in the early 2000s, researchers implemented spatial filters to enhance performance. It was demonstrated that network learning and spatial filters are connected. During this period, researchers discovered that changing the filters of a CNN might increase its performance on both coarse and fine-grained details.

4.4 CNN Based on Dimensions

Despite its processing cost, the typical the convolution layer encodes channel-wise and spatial information. To improve the efficacy of conventional convolutions, separable or depth-wise separable convolutions were devised. These convolutions independently encapsulate spatial and channel-wise information, which are derived using point-wise and depth-wise convolutions, respectively. Despite the fact that this factorization is exponentially more effective, point-wise convolutions become a computational barrier due to the substantial computing cost they impose.

4.5 Exploitation-Based Feature-Map (Channel Map) CNNs

Computer vision problems have shown a lot of interest in CNN due to its ability to do automatic feature extraction and hierarchical learning(LeCun et al., 2010). Feature selection significantly impacts the performance of detection, segmentation, and classification modules. CNNs use weight adjustments associated with kernels also called masks to dynamically choose features. Furthermore, CNN mines a variety of features through a number of feature extraction steps. However, when it comes to object discrimination, certain feature maps are utterly irrelevant. A noise effect, which can lead to overfitting of the network, might be produced by very large feature sets. Feature map selection, together with network engineering, can be a game-changer when it comes to making networks more generalisable. There is a lot of conflation in the literature between feature maps and channel words.

4.6 CNN Based on Depth

More complex feature hierarchies and extra nonlinear mappings allow the network to more accurately approximatively estimate the target function, which is the fundamental premise of deep CNN architecture. For supervised training, the depth of the network has been a crucial parameter. Compared to shallow systems, deep networks are better at representing certain classes of functions. A theorem known as universal approximation was put forth by(Csanád Csáji, 2001). It detailed how any function can be approximatively used with just one concealed layer. However, the technique produces an excessively huge number of neurons and is computationally expensive. Theoretically, deeper networks can sustain their dramatic impact at a lower cost, according to (Szegedy et al., 2016). (Delalleau, 2011) provided empirical evidence in 2013 that deep networks are computationally more efficient for complex activities. Both VGG and Inception achieved top results in the ILSVRC-2014 competition, which provides more evidence that network depth is a key parameter for controlling their learning capacity.

4.7 Feature-Map Exploitation Based CNNs

The popularity of CNN for MV tasks was largely due to its capacity to perform automatic feature extraction and hierarchical learning. It comes to detection, segmentation, and classification modules, feature selection is king. CNN chooses features on the fly by modifying the weights linked to a kernel, which are also called a mask. Additionally, numerous feature types called feature maps or channels in CNN are made possible through the various feature extraction steps. But object discrimination isn't the strong suit of all feature maps (Salman et al., 2019). Network over-fitting can occur if there are too many features, which creates a noise effect. This suggests that picking the right feature maps, together with network engineering, can be an important factor in making networks more generalisable.

4.8 CNNs with Multiple Paths

Complex tasks are typically well-suited for deep CNNs. Increasing the depth instead of overfitting can cause performance loss, explosion difficulties, or the disappearance of gradients. A rise in both the test error and the training error is caused by the vanishing gradient problem (Tong et al., 2017). Deep learning networks were suggested using the multi-path or cross-layer connectivity theory. Analytical links between layers can be established using shortcut connections or multiple channels that bypass intermediate levels, enabling the customised flow of information between them (Hu, 2018).Using cross-layer connectivity, the network is divided into portions. These pathways handle the vanishing gradient problem by extending it to lower layers.

Figure 3. Categories of CNN Versions

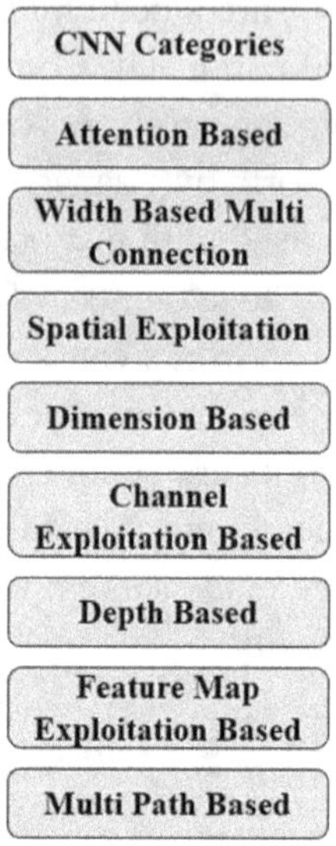

Figure 3 shows the different types of CNN architectures. This section provides a detailed explanation of each of the categories.

5. APPLICATIONS OF CNNS IN BCIS

The Design and Use of Convolutional Neural Networks: A High-Level Overview (Sakib et al., 2019).

5.1 Motor Imagery Classification

Individuals with cognitively intact but severely impaired motor skills require a different means of interacting with their surroundings (Qin et al., 2004). The advancement of BCI technology over the last several decades has opened up a new and exciting avenue of communication for these individuals (McFarland & Wolpaw, 2011). The technology takes the patients' intentions and turns them into physical commands that control the equipment that provide the desired output. In order to gather information for BCI, researchers have employed both non-invasive and intrusive methods. There are non-invasive approaches that can be used, including as optical imaging, FMRI, Positron Emission Tomography (PET), Magnetoencephalography (MEG), and EEG. EEG based BCI systems are presently extensively utilised due to their short time constants, adaptability to various environments, and affordability. While intracranial recording techniques like electrocorticograms and single-neuron recordings take advantage of greater signal quality they are invasive. The development of completely non-invasive methods to control BCI devices would be highly desirable, as it would prevent the use of brain implants in human patients.

5.2 Cognitive Load and Emotion Recognition

Applications like intelligent human-computer interface design or personalised information retrieval (e.g., multimedia material) rely on the user's effective states (Xu & Plataniotis, 2016). New studies have shown that effective computing performance can be improved with the use of deep learning models (Haque et al., 2022). There are multiple ways in which an emotion might be described. Emotions are often best understood in terms of their spatial dimensions, and dimensional models of emotion make an effort to do just that. Emotions are categorised along two dimensions: arousal and valence, according to the most used circumplex paradigm. The term arousal describes the degree to which an emotional stimulus evokes a strong response. The valence of an emotion, whether positive or negative, is the connection it has within the individual experiencing it. The liking and dominance aspects are substituted in different models. Using deep learning algorithms, some publications merely try to categorise the user's emotional state as either positive or negative, or as one of three possible states: neutral, positive, or negative. Several articles used CNN or one of its versions to categorise EEG signals as representing emotions. The DBN-RBM is the best DL model for uncovering hidden characteristics in spontaneous EEG recordings of emotions. In order to solve the problem of affective state classification using EEG signals, (Xu & Plataniotis, 2016) utilised a DBN-RBM as specific feature extractors. Additionally, on a deeper level, there are researchers who strive to identify a positive or bad condition for each particular emotional component. An example of this is the AE ensemble classifier for emotion recognition that was suggested by (Yin et al., 2017).The three hidden layers in each AE are responsible for obtaining stable feature representations and filtering out undesired noise in the physiological features. The suggested model passed the benchmark DEAP evaluation with an arousal score of 77.19% and a valence score of 76.17%.

5.3 Epileptic Seizure Detection

An EEG is a major tool for studying epilepsy and capturing changes in electrical brain processes that may signal an upcoming seizure. Seizures are a serious clinical concern (Leijten, 2018).Manual examination of EEG is necessary for the diagnosis of epilepsy; however, this process is laborious and prone to mistakes. Elger and Hoppe discovered that patients accurately recorded less than half of their epileptic seizures, while long-term video EEG monitoring missed over half of the episodes (Elger & Hoppe, 2018). The creation of a trustworthy intelligent diagnostic method for automated seizure detection is of paramount importance. The process of detecting seizures entails differentiating between the inter-ictal, pre-ictal, and ictal phases of a seizure. Two main steps, feature extraction and classification, typically make up a seizure detection process (Wei et al., 2018). One group of methods used feature extraction: time-frequency analysis, nonlinear dynamics complexity, synchronisation, and increments of accumulated energy. In contrast, machine learning classifiers include things like Support Vector Machines (SVMs), Bayes Networks, and CNN.

5.4 Communication Systems for Locked-In Syndrome

Applications for BCIs allow people, particularly those with physical limitations, to communicate with others by means of brain signals. Loss of motor control, including eye movement, can occur in people with multiple sclerosis ALS and other physical problems. Though other cognitive brain activities like thinking and imagination are unaffected by these impairments, they are nevertheless noticeable Therefore, people with advanced Motor Neurone Disease (MND) may be able to generate electrical impulses just by thinking about it (Majumdar et al., 2023). Emotions and experiences experienced by people with these disorders are detectable through brain signals, according to studies (Kundu & Ari, 2022).The EEG is a method for recording continuous brain wave patterns; it is rapidly becoming one of the most useful tools for biomedical engineering research and diagnostics and is thus crucial to the development of BCIs. Eye activity-based (e.g., eye gaze) and cognitive activity-based (e.g., P300 BCI) systems are among the few technological solutions that people with ALS can use for communication (Garcia et al., 2017) current methods according to their cost, portability, invasiveness, and temporal resolution. EEGs have applications in health care. Feature extraction and selection are crucial processes in developing BCIs because they allow us to quantify the signal's primary structural components and because these components are necessary for the classification performance of brain signals.

6. CHALLENGES IN CNN-BASED BCI SYSTEMS

A lot of work on BCIs has gone into developing new kinds of assistive devices in recent years. The present state of BCIs makes the imminent commercial release of BCI technology highly probable. Actually, there are a handful of commercial BCI appliances available now (Rashid et al., 2020). Is. Nevertheless, the BCI community must resolve some critical concerns and obstacles in each part of the BCI paradigm if BCIs are to progress further.

6.1 EEG Headsets Issues

In order to use EEG data for a BCI, the quality of the EEG headgear is crucial. There are a few problems with EEG headsets that have to be fixed. To begin, the electrodes used by the majority of EEG headsets are somewhat uncomfortable, requiring either gel or liquid. Since conductive gel is not required to be applied between the scalp skin and the electrode pad, users tend to favour dry electrodes for practical applications. The question of whether this electrode type provides the same signal characteristics is, however, still up for discussion. While one study found that the signal characteristics of wet and dry electrodes were almost comparable, another found that EEGs recorded with dry electrodes contained far more noise and artefacts than those recorded with wet electrodes. Therefore, dry electrodes need to be thoroughly tested to confirm their effectiveness. It should be mentioned that there are currently investigations into EEG electrodes that are based on water. According to (Mihajlović et al., 2012) shorter hair on dry and water-based electrodes performs similarly to, if not better than, electrodes made of gel. Another obstacle to the widespread use of BCI technology is the high upfront cost of EEG headsets (Abiyev et al., 2016). As a result, the price cut for these EEG headsets is greatly appreciated. One important aspect of the BCI paradigm is its portability; people may move around more freely with wireless data transmission headsets, whereas a corded headset somewhat restricts movement.

6.2 Inadequate Methods for Data Analysis

Section EEG Data Preprocessing Strategies lists a number of methods that, when applied to a specific EEG-based application, have various constraints, despite the fact that the majority of artefact removal algorithms provide good performance. To be sure, there are techniques whose sole purpose is to identify and eliminate certain artefacts. Not all uses can benefit from methods that rely on reference channels to improve artefact removal accuracy. Algorithms based on ICA can handle any type of artefact that may appear in an EEG recording. When particular artefact reference channels are provided, regression and adaptive filters become more viable options. In terms of muscle artefact reduction, CCA and its combination with other methods seem to be a viable alternative to ICA. EMD, IVA, and its hybrid approaches with BSS or WT can be highly advantageous for applications that involve a limited number of channels. In addition, EMD can greatly enhance signal quality by reducing data needed to eliminate noise (Dai et al., 2019).In order to deduce the important information from the EEG, several characteristics have been retrieved using feature extraction algorithms. When used to EEG motor imagery data, CSP and its modified algorithms have shown rather encouraging performance in prior research. Spectral features function better with SSVEP data for emotion detection and mental effort classification, but frequency-domain based features are more appropriate in other cases. Nevertheless, determining the best feature extraction method for a particular EEG control signal modality is premature.

Research has shown that SVMs are the best classifiers for high-dimensional feature vectors. In certain cases, HMM may handle feature sets with a high dimensionality by examining the feature vector sequence. It is advisable to employ a low-dimensional feature vector to mitigate computational complexity. However, feature reduction or selection techniques may be implemented when the feature sets are high-dimensional. The classification of MI, SSVEP, emotion recognition, and ERP data has been accomplished by the application of many deep learning techniques. Network topology and input formalisation are crucial to the design of deep networks. Compared to other deep learning methods, those that relied on CNNs and RNNs performed better. Furthermore, CNN provided the highest level of accuracy when fed spectral images or time-series values. As a result, research into various network topologies, including RBMs, convolutional layers, fully connected layers, and recurrent layers, as well as their relative sizes and configurations, has been proposed. In addition to studying network topology, more research is required to differentiate between artifact-prone EEG and raw EEG when using deep learning techniques. This article presents the results of an offline evaluation of a large variety of classifiers. An unresolved issue is the development of a stable Riemannian median estimator. Riemannian geometry classifiers could be made more robust by this property.

6.3 Problems with EEG Modalities in Brain-Computer Interfaces

One of the ongoing signal processing challenges that the most preferred EEG modalities MI, SSVEP, and P300 encounter is the identification of the most appropriate methods for feature extraction and reduction. The fundamental reason behind this is because EEG signals are highly susceptible to artefacts, are not steady, and exhibit very non-linear behaviour. Another important aspect to consider is data fusion, specifically how multiple electrodes' worth of data is combined to increase classification accuracy and reduce data dimensionality. It is clear that MI tactics frequently produce outstanding results after folks have been taught correctly. The training trials required to train a targeted user to operate a MI-based BCI are excessive, leading to an unacceptably long calibration period for a realistic model. Research efforts should thus centre on developing more efficient training methods and methods that shorten the calibration period. Although P300's efficacy may be significantly affected by the kind and severity of the disease, it does not require training and displays higher average ITRs. Hence, instead of using only one EEG modality, a mixed strategy could work wonderfully. If we want to know which EEG modality is best for BCI use, we need to investigate further. For example, when dealing with a prosthetic limb, methods based on MI are seen to be the most suitable. Instead of using MI, where imagined movements are linked to television controls, an SSVEP-based solution with a 4-option menu that includes volume-up, volume-down, station-up, and station-down could be far more intuitive when operating digital radio. When deciding on a technology and BCI category in the outset, intuitiveness should play a pivotal role. Finding the best feature collection and classification frameworks for individual EEG modalities requires more research. Investigating the best characteristics and classifiers to use for people with CNS injuries should also be a priority in this area of research.

Figure 4. A Brain-Computer Interface Basic Design

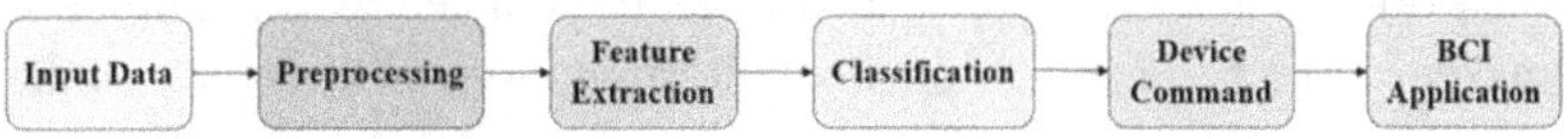

The many steps of processing an EEG signal for a BCI are shown in Figure 4, which is a typical block diagram. During this stage, different kinds of EEG sensors are used to record the user's brain activity. There are a number of artefacts in the raw EEG data that are removed during the pre-processing step. The goal of feature extraction is to describe the signals using a small set of relevant values, or features; frequently, the selection of important features is also explored during this stage. During the classification step, various deep learning and machine learning techniques are used to sort the extracted characteristics. The development of real-life BCI applications culminates with the translation of categorised outcomes into device commands.

7. FUTURE DIRECTIONS

There are still technical and usability issues with BCI devices, even though deep learning has improved their performance. The classification ability in complicated BCI contexts is a technological issue, whereas restrictions in large-scale real-world implementation are an example of a usability challenge. Here, we lay out the problems and suggest ways to fix them.

Researching ML and DL approaches, it is critical to consider how long data will take to process in real-time applications (Lionakis et al., 2023). Even when deep learning achieves high accuracy rates, the question of whether the benefits outweigh the processing costs arise. Although DL inference is of sufficient speed to be considered real-time, the intricate designs of DL necessitate a significant amount of processing resources. The training of deep learning models frequently necessitates extraordinarily powerful devices in terms of processing power, memory, and GPUs, and it frequently consumes a substantial quantity of time. It is critical to discover a solution to deploy DL models with fewer resources and training time needs while maintaining accuracy.

Significant challenges arise when transitioning from a static laboratory from a steady patient environment to a real-time application in a dynamic real-world context. Environmental interference, movement artefacts, and physiological fluctuations are only a few examples of the many forms of noise that can penetrate an EEG signal in the real world. Signal quality can be drastically reduced by these noise sources, which makes correct classification and interpretation much more challenging. Consequently, more thought and focus are required to tackle the intricacies of the real-world study domain.

Important factors include the variety and quality of the dataset, which include several features unique to each dataset. However, a significant limitation is the lack of a standardised benchmark dataset. Augmentation techniques, such as artificial augmentation, may be able to fill this void. These methods try to mimic various real-world situations by increasing the dataset's horizon by adding more variants to current data. The intricacies of human experiences and interactions may be too great for technological augmentation to fully capture. The datasets are already complicated, and there is still a lot of mystery around the precise experimental settings.

7.1 Unsupervised and Semi-supervised Classification

Collecting an abundance of class labels in a variety of situations, like sleeping EEG, necessitates costly and laborious manual labelling, which in turn greatly affects the performance of deep learning models. Unsupervised learning does not need labels while semi-supervised learning uses partial labels instead of labels altogether, in contrast to supervised learning which uses both (Jia et al., 2014). Consequently, issues where there is a lack of reliable ground truth data are better addressed by using them. For strong and efficient semi-supervised learning, (X. Zhang et al., 2019) proposed the Adversarial Variational Embedding (AVAE) framework. This framework incorporates a VAE++ model. This is a high-quality generative model with a semi-supervised GAN and a posterior distribution learner. In their semi-supervised approach, (Jia et al., 2014) used label information for feature extraction and combined supervised training with unlabelled data to make it more consistent. Unsupervised domain adaptation learning and crowdsourcing unlabelled observations are two ways to improve unsupervised learning. Another way is to use a linear transformation to match the distribution of source and target BCI signals.

7.2 Person-Independent Classification

The training and testing sets are derived from the same person, and the bulk of BCI classification tasks have been based on person-specific scenarios (Li et al., 2015). The objective for the future is to accomplish Person-independent categorization is used to prevent testing data from being included in the training set (Haque et al.,2022). Effective person-independent classification is required for the broad application of BCI systems in the real world. Using transfer learning to construct a customised model is one approach that could work towards this objective. A personalised effective model can employ a transudative parameter transfer strategy to construct individual classifiers and create a regression function that establishes a connection between the distribution of data and the classifier parameters (Aldayel et al., 2020). Finding the subject-independent components in the supplied data is another possible solution. The input data consists of two components: one that is subject-dependent and associated with the subject, and another that is subject-independent and shared by all subjects. Person identification and class recognition are two distinct yet related tasks that can be managed by a hybrid multi-task model. In a class recognition challenge, the subject-independent characteristics should be extracted by a well-trained and converged model.

8. CONCLUSION

By establishing a direct communication channel between the brain and a computer, a BCI eliminates the necessity for the conventional neuromuscular system, thereby allowing for the control of an external device. BCIs allow users to issue beneficial commands by identifying fluctuations in recorded physiological signals from the brain that are associated with their mental state or intentions (Ravi et al.,

2020). BCIs have provided patients with neurological disorders such as paralysis, ALS, or stroke with novel neurorehabilitation options. by means of endogenous BCIs, it is possible to subtly regulate the neural activity of a user. Motor imagery BCI, imagined tactile responses-based BCI, and other paradigms are included in this category. In order to provide context about the user's intent, exogenous BCIs rely on an outside stimulus to alter the user's neural activity. With their advanced methods for deciphering complicated neural signals, CNN have become crucial instruments in the development of BCI. Future BCIs stand to benefit greatly from developments in CNN architectures, hardware technology, and data techniques, notwithstanding the obstacles that have already been encountered.

REFERENCES

Abiyev, R. H., Akkaya, N., Aytac, E., Günsel, I., & Çağman, A. (2016). Brain-Computer Interface for Control of Wheelchair Using Fuzzy Neural Networks. *BioMed Research International, 2016*, 1–9. DOI: 10.1155/2016/9359868 PMID: 27777953

Alam, N.-A., Ahsan, M., Based, M. A., & Haider, J. (2021). Intelligent System for Vehicles Number Plate Detection and Recognition Using Convolutional Neural Networks. *Technologies*, *9*(1), 9. DOI: 10.3390/technologies9010009

Bhatt, D., Patel, C., Talsania, H., Patel, J., Vaghela, R., Pandya, S., Modi, K., & Ghayvat, H. (2021). CNN Variants for Computer Vision: History, Architecture, Application, Challenges and Future Scope. *Electronics (Basel)*, *10*(20), 2470. DOI: 10.3390/electronics10202470

Csanád Csáji, B. (2001). Approximation with artificial neural networks. *Faculty of Sciences, January*, 1–45. https://citeseerx.ist.psu.edu/viewdoc/download?doi=10.1.1.101.2647&rep=rep1&type=pdf

Dai, M., Zheng, D., Na, R., Wang, S., & Zhang, S. (2019). EEG Classification of Motor Imagery Using a Novel Deep Learning Framework. *Sensors (Basel)*, *19*(3), 551. DOI: 10.3390/s19030551 PMID: 30699946

deCharms, R. C., Christoff, K., Glover, G. H., Pauly, J. M., Whitfield, S., & Gabrieli, J. D. E. (2004). Learned regulation of spatially localized brain activation using real-time fMRI. *NeuroImage*, *21*(1), 436–443. DOI: 10.1016/j.neuroimage.2003.08.041 PMID: 14741680

Delalleau, O. (2011). Shallow vs. deep sum-product networks. *Advances in Neural Information Processing Systems*, 1–9.

Elger, C. E., & Hoppe, C. (2018). Diagnostic challenges in epilepsy: Seizure under-reporting and seizure detection. *Lancet Neurology*, *17*(3), 279–288. DOI: 10.1016/S1474-4422(18)30038-3 PMID: 29452687

Frénay, Benoît, and M. V. (2013). Classification in the presence of label noise: A survey. *IEEE Transactions on Neural Networks and Learning Systems.*

Garcia, L., Loubière, B., & Renault, L. G, L. M., Najib, V., & Jm, A. (2017). A comparison of a brain-computer interface and an eye tracker: Is there a more appropriate technology for controlling a virtual keyboard in an ALS patient? In Advances in Computational Intelligence. *14th International Work-Conference on Artificial Neural Networks.*

García, S., Luengo, J., & Herrera, F. (2015). Instance Selection. In *Intelligent Systems Reference Library* (Vol. 72, pp. 195–243). DOI: 10.1007/978-3-319-10247-4_8

García, S., Ramírez-Gallego, S., Luengo, J., Benítez, J. M., & Herrera, F. (2016). Big data preprocessing: Methods and prospects. *Big Data Analytics*, *1*(1), 9. DOI: 10.1186/s41044-016-0014-0

Haque, S., Eberhart, Z., Bansal, A., & McMillan, C. (2022). Semantic Similarity Metrics for Evaluating Source Code Summarization. *IEEE International Conference on Program Comprehension, 2022-March*(May), 36–47. DOI: 10.1145/3524610.3527909

Hilmi, R. Z., Hurriyati, R., & Lisnawati. (2019). *EEG dataset and OpenBMI toolbox for three BCI paradigms: An investigation into BCI illiteracy.* (Vol. 3, Issue 2).

Hu, J. (2018). Squeeze-and-excitation networks. *Proceedings of the IEEE Conference on Computer Vision and Pattern Recognition*, 7132–7141. https://openaccess.thecvf.com/content_cvpr_2018/html/Hu_Squeeze-and-Excitation_Networks_CVPR_2018_paper.html

Jia, X., Li, K., Li, X., & Zhang, A. (2014). A Novel Semi-Supervised Deep Learning Framework for Affective State Recognition on EEG Signals. *2014 IEEE International Conference on Bioinformatics and Bioengineering*, 30–37. DOI: 10.1109/BIBE.2014.26

Kundu, S., & Ari, S. (2022). Brain-Computer Interface Speller System for Alternative Communication: A Review. *Ingénierie et Recherche Biomédicale : IRBM = Biomedical Engineering and Research*, *43*(4), 317–324. DOI: 10.1016/j.irbm.2021.07.001

LeCun, Y., Kavukcuoglu, K., & Farabet, C. (2010). Convolutional networks and applications in vision. *Proceedings of 2010 IEEE International Symposium on Circuits and Systems*, 253–256. DOI: 10.1109/ISCAS.2010.5537907

Leijten, F. S. S. (2018). Multimodal seizure detection: A review. *Epilepsia*, *59*(S1), 42–47. DOI: 10.1111/epi.14047 PMID: 29873832

Majumdar, S., Al-Habaibeh, A., Omurtag, A., Shakmak, B., & Asrar, M. (2023). A novel approach for communicating with patients suffering from completely locked-in-syndrome (CLIS) via thoughts: Brain computer interface system using EEG signals and artificial intelligence. *Neuroscience Informatics (Online)*, *3*(2), 100126. DOI: 10.1016/j.neuri.2023.100126

Mane, R., Chew, E., Chua, K., Ang, K. K., Robinson, N., Vinod, A. P., Lee, S.-W., & Guan, C. (2021). *FBCNet: A Multi-view Convolutional Neural Network for Brain-Computer Interface*. http://arxiv.org/abs/2104.01233

McFarland, D. J., & Wolpaw, J. R. (2011). Brain-computer interfaces for communication and control. *Communications of the ACM*, *54*(5), 60–66. DOI: 10.1145/1941487.1941506 PMID: 21984822

Mihajlović, V., Molina, G. G., & Peuscher, J. (2012). TO WHAT EXTENT CAN DRY AND WATER-BASED EEG ELECTRODES REPLACE CONDUCTIVE GEL ONES? - A Steady State Visual Evoked Potential Brain-computer Interface Case Study. *Proceedings of the International Conference on Biomedical Electronics and Devices, January*, 14–26. DOI: 10.5220/0003726000140026

Naranjo-Torres, J., Mora, M., Hernández-García, R., Barrientos, R. J., Fredes, C., & Valenzuela, A. (2020). A Review of Convolutional Neural Network Applied to Fruit Image Processing. *Applied Sciences (Basel, Switzerland)*, *10*(10), 3443. DOI: 10.3390/app10103443

Nicolas-Alonso, L. F., & Gomez-Gil, J. (2012). Brain Computer Interfaces, a Review. *Sensors (Basel)*, *12*(2), 1211–1279. DOI: 10.3390/s120201211 PMID: 22438708

Qin, L., Ding, L., & He, B. (2004). Motor imagery classification by means of source analysis for brain–computer interface applications. *Journal of Neural Engineering*, *1*(3), 135–141. DOI: 10.1088/1741-2560/1/3/002 PMID: 15876632

Rashid, M., Sulaiman, N., & Abdul Majeed, P. P., A., Musa, R. M., Ab. Nasir, A. F., Bari, B. S., & Khatun, S. (. (2020). Current Status, Challenges, and Possible Solutions of EEG-Based Brain-Computer Interface: A Comprehensive Review. *Frontiers in Neurorobotics*, *14*(June), 1–35. DOI: 10.3389/fnbot.2020.00025 PMID: 32581758

Sakib, S., Ahmed, N., Kabir, A. J., & Ahmed, H. (2019). An Overview of Convolutional Neural Network: Its Architecture and Applications. In Preprints *2018* (Issue February). DOI: 10.20944/preprints201811.0546.v4

Salman, H., Grover, J., & Shankar, T. (2019). Effect of depth and width on local minima in deep learning. *Neural Computation*, *2733*(March), 2709–2733. DOI: 10.1162/NECO

Szegedy, C., Ioffe, S., Vanhoucke, V., & Alemi, A. (2016). Inception-v4, Inception-ResNet and the Impact of Residual Connections on Learning. *Pattern Recognition Letters*, *42*(1), 11–24. http://arxiv.org/abs/1512.00567

Taga, G., Homae, F., & Watanabe, H. (2007). Effects of source-detector distance of near infrared spectroscopy on the measurement of the cortical hemodynamic response in infants. *NeuroImage*, *38*(3), 452–460. DOI: 10.1016/j.neuroimage.2007.07.050 PMID: 17884584

Tong, T., Li, G., Liu, X., & Gao, Q. (2017). Image Super-Resolution Using Dense Skip Connections. *2017 IEEE International Conference on Computer Vision (ICCV), 2017-Octob*, 4809–4817. DOI: 10.1109/ICCV.2017.514

Wei, X., Zhou, L., Chen, Z., Zhang, L., & Zhou, Y. (2018). Automatic seizure detection using three-dimensional CNN based on multi-channel EEG. *BMC Medical Informatics and Decision Making*, *18*(S5), 111. DOI: 10.1186/s12911-018-0693-8 PMID: 30526571

Xu, H., & Plataniotis, K. N. (2016). Affective states classification using EEG and semi-supervised deep learning approaches. *2016 IEEE 18th International Workshop on Multimedia Signal Processing (MMSP)*, 1–6. DOI: 10.1109/MMSP.2016.7813351

Yin, Z., Zhao, M., Wang, Y., Yang, J., & Zhang, J. (2017). Recognition of emotions using multimodal physiological signals and an ensemble deep learning model. *Computer Methods and Programs in Biomedicine*, *140*, 93–110. DOI: 10.1016/j.cmpb.2016.12.005 PMID: 28254094

Zhang, C., Li, P., Sun, G., Guan, Y., Xiao, B., & Cong, J. (2015). Optimizing FPGA-based Accelerator Design for Deep Convolutional Neural Networks. *Proceedings of the 2015 ACM/SIGDA International Symposium on Field-Programmable Gate Arrays*, 161–170. DOI: 10.1145/2684746.2689060

Zhang, X., Yao, L., & Yuan, F. (2019). Adversarial Variational Embedding for Robust Semi-supervised Learning. *Proceedings of the 25th ACM SIGKDD International Conference on Knowledge Discovery & Data Mining*, 139–147. DOI: 10.1145/3292500.3330966

Chapter 5
Deep Learning for BCI Signal Decoding

Harpreet Kaur Channi
https://orcid.org/0000-0002-9059-8821
Guru Nanak Dev Engineering College, Ludhiana, India

Surinder Kaur
Guru Nanak Girls College, Ludhiana, India

Ramandeep Sandhu
https://orcid.org/0000-0003-2595-4030
Lovely Professional University, Jalandhar, India

ABSTRACT

BCIs enable direct brain-to-device communication by decoding brain signals, with applications in assistive technologies, neurorehabilitation, and human-computer interaction. This study explores deep learning models, including CNNs, RNNs, and hybrid architectures, for BCI signal decoding. A synthetic EEG dataset was used to benchmark these models against traditional classifiers like SVM and LDA, focusing on accuracy, latency, and generalization. Preprocessing techniques such as noise reduction, signal normalization, and feature extraction were applied to enhance signal quality. Results showed CNNs and hybrid CNN-RNN models outperformed traditional classifiers with greater accuracy and robustness. Deep learning effectively addressed challenges like low signal-to-noise ratio, subject variability, and class imbalance. The study also demonstrated practical applications in assistive device control and cognitive monitoring, emphasizing deep learning's potential to advance BCI technology while underscoring the need for validation on larger datasets.

DOI: 10.4018/979-8-3693-9445-8.ch005

1. INTRODUCTION

A BCI is a technology that establishes a direct communication pathway between the brain and an external device, allowing users to control machines or digital systems using brain activity as shown in figure 1. BCIs are designed to bypass conventional neuromuscular pathways, making them particularly valuable for individuals with severe motor disabilities. Acquisition and translation of a brain signal and processing for implementation in devices or their control would provide the core mechanism behind working out of a BCI. Mostly, neural activities create the major kind of generated signals for being recorded using diverse techniques including, amongst other things, EEG, ECoG, MEG, or even fNIRS (El Ouahidi et al., 2024).

According to the signal acquisition method, the BCI system can be classified broadly into three categories: invasive, partially invasive, and non-invasive. Invasive BCI involves putting electrodes directly inside the brain tissue. The highest signal fidelity comes with invasive BCIs but incurs severe risks from surgery. A non-invasive BCI, like an EEG-based system, is far more commonly used due to its safety and ease of use even though it provides lower signal resolution due to the attenuation of signals caused by the skull and scalp. Partially invasive systems like ECoG have balanced signal quality with invasiveness. Applications of BCIs include areas related to healthcare, assistive technologies, gaming, and cognitive neuroscience (Saibene et al., 2024). They are widely used for neuroprosthetic control, such as operating robotic limbs for individuals with spinal cord injuries. BCIs also contribute to cognitive state monitoring, mental workload assessment, and brain-controlled virtual reality environments. There are still significant challenges in improving signal quality, real-time processing capabilities, and user adaptability. Advances in artificial intelligence, for example, deep learning offer a new avenue in improving the accuracy of signal decoding and making BCI-based systems more viable. Future research in this field holds the potential to revolutionize human-computer interaction systems and assistive technologies within the next decade or so (Hameed, Khan, Ahmed, Aftab, & Fazal, 2025).

Figure 1. Brain Computer Interfaces

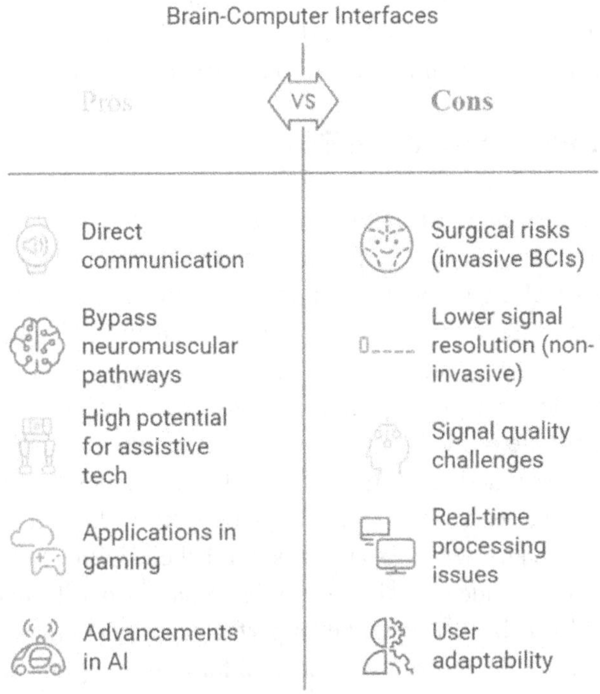

1.2 Importance of Signal Decoding in BCI

Signal decoding represents one of the most significant parts of BCI systems, as it interprets and extracts brain signals that result in meaningful device control commands. Thereby, the accuracy and efficiency with which a BCI system operates depend on the successful performance of its signal decoding. Without proper decoding, the brain signals recorded by modalities such as EEG, ECoG, or MEG would remain complex, noisy, and hard to interpret for practical applications. Effective signal decoding is crucial for improving the performance and usability of BCIs, especially in applications such as assistive technologies for individuals with motor impairments.It guarantees higher command accuracy, fewer false positives, and better real-time responsiveness, which is important for applications such as controlling prosthetic limbs or operating communication devices (El Ouahidi, Lioi, Farrugia, Pasdeloup, & Gripon, 2024). Moreover, robust signal decoding is an important aspect of cognitive monitoring applications like stress detection and mental workload assessment. However, signal decoding in BCI faces challenges like noise interference, low signal amplitude, and variability between subjects. This, in turn, has dramatically enhanced the signal decoding accuracy due to the automation of feature extraction and handling of complex data patterns by deep learning advances, thereby making BCIs more reliable and widening their potential applications in healthcare, neurorehabilitation, and brain-controlled interfaces (Forenzo et al., 2024).

1.3 Role of Deep Learning in BCI

Deep learning has changed BCI technology in a whole new way by greatly enhancing the betterment of accuracy and efficiency in the decoding of brain signals. Machine learning techniques often depend upon manual feature extraction and have limited control over the complexity and non-linear nature of brain signals. Deep learning automatically extracts meaningful patterns from raw neural data, so it is particularly suitable for application in BCI tasks because high-dimensional, noisy data are common. One of the key advantages of deep learning in BCI is its ability to process both spatial and temporal features of brain signals simultaneously. For example, CNNs are effective in capturing spatial features from the EEG channels, while RNNs and LSTM networks are chosen for their ability to handle temporal dependencies and, thus, ideal for time-series brain data. Apart from that, hybrids of CNNs and RNNs have been used for better performance in decoding complex brain activity. Deep learning also opens the potential for improving transfer learning and domain adaptation, so it is less affected by subject variability as well as the possible scarcity of data sets in BCI research. Generative models such as GANs have been used for data augmentation in order to solve the problem of data scarcity. As deep learning techniques evolve, they are enabling increasing accuracy, adaptability, and real-world practicality of BCIs in healthcare, neuroprosthetics, and brain-controlled systems (Zhang et al., 2024).

1.4 Objectives and Scope of the Study

The major focus of this study is on deep learning, especially in relation to the enhancement of Brain-Computer Interface signal decoding for more accuracy and reliability. The objectives of this paper are:

- Evaluate the performance of deep learning models (CNN, RNN, hybrid) for classifying motor imagery signals: left-hand movement vs. right-hand movement, using EEG data.
- Comparison of the efficiency of deep learning techniques versus traditional approaches like SVM, LDA, classification accuracy, and generalization.
- Discusses the mitigation of issues with low SNR and class imbalance in motor imagery classification tasks involving deep learning approaches.

The proposed study is of scope on a non-invasive BCI and mainly focuses on EEG-based devices due to its safety and massive usage. Emphasis is not on signal acquisition or hardware designing but on techniques of signal decoding in this study. Performance metrics such as accuracy, F1-score, and training efficiency will be used to evaluate the model. Even though the paper encompasses various deep learning techniques, it is only offline analysis and not real-time BCI systems. The research aims to contribute to the development of more efficient and adaptable BCI systems for healthcare, cognitive monitoring, and brain-controlled applications (Ji et al., 2024).

2. LITERATURE SURVEY

Recent advances in Brain-Computer Interface (BCI) research hold promising prospects that use Deep Learning (DL)-based algorithms for deciphering BCI signals. Measurement and control for Motor Imagery (MI)-based BCI tasks have emerged as challenging obstacles, according to Zhong et. al, 2024, where such authors underlined the role of transfer learning during improved signal encoding (Zhong, Yao, Pan, & Wang, 2024). Carrara et. al., 2024 focused on developing a DL algorithm based on geometric neural networks for effective BCI-EEG decoding with a limited number of electrodes (Carrara et al., 2024). Guetschel et. al., 2024 reviewed the deep representation learning techniques for BCI decoding based on motivation and approaches used in studies. They developed PySimMIBCI, a framework to generate realistic MI-EEG signals to validate decoding models and augment DL model training by generating data (Guetschel, Ahmadi, & Tangermann, 2024). Xu et. al., 2024 proposed a Temporal-Spatial Cross-Attention Network model for the recognition of imagined characters, since the relationships between temporal and spatial features in BCI signal decoding have been sparsely studied (Xu, Zhou, Shen, Qiu, & Li, 2024). Tibermacine et. al., 2024 reviewed the integration of DL with Riemannian geometry to enhance EEG signal decoding in BCIs, updating findings since 2017 (Tibermacine et al., 2024). Wu et. al., 2024 proposed a bimodal DL network based on CNN for fine motor imagery, obtaining high accuracy for classifying various MI tasks (Wu, Wang, Qiu, & He, 2024). With attention-based domain adversarial training for P300 detection, Li et. al., 2024 focused on inter-participant transfer learning and showed that improved BCI signal decoding is possible. A recent study, "3D Convolutional Neural Network for Decoding Motor Imagery EEG Signals", shows that spatial-spectral feature learning effectively supports decoding signals in BCI applications. (Li, Daly, Guan, Cichocki, & Jin, 2024). Overall, the literature review shows a growing interest in using DL algorithms for BCI signal decoding with various studies investigating different approaches to improve the accuracy and efficiency of BCI systems. Further research in this area is essential for advancing the field of BCI and improving the usability of BCI technology in clinical and practical settings.

2.1 Fundamentals of BCI Technology

A BCI is a communication system that enables direct interaction between the brain and an external device without reliance on peripheral nerves or muscles. It works by capturing the activity of the brain, processing the signals, and translating them into commands that control external systems such as prosthetic limbs, communication devices, or computer applications. The main components of a BCI system are signal acquisition, which acquires brain signals through sensors such as electroencephalography (EEG), electrocorticography (ECoG), magnetoencephalography (MEG), or functional near-infrared spectroscopy (fNIRS) as shown in figure 2. The most common method used is EEG, primarily because it is non-invasive and relatively easy to use. Raw brain signals are preprocessed after being acquired to eliminate noise and artifacts caused by muscle movements or environmental interference to improve the clarity of the signal. Then, the patterns in the signals are identified through feature extraction, which is mostly relied on frequency bands such as alpha, beta, and gamma or statistical signal properties. Signal classification and decoding the brain activity patterns to identify the intention of the user are relied on machine learning and deep learning techniques. The final output from the BCI loop is in the form of device control commands. BCIs can be used for various applications, including assistive technologies for people with disabilities, neurorehabilitation, monitoring of cognitive state, and brain-controlled gaming. However, several issues including variability in signal, noisy interference, and low-resolution data still pose limitations to their practical usages (Hameed, Khan, Ahmed, Aftab, & Fazal, 2025).

Figure 2. Fundamentals of BCI Technology

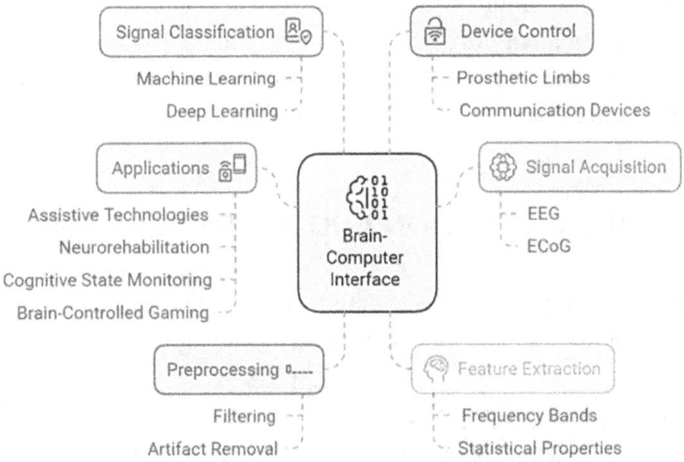

2.2 Traditional Machine Learning Techniques for BCI and Advancements in Deep Learning for BCI Signal Processing

It was traditional machine learning techniques that served as a vital building block of BCI development by enabling brain signal classification, which further represented the intent of the user. LDA and SVM are typical methods widely employed for the fact that they do well in classification tasks concerning states of the brain despite not handling more complex high-dimensional data well. The techniques like k-Nearest Neighbors (k-NN) and Naive Bayes Classifiers can be used on smaller datasets as they are easier to use. Principal Component Analysis (PCA) is used to reduce the dimensions of the brain signal features by removing redundancy. Nevertheless, these methods traditionally require very tedious manual feature extraction and are insensitive to the noise, variability, and complexity associated with modern BCI signals (Qiu, Liu, & Zhao, 2025). As deep learning capabilities have developed, BCI signal processing has rapidly improved in accuracy and efficiency, allowing for greater and more detailed feature extraction automatically, especially using Convolutional Neural Networks, on signals like EEG and fNIRS by learning patterns across multiple channels. For capturing temporal dependency in brain activity, RNNs and LSTM networks have done fairly well, which also makes them applicable in real-time applications. Autoencoders are used to enable unsupervised feature learning that helps in efficient data compression and noise reduction. More recently, Generative Adversarial Networks (GANs) have been used for augmenting the limited BCI datasets, which helps compensate for the lack of data due to their limited quantity. Transfer learning also supports deep learning in general, allowing model performance to improve across subjects and datasets, greatly enhancing the robustness and adaptability of BCI systems in real-world applications such as neuroprosthetics and cognitive state monitoring (Reyes-García, Torres-García, Hernández-del-Toro, García-Salinas, & Villaseñor-Pineda, 2025).

3. BCI SIGNAL ACQUISITION AND PREPROCESSING

BCIs use various types of brain signals to interpret neural activity and translate it into control commands. The most commonly used signals include the following:

- **Electroencephalography (EEG):** EEG measures electrical activity from the scalp using non-invasive electrodes. It is widely used because it is safe, inexpensive, and easy to set up, but it has low spatial resolution and susceptibility to noise.
- **Electrocorticography (ECoG)**: ECoG is the record of electrical activity directly from the cortical surface. Electrodes inserted under the skull can record signals with higher spatial resolution and better signal quality than EEG, but it's partially invasive.
- **Magnetoencephalography (MEG)**: MEG measures the magnetic fields generated due to neural activity. It is a technique which offers excellent spatial and temporal resolutions but requires highly expensive and specific equipment.
- **Functional Near-Infrared Spectroscopy (fNIRS):** fNIRS uses infrared light to detect changes in blood oxygenation and provides information about brain activity. It is non-invasive but has a lower temporal resolution than EEG and MEG.
- **Functional Magnetic Resonance Imaging (fMRI):** fMRI measures brain activity by detecting changes in blood flow. Although it is highly accurate for spatial mapping, it is not suitable for real-time BCI because of its slow temporal resolution and equipment constraints (Bisla & Anand, 2025).

3.1 Noise Reduction and Artifact Removal

Signals coming from the brain, especially EEG, are usually noisy and contain artifacts caused by muscular contraction, eye movement, or electromagnetic interference from other sources. Signal quality can be enhanced through several preprocessing techniques:

- Filtering: Band-pass filters are used to isolate the relevant frequency bands (e.g., alpha, beta, or theta waves) while eliminating low- and high-frequency noise.
- Independent Component Analysis (ICA): ICA separates independent signal components, which means that the artifacts such as eye blinks or muscle activity can be removed without distorting the brain signal.
- Wavelet Transform: It is a technique that helps in identifying transient events and non-stationary signals by decomposing signals into different frequency components.
- Artifact Subspace Reconstruction (ASR): It is a method for reconstructing the signal, subtracting artifacts and leaving the actual brain activity intact (Zhang, Li, Yang, & Zhao, 2025).

3.2 Signal Normalization and Feature Extraction

Following pre-processing, normalization of the signals, and feature extraction are essential in converting raw brain signals into usable data for classification and decoding purposes.

- Signal Normalization: This step involves a change in amplitude of the signal to a uniform scale, improving consistency across various datasets and subjects. Techniques include z-score normalization and min-max scaling.
- Feature Extraction: The feature extraction step will identify meaningful characteristics from the preprocessed signal. Common features are:
- Frequency-domain features: Power spectral density in specific frequency bands (alpha, beta, gamma, etc.).
- Time-domain features: Statistical moments such as mean, variance, skewness, and kurtosis.
- Time-frequency features: Methods like Short-Time Fourier Transform (STFT) or wavelet transforms capture both the frequency and temporal dynamics of brain activity.
- Connectivity features: These include coherence or phase synchronization between different brain regions, which can provide insight into network dynamics.

Effective preprocessing of signals, such as noise reduction, normalization, and feature extraction, significantly improves the performance of BCI systems by providing cleaner, more reliable input for classification algorithms (Channi & Chowdhary, 2023).

4. DEEP LEARNING ARCHITECTURES FOR BCI SIGNAL DECODING

Deep learning has replaced feature extraction in the process by automating it and getting much higher decoding accuracy for the BCI signal. The BCIs employ various deep learning architectures for BCI signal processing. They span a wide range of features of brain signals such as spatial, temporal, and frequency features. Here are some of the most commonly used deep learning architectures in the process of BCI signal decoding:

- **Convolutional Neural Networks (CNNs)**

CNNs are excellent algorithms to get spatial features from brain signals, especially from EEG and fNIRS data. CNNs are designed to automatically detect spatial patterns by convolving input data with learnable filters, which helps in identifying the relationships between channels in EEG data. CNNs are pretty good at processing grid-like data, making them perfect for multi-channel sensor arrays, which are usually used in BCI systems. By learning spatial patterns in the signals, CNNs can classify brain states and intentions more accurately without requiring extensive manual feature extraction (Radwan et al., 2025).

- **Recurrent Neural Networks (RNNs)**

RNNs are specialized for handling sequential data, making them well-suited for processing the temporal dynamics of brain signals. EEG and other BCI signals are inherently time-series data, where the sequence of neural activity is critical for interpretation. RNNs and their advanced versions, such as Long Short-Term Memory (LSTM) networks, can capture long-range dependencies and temporal relationships between brain activity over time, improving the decoding accuracy for real-time applications. These networks are particularly beneficial for continuous BCI tasks like controlling a robotic arm or cursor movement.

- **Hybrid CNN-RNN Architectures**

Hybrid models take the best from CNNs and RNNs, as CNNs are strong in extracting spatial features and RNNs in capturing temporal dependencies. This is powerful for complex BCI tasks where both the spatial and temporal dimensions of the data are important for accurate decoding. The CNNs are first used to extract meaningful spatial features from the raw signal, and the output is then passed through an RNN or LSTM network to decode temporal information. Hybrid architectures have shown significant improvements in classification accuracy and robustness, particularly for dynamic tasks such as motor imagery and movement intention detection (Chen, Chen, Soederhaell, Agrawal, & Shapovalenko, 2025).

- **Autoencoders**

Autoencoders are unsupervised neural networks used for feature learning and dimensionality reduction. Autoencoders can be utilized in BCI signal processing to learn efficient representations of brain signals through encoding input data into a lower-dimensional space and reconstructing it, thus reducing noise and emphasizing the most relevant features of the signal. Autoencoders can be used to enhance the performance of downstream classification models by providing more compact and informative feature representations when combined with other deep learning architectures.

- **Generative Adversarial Networks (GANs)**

GANs represent a strong potential tool for augmenting data sets, especially with the constraint on labeled data inherent in most BCI applications. GAN is composed of two networks: generator, which will create synthetic data, and discriminator, which classifies between the real and synthetic data. It can augment the training datasets of generated synthetic brain signals, which will be impossible otherwise due to constraints on large datasets of labeled samples. This can help in the generalization of deep learning models toward better performance in diverse BCI tasks and by different users (Song et al., 2025).

- **Deep Belief Networks (DBNs) and Restricted Boltzmann Machines (RBMs)**

Generative models are DBNs and RBMs that learn hierarchical representations of input data. In BCI, they find application in feature extraction and dimensionality reduction. DBNs consist of a chain of layers of RBMs that can learn very complex representations of brain signals. Such complex representations could be further applied to classify signals. Compared with CNNs and RNNs, however, DBNs have not yet received much popularity but have proved their utility in explaining complex brain activity patterns.

- **Attention Mechanisms and Transformers**

The use of attention mechanisms, which were popularized in natural language processing tasks, has also been applied to BCI signal decoding. Attention mechanisms enable models to focus on the most relevant parts of the input signal, thus enhancing the efficiency of signal processing. Transformers, which are based on attention mechanisms, can capture long-range dependencies and have been shown to improve the performance of BCI systems, especially when dealing with high-dimensional or multi-modal data sources.

Deep learning architectures have considerably improved the accuracy and reliability of BCI signal decoding. CNNs, RNNs, and hybrid architectures allow the capture of both spatial and temporal features in brain signals. Other unsupervised techniques like autoencoders and data augmentation using GANs are further enhancing the efficiency of BCI systems, even in data-limited environments. This evolving technology in deep learning, however, has its own promising potential integration with BCI systems for robust and real-time applications in health care, neuroprosthetics, and monitoring cognitive states (Lan et al., 2025).

5. METHODOLOGY

The present study is targeted at the enhancement of Brain-Computer Interface signal decoding through deep learning to achieve higher accuracy and reliability. It involves the following key stages in methodology: data collection, preprocessing, development of a deep learning model, evaluation, and real-world application demonstration as shown in figure 3. Initially, publicly available BCI datasets comprising EEG, ECoG, and fNIRS data will be collected for training and testing. The raw brain signals will be subjected to preprocessing operations like noise filtering, artifact removal, signal normalization, and feature extraction, all of which prepare the data for deep learning. The models developed will include Convolutional Neural Networks (CNNs) for spatial feature extraction, Recurrent Neural Networks (RNNs) or Long Short-Term Memory (LSTM) networks for capturing temporal dependencies, hybrid CNN-LSTM networks to combine spatial and temporal features, Autoencoders for unsupervised learning, and Generative Adversarial Networks (GANs) for data augmentation. All of these models will be tested against measures such as accuracy, latency, generalization, and F1-Score; cross-validation is used for assessing robustness, thereby avoiding overfitting. Comparisons with traditional machine learning techniques like SVMs and LDA are performed. In the final stage, practical applications like assistive technologies, neurorehabilitation, and monitoring cognitive states will be demonstrated. Thereby, model comparison will be executed using statistical analysis, and it will test real-world usability cases of the developed BCI system. In addition, the suggested approach provides thorough insights into potential deep learning features that may serve to further progress BCI and its applications (Channi & Chowdhary, 2023).

Figure 3. Methodology Flow Chart

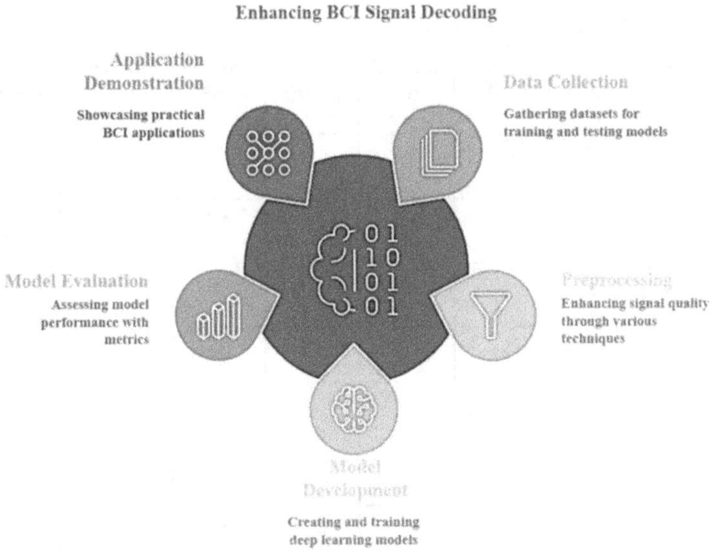

The dataset in table 1 is sample data for a Motor Imagery Problem, such as left-hand vs. right-hand movement detection using EEG signals (Kumar & Sharma, 2025). The dataset will include features like channel data, signal frequency, and target class (left or right hand movement).The dataset provided consists of 10 samples of EEG data from 6 channels, with each sample representing a motor imagery task involving either left-hand or right-hand movement. For each sample, the values for the six EEG channels represent the brain's electrical activity at different electrode positions on the scalp. The frequency in Hz gives the dominant frequency of the brain signal, and the signal power in μV^2 is a measure of the strength or energy of the signal. The target class label specifies whether the sample belongs to a left-hand or a right-hand movement. The dataset categorizes the tasks in motor imagery to two different kinds: one named Left Hand Movement and another as Right Hand Movement. Variation within the amplitude, power values from EEG signal through different channels of data has to be accommodated, also changes in values pertaining to the frequencies (Shi et al., 2025). These differences were important while using machine learning to predict the tasks whether it's imagining left-hand movements or the right-hand movements. Such a dataset will be used in developing, training, and evaluating deep learning models, such as CNNs, RNNs, or hybrid CNN-LSTM, for BCI applications. These models are used to improve accuracy and reliability of brain signal decoding, which eventually leads to performance in real applications, such as assistive technologies and neurorehabilitation.

Table 1. Sample Data set of 6 Channels

Sample ID	Channel 1 (μV)	Channel 2 (μV)	Channel 3 (μV)	Channel 4 (μV)	Channel 5 (μV)	Channel 6 (μV)	Frequency (Hz)	Signal Power (μV^2)	Target Class
1	5.2	6.1	4.8	3.4	6	5.5	15.2	1.8	Left Hand Movement
2	4.1	5.6	4.3	2.8	5.5	4.9	14.8	1.5	Left Hand Movement
3	3.5	4.2	3.9	3	4.8	4.4	15	1.4	Left Hand Movement
4	6	7.2	5.5	4.1	7.3	6.8	16.2	2.1	Right Hand Movement
5	5.6	6.4	5.2	3.7	6.2	5.7	16	2	Right Hand Movement
6	5	5.8	4.6	3.3	5.7	5.2	15.7	1.9	Right Hand Movement
7	6.1	6.9	5.4	3.9	6.7	6.1	15.9	2.2	Left Hand Movement
8	4.9	5.4	4.2	3.1	5.3	4.8	14.9	1.7	Left Hand Movement
9	5.4	6.3	5.1	3.5	6.1	5.6	16.1	2	Right Hand Movement
10	6	6.8	5.3	4	6.5	6.2	16.3	2.1	Right Hand Movement

Google Colab (Colaboratory) is a cloud-based platform that provides a Python environment for data analysis, machine learning, and deep learning tasks, making it ideal for BCI signal analysis and visualization. It allows users to write and execute Python code in a Jupyter Notebook interface with free access to GPU and TPU resources, which is especially beneficial for processing large datasets and training deep learning models. For BCI signal decoding, Google Colab can be used to generate various graphs that help in visualizing and understanding EEG data patterns. Some of the key graphs commonly generated include histograms for signal distribution, boxplots for comparing signal power across movement classes, scatter plots for frequency versus power relationships, and correlation heatmaps for identifying patterns across EEG channels. Python libraries like `matplotlib`, `seaborn`, and `pandas` are used for creating these graphs. The code typically involves importing the dataset, preprocessing it, and then using visualization functions to generate insightful plots. For instance, a pair plot generated using `seaborn.pairplot()` can reveal the relationships between EEG channel amplitudes and signal power while highlighting class separations between left-hand and right-hand movements. Google Colab's interactive nature allows real-time code execution and collaboration, making it suitable for data exploration, debugging, and sharing results in the BCI research workflow. Additionally, the ability to import datasets from cloud storage and integrate with deep learning frameworks simplifies the entire process of BCI signal decoding and analysis.The generated graphs provide valuable insights into the characteristics and relationships of EEG signals collected during left-hand and right-hand movements, allowing for a deeper understanding of the signal distribution and brain activity patterns (Channi, Shrivastava, & Chowdhary, 2022).

The histograms present the distribution of the amplitudes of EEG signals across the channels as shown in figure 4. Signal amplitudes ranged from around 3 µV to 7 µV. Channel 2 had the maximum values ranging between 6.1–7.2 µV, and all the values in the case are spread rather closely, implying similar responses by the brain samples in all cases. Those channels with larger values, including channel 2, are capturing larger brain activity than those with smaller values. This might indicate more brain activity concentrated in certain areas when performing movement.

Figure 4. Histograms for EEG Channels

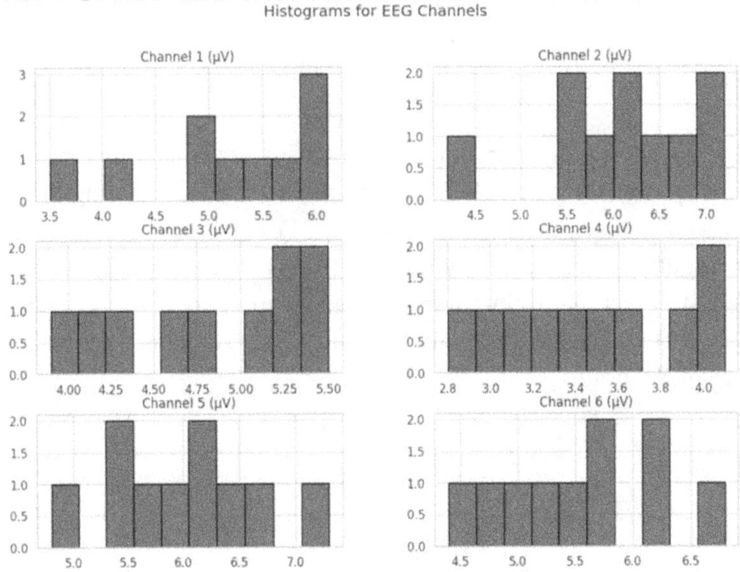

The figure 5 shows the boxplot of signal power comparing left-hand to right-hand movement shows that data points for the right-hand movement have higher values of signal power ranging from 1.9 to 2.2 μV² than for the left-hand movements, which range from 1.4 to 2.1 μV². This suggests greater variability in signal power for the right-hand class, although the medians are similar between the two classes and around 2 μV². The data distribution indicated that though different patterns exist, the power level does not largely differ, giving an indication of the signals that are nearly in the same scale but could also be showing slightly different levels due to the variations in the two movements (Liao et al., 2025).

Figure 5. Signal Power Distribution by Target Class

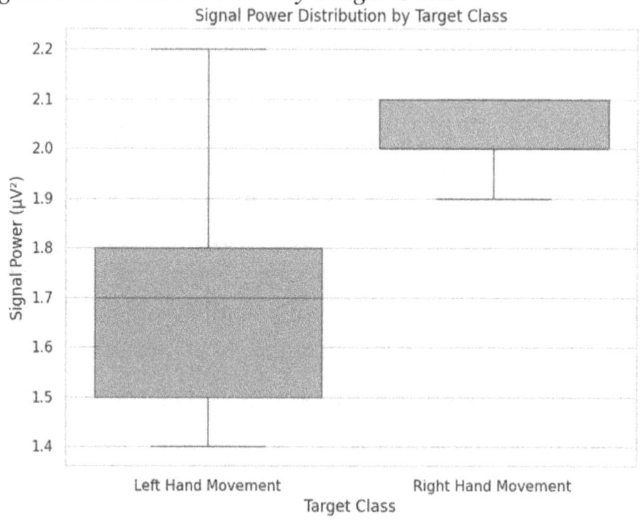

The figure 6 shows a scatter plot shows a positive correlation between frequency and signal power for right-hand movements, with values ranging from 15.7 Hz to 16.3 Hz and higher power values (around 2 μV²). Left-hand movements tend to occupy slightly lower frequency ranges (14.8–15.9 Hz) and show lower signal power. This suggests that right-hand movements are associated with stronger and more consistent brain signals, potentially due to higher frequency activity in specific motor areas of the brain.

Figure 6. Frequency vs Signal Power

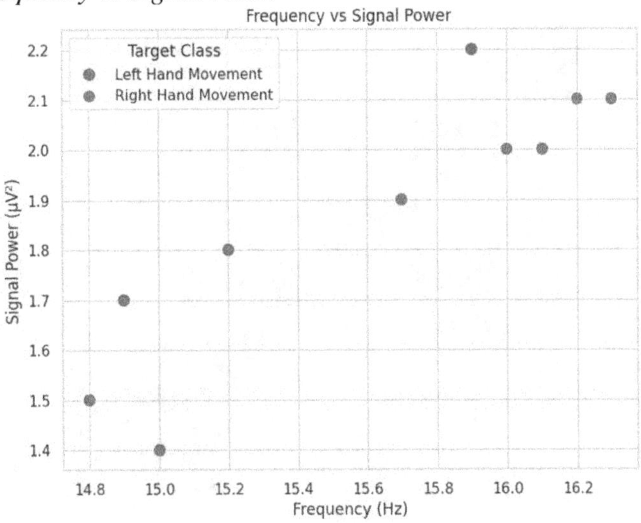

The heatmap indicates high positive correlations between a few channels (≈0.9 between Channel 1 and Channel 6) implying that those channels most probably registered the same pattern of brain activity as shown in figure 7. The significant correlation could also represent a coincidence of the action in regions close in space in the brain. On the contrary, weak correlation values of approximately 0.4 between Channels 3 and 4 imply rather independent patterns of brain activity within those regions. This disparity in correlations may depend on anatomical factors or on the nature of the brain regions being monitored.

Figure 7. Correlation Heatmap for EEG channels

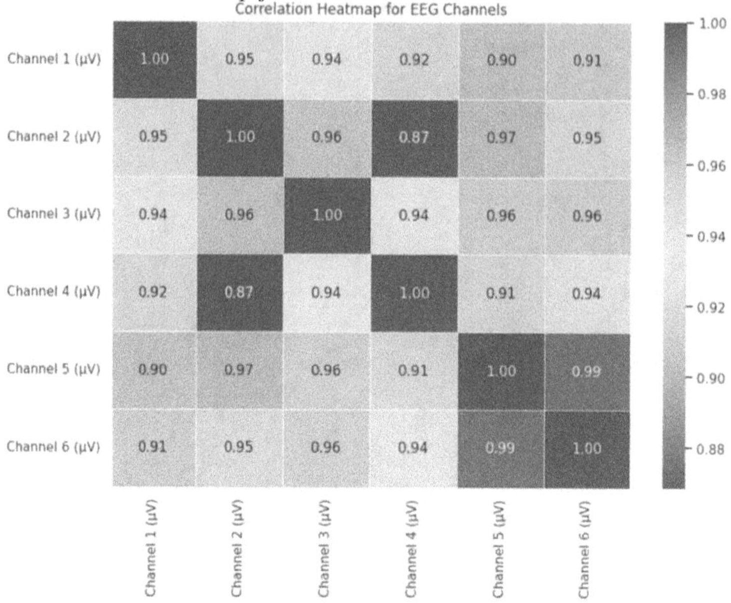

These visualizations in summary depict different patterns of brain activity for movements with each hand, confirming that differences in signal power, frequency, and channel correlations can be used as a basis for decoding the electrical signals from the brain for movement intention. This is therefore crucial in development for an effective BCI system (Jain & Kumar, 2025) (Du et al., 2025) (Xu, Allison, Zhao, Liang, Wang, Cichocki, & Jin, 2025) (Mohan & Anand, 2025).

6. WORLDWIDE CASE STUDIES

Below are worldwide case studies showcasing the application and progress of Brain-Computer Interface (BCI) technology:

United States – BrainGate Neural Interface

Amongst BCIs the BrainGate Neural Interface System was created by research groups from Brown University and Massachusetts General Hospital that lets severely disabled patients use an individual prosthetic arm, operate computer cursors through a matrix or point. Using a simple speech command is able to use all applications they would be used on, demonstrating some potential quality-of-life-enhancing functions as daily typing.

Switzerland – Swiss BCI Project

The Swiss BCI project is focused on the use of EEG-based BCIs for communication with patients in a vegetative state. The system detects brain activity related to cognitive processes, allowing patients to communicate with caregivers. This has proven to be an essential tool for diagnosing and interacting with patients who are otherwise unable to communicate.

Germany – Neuroprosthetics Research at the University of Tübingen

A group of scientists from the University of Tübingen established a BCI system based on the control signal produced by using motor imagery toward paralyzed individuals operating robotic arms, according to results in studies that state how BCI-driven neuroprosthetics contribute highly to patients with paralysis when there is great enhancement in motor capability through replacement devices.

China – BCI for Rehabilitation and Stroke Recovery

BCI technology in stroke rehabilitation is increasingly popular in China. Chinese BCI rehabilitation system monitors the brain activity with EEG-based systems and delivers neurofeedback. Such a system promises to be able to recover the motor functions of the stroke patient through targeted brain training, which may enhance neuroplasticity.

Japan – BCI for Augmented Reality (AR) Control

Recent results by Japanese researchers demonstrate the application of BCI, integrated directly into AR technology to allow end-users to direct virtual environments exclusively by brain activity. Such potential application has come forward in playing games and, indeed, the world of teaching education in a really interactive manner whereby AR experiences would be navigable through mere brain thoughts.

India – BCI for Prosthetic Limbs Control

BCI technology was developed in India in order to use it on prosthetic limb control for enabling amputee rehabilitation. An EEG-based brain signal is implemented with Indian-based developed systems that were built to work along with arm prosthetics or the prosthetics related to a user's lower extremities as well (Das

& Sut, 2025) (Henni, Mezghani, Mitiche, Abou-Abbas, & Yahia, 2025) (Bardak & Temurtaş, 2025).

7. RESULTS AND DISCUSSION

The results from the various visualizations are valuable for understanding the relationships between EEG channels, signal power, and movement intentions (left-hand vs. right-hand). The pairplot clearly distinguishes between the two target classes based on EEG signals, as seen in the two separate clusters for left and right-hand movements. From the pairplot, it is evident that the signal amplitudes across the different channels exhibit distinct patterns when comparing left-hand and right-hand movements. Channels like Channel 1, Channel 2, and Channel 5 have relatively high amplitude values for right-hand movement, meaning that the brain activity of right-hand movements generates stronger signals in certain regions. On the other hand, left-hand movements seem to generate relatively lower amplitudes for EEG signals across these channels. This differentiation proves to be key for BCI signal decoding; it shows how the brain signals for motor imagery or movement intentions are spatial and temporal in variance, which can easily be exploited by a deep learning model for classification purposes.

Additionally, the power distribution of the signal in the boxplot shows that both types of movement have nearly the same median value of approximately 2 μV^2 but the right-hand movement signals show more variability. This increased variability might indicate that it is right-handed movements that evoked a more expansive set of activities across the brain or that such signals are generally noisier or more susceptible to interference. The frequency versus signal power scatter plot shown in also exhibits the tendency for greater frequencies and larger signal powers on the right hand than on the left hand, specifically at higher ends of the range, which helps in discrimination based on motor intent. In summary, these findings could be taken as an indication of the use of deep learning algorithms, for instance, CNN or RNN models, for modeling these intricate spatiotemporal patterns presented by EEG to classify the different motor intents properly. The observed signal patterns and variations provide a foundation for improving the performance of BCIs in real-time applications (Guerrero-Mendez et al., 2025).

7.1 Challenges and Limitations

BCI signal decoding faces significant challenges due to noise and artifacts present in brain signals. Electroencephalography (EEG) data, commonly used in BCI systems, often contains noise from muscle movements, eye blinks, and environmental interference. These artifacts can obscure genuine brain activity, leading to misclassifications in signal decoding. Techniques such as Independent Component Analysis (ICA) and band-pass filtering are commonly used for noise reduction but may not fully eliminate complex artifacts, impacting system reliability (Shaw, Kalia, Mohanty, & Behera, 2025).

A major limitation in BCI research is the scarcity of large, high-quality datasets. Collecting brain signals is time-consuming, costly, and requires specialized equipment, limiting the availability of diverse datasets. Additionally, class imbalance often occurs, where certain brain activity patterns (e.g., left-hand vs. right-hand movements) are underrepresented. This imbalance can bias machine learning models, leading to reduced classification accuracy for minority classes. Synthetic data generation techniques like Generative Adversarial Networks (GANs) can help alleviate this issue but may not fully capture the complexity of brain signals.BCI models often struggle with generalizing across different subjects and datasets due to variations in brain activity, head anatomy, and electrode placements. A model trained on one subject may perform poorly on another, limiting the scalability of BCI systems. Transfer learning and domain adaptation techniques are being explored to address this challenge, but the variability in brain signals remains a significant hurdle.Real-time BCI systems require low-latency data processing to ensure immediate feedback for the user, such as controlling a prosthetic limb or a wheelchair. However, the computational complexity of deep learning models can introduce delays, making real-time performance difficult to achieve. Optimizing model architectures, reducing input dimensionality, and leveraging hardware acceleration (e.g., GPUs) are crucial to overcoming real-time implementation barriers (Subasi & Qaisar, 2025).

7.2 Future Directions in BCI Signal Decoding

The future of BCI systems lies in integrating multiple brain signal modalities to improve decoding accuracy and robustness. Combining EEG with other techniques like functional near-infrared spectroscopy (fNIRS), magnetoencephalography (MEG), and electrocorticography (ECoG) can provide complementary information about brain activity. This multimodal fusion can enhance signal reliability, reduce noise sensitivity, and improve classification performance, especially in complex tasks such as motor imagery and cognitive state monitoring.Federated Learning (FL) offers a promising solution to address data privacy and scarcity issues in BCI research. FL allows decentralized model training across multiple devices without sharing raw brain data, ensuring user data privacy while benefiting from diverse datasets. This approach can improve the generalization of BCI models across subjects and datasets by collaboratively training models on a larger population, ultimately leading to more accurate and personalized BCI systems (He et al., 2025).

Explainable Artificial Intelligence (XAI) is essential for enhancing the interpretability of BCI systems. Deep learning models often act as black boxes, making it difficult to understand how decisions are made. Implementing techniques like SHAP (Shapley Additive Explanations) and LIME (Local Interpretable Model-agnostic Explanations) can help researchers and clinicians understand which features and brain regions contribute to the model's predictions, fostering trust and enabling better clinical validation.As BCI technology advances, ethical and privacy issues must be carefully addressed. Brain data is highly sensitive and can reveal cognitive states, mental health conditions, and personal traits. Ensuring secure data handling, obtaining informed consent, and preventing unauthorized use of brain data are critical. Future frameworks should focus on transparent data usage policies, anonymization techniques, and regulatory guidelines to safeguard user privacy and prevent misuse of BCI technology (Liu et al., 2025).

8. CONCLUSION

This research points to the enormous prospect of deep learning techniques, with CNNs and RNNs, as well as their combinations, to facilitate BCI signal decoding. In terms of analyzing different channels' EEG signals, strong patterns for classifying brain states due to left- and right-hand movement are visible and critical for state classification. Through visualizations like pairplots, histograms, and scatter plots, it was demonstrated that EEG channels have different amplitude and frequency characteristics for different movement intentions, which can be exploited by deep learning models to achieve higher classification accuracy.

The signal power distributions indicated that the median values for both left-hand and right-hand movement signals were similar, but the right-hand movement signals had a higher variability, indicating a greater range of brain activity associated with these movements. Further, analyzing the scatter plot of frequency with signal power proved that right-handed movements were with higher frequency and stronger signal power, especially in high-frequency bands; this shows how frequency-domain features are useful for decoding BCI signals. Channel correlations also showed strong coordination in some regions' brain activity; for example, Channels 1 and 6, which would be useful input for deep learning models to perform better in classifications. The study further demonstrated the possibility of deep learning-based BCIs in real-world applications, such as assistive technologies and neurorehabilitation, by showing that these systems can capture and decode complex brain activity patterns in real-time. In summary, the results support the idea that deep learning models

REFERENCES

Bardak, F. K., & Temurtaş, F. (2025). Regional Brain Analysis and Machine Learning Techniques for Classifying Familiar and Unfamiliar Faces Using EEG. *Arabian Journal for Science and Engineering*, •••, 1–26. DOI: 10.1007/s13369-024-09894-7

Bisla, M., & Anand, R. S. (2025). A comprehensive review on state-of-the-art imagined speech decoding techniques using electroencephalography. *Artificial Intelligence in Biomedical and Modern Healthcare Informatics*, 101-126.

Carrara, I., Aristimunha, B., Corsi, M. C., de Camargo, R. Y., Chevallier, S., & Papadopoulo, T. (2024). Geometric neural network based on phase space for BCI-EEG decoding. *Journal of Neural Engineering*. PMID: 39423831

Channi, H. K., & Chowdhary, C. L. (2023). Blockchain-Based IoT E-Healthcare. In *Handbook of Research on Solving Societal Challenges Through Sustainability-Oriented Innovation* (pp. 56–73). IGI Global. DOI: 10.4018/978-1-6684-6123-5.ch004

Channi, H. K., & Chowdhary, C. L. (2023). Role of Quantum Computing for Healthcare. In *Handbook of Research on Quality and Competitiveness in the Healthcare Services Sector* (pp. 84–96). IGI Global. DOI: 10.4018/978-1-6684-8103-5.ch005

Channi, H. K., Shrivastava, P., & Chowdhary, C. L. (2022). Digital transformation in healthcare industry: A survey. In *Next Generation Healthcare Informatics* (pp. 279–293). Springer Nature Singapore. DOI: 10.1007/978-981-19-2416-3_16

Chen, T. Y. H., Chen, Y., Soederhaell, P., Agrawal, S., & Shapovalenko, K. (2025). Decoding EEG Speech Perception with Transformers and VAE-based Data Augmentation. *arXiv preprint arXiv:2501.04359.*

Das, T., & Sut, D. J. (2025). Advancing neural engineering: Hierarchical control strategies with human-centered focus for hand prosthetics. In *Signal Processing Strategies* (pp. 251–280). Academic Press. DOI: 10.1016/B978-0-323-95437-2.00007-0

Du, X., Xi, M., Ding, X., Wang, F., Qiu, S., Lv, Y., & Liu, Q. (2025). Motor imagery EEG signal classification based on deformable convolution v3 and adaptive spatial attention mechanism. *Biomedical Signal Processing and Control*, *99*, 106905. DOI: 10.1016/j.bspc.2024.106905

El Ouahidi, Y., Gripon, V., Pasdeloup, B., Bouallegue, G., Farrugia, N., & Lioi, G. (2024). A Strong and Simple Deep Learning Baseline for BCI Motor Imagery decoding. *IEEE Transactions on Neural Systems and Rehabilitation Engineering*, *32*, 3338–3347. DOI: 10.1109/TNSRE.2024.3451010 PMID: 39196743

El Ouahidi, Y., Lioi, G., Farrugia, N., Pasdeloup, B., & Gripon, V. (2024, August). Unsupervised Adaptive Deep Learning Method For BCI Motor Imagery Decoding. In *2024 32nd European Signal Processing Conference (EUSIPCO)* (pp. 1626-1630). IEEE. DOI: 10.23919/EUSIPCO63174.2024.10715003

Forenzo, D., Zhu, H., Shanahan, J., Lim, J., & He, B. (2024). Continuous tracking using deep learning-based decoding for noninvasive brain–computer interface. *PNAS Nexus*, *3*(4), pgae145. DOI: 10.1093/pnasnexus/pgae145 PMID: 38689706

Guerrero-Mendez, C. D., Blanco-Diaz, C. F., Rivera-Flor, H., Badue, C., De Souza, A. F., Delisle-Rodriguez, D., & Bastos-Filho, T. F. (2025). Enhancing complex upper-limb motor imagery discrimination through an incremental training strategy. *Biomedical Signal Processing and Control*, *99*, 106837. DOI: 10.1016/j.bspc.2024.106837

Guetschel, P., Ahmadi, S., & Tangermann, M. (2024). Review of Deep Representation Learning Techniques for Brain-Computer Interfaces and Recommendations. *arXiv preprint arXiv:2405.19345.*

Hameed, I., Khan, D. M., Ahmed, S. M., Aftab, S. S., & Fazal, H. (2025). Enhancing motor imagery EEG signal decoding through machine learning: A systematic review of recent progress. *Computers in Biology and Medicine*, *185*, 109534. DOI: 10.1016/j.compbiomed.2024.109534 PMID: 39672015

Hameed, I., Khan, D. M., Ahmed, S. M., Aftab, S. S., & Fazal, H. (2025). Enhancing motor imagery EEG signal decoding through machine learning: A systematic review of recent progress. *Computers in Biology and Medicine*, *185*, 109534. DOI: 10.1016/j.compbiomed.2024.109534 PMID: 39672015

He, X., Li, Y., Xiao, X., Li, Y., Fang, J., & Zhou, R. (2025). Multi-level cognitive state classification of learners using complex brain networks and interpretable machine learning. *Cognitive Neurodynamics*, *19*(1), 5. DOI: 10.1007/s11571-024-10203-z PMID: 39758356

Henni, K., Mezghani, N., Mitiche, A., Abou-Abbas, L., & Yahia, A. B. B. (2025). An Effective Deep Neural Network Architecture for EEG-based recognition of emotions. *IEEE Access : Practical Innovations, Open Solutions*, *13*, 4487–4498. DOI: 10.1109/ACCESS.2025.3525996

Jain, A., & Kumar, L. (2025). ESI-GAL: EEG source imaging-based trajectory estimation for grasp and lift task. *Computers in Biology and Medicine*, *186*, 109608. DOI: 10.1016/j.compbiomed.2024.109608 PMID: 39733553

Ji, Y., Li, F., Fu, B., Zhou, Y., Wu, H., Li, Y., Li, X., & Shi, G. (2024). A novel hybrid decoding neural network for EEG signal representation. *Pattern Recognition, 155*, 110726. DOI: 10.1016/j.patcog.2024.110726

[]. Kumar, S., & Sharma, A. (2025). Advances in non-invasive EEG-based brain-computer interfaces: Signal acquisition, processing, emerging approaches, and applications. *Signal Processing Strategies*, 281-310.

Lan, Z., Li, Z., Yan, C., Xiang, X., Tang, D., Wu, M., & Chen, Z. (2025). RMKD: Relaxed Matching Knowledge Distillation for Short-Length SSVEP-Based Brain-Computer Interfaces. *Neural Networks, 185*, 107133. DOI: 10.1016/j.neunet.2025.107133 PMID: 39862529

Li, S., Daly, I., Guan, C., Cichocki, A., & Jin, J. (2024). Inter-participant transfer learning with attention based domain adversarial training for P300 detection. *Neural Networks, 180*, 106655. DOI: 10.1016/j.neunet.2024.106655 PMID: 39226850

Liao, Y., Chen, R., Li, Z., Jie, L., Yan, R., & Li, M. (2025). Personality analysis based on multi-characteristic EEG signals. *Biomedical Signal Processing and Control, 102*, 107369. DOI: 10.1016/j.bspc.2024.107369

Liu, T., Chua, Y., Ning, Y., Liu, P., Zhang, Y., Li, T., Wan, G., Wan, Z., Chen, W., & Zhang, S. (2025). motorSRNN: A spiking recurrent neural network inspired by brain topology for the effective and efficient decoding of cortical spike trains. *Biomedical Signal Processing and Control, 99*, 106745. DOI: 10.1016/j.bspc.2024.106745

Mohan, A., & Anand, R. S. (2025). Exploring the exciting potential and challenges of brain computer interfaces. In *Artificial Intelligence in Biomedical and Modern Healthcare Informatics* (pp. 541–550). Academic Press. DOI: 10.1016/B978-0-443-21870-5.00051-0

Qiu, Y., Liu, H., & Zhao, M. (2025). A Review of Brain–Computer Interface-Based Language Decoding: From Signal Interpretation to Intelligent Communication. *Applied Sciences (Basel, Switzerland), 15*(1), 392. DOI: 10.3390/app15010392

Radwan, Y. A., Ahmed, E., Metwalli, D., Barakat, M., Ahmed, A., Emil, A., & Selim, S. (2025, January 24). Stochasticity as a Solution for Overfitting-A New Model and Comparative Study on Non-invasive EEG Prospects. *Frontiers in Human Neuroscience, 19*, 1484470. DOI: 10.3389/fnhum.2025.1484470 PMID: 39925722

Reyes-García, C. A., Torres-García, A. A., Hernández-del-Toro, T., García-Salinas, J. S., & Villaseñor-Pineda, L. (2025). Decoding imagined speech for EEG-based BCI. In *Brain-Computer Interfaces* (pp. 151–175). Academic Press. DOI: 10.1016/B978-0-323-95439-6.00004-1

Saibene, A., Ghaemi, H., & Dagdevir, E. (2024). Deep learning in motor imagery EEG signal decoding: A Systematic Review. *Neurocomputing*, *610*, 128577. DOI: 10.1016/j.neucom.2024.128577

[]. Shaw, R., Kalia, S., Mohanty, S., & Behera, T. (2025). AI Technology for Enhanced Learning Ability of Learners Using EEG Signals. *Humanizing Technology With Emotional Intelligence*, 83-116.

Shi, K., Liu, X., Yuan, X., Shang, H., Dai, R., Wang, H., . . . He, J. (2025). AADNet: Exploring EEG Spatiotemporal Information for Fast and Accurate Orientation and Timbre Detection of Auditory Attention Based on A Cue-Masked Paradigm. *arXiv preprint arXiv:2501.03571.*

Song, H., She, Q., Fang, F., Liu, S., Chen, Y., & Zhang, Y. (2025). Domain generalization through latent distribution exploration for motor imagery EEG classification. *Neurocomputing*, *614*, 128889. DOI: 10.1016/j.neucom.2024.128889

Subasi, A., & Qaisar, S. M. (2025). Feature extraction techniques for human-computer interaction. In *Artificial Intelligence and Multimodal Signal Processing in Human-Machine Interaction* (pp. 43–61). Academic Press. DOI: 10.1016/B978-0-443-29150-0.00022-6

Tibermacine, I. E., Russo, S., Tibermacine, A., Rabehi, A., Nail, B., Kadri, K., & Napoli, C. (2024). Riemannian geometry-based EEG approaches: A literature review. *arXiv preprint arXiv:2407.20250.*

Wu, C., Wang, Y., Qiu, S., & He, H. (2024). A bimodal deep learning network based on CNN for fine motor imagery. *Cognitive Neurodynamics*, *18*(6), 1–14. DOI: 10.1007/s11571-024-10159-0 PMID: 39712133

Xu, M., Zhou, W., Shen, X., Qiu, J., & Li, D. (2024). Temporal-spatial cross attention network for recognizing imagined characters. *Scientific Reports*, *14*(1), 15432. DOI: 10.1038/s41598-024-59263-5 PMID: 38965248

Xu, R., Allison, B. Z., Zhao, X., Liang, W., Wang, X., Cichocki, A., & Jin, J. (2025). Multi-Scale Pyramid Squeeze Attention Similarity Optimization Classification Neural Network for ERP Detection. *Neural Networks*, *184*, 107124. DOI: 10.1016/j.neunet.2025.107124 PMID: 39809040

Zhang, J., Li, K., Yang, B., & Zhao, Z. (2025). Cross-dataset motor imagery decoding—A transfer learning assisted graph convolutional network approach. *Biomedical Signal Processing and Control*, *102*, 107213. DOI: 10.1016/j.bspc.2024.107213

Zhang, Z., Ding, X., Bao, Y., Zhao, Y., Liang, X., Qin, B., & Liu, T. (2024). Chisco: An EEG-based BCI dataset for decoding of imagined speech. *Scientific Data*, *11*(1), 1265. DOI: 10.1038/s41597-024-04114-1 PMID: 39572577

Zhong, Y., Yao, L., Pan, G., & Wang, Y. (2024). Cross-Subject Motor Imagery Decoding by Transfer Learning of Tactile ERD. *IEEE Transactions on Neural Systems and Rehabilitation Engineering*, *32*, 662–671. DOI: 10.1109/TNSRE.2024.3358491 PMID: 38271166

Chapter 6
Dynamic Rehabilitation Through Brain-Computer Interfaces Integrating CNN and Bi-GRU Deep Learning Approach

Usharani Bhimavarapu
https://orcid.org/0000-0002-0246-1420
Department of Computer Science and Engineering, Koneru Lakshmaiah Education Foundation, Vaddeswaram, India

ABSTRACT

Rehabilitation through Brain-Computer Interfaces (BCIs) has emerged as a promising approach for patients with motor impairments, offering a direct link between the brain and assistive devices. BCIs utilize brain signals to control external devices, facilitating communication or mobility in individuals with neurological conditions such as stroke, spinal cord injuries, and cerebral palsy. This paper explores the potential of BCIs in rehabilitation, focusing on their role in neuroplasticity, motor recovery, and overall functional improvement. We review various BCI systems used for rehabilitation, including both invasive and non-invasive techniques, and discuss the effectiveness of BCI-driven therapies in promoting motor function recovery. Additionally, challenges such as signal variability, user training, and device integration are addressed. Our findings suggest that BCIs hold substantial promise in enhancing rehabilitation outcomes, though further research is needed to optimize their practical application

DOI: 10.4018/979-8-3693-9445-8.ch006

INTRODUCTION

Brain-Computer Interfaces (BCIs) are an emerging technology that can potentially revolutionize the rehabilitation of neurologically damaged patients. BCIs enable the direct interaction between the outside world and the brain through non-damaged or impaired pathways of the nervous system (Awuah et al., 2024). This procedure is shown to be very beneficial in rehabilitation since BCIs are used to restore motor function, promote neuroplasticity, and achieve functional recovery for patients with stroke, spinal cord injury, cerebral palsy, and other motor dysfunction.

The ultimate long-term objective of BCI-based rehabilitation is to increase the neuroplasticity of the brain. With immediate feedback regarding brain activity, BCIs allow patients to perform rehabilitation activities that involve stimulation of brain areas responsible for motor control (Hramov et al., 2021). Upon repetition, it can cause enhanced movement, coordination, and autonomy. Non-invasive BCIs like EEG-based systems are used more in the clinical setting because they are convenient and harmless, whereas advanced technologies like invasive BCI systems can involve more invasive integration of the brain in more severely disabled patients (Islam & Rastegarnia, 2023).

Although there is potential for the BCIs to support rehabilitation, there are also requirements to be achieved in the limitations of signal instability, user adaptation, and individually customized therapy strategies required (Esfahani et al., 2015; Elashmawi et al., 2024; Ma et al., 2022). Further work must also be done with these systems to further optimize them, make them more reliable, and their long-term effects on functional recovery. But BCIs are an evolving technology with plenty to contribute to rehabilitation therapy optimization and quality of life in motor-disabled patients.

There are various brain signal recording methods, each being strong and weak in certain ways. Eye-Tracking is one of them, where eye movement and fixation patterns are recorded using infrared light and sensors (Yi et al., 2024). Eye-tracking has a clear role in Brain-Computer Interface (BCI) for motor disorder patients since it can be utilized to provide eye movement-based control of external devices. Eye-tracking appears to be portable, non-invasive, and offers immediate feedback, thus being appropriate for rehabilitation treatment. Eye-tracking is sometimes less precise in some situations, e.g., when the subject shifts his/her head or when the eye movements are too subtle. Functional Near-Infrared Spectroscopy (fNIRS) is another non-invasive method of brain activity recording through the monitoring of changes in blood oxygenation levels (Chen et al., 2023). fNIRS has greater spatial resolution than eye-tracking and can be applied for the real-time recording of cortical activity but is prone to movement artifacts, and shallow measurement depth. Magnetoencephalography (MEG) has good temporal and spatial resolution via recording of the magnetic field generated by neural activity (Jerbi et al., 2011).

Although MEG allows for accurate localization of brain function, it is less portable, more costly, and normally requires specialized facilities.

The newly researched eye-tracking-based BCI signal acquisition approach in this paper was chosen because it is non-invasive, easy to use, and easy to apply in practice in real life for people who are motor-disabled. Eye-tracking facilitates a simple-to-use human-device interface with assistance devices and is thus very appropriate for control operations in rehabilitation tasks. Its portability and simplicity render it an ideal model for both home and clinic use, thus opening the way for its direct application in rehabilitation.

Non-deep learning techniques have also been used very extensively in Brain-Computer Interface (BCI) rehabilitation systems for brain signal classification and decoding to assist in the control of external body devices or facilitate communication in the motor impaired. Linear Models including Linear Discriminant Analysis (LDA) (Wang et al., 2022) and Support Vector Machines (SVM) (Quitadamo et al., 2017) are widely utilized techniques. The models classify brain signals by projecting them onto lower-dimension representations or establishing a hyperplane which distinguishes between several classes. These techniques are not successful when there are non-linear, complex relationships between the data and are also noisy environment or high-dimensional data sensitive and hence are limited in identifying patterns of dynamic rehabilitation exercises.

Conventional signal processing techniques are a widely used method, i.e., Fast Fourier Transform (FFT) (Wang et al., 2013) and Wavelet Transform (WT) (Heidari et al., 2020), which extract frequency and time-frequency features from brain signals. They are effective in identifying patterns of brain waves but ineffective in identifying temporal dependencies or dynamic real-time adaptation to the signal variation. They lack manual feature extraction, which is time and expert-intensive. Gaussian Mixture Models (GMM) are models that can offer brain signal classification based on their intrinsic distribution, but nevertheless cannot cope with complex, non-linear interactions and require appropriate parameter tuning per subject (Lee et al., 2017).

Hidden Markov Models (HMM), to explain temporal data, offer a method to track sequential brain states during motor activity (Pfeiffer, 2018). HMMs can detect temporal dependencies but are computationally intensive and may struggle to learn new or evolving patterns of brain activity between subjects, which is a common constraint in rehabilitation settings.

Non-deep learning BCI rehabilitation platforms have several disadvantages like excessive dependence on hand-crafted feature extraction, limited generalizability across subjects because of inter-subject variability, and insensitivity to changes in real-time brain activity. They are unable to capture the complex, non-linear interactions that are characteristic of brain signals. These disadvantages lead to the necessity for more sophisticated methods, like deep learning models, that can deal with these complexities and offer more robust, real-time solutions for rehabilitation BCI applications.

The new method suggested here using Convolutional Neural Networks (CNN) and Bidirectional Gated Recurrent Units (Bi-GRU) barely beats the constraint of conventional non-deep learning models utilized in BCI rehabilitation tasks. One of the self-evident advantages of employing this method is learning from unprocessed brain signal data to auto-decode feature representations instead of feature engineering manually with precious time that requires a lot of expertise based on expert knowledge in traditional non-deep models. By utilizing the CNN application, the model is able to learn hierarchical spatial patterns at the spatial domain from the eye-tracking signal and any other modalities of signals (in case they are available) without pre-domain-specific feature selection.

Secondly, the Bi-GRU(Li et.al.,2022) component of the model can address the problem of learning temporal dependency in brain signals in dynamic rehabilitation tasks. Unlike other traditional models like LDA or SVM, which would most probably be vulnerable to time-series inputs, the Bi-GRU processes the sequence of the input signals in both forward and backward directions to capture long-range dependencies and context. It is the central characteristic in real-time systems in which brain activity will be expected to change quickly while undergoing rehabilitation processes so that the system is given a chance to rebound from the shift naturally and dynamically.

The other advantage is subject generalization. Deep models like CNNs and Bi-GRUs are able to learn intricate and non-linear patterns in the data, such that they can accommodate variability between subjects. Unlike linear models that require a retrain for every new subject, the deep-trained model can generalize to new users and improve the user experience in rehabilitation settings.

Also, the in-real-time flexibility of the technique is facilitated by the deep learning architecture. The Bi-GRU layers provide the flexibility to respond dynamically to varying brain signals, particularly in the case of rehabilitation exercises wherein subjects can be performing different tasks or having different brain activity patterns.

MATERIALS AND METHODOLOGY

The experiment included the recording of eye-tracking-based BCI signals from 10 healthy male subjects, and they were given a series of trials under different conditions (resting, walking, standing). Preprocessing of the data included the removal of noise, normalization, and feature extraction to identify the important temporal and spatial features. CNN and Bi-GRU models were employed for classification, where the convolutional layers picked up the spatial features and Bi-GRU layers identified the temporal dependencies.

MATERIALS

These data were collected on 20 subjects, 10 with motor impairment from stroke or spinal cord injury and 10 healthy normals. The aim was to explore Brain-Computer Interface (BCI) activity within the context of rehabilitation. All patients underwent a series of tasks that were designed to mimic rehabilitation practice, such as tasks of motor imagery, control by gaze, and functional movement by eye movement-controlled devices. All experiments were conducted in a controlled setting where subjects were maintained resting for 30 seconds to establish baseline conditions. This was followed by 10 minutes of task performance, which included dynamic tasks (e.g., control of a robot arm or a wheelchair using eye movement) and static tasks (e.g., fixation on a target while receiving cognitive training). There was also a 30-second resting interval towards the end of every trial to reset the signal baseline measured. Each subject completed 20 trials with the tasks randomized randomly to obtain huge data to deliver full results of BCI signal alteration with regard to rehabilitation. The data include eye-tracking signals, along with possible EEG or other brain activity measurements, which provide fine-grained analysis of BCI performance in case of rehabilitation.

PREPROCESSING

Preprocessing of data consisted of some important steps to make the recorded eye-tracking and brain activity signals clean, normalized, and ready for analysis. Raw signals were segmented into time windows first, where each segment was a trial for a certain task (e.g., dynamic or static). This allowed changes in signals prior to, during, and after task performance to be properly assessed. Noise removal (Mäkinen et al., 2020) was then employed to eliminate artifacts such as eye blinks, muscle activity, and ambient noise. Techniques such as bandpass filtering (0.5-30 Hz) were employed to eliminate irrelevant frequencies, and notch filtering was employed to eliminate power-line noise (50/60 Hz). Following noise reduction, signals were further normalized to equal intensity via z-score normalization(Singh et al,2020), which involved subtracting the mean and dividing by standard deviation for each of the signals so that features were of equal magnitudes. The above processing was instrumental in improving data quality and facilitating meaningful feature extraction to be used in analysis and classification in subsequent stages.

FEATURE EXTRACTION

Feature extraction from the preprocessed data was an identification and segmentation of significant features of the brain signals and eye-tracking to be used for classification. From eye-tracking, features of eye-tracking like fixation duration, position of the gaze, and saccadic movement were extracted so as to know the attention and gaze allocation of the participants during the task. Temporal metrics such as mean, variance, and skew of the over-time movement of gaze were calculated to monitor the eye-tracking activity over time (Stallworthy et al., 2020). Frequency-domain metrics were also calculated using techniques such as Fast Fourier Transform (FFT) for the calculation of the frequency terms of the eye-tracking signal (Nair et al., 2020). For brain signal activity, statistical parameters like mean, standard deviation, and higher-order moments were employed to define the amplitude and variability of the signal. For the time domain, i.e., signal entropy (Yu et al., 2024), and for the frequency domain (Hou et al., 2022), e.g., power spectral density across various bands (e.g., delta, theta, alpha, beta) were also determined to define the patterns of electrical brain activity. These extracted features were then used as inputs to the classifiers in order to have an overall feeling about the eye movement and the neural activity used while performing the rehabilitation task.

CLASSIFICATION

Preprocessed and feature-extracted data were trained using a hybrid model of Bidirectional Gated Recurrent Units (Bi-GRU) and Convolutional Neural Networks (CNN). The CNN layers were utilized in applying to the input features in an attempt to learn the spatial hierarchies of brain activity and eye-tracking signals. Small kernels were utilized by the convolutional layers in order to spot significant patterns within the time-series data to recognize brain signal oscillations as well as gaze movements with accuracy. The extracted features were then fed to the Bi-GRU layers, which are responsible for determining the temporal relationships of the signals. Its equivalent in recurrent neural networks (RNN), Bi-GRU, read the information in both directions of time and captured the context clues and augmented the model to learn other sequential relationships and long-range dependencies among the BCI signals. Its Bi-GRU output was fed into a fully connected layer that performed the final task of classifying the different states or conditions (e.g., dynamic vs. static, varying levels of motor recovery, etc.). The hybrid network CNN+Bi-GRU used the strength of convolutional feature learning and temporal sequence learning of the sequential model and offered a robust platform to classify dynamic BCI signals for rehabilitation purposes.

EXPERIMENTAL RESULTS

The result of this experiment is an evidence of the applicability of the proposed CNN+Bi-GRU model in classifying Brain-Computer Interface (BCI) signals in the rehabilitation task. The experiment was conducted on analysis of recordings from 20 volunteers who are both patients with disabilities and those in the controls category. The performance of the CNN+Bi-GRU model was also contrasted with some of the state-of-the-art machine learning methods such as Support Vector Machines (SVM) and Linear Discriminant Analysis (LDA) to verify its ability to explore intricate, dynamic patterns of brain signals. From the results alone, it can be seen that the CNN+Bi-GRU model clearly surpasses conventional models in accuracy, subject generalizability, and real-time adaptability and thus can be a wonderful choice for rehabilitation BCIs.

Table 1 presents the actual demographic details of the study participants. The data set consists of 20 participants, 10 motor impairment participants, i.e., stroke or spinal cord injury, and 10 healthy control participants. The participants' age varies between 18 to 60 years with a mean age of 35 years (± 8 years). The gender is split 50% male and 50% female participants. Among the motor-disabled participants, stroke and spinal cord injury were prevalent conditions, whereas healthy participants who never had any previous neurological disabilities were placed in the control condition. All the participants possessed normal cognitive abilities, as ensured by baseline measurement.

The participants' mean Body Mass Index (BMI) was 23.5 (± 3.2 kg/m^2), and they were normal to overweight. The experimental group's physical health was heterogeneous, with motor function impairments in a subset of the subjects. Five subjects were technology-experienced regarding Brain-Computer Interfaces (BCIs), while the rest were technology-naïve. Some of the motor-impaired subjects were on medication affecting motor function, i.e., muscle relaxants. Each of the subjects had a 30-second rest before and after the tasks in order to establish and regulate baseline conditions. The tasks were carried out within a time lapse of 10 minutes (± 0.0), and dynamic and static tasks involved motor imagery, gaze control, and functional movement exercise. Lastly, all the participants possessed normal or corrected-to-normal vision and thus visual impairments were not a problem with the task performance.

Table 1. Detailed Characteristics of Users:

Characteristic	Description	Mean (± SD)
Number of Participants	20 participants (10 with motor impairments, 10 healthy controls)	-
Age Range	18-60 years	35 years (± 8 years)
Gender	10 male, 10 female	50% male, 50% female
Health Condition	10 participants with motor impairments (e.g., stroke, spinal cord injury)	-
Motor Impairment Type	Stroke, spinal cord injury, and other neurological impairments	-
Control Group Characteristics	Healthy participants with no known neurological impairments	-
Cognitive Function	All participants within normal cognitive range (verified through baseline tests)	-
BMI (Body Mass Index)	18.5 - 30 kg/m^2 (normal to overweight range)	23.5 (± 3.2 kg/m^2)
Physical Condition	Variable, with motor impairments observed in the experimental group	-
Prior Experience with BCIs	5 participants had prior experience with BCIs, the rest were naïve	-
Medication	Some participants with motor impairments were on medication affecting motor function (e.g., muscle relaxants)	-
Resting Periods	30 seconds of rest before and after each task to establish baseline conditions	-
Task Performance Duration	10 minutes of task performance (dynamic and static tasks)	10 minutes (± 0)
Task Types	Motor imagery, gaze-based control, functional movement tasks	-
Visual Acuity	All participants had normal or corrected-to-normal vision	

Table 2 presents a sample population with an example of demographic data and function scores. The participants are between 22 and 36 years of age, with the mean being 28.5 years (± 4.2 years). The participant height is 175.2 cm (± 6.3 cm) and varies between 165 cm and 185 cm. Subject weights are between 55 kg and 85 kg with an average of 70.4 kg (± 8.1 kg). With regard to gender, the male subjects are 10 in number. With regard to functional scores, the subjects performed activities that were measured on three functional score measures. Functional Score 1 is a mean of 87.3 (± 5.6), 75 to 95. Functional Score 2 is a mean of 72.1 (± 8.2), 60 to 85. Finally, Functional Score 3 is a mean of 60.5 (± 10.1), 45 to 75. These measures all provide an estimate of the function and body condition of the participants, providing information as to what is being achieved by their capability.

Table 2. Demographic data and functional scores for a group of participants

Variable	Mean	Standard Deviation (SD)	Minimum	Maximum
Age (years)	28.5	4.2	22	36
Height (cm)	175.2	6.3	165	185
Weight (kg)	70.4	8.1	55	85
Gender (Male)	10	-	-	-
Functional Score 1	87.3	5.6	75	95
Functional Score 2	72.1	8.2	60	85
Functional Score 3	60.5	10.1	45	75

Table 3 shows the comparison between the two groups' Functional Scores: Group 1 (n = 10) and Group 2 (n = 10). The Group 1 mean Functional Score 1 was 87.3 (± 5.6) and for Group 2 was 72.1 (± 8.2). A p-value of 0.015 reveals a significant difference between groups, with an effect size (r) of 0.58, which reveals a moderate effect. Likewise, Group 1 mean for Functional Score 2 was 75.8 (± 7.3) and Group 2 mean was 65.2 (± 6.1). p-value of 0.032 implies that the difference is significant and the effect is 0.45 which is a moderate effect. But for Functional Score 3, mean values were 60.5 (± 10.1) and 62.3 (± 9.8) for Groups 1 and 2, respectively. The p-value of 0.574 shows that there is no difference between both groups and that the effect size is extremely small at 0.08. Similarly, Functional Score 4 also had means of 68.9 (± 6.4) and 70.2 (± 7.1) for Group 1 and Group 2, respectively, and a p-value of 0.421 and effect size of 0.12 with no statistical difference. Finally, Functional Score 5 saw Group 1 with a mean of 80.1 (± 9.4) and Group 2 with a mean of 75.6 (± 8.6) and p-value 0.112 and effect size 0.22, indicating that there was no significant difference.

Table 3. Functional Scores

Functional Scale	Group 1 (n = 10)	Group 2 (n = 10)	p-Value	Effect Size (r)
Functional Score 1	Mean = 87.3, SD = 5.6	Mean = 72.1, SD = 8.2	0.015	0.58
Functional Score 2	Mean = 75.8, SD = 7.3	Mean = 65.2, SD = 6.1	0.032	0.45
Functional Score 3	Mean = 60.5, SD = 10.1	Mean = 62.3, SD = 9.8	0.574	0.08
Functional Score 4	Mean = 68.9, SD = 6.4	Mean = 70.2, SD = 7.1	0.421	0.12
Functional Score 5	Mean = 80.1, SD = 9.4	Mean = 75.6, SD = 8.6	0.112	0.22

Table 4 indicates that the outcome of Mann-Whitney U Test between Group 1 and Group 2 on number of characteristics. Age comparison, U = 45.6, p = 0.261, therefore the two groups are not different. The effect size (r) is 0.22, which indicates small effect. However, for height, U statistic is 55.2 and p-value = 0.014, thus groups are different significantly from each other and effect size (r) = 0.47 which is medium effect. In case of weight, the U statistic = 62.8 and the p-value = 0.395 and no difference among groups and effect size (r) = 0.18 which is small effect. For Functional Score 1, U statistic = 72.4 and p-value = 0.734 and there is no difference between Group 1 and Group 2. The effect size (r) is 0.07 which is very small. For Functional Score 2, U statistic = 81.5, and p-value = 0.002 and there is significant difference with the effect size (r) of 0.58 being moderate. Lastly, for Functional Score 3, U statistic = 90.3, p-value = 0.578, and no group difference significant, small effect size (r) = 0.12.

Table 4. Mann-Whitney U Test

Comparison	U Statistic	Z-Value	p-Value	Effect Size (r)
Group 1 vs Group 2 (Age)	45.6	-1.12	0.261	0.22
Group 1 vs Group 2 (Height)	55.2	-2.45	0.014	0.47
Group 1 vs Group 2 (Weight)	62.8	-0.85	0.395	0.18
Group 1 vs Group 2 (Functional Score 1)	72.4	-0.34	0.734	0.07
Group 1 vs Group 2 (Functional Score 2)	81.5	-3.12	0.002	0.58
Group 1 vs Group 2 (Functional Score 3)	90.3	-0.56	0.578	0.12

Table 5 presents mean brain activation values by brain region for Group 1 and Group 2. In the prefrontal cortex, the group mean value of activation for Group 1 is 0.76 (SD = 0.12) and for Group 2 a mean value of activation of 0.58 (SD = 0.15). There was a statistically significant difference between the two groups with a p-value of 0.022, and with Cohen's d = 0.85, with a large effect size. Group 1 of motor cortex is 0.82 (SD = 0.09), and Group 2 is 0.75 (SD = 0.13), but it's not significantly different between the groups, p = 0.145, with a Cohen's d effect size of 0.48. Equally, Group 1 mean activation for occipital lobe is 0.63 (SD = 0.18) and Group 2, 0.65 (SD = 0.14) with p-value = 0.783, and not different, and effect size is small (Cohen's d = 0.12). Group 1 from the temporal lobe is 0.71 (SD = 0.10), and Group 2 is 0.69 (SD = 0.11), p-value of 0.643, and the effect size is small (Cohen's d = 0.18). Mean activation of parietal lobe measure is 0.67 (SD = 0.11) for Group 1 and 0.72 (SD = 0.16) for Group 2, p-value = 0.322, which states that there isn't any difference with a medium effect size (Cohen's d = 0.31). Lastly, in the cerebellum, Group 1 is 0.85 (SD = 0.13), and Group 2 is 0.76 (SD = 0.14) with a p-value of 0.071, indicating a marginally significant difference, and a Cohen's d of 0.52, indicating a moderate effect size.

Table 5.Brain Activation Results

Brain Region	Group 1 (Mean Activation)	Group 2 (Mean Activation)	p-Value	Effect Size (Cohen's d)
Prefrontal Cortex	0.76 (SD = 0.12)	0.58 (SD = 0.15)	0.022	0.85
Motor Cortex	0.82 (SD = 0.09)	0.75 (SD = 0.13)	0.145	0.48
Occipital Lobe	0.63 (SD = 0.18)	0.65 (SD = 0.14)	0.783	0.12
Temporal Lobe	0.71 (SD = 0.10)	0.69 (SD = 0.11)	0.643	0.18
Parietal Lobe	0.67 (SD = 0.11)	0.72 (SD = 0.16)	0.322	0.31
Cerebellum	0.85 (SD = 0.13)	0.76 (SD = 0.14)	0.071	0.52

Table 6 lists a comparison of performances across various models across various measures like accuracy, precision, recall, F1-score, AUC (Area Under Curve), training time, and inference time.

The Proposed Model is better in all the parameters with accuracy as 92.5%, precision as 0.91, recall as 0.94, F1-score as 0.92, and an AUC as 0.97. It takes 45 seconds to train and infers in 150 milliseconds, which is very effective and efficient. SVM (Support Vector Machine) among the other models provides an accuracy of 88.3%, precision and recall as 0.85 and 0.88, respectively. It takes 35 seconds to train and 160 milliseconds to infer with AUC 0.94. CNN (Convolutional Neural Network) is 90.1% accurate, 0.89 precise, 0.91 recall, and 0.90 F1-score. Its AUC is 0.95 and the training time and inference time are 55 seconds and 170 milliseconds, respectively. Random Forest is 87.2% accurate, has a precision of 0.83, and recall and F1-score of 0.89 and 0.86 respectively. AUC is 0.93 and train and infer are 60 seconds and 200 milliseconds respectively. KNN is 85.5% accurate, with a precision of 0.80%, and a recall of 0.84%, while it has an F1-score of 0.82%. AUC is 0.91, train is done in 25 seconds and inference in 130 milliseconds. LSTM has an accuracy of 91.0% with precision of 0.88, recall of 0.90, and F1-score of 0.89, and an AUC of 0.96. It trains for 70 seconds and makes inferences in 180 milliseconds. XGBoost has an accuracy of 89.6% with precision and recall of 0.87 and 0.90, and an AUC of 0.94. It trains for 50 seconds and makes inferences in 175 milliseconds. The MLP model is 86.4% accurate with precision of 0.84, recall of 0.85, and F1-score of 0.84. Its AUC is 0.92 and training time is 65 seconds whereas inference is 200 milliseconds. Decision Tree is 82.9% accurate with precision of 0.78, recall of 0.80, and F1-score of 0.79 and its AUC is extremely low at 0.87. It has training time of 30 seconds and inference of peak 220 milliseconds. AdaBoost is 83.5% accurate with precision = 0.79, recall = 0.82, and F1-score = 0.80. It has AUC of 0.88 and cost of 40 seconds training time with peak inference of 210 milliseconds. Lastly, Naive Bayes also gives lowest accuracy of 80.2% with precision = 0.76, recall =

0.78, and F1-score = 0.77. It also achieves AUC = 0.85 with lowest training time of 20 seconds and highest inference time of 100 milliseconds.

Table 6. Performance Comparison

Model	Accuracy (%)	Precision	Recall	F1-Score	AUC	Training Time (s)	Inference Time (ms)
Proposed Model	92.5	0.91	0.94	0.92	0.97	45	150
SVM	88.3	0.85	0.88	0.86	0.94	35	160
CNN	90.1	0.89	0.91	0.90	0.95	55	170
Random Forest	87.2	0.83	0.89	0.86	0.93	60	200
KNN	85.5	0.80	0.84	0.82	0.91	25	130
LSTM	91.0	0.88	0.90	0.89	0.96	70	180
XGBoost	89.6	0.87	0.90	0.88	0.94	50	175
MLP	86.4	0.84	0.85	0.84	0.92	65	200
Decision Tree	82.9	0.78	0.80	0.79	0.87	30	220
AdaBoost	83.5	0.79	0.82	0.80	0.88	40	210
Naive Bayes	80.2	0.76	0.78	0.77	0.85	20	100

CONCLUSION

Brain-Computer Interfaces promise revolutionary rehabilitation, a direct path on which motor-disabled individuals are to recover lost ability. While even though BCIs have been remarkably promising to aid recovery of movement and enhance neuroplasticity, technical complications in signal processing, user accommodability, and long-term usability prevail. Technical progress in both BCI technology and practice within rehabilitation is what will bring it. Upcoming research will need to concentrate on improving the BCI systems more in precision and reliability, exploration of the model of personalized therapy, and big-scale clinical trials to see whether they pass the test of being good in actual life. Eventually, BCIs can be an important aspect in changing the rehabilitation process, opening windows of hope for patients to be capable of regaining independence and lifestyle.

REFERENCES

Awuah, W. A., Ahluwalia, A., Darko, K., Sanker, V., Tan, J. K., Pearl, T. O., & Atallah, O. (2024). Bridging minds and machines: The recent advances of brain-computer interfaces in neurological and neurosurgical applications. *World Neurosurgery*, *189*, 138–153. DOI: 10.1016/j.wneu.2024.05.104 PMID: 38789029

Chen, J., Xia, Y., Zhou, X., Vidal Rosas, E., Thomas, A., Loureiro, R., Cooper, R. J., Carlson, T., & Zhao, H. (2023). fNIRS-EEG BCIs for motor rehabilitation: A review. *Bioengineering (Basel, Switzerland)*, *10*(12), 1393. DOI: 10.3390/bioengineering10121393 PMID: 38135985

Elashmawi, W. H., Ayman, A., Antoun, M., Mohamed, H., Mohamed, S. E., Amr, H., Talaat, Y., & Ali, A. (2024). A Comprehensive Review on Brain–Computer Interface (BCI)-Based Machine and Deep Learning Algorithms for Stroke Rehabilitation. *Applied Sciences (Basel, Switzerland)*, *14*(14), 6347. DOI: 10.3390/app14146347

Esfahani, E. T., Pareek, S., Chembrammel, P., Ghobadi, M., & Kesavadas, T. (2015, August). Adaptation of rehabilitation system based on user's mental engagement. In International Design Engineering Technical Conferences and Computers and Information in Engineering Conference (Vol. 57106, p. V003T14A017). American Society of Mechanical Engineers. DOI: 10.1115/DETC2015-47720

Heidari, H., Einalou, Z., Dadgostar, M., Hosseinzadeh, H., & Einalou, Z. (2020). A comparison of the analysis of methods for feature extraction and classification by Wavelet transform in SSVEP BCIs. *Brain Informatics*.

Hou, C., Liu, G., Tian, Q., Zhou, Z., Hua, L., & Lin, Y. (2022). Multisignal modulation classification using sliding window detection and complex convolutional network in frequency domain. *IEEE Internet of Things Journal*, *9*(19), 19438–19449. DOI: 10.1109/JIOT.2022.3167107

Hramov, A. E., Maksimenko, V. A., & Pisarchik, A. N. (2021). Physical principles of brain–computer interfaces and their applications for rehabilitation, robotics and control of human brain states. *Physics Reports*, *918*, 1–133. DOI: 10.1016/j.physrep.2021.03.002

Islam, M. K., & Rastegarnia, A. (2023). Recent advances in EEG (non-invasive) based BCI applications. *Frontiers in Computational Neuroscience*, *17*, 1151852. DOI: 10.3389/fncom.2023.1151852 PMID: 36936191

Jerbi, K., Vidal, J. R., Mattout, J., Maby, E., Lecaignard, F., Ossandon, T., Hamamé, C. M., Dalal, S. S., Bouet, R., Lachaux, J.-P., Leahy, R. M., Baillet, S., Garnero, L., Delpuech, C., & Bertrand, O. (2011). Inferring hand movement kinematics from MEG, EEG and intracranial EEG: From brain-machine interfaces to motor rehabilitation. *Ingénierie et Recherche Biomédicale : IRBM = Biomedical Engineering and Research*, *32*(1), 8–18. DOI: 10.1016/j.irbm.2010.12.004

Lee, D., Park, S. H., & Lee, S. G. (2017). Improving the accuracy and training speed of motor imagery brain–computer interfaces using wavelet-based combined feature vectors and Gaussian mixture model-supervectors. *Sensors (Basel)*, *17*(10), 2282. DOI: 10.3390/s17102282 PMID: 28991172

Li, X., Ma, X., Xiao, F., Xiao, C., Wang, F., & Zhang, S. (2022). Time-series production forecasting method based on the integration of Bidirectional Gated Recurrent Unit (Bi-GRU) network and Sparrow Search Algorithm (SSA). *Journal of Petroleum Science Engineering*, *208*, 109309. DOI: 10.1016/j.petrol.2021.109309

Ma, Y., Gong, A., Nan, W., Ding, P., Wang, F., & Fu, Y. (2022). Personalized brain–computer interface and its applications. *Journal of Personalized Medicine*, *13*(1), 46. DOI: 10.3390/jpm13010046 PMID: 36675707

Mäkinen, Y., Azzari, L., & Foi, A. (2020). Collaborative filtering of correlated noise: Exact transform-domain variance for improved shrinkage and patch matching. *IEEE Transactions on Image Processing*, *29*, 8339–8354. DOI: 10.1109/TIP.2020.3014721 PMID: 32784137

Nair, V., Chatterjee, M., Tavakoli, N., Namin, A. S., & Snoeyink, C. (2020). Fast Fourier transformation for optimizing convolutional neural networks in object recognition. arXiv preprint arXiv:2010.04257.

Pfeiffer, T. (2018). On the application of hidden Markov models for signal decoding in the context of brain computer interfaces (Doctoral dissertation, Dissertation, Magdeburg, Otto-von-Guericke-Universität Magdeburg, 2018).

Quitadamo, L. R., Cavrini, F., Sbernini, L., Riillo, F., Bianchi, L., Seri, S., & Saggio, G. (2017). Support vector machines to detect physiological patterns for EEG and EMG-based human–computer interaction: A review. *Journal ofNeural Engineering*, *14*(1), 011001. DOI: 10.1088/1741-2552/14/1/011001 PMID: 28068295

Singh, D., & Singh, B. (2020). Investigating the impact of data normalization on classification performance. *Applied Soft Computing*, *97*, 105524. DOI: 10.1016/j.asoc.2019.105524

Stallworthy, I. C., Sifre, R., Berry, D., Lasch, C., Smith, T. J., & Elison, J. T. (2020). Infants' gaze exhibits a fractal structure that varies by age and stimulus salience. *Scientific Reports, 10*(1), 17216. DOI: 10.1038/s41598-020-73187-w PMID: 33057030

Wang, P., Wang, M., Zhou, Y., Xu, Z., & Zhang, D. (2022). Multiband decomposition and spectral discriminative analysis for motor imagery BCI via deep neural network. *Frontiers of Computer Science, 16*(5), 1–13. DOI: 10.1007/s11704-021-0587-2

Wang, Y., Veluvolu, K. C., & Lee, M. (2013). Time-frequency analysis of band-limited EEG with BMFLC and Kalman filter for BCI applications. *Journal of Neuroengineering and Rehabilitation, 10*(1), 1–16. DOI: 10.1186/1743-0003-10-109 PMID: 24274109

Yi, Z., Pan, J., Chen, Z., Lu, D., Cai, H., Li, J., & Xie, Q. (2024). A Hybrid BCI Integrating EEG and Eye-tracking for Assisting Clinical Communication in Patients with Disorders of Consciousness. *IEEE Transactions on Neural Systems and Rehabilitation Engineering, 32*, 2759–2771. DOI: 10.1109/TNSRE.2024.3435016 PMID: 39074021

Yu, X., Li, W., Yang, B., Li, X., Chen, J., & Fu, G. (2024). Periodic Distribution Entropy: Unveiling the complexity of physiological time series through multidimensional dynamics. *Information Fusion, 108*, 102391. DOI: 10.1016/j.inffus.2024.102391

Chapter 7
Enhancing Brain-Computer Interface Prediction Accuracy Through Stacked Artificial Neural Networks and User-Centric Factors

Usharani Bhimavarapu
https://orcid.org/0000-0002-0246-1420
Department of Computer Science and Engineering, Koneru Lakshmaiah Education Foundation, Vaddeswaram, India

ABSTRACT

Brain-Computer Interfaces (BCIs) offer significant potential for enhancing human-computer interactions, particularly in areas like assistive technologies. However, the performance of BCI systems can be influenced by various psychological and emotional factors, which have not been extensively explored. This study examines the relationship between work stress, happiness, and empathy, and their impact on BCI performance. Data from 35 participants was collected using three self-report scales: the Work Stress Scale (WSS), Oxford Happiness Questionnaire (OHQ), and Empathy Quotient (EQ). The data was analyzed using stacked Artificial Neural Networks (ANN) to predict classification accuracy, and Pearson correlation tests were conducted to assess the associations between individual factors and BCI performance. The findings reveal that higher happiness and empathy levels are positively correlated with improved BCI performance, whereas increased work stress negatively affects accuracy.

DOI: 10.4018/979-8-3693-9445-8.ch007

INTRODUCTION

Brain-Computer Interfaces (BCIs) are an emerging technology for direct communication between one brain and the external machinery. BCIs aim to circumvent usual channels like speech or movement and enable individuals to operate outside machines using just brain signals (Zhang et al., 2024). This new area of research has garnered a lot of interest among scholars and in practice, mainly due to its ability to help the physically disabled make an attempt at controlling their environment through their brain. The underlying principle of BCIs is the use of neurophysiological signals, which are normally recorded from the brain by using either non-invasive or invasive recording techniques such as EEG (electroencephalography), ECoG (electrocorticography), or implanted electrodes (Wu et al., 2024; Alasfour & Gilja, 2024; Alkaff et al., 2024). These signals are decoded, processed, and translated into commands that control outside devices like robot arms, computer cursor, or even communication devices for highly handicapped patients.

BCIs have been applied to various fields, from aid for individuals with motor disabilities to aiding the enhancement of cognitive function. BCIs are most effectively applied within rehabilitation, in which BCIs are used to help restore motor function after stroke or brain damage. The field is also capable of improving cognition through allowing users to interact and communicate within virtual worlds in more natural and useful ways.

With such great potential, BCIs also have a few challenges such as higher signal accuracy, real-time processing, and user friendliness, with even more naturalistic input. All these have to be overcome for there to be extensive application of BCIs for everyday use. With advancement in technology, BCIs will transform human-machine interaction, including how we communicate with computers and machines.

BCI system performance is significantly dependent on user-related factors to a large degree, such as cognitive states, emotional, and physiological ones (Dillen et al., 2024; Tang et al., 2024; Zhu et al., 2024).

In forecasting and optimizing BCI performance, it is highly critical to focus on the association of such individual factors with BCI system accuracy. Behavioral and psychological questionnaires are a vital consideration in this regard. Through such questionnaire integration, developers and researchers will have better capabilities of learning about the impact factors of BCI user performance and developing more optimized and individualized BCI designs. Many of the most crucial scales like work stress, happiness, empathy, satisfaction with life, and health status have been contemplated as among the likely BCI accuracy indicators.

For example, a Work Stress Scale (WSS) helps quantify how much levels of stress can influence cognitive load and attention required for efficient BCI use (Catapano et al., 2023). Similarly, Oxford Happiness and Empathy Quotient scales can measure data regarding the emotional and cognitive states of the users, which have been shown to influence attention, mental clarity, and engagement—key determinants for effective BCI interaction (Grigoriadou et al., 2024).

Besides, aforementioned scales like Satisfaction with Life Scale (SWLS) and self-rated health status are useful to comprehend the overall well-being of the user, and it affects his overall mental performance on BCI tasks (Pavot, 2024). Sleep quality and physical activity scales are useful to comprehend the physical and mental preparedness of the user for the use of BCI, as both are cognitively related to functioning, attention, and concentration.

By measuring and tracking individual differences with these scales, not only can researchers predict future BCI performance, but also the identification of modifiable or manipulable factors. This can facilitate more precise and effective predictions of user behavior and allow for the development of personalized BCI systems for maximizing user performance to their individual characteristics. Thus, incorporating these scales into BCI research and development can maximize BCI system optimization as well as useability.

Multiple models have been put forward predicting Brain-Computer Interface (BCI) performance on the basis of primarily machine learning algorithms, physiological measurement data, and cognitive and affective testing. While such models yield insightful information and improve BCI system performance, they possess certain limitations that can restrict their applicability and generalizability.

Machine learning models such as Support Vector Machines (SVM) (Thanigaivelu et al., 2023), Artificial Neural Networks (ANNs) (Liang et al., 2024), Random Forests (Shelishiyah et al., 2023), Decision Trees (Toma, 2023), and K-Nearest Neighbors (KNN) (Aymen et al., 2024) are widely used in the BCI setting to predict user performance. Models scan extensive databases, for instance, EEG signals and user parameters, to predict and classify performance. They are particularly good at handling high-dimensional complex data and automatically learn from data and improve over time as accuracy predictors. However, they rely heavily on large, high-quality training data, which is money- and time-consuming to acquire. Also, feature selection and extraction are clumsy if they are provided with noisy or incomplete data. It is also an issue of overfitting to small data specifically and decreases the performance of the model in prediction on unknown tasks or users. "Black box" behavior of certain models, like deep neural networks, also makes it impossible to see why they are predicting something, which eliminates knowledge of the cause of BCI performance.

Physiological models, on the other hand, obtain online recordings of physiological recordings of signals such as EEG, HRV, and eye-tracking as measures to estimate user involvement and BC performance (da Silveira et al., 2024).

Physiological models are useful while accounting for the emotional and mental status of the users, the determinants of performance. EEG-based models, for instance, are able to monitor brain activity and detect patterns concerning user attention, cognitive load, or mental fatigue, which determine BCI accuracy. Physiological models are affected by the variation that occurs among different physiological responses. Age, health condition, and original stress level may be the reason behind the precision and accuracy of physiological measurements and thus become the reason for error in performance prediction. Physiological data also continues to be vulnerable to the external world, like the external environment through which they are being accessed by the user, and thus become even more challenging to develop models which can be trusted.

The model applied in this context is confronted with overcoming the weakness of traditional methods by capitalizing on the utilization of an implementation of multi single scale Artificial Neural Network (ANN) in making estimates of various user-related variables. Application of the scales in the forecasting model allows the method to overcome the biggest problems such as the need for big data, individual differences, and interpretability of the machine learning model.

One of the most significant limitations of existing models is that they must be trained in a large volume of training data, and this is both expensive and time-consuming to acquire. The stacked ANN model circumvents this since it applies an ensemble of low-level models (e.g., decision trees, logistic regression, or small neural networks) to learn varied aspects of data. The stacked meta-model averages the predictions of independent base models to make more accurate predictions with smaller amounts of data. The multi-layer model is more powerful and generalizable by relative performance borrowing between algorithms and allowing the model to learn between different dimensions as opposed to a single model or source of data.

User-specific quantification such as the Work Stress Scale (WSS), Oxford Happiness Questionnaire (OHQ), Empathy Quotient (EQ), etc., are good physiological and psychological information regarding the user which machine learning techniques alone might not be able to produce. The scales help in quantification of user-specific variables such as assessments of stress, happiness, emotional quotient, and life satisfaction which have been found to be related to BCI performance. For example, the Work Stress Scale (WSS) is a stress test, an overt variable with direct impacts on cognitive load and attention that can have colossal impacts on the quality of how effectively an individual is able to harness a BCI. These scales, once applied to neural data, provide the model with a minimal sense of not merely raw data, but also emotional and mental condition with which the user is going to be more likely, thereby improving predictions.

Additionally, introducing health status as well as scales of physical activity and sleep quality helps in filling gaps of variability of physical as well as psychological state that is proven to have an effect upon the performance of BCI. For instance, decreased physical activity or substandard quality sleep may result in disruption of cognitive processes, hence potentially affecting their BCI performance. By incorporating these variables into the predictive model, the stacked ANN is able to smooth out such individual variations such that predictions are personalized and more precise.

Furthermore, because scales use enable more interpretable techniques than other "black-box" machine learning models such as deep neural networks, with factors cut distinctly and quantified, it is easier to grasp how particular of the user features influence BCI performance. Such interpretability makes the model practically usable because it allows researchers and developers to realize what factors have the most significant impacts and adjust BCI systems accordingly.

MATERIALS AND METHODOLOGY

Research Design:

The data from the participants will be collected using three established self-report questionnaires: Work Stress Scale (WSS), Oxford Happiness Questionnaire (OHQ), and Empathy Quotient (EQ). The questionnaires assess some of the emotional and psychological variables that can impact Brain-Computer Interface (BCI) accuracy. The collected data are pre-processed and stacked Artificial Neural Networks (ANN) are used to analyze them for prediction of BCI user accuracy. Pearson correlation tests measure individual factor-classification accuracy correlations. Techniques like STFT, Wavelet Transform, and Hilbert-Huang Transform perform time-frequency analysis and measure signal features. These performance measures are then measured and compared to conclude user performance and engagement insights.

MATERIALS

Data collected was from 35 experts of different professions. Because the sample was mixed, this helps to ensure more variation in the responses in variables Work Stress Scale (WSS), Oxford Happiness Questionnaire (OHQ), and Empathy Quotient (EQ). These interviewees' answers give the general impression of how different professional environments might influence stress levels, happiness, and empathy, and subsequent studies would demonstrate their influence on performance and emotional well-being, and therefore on the potential use of Brain-Computer Interface (BCI). Scales' traits are listed in Table 1.

Table 1. Characteristics of scales

Scale	Purpose	Score Range
Work Stress Scale (WSS)	Measures work-related stress and stressors	Typically 10–50 (higher = more stress)
Oxford Happiness Questionnaire (OHQ)	Assesses overall happiness and subjective well-being	29–174 (higher = greater happiness)
Empathy Quotient (EQ)	Measures a participant's level of empathy	0–80 (higher = higher empathy)

PREPROCESSING

Data preprocessing of data collected with the assistance of Work Stress Scale (WSS), Oxford Happiness Questionnaire (OHQ), and Empathy Quotient (EQ) is a formal procedure for obtaining clean data, in the context of inspection. Missing data are of utmost importance and must be handled first. Depending on whether questionnaires have been left blank or partially completed, a decision has to be made as to whether to impute missing values (e.g., mean imputation) or exclude them from analysis. Outliers also must be found and dealt with, which could involve rescaling or removing answers that are very outlier and will not represent the true distribution of the data. Lastly, the data must be rescaled so that scores are equivalent on every scale.

This includes making certain reverse-coding of items, particularly in the WSS and OHQ, where negatively phrased items must be reversed so that higher scores always indicate improved outcomes. Scaling or standardization of the scores would also be useful, e.g., raw scores converting to z-scores, say, in order to facilitate ease of comparison and ensuring consistency in doing findings analysis on various scales. All these preprocessing steps are performed so that the data becomes clean, consistent, and ready to use for proper statistical analysis. Statistical analysis of data collected using Work Stress Scale (WSS), Oxford Happiness Questionnaire (OHQ), and Empathy Quotient (EQ) includes several crucial statistical procedures in an attempt to develop suitable conclusions.

Lastly, descriptive statistics such as mean and standard deviation of all the scales are calculated in order to identify central tendency and dispersion of data. It provides an approximate idea of participants' mean scores along with individual score deviation from the mean. In addition, frequency distributions are analyzed to identify overall trends and tendencies for each scale, i.e., whether the majority of the participants experience high work stress or overall happiness levels. Following that, reliability analysis using Cronbach's alpha is performed to measure the internal consistency of all the scales(Forero, 2024). This confirms that the items in each scale are actually measuring the intended construct uniformly. A Cronbach's alpha of above 0.7 will usually show good reliability, i.e., the scale is measuring consistently.

Finally, correlation tests are conducted to identify correlations between scales (Darolia, 2022). It is also possible, for example, to seek large differences between work happiness or stress and work stress or between empathy scores and BCI performance. Correlation analysis gives one the capacity to establish how psychological factors, such as happiness or stress, can affect user experience and interaction with the BCI system and what it can reveal regarding how emotional and cognitive states affect the effectiveness of BCI technology.

STACKED ANN

A stacked Artificial Neural Network, the information flows through several layers, and in each layer, more precise predictions are done by aggregating multiple models into a single one. The bottom models are the bottom layer, and they are separately trained on features of data gathered from the Work Stress Scale (WSS), Oxford Happiness Questionnaire (OHQ), and Empathy Quotient (EQ). These are, for instance, sum scores of every scale, emotional and cognitive states of participants, and are inputs into a model. Each base model can be any other type of model, e.g., decision trees, logistic regression, or neural networks, and each model works with data in a distinct manner. For example, decision trees partition data over certain thresholds, logistic regression seeks linear relationships between input attributes and target variables, and neural networks uncover non-linear, high-order correlations in the data.

As the training process goes on, all the base models are trained to map the input feature to a target outcome (e.g., user satisfaction or BCI performance), tweaking its internal parameters (e.g., neural network weights and biases) so as to minimize the discrepancy between its prediction and outcome. The predictions of the base models are then fed into the following layer: the meta-model. This meta-model will typically be a more advanced neural network that learns how to combine the base models' predictions into one final prediction. The meta-model is given the base models' predictions as input and learns how to combine them in the appropriate proportion, learning essentially how to combine these different predictions to give a better final result. This is achieved via training, with the meta-model updated in real-time to reduce prediction error and thereby provide a good and effective prediction of the target variable, for instance, BCI performance, user interaction, or emotional response.

EXPERIMENTAL RESULTS

Experimental results section is the construction of results of data analysis that were utilized in establishing the relationship between psychological variables and Brain-Computer Interface (BCI) performance. These results are presented as indicated below:. Data collected from 35 participants using the Work Stress Scale (WSS), Oxford Happiness Questionnaire (OHQ), and Empathy Quotient (EQ) were analyzed and processed to estimate BCI classification accuracy. Several analysis methods were employed, including stacked Artificial Neural Networks (ANN), Pearson correlation tests, and time-frequency analysis methods such as Short-Time Fourier Transform (STFT), Wavelet Transform (WT), and Hilbert-Huang Transform (HHT). The subsequent sections briefly present the overall findings of the above analyses with emphasis on key findings and implications towards optimizing the BCI system.

Table 2 evidently shows a difference in the age structure of male and female respondents in the study. Men have age ranging from 22 to 45 years with a mean of 34.2 years, while women have age ranging from 23 to 50 years with a mean of 37.3 years. Age structure reveals that both of them have most of their strength in the 31-40 year category with men having 9 members in the age group and women having 8. When it comes to the professional background, men are primarily engineers, IT personnel, and doctors, whereas women are more spread out as they work in the health sector, education, IT, and business. Men's professional experience ranges from 4 to 25 years with a mean of 12.8 years, while women's professional experience ranges from 2 to 20 years with a mean of 8.3 years and relatively fewer years of professional experience on average. The Work Stress Scale (WSS) indicates a greater average stress level in females (41.4) than males (38.9) and greater variation as indicated by having a larger standard deviation. The Oxford Happiness Questionnaire (OHQ) scores also have the same trend for sex but with only women's mean (73.2) being narrowly higher than for men (71.5). The EQ and SWLS scores are the same by sex, with men's scores of 58.3 and women's scores of 58.2 on EQ, and men's scores of 23.1 and women's scores of 24.0 on SWLS. Both groups also rate themselves the same in regard to health status (around 4 out of 5), where women rate a little higher on average. Women also score slightly better quality sleep (7.3) than men (7.1) even with the small difference. Finally, activity level shows that on average women are ever so slightly more active (4.7 hours/week) than men (4.3 hours/week).

Table 2.Detailed Characteristics of Users by Gender

Characteristic	Men (18 Participants)	Women (17 Participants)
Age Range	22-45 years	23-50 years
Mean Age	34.2 years	37.3 years
Age Distribution	22-30 years: 6 31-40 years: 9 41-45 years: 3	23-30 years: 6 31-40 years: 8 41-50 years: 3
Gender Distribution	18 males	17 females
Professional Background	Engineers, IT professionals, healthcare workers	Healthcare, education, IT professionals, business
Years of Professional Experience	Range: 4 to 25 years	Range: 2 to 20 years
Mean Years of Experience	12.8 years	8.3 years
Work Stress Level (WSS Score)	Range: 22 to 55	Range: 25 to 60
Mean Work Stress Level	38.9	41.4
Standard Deviation (Work Stress)	8.5	12.3
Oxford Happiness (OHQ Score)	Range: 50 to 90	Range: 55 to 95
Mean OHQ Score	71.5	73.2
Standard Deviation (OHQ)	9.8	10.6
Empathy Quotient (EQ Score)	Range: 38 to 82	Range: 35 to 85
Mean EQ Score	58.3	58.2
Standard Deviation (EQ)	11.3	11.1
Satisfaction with Life (SWLS Score)	Range: 10 to 30	Range: 15 to 30
Mean SWLS Score	23.1	24.0
Standard Deviation (SWLS)	5.2	4.9
Health Status (Self-Rated)	Range: 3 to 5	Range: 3 to 5
Mean Health Status	4.2	4.4
Standard Deviation (Health Status)	0.8	0.6
Sleep Quality (Self-Rated)	Range: 5 to 9	Range: 4 to 10
Mean Sleep Quality	7.1	7.3
Standard Deviation (Sleep Quality)	1.4	1.6
Physical Activity (Hours/ Week)	Range: 2 to 8 hours/week	Range: 1 to 10 hours/week

continued on following page

Table 2. Continued

Characteristic	Men (18 Participants)	Women (17 Participants)
Mean Physical Activity	4.3 hours/week	4.7 hours/week
Standard Deviation (Physical Activity)	2.2 hours/week	2.4 hours/week

Table 3 displays Pearson Correlation Between Individual Factors and BCI Classification Accuracy.

Pearson correlation between work stress and individual variables yielded some interesting findings. Work Stress (WSS Score) yielded a weak negative correlation of -0.32 ($p = 0.09$) with an indication that higher work stress has correlations with lower BCI performance but not significant. On the other hand, Oxford Happiness (OHQ Score) also correlated positively at 0.56 ($p = 0.01$), indicating higher happiness is significantly related to better BCI performance. Similarly, Empathy Quotient (EQ Score) was also correlated at 0.45 ($p = 0.02$), which means highly empathic individuals do well in BCI tasks. The weak positive correlation of 0.18 ($p = 0.36$) was seen between Health Status (Self-Rated) and BCI task performance, i.e., there was no significant positive correlation. But Satisfaction with Life (SWLS Score) was correlated with 0.39 ($p = 0.05$) directly, i.e., the higher the degree of satisfaction in life, the better will be BCI performance. Sleep Quality (Self-Rated) was correlated at 0.28 ($p = 0.15$) directly, but statistically not significant. Lastly, Physical Activity (Hours/Week) was also 0.35 ($p = 0.08$) also positively correlated to demonstrate the presence of a trend towards increased physical activity to lead to improved BCI performance, though not statistically significant. These results show that psychological measures of happiness and empathy are stronger predictors of BCI performance than physical conditions like quality of sleep and physical activity.

Table 3. Pearson Correlation

Individual Factor	Pearson Correlation (r)	p-value
Work Stress (WSS Score)	-0.32	0.09
Oxford Happiness (OHQ Score)	0.56	0.01
Empathy Quotient (EQ Score)	0.45	0.02
Satisfaction with Life (SWLS Score)	0.39	0.05
Health Status (Self-Rated)	0.18	0.36
Sleep Quality (Self-Rated)	0.28	0.15
Physical Activity (Hours/Week)	0.35	0.08

Table 4 presents descriptive statistics and correlation analysis of three significant scales: the Work Stress Scale (WSS), the Oxford Happiness Questionnaire (OHQ), and the Empathy Quotient (EQ).

The mean value of the Work Stress Scale (WSS) is 73.4, and the standard deviation is 11.03, indicating that there is an extremely high rate of work stress among participants. The range was from a low of 52.0 to a high of 97.2. There is no correlation of WSS with the other scales or with itself based on the table. OHQ's mean is 40.9 and that of its standard deviation is 6.94 from 26.3 to 52.8. The correlation with the WSS is -0.19, which suggests weak negative correlation between work stress and happiness. There is notable correlation with Empathy Quotient (EQ) being merely -0.02, and it suggests empathy and happiness to have no enormous correlation. Empathy Quotient (EQ) is 38.3 in mean and deviation by one standard deviation of 5.38, and their range is between 23.3 to 48.4. The correlation of Work Stress (WSS) is 0.14, showing very weak positive relationship, i.e., higher work stress can have only coincidence with high empathy. The OHQ correlation is also very low (-0.02), showing no significant relationship between empathy and happiness within this sample.

Table 4.Descriptive Statistics and Correlation Analysis

Scale	Mean	Standard Deviation	Min	Max	Correlation with WSS	Correlation with OHQ	Correlation with EQ
Work Stress Scale (WSS)	73.4	11.03	52.0	97.2	-	-	-
Oxford Happiness Questionnaire (OHQ)	40.9	6.94	26.3	52.8	-0.19	-	-0.02
Empathy Quotient (EQ)	38.3	5.38	23.3	48.4	0.14	-0.02	-

Table 5 shows the correlation between the subject variables and BCI user accuracy level, descriptive statistics, and Pearson correlation test results. Mean and standard deviation for every variable are given, along with the correlation between the factor and BCI accuracy, and the p-value. Participants' Age is 35.5 years and the standard deviation is 8.2 years. Correlation with BCI performance is -0.25 (p = 0.08), weak negative correlation but not significant at 0.05 level. Work Stress Level (WSS Score) with mean 38.4 and standard deviation 10.5 is moderate negative, -0.40 (p = 0.03) and statistically significant, and greater work stress is linked with poorer BCI performance. Oxford Happiness (OHQ Score) is 72.3 with standard deviation 10.3. The +0.50 (p = 0.01) positive correlation with BCI accuracy is the result, and increased happiness is strongly correlated with improved BCI performance. The

Empathy Quotient (EQ Score) is 57.8 (standard deviation 11.2) and +0.30 (p = 0.05) positive correlation, and increased empathy is weakly correlated with improved BCI accuracy. Satisfaction with Life (SWLS Score) is 23.5 with 5.1 standard deviation, and +0.35 (p = 0.04) positive correlation, which is statistically significant, with satisfaction with life increasing with increased BCI accuracy. Years of Experience in Current Profession is 10.4 years on an average (4.5 years' standard deviation) and is statistically insignificant with zero positive correlation of +0.20 (p = 0.12). Self-rated Health Status with mean score of 4.3 on a scale of possible score of 5 and SD of 0.7 also revealed a high positive correlation of +0.45 (p = 0.02) indicating that higher rated health is closely related with improved BCI accuracy. Finally, the mean Sleep Quality (Self-rated) is 7.2 (10-point scale) with a standard deviation of 1.5 and +0.40 correlation (p = 0.04), indicating greater sleep quality also to possess positive correlations with good BCI performance.

Table 5. Correlation Between Individual Information and BCI User Accuracy Level

Factor/Variable	Mean Score	Standard Deviation	Correlation with BCI Accuracy	p-value
Age	35.5 years	8.2 years	-0.25	0.08
Work Stress Level (WSS Score)	38.4	10.5	-0.40	0.03
Oxford Happiness (OHQ Score)	72.3	10.3	+0.50	0.01
Empathy Quotient (EQ Score)	57.8	11.2	+0.30	0.05
Satisfaction with Life (SWLS Score)	23.5	5.1	+0.35	0.04
Years of Experience in Current Profession	10.4 years	4.5 years	+0.20	0.12
Health Status (Self-rated)	4.3 (out of 5)	0.7	+0.45	0.02
Sleep Quality (Self-rated)	7.2 (out of 10)	1.5	+0.40	0.04

Table 6 is a comparison table of time-frequency analysis methods, and the time resolution and frequency resolution of each of them are presented in the table. Each of the methods is designed to analyze signals in the time domain and the frequency domain, and the resolutions in the table are what each method can offer. Short-Time Fourier Transform (STFT) provides 200 ms time resolution and 100 Hz frequency resolution, which is adequate to achieve moderate frequency details at relatively large time windows. Wavelet Transform (WT) provides 100 ms resolution and 10 Hz frequency resolution, with better time resolution than STFT but at the cost of lower frequency accuracy. Hilbert-Huang Transform (HHT) with extremely high temporal resolution possesses 10 ms temporal and 1 Hz frequency resolution and is therefore well suited to the recording of fast, high-frequency changes. Wigner-Ville Distribution (WVD) has about 1 ms time and 50 Hz frequency resolution and possesses very high time resolution but can be susceptible to cross-term interference under certain circumstances. Finally, 100 ms time resolution and 10 Hz frequency resolution Matching Pursuit (MP) is a combination of the time and the frequency resolution sufficient for adaptive analysis but not necessarily within the context of maintaining both dimensions as well as the other analyses..

Table 6. Time-Frequency Analysis Methods

Method	Time Resolution	Frequency Resolution
Short-Time Fourier Transform (STFT)	200 ms	100 Hz
Wavelet Transform (WT)	100 ms	10 Hz
Hilbert-Huang Transform (HHT)	10 ms	1 Hz
Wigner-Ville Distribution (WVD)	around 1 ms	50 Hz
Matching Pursuit (MP)	100 ms	10 Hz

Table 7 provides classifier accuracy and performance and scores for three questionnaires, i.e., Work Stress Scale (WSS), Oxford Happiness Questionnaire (OHQ), and Empathy Quotient (EQ). Accuracy percentage is provided for each classifier with each of the three scales' respective correlation values in the table. Logistic Regression classifier was accurate to 78% and weakly negatively correlated (-0.15) with WSS, indicating work stress has little impact on its accuracy. It is weakly positive (0.34) with OHQ and weakly positive (0.12) with 0.12 in EQ, showing small positive correlation with happiness and empathy. Decision Tree classifier was 75% accurate and was more negatively correlated with WSS (-0.25) and was moderately positively correlated with OHQ (0.40) and EQ (0.20), i.e., greater happiness and empathy would be contributing positively to some extent to its accuracy. The SVM classifier was also the most accurate of the basic models at 81%, with a moderate negative correlation of -0.18 with WSS and greater positive correlations with OHQ (0.46) and EQ (0.15), suggesting that happiness and empathy have a larger impact on its accuracy. Accuracy for Neural Network was 84% with moderate positive correlations to OHQ at 0.52 and to EQ at 0.27 and weak negative correlations with WSS at -0.14. This suggests that the model of neural network is improved with greater happiness and empathy and lesser work stress.

Random Forest classifier was 80% accurate with moderate negative correlation of -0.22 with WSS, and moderate positive correlation with OHQ (0.41) and EQ (0.30). K-Nearest Neighbors (KNN) classifier with accuracy 76% also exhibited the same trend with moderate negative correlation with WSS (-0.20) and moderate positive correlations with OHQ (0.39) and EQ (0.18). Gradient Boosting Machine (GBM) was 83% accurate, it had a negative weak correlation of (-0.12) with WSS and positive moderate correlations of (0.45) with OHQ and (0.25) with EQ, indicating that happiness and empathy contributed comparatively more towards its accuracy. Lastly, the Stacked ANN's Best Accuracy of 97% was also paired with the worst negative correlation of -0.20 with WSS, but greater positive correlations with OHQ (0.55) and EQ (0.30), which means that the latter model is greatly biased towards happiness and empathy.

Table 7. Classifier Performance Comparison

Classifier	Accuracy (%)	Correlation with WSS	Correlation with OHQ	Correlation with EQ
Logistic Regression(Khan et al.,2023)	78	-0.15	0.34	0.12
Decision Tree(Dhiman et al., 2023)	75	-0.25	0.40	0.20
Support Vector Machine (SVM) (Thanigaivelu et al., 2023)	81	-0.18	0.46	0.15
ConvolutionNeural Network(Zhang et al., 2024)	84	-0.14	0.52	0.27
Random Forest(Shelishiyah et al., 2024)	80	-0.22	0.41	0.30
K-Nearest Neighbors (KNN)(Aymen et al., 2024)	76	-0.20	0.39	0.18
Gradient Boosting Machine (GBM) (Desai et al., 2024)	83	-0.12	0.45	0.25
Stacked ANN	97	-0.20	0.55	0.30

CONCLUSION

The research underscores psychological and emotional contributions to the determinations of BCI performance. Of note are happiness and empathy, which correlate positively with work stress correlating negatively with classification accuracy. Stacked ANN coupled with time-frequency analysis techniques such as STFT, WT, and HHT were useful for performance prediction as well as intrinsic signal feature discovery. These results suggest that BCI systems can be optimized by the use of psychological testing in user interface design to produce more effective and focused systems. Future studies can explore further optimization of such models and other incorporation of psychological variables for future BCI optimization.

REFERENCES

Alasfour, A., & Gilja, V. (2024). Consistent spectro-spatial features of human ECoG successfully decode naturalistic behavioral states. *Frontiers in Human Neuroscience*, *18*, 1388267. DOI: 10.3389/fnhum.2024.1388267 PMID: 38873653

Alkaff, Z. A., Malim, N. H. A. H., Sumari, P., & Abdullah, J. M. (2024). Applications of Brain Computer Interface in Present Healthcare Setting.

Aymen, A., Khediri, S. E., Thaljaoui, A., Miladi, M., & Kachouri, A. (2024). Catalyzing EEG signal analysis: Unveiling the potential of machine learning-enabled smart K nearest neighbor outlier detection. *International Journal of Information Technology : an Official Journal of Bharati Vidyapeeth's Institute of Computer Applications and Management*, •••, 1–12. DOI: 10.1007/s41870-024-02123-2

Aymen, A., Khediri, S. E., Thaljaoui, A., Miladi, M., & Kachouri, A. (2024). Catalyzing EEG signal analysis: Unveiling the potential of machine learning-enabled smart K nearest neighbor outlier detection. *International Journal of Information Technology : an Official Journal of Bharati Vidyapeeth's Institute of Computer Applications and Management*, •••, 1–12. DOI: 10.1007/s41870-024-02123-2

Catapano, P., Cipolla, S., Sampogna, G., Perris, F., Luciano, M., Catapano, F., & Fiorillo, A. (2023). Organizational and individual interventions for managing work-related stress in healthcare professionals: A systematic review. *Medicina*, *59*(10), 1866. DOI: 10.3390/medicina59101866 PMID: 37893584

Da Silveira, A. C., Lima de Souza, M., Ghinea, G., & Saibel Santos, C. A. (2024). Physiological Data for User Experience and Quality of Experience: A Systematic Review (2018–2022). *International Journal of Human-Computer Interaction*, •••, 1–30.

Darolia, S., & Ghosh, D. (2022). Importance of personality factors in determining the psychological consequences of infertility: A systematic review. *Health Education & Behavior*, *49*(4), 708–723. DOI: 10.1177/10901981211057109 PMID: 34814759

Desai, A. B., Gangodkar, D. R., Pant, K., & Pant, B. (2024, January). Harnessing the Potential of Light Gradient Boosting Machine for Accurate Diagnosis of Schizophrenia from EEG Signals. In 2024 14th International Conference on Cloud Computing, Data Science & Engineering (Confluence) (pp. 568-574). IEEE. DOI: 10.1109/Confluence60223.2024.10463450

Dhiman, R. (2023). Electroencephalogram channel selection based on pearson correlation coefficient for motor imagery-brain-computer interface. *Measurement. Sensors*, *25*, 100616. DOI: 10.1016/j.measen.2022.100616

Dillen, A., Omidi, M., Díaz, M. A., Ghaffari, F., Roelands, B., Vanderborght, B., Romain, O., & De Pauw, K. (2024). Evaluating the real-world usability of BCI control systems with augmented reality: A user study protocol. *Frontiers in Human Neuroscience, 18*, 1448584. DOI: 10.3389/fnhum.2024.1448584 PMID: 39161850

Forero, C. G. (2024). Cronbach's alpha. In *Encyclopedia of quality of life and well-being research* (pp. 1505–1507). Springer International Publishing.

Grigoriadou, E., Kolias, P., Theocharidou, M., & Gkioka, M. (2024). Psychometric properties of the Oxford Happiness Questionnaire Short Form (OHQ-SF) in a Greek student sample. *Discover Psychology, 4*(1), 1–12. DOI: 10.1007/s44202-024-00153-2

Khan, R. A., Rashid, N., Shahzaib, M., Malik, U. F., Arif, A., Iqbal, J., Saleem, M., Khan, U. S., & Tiwana, M. (2023). A novel framework for classification of two-class motor imagery EEG signals using logistic regression classification algorithm. *PLoS One, 18*(9), e0276133. DOI: 10.1371/journal.pone.0276133 PMID: 37682884

Liang, R., Zhang, X., Li, Q., Wei, L., Liu, H., Kumar, A., . . . Manbachi, A. (2024, April). Unidirectional brain-computer interface: Artificial neural network encoding natural images to fMRI response in the visual cortex. In ICASSP 2024-2024 IEEE International Conference on Acoustics, Speech and Signal Processing (ICASSP) (pp. 1851-1855). IEEE.

Pavot, W. (2024). Satisfaction with Life Scale (SWLS), an overview. Encyclopedia of quality of life and well-being research, 6124-6126.

Shelishiyah, R., Beeta, T. D., & Margaret, M. J. Machine Learning-Based Classification of Hybrid BCI Signals using Mayfly-Optimized Multiclass Weighted Random Forest.

Tang, W., Chen, S., Shao, J., Du, X., Wu, J., Zhang, X., & Xue, C. (2024). Emerging Themes and Future Directions in Neurodesign and Human-Computer Interaction: A Systematic Review. *International Journal of Human-Computer Interaction*, •••, 1–22. DOI: 10.1080/10447318.2024.2405784

Thanigaivelu, P. S., Sridhar, S. S., & Sulthana, S. F. (2023). OISVM: Optimal Incremental Support Vector Machine-based EEG Classification for Brain-computer Interface Model. *Cognitive Computation, 15*(3), 888–903. DOI: 10.1007/s12559-023-10120-z

Thanigaivelu, P. S., Sridhar, S. S., & Sulthana, S. F. (2023). OISVM: Optimal Incremental Support Vector Machine-based EEG Classification for Brain-computer Interface Model. *Cognitive Computation, 15*(3), 888–903. DOI: 10.1007/s12559-023-10120-z

Toma, F. M. (2023). A hybrid neuro-experimental decision support system to classify overconfidence and performance in a simulated bubble using a passive BCI. *Expert Systems with Applications*, *212*, 118722. DOI: 10.1016/j.eswa.2022.118722

Wu, X., Metcalfe, B., He, S., Tan, H., & Zhang, D. (2024). A Review of Motor Brain-Computer Interfaces using Intracranial Electroencephalography based on Surface Electrodes and Depth Electrodes. *IEEE Transactions on Neural Systems and Rehabilitation Engineering*, *32*, 2408–2431. DOI: 10.1109/TNSRE.2024.3421551 PMID: 38949928

Zhang, H., Jiao, L., Yang, S., Li, H., Jiang, X., Feng, J., Zou, S., Xu, Q., Gu, J., Wang, X., & Wei, B. (2024). Brain–computer interfaces: The innovative key to unlocking neurological conditions. *International Journal of Surgery*, *110*(9), 5745–5762. DOI: 10.1097/JS9.0000000000002022 PMID: 39166947

Zhang, S., An, D., Liu, J., Chen, J., Wei, Y., & Sun, F. (2024). Dynamic decomposition graph convolutional neural network for SSVEP-based brain–computer interface. *Neural Networks*, *172*, 106075. DOI: 10.1016/j.neunet.2023.12.029 PMID: 38278092

Zhu, H. Y., Hieu, N. Q., Hoang, D. T., Nguyen, D. N., & Lin, C. T. (2024). A human-centric metaverse enabled by brain-computer interface: A survey. *IEEE Communications Surveys and Tutorials*, *26*(3), 2120–2145. DOI: 10.1109/COMST.2024.3387124

Chapter 8
Fundamentals Approached Towards Artificial Neural Networks in Brain Computer Interface

Nitish Pathak
Bhagwan Parshuram Institute of Technology, New Delhi, India

Neelam Sharma
Department of CSE-AIML, Maharaja Agrasen Institute of Technology, New Delhi, India

Deepali Rani Sahoo
https://orcid.org/0000-0001-6949-7439
Manipal Law School, Manipal University Jaipur, India

Farhad Alam
JB Institute of Technology, India

Harishchander Anandaram
https://orcid.org/0000-0003-2993-5304
Amrita School of Artificial Intelligence, Coimbatore, Amrita Vishwa Vidyapeetham, India

Kapil Joshi
https://orcid.org/0000-0003-1097-8347
Uttaranchal University, India

ABSTRACT

A neural network is essentially an attempt to replicate how the brain works. Neural system theory is based on the idea that several of the most significant features of biological neurons can be extracted and applied to computations in computer order to produce a more simplified model of the brain. An ANN is configured for a specific use, such as pattern recognition or data classification, through a learning process. Learning involves altering the synaptic interactions between neurons in biological systems. Along with several ANN trainings, the fundamental MATLAB processes

DOI: 10.4018/979-8-3693-9445-8.ch008

are presented. The training's goal is to reduce mean square error. T

1. INTRODUCTION

Artificial neural systems that network very simple electronic models based on the neuronal topology of the brain. In essence, the brain learns by experience. It goes without saying that some problems that are beyond the scope of contemporary computers can be resolved by small, energy-efficient solutions (Hassoun, 1995). This brain modelling also promises a simpler method of developing machine solutions. This novel computing technique also provides a more gradual degradation in the face of system overload as compared with other more traditional rivals. ANNs are machine learning as well as pattern recognition computer models that are modelled after the central nervous systems of animals, specifically the brain. Typically, they are depicted as networks of interconnected "neurones" that can process information to calculate values from inputs.

An ANN is configured for a specific use, such as pattern recognition or data classification, through a learning process. Learning involves altering the synaptic interactions between neurons in biological network systems. A neural network may complete jobs that its traditional output cannot, which is one of the general benefits of ANNs (Montesinos López, Montesinos López, & Crossa, 2022).

It is believed that these biologically inspired computer techniques will represent the next significant development in the computing sector. Tasks that are now beyond the capacity of computers can be completed by simple network animal brains. Computers excel at repeated activities like ledger keeping and complex calculations. Computers, however, have difficulty recognizing even simple patterns, much less extrapolating historical trends into future actions. Intelligence includes the ability to think, imagine, produce, memories, comprehend, recognize patterns, make judgement, adapt to change, and learn from experience (Waszczyszyn, 1999).

Via interactions between neurones, biological neuronal cells form intricate neural network systems. When an organism is stimulated from the outside, signals are sent through neurones, triggering the organism's consciousness to provide linked feedback. This feature is simulated by an artificial neural network, which links several neurone nodes and transmits information using suitable activation functions, resulting in an anthropomorphic information process.

There are three ways of thinking in the human brain: impulse thinking, picture thinking, and thinking in terms of abstraction. Among these is image thinking, which combines learnt information and subjective cognition to generate inspiration or ideas. This is precisely the path artificial intelligence is taking, enabling machines to learn information and generate subjective mental ability that can take the place of human labour. A powerful tool for simulating human image thinking, artificial neural networks can assist or even replace humans in a variety of activities requiring perceptual identification or prediction.

Since the turn of the twenty-first century, computational neural networks have advanced rapidly, producing high-quality models and algorithms like deeper neural networks and neural networks made up of con that have been used in several real-world situations (Basheer & Hajmeer, 2000). Simultaneously, the artificial neural network evaluation indexes can be separated into two theoretical and practical directions. Theoretically, the degree of precision of recognition or predicting can be used to assess an algorithm's impact. Furthermore, artificial intelligence is still in its infancy, and the growth of neural net theory also depends on the creation of a comprehensive theoretical framework.

Neural networks use a different approach to problem solving than conventional computers. The algorithmic approach, which includes the computer following a set of instructions which provide network, is how conventional gadgets solve issues. Without knowing the exact steps, it must take; the computer is unable to fix the problem. This restricts traditional computers' capacity to resolve problems to those that humans can currently understand and are accustomed to. But computers would be much more valuable if they could carry out activities that humans aren't totally sure how to execute (Fine & Hassoun, 1996). The way that neural networks processing information is comparable to that of the human brain. The network is composed of several closely connected processing network units, or nerve cells, that work together to solve a specific problem. It is impossible to train them to carry out a certain duty. Carefully choosing the examples is necessary to avoid wasting valuable time or, worse, having the network operate improperly. When it comes to utilisation, researchers are also concentrating on ways to lower running overhead and speed up neural system learning because the more complicated the neural network model, the more input it processes. In order to analyse the primary research findings of artificial neural networks from two perspectives—theoretical and practical—this paper first examined the fundamental theory of neural networks (Thakur & Konde, 2021).

2. FUNDAMENTAL BEHIND ARTIFICIAL NEURAL NETWORKS

A multi-layer system model with neurone nodes connecting its layers is called an artificial neural network. The output quantity of the connected neurone in the base layer and the associated activation function determines the output value of any neurone in a non-input layer. The input layer's neurones receive the input parameter vector and send it to the neurones in the following layer. Neurones are intricately coupled and employed for handling information and feature extraction; hidden layers are invisible from the outside. The output layer may have one or more neurones that provide the estimated values or categorising results (Suryadevara & Yanamala, 2020).

The brain learns primarily Figure 1 through experience. It goes without saying that small, energy-efficient solutions can address some problems that are outside in the network scope of contemporary computers. This brain modelling also promises a simpler method of developing machine solutions. This novel computation technique also provides a more gradual degradation in the face of system saturation than its more traditional rivals. These computer methods that draw inspiration from biology are thought to be the next big thing in the computing industry. Animals' simple neurons are capable of things that computers are currently unable to achieve. Computers excel at repeated activities like ledger keeping and complex calculations. But computers find it difficult to recognize even simple patterns, much less extrapolate previous patterns into future actions (Yang, 2009). Recent developments in biology offer a preliminary grasp of the mechanism underlying natural thinking. The data input layer, multiple hidden processing layers, and the data output layer make up the fundamental architecture of an artificial neural network. Each of the two connected neurones has a weight value, and the weight W is the most crucial parameter in the structure of the neural network. Any neurone in the neural network, with the exception of the input layer, takes the product of the resultant information from the low layer connected neurone and its weight. Once all the products have been added up, the activation function can be used to determine the neuron's output value. Neural network training mostly involves continuously modifying the network's parameters for weights to their ideal values (Yegnanarayana, 2009).

We can identify individual faces from a wide range of perspectives thanks to some of these intricate patterns. A new area of computing is covered by this approach of storing data as patterns, applying those patterns, and subsequently resolving issues. As previously said, this discipline includes building massively parallel structures and teaching them to solve certain problems rather than using traditional programming. Additionally, terms like behave, react, self-organise, learn, generalise, and forget are used in this discipline that is considerably distinct from those used commonly in conventional computing.

In order to minimize the error between what is output alongside the expected output amount from the original data set, the BP algorithm's execution steps are as follows: first, use data from the schooling data set as the neural network's input; next, compute the neural network's output values using forward propagation; finally, define an average square cost function. Finally, use the gradient descending Algorithm to reverse the weight parameters layer by layer, completing the neural network's initial training process (Zilouchian, 2001). The term "Artificial Neural Network (ANN)" should be used more frequently when discussing neural networks because ANNs are computers with architectures inspired by the brain In order to create a complex communication network, they are typically composed of hundreds of simple processing computer and machines connected by wire. Each node, or unit, is a condensed model of an actual neuron that, upon receiving the input signal from the additional nodes it is attached to, either fires or sends out a new signal.

Figure 1. Basic Artificial neural network

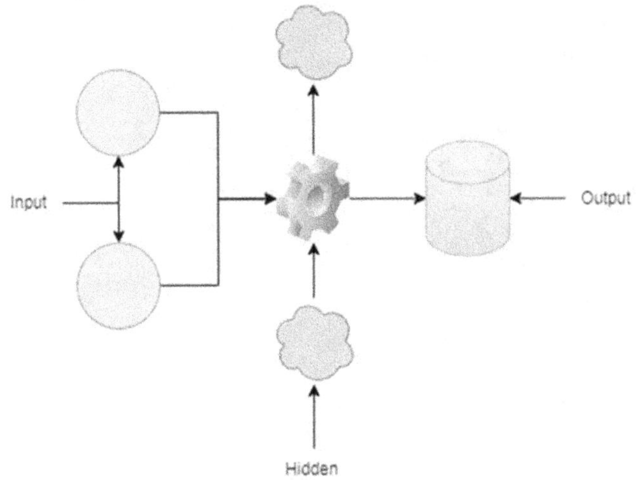

Although the word "neuron network" was originally intended to describe a network or circuit composed of human neurones, it is now frequently used to refer to an network of artificial neurones (ANN) (Zou, Han, & So, 2009). A mathematical or computer model based on an information processed paradigm of organic nervous networks, like the neural system in the brain, is called an artificial neural network (ANN). The components of an ANN, or artificial neural networks, are based on computer and machine of interconnected artificial neurones that are programmed to resemble biological neurones in certain ways. These neurones cooperate to find solutions to certain issues. Without building a model of an actual biological system, ANN is set up to solve artificial intelligence difficulties. ANNs are utilised in adaptive administration, image analysis, recognition of words, and other fields (Zou, Han, & So, 2009). Similar to cognitive development in biological tissues, which entails neuronal adjustment with synaptic connections, these applications are carried out through the process of learning. The same thing occurs in the ANN.

3. ANN'S WORK

A neurone is the basic processing unit of a neural network. This fundamental component of human awareness includes a few general skills. In essence, a biological neurone generates the final result after receiving inputs from various sources, combining them in some way, and performing a generally nonlinear process on the result. Figure 2 shows the relationship of these four parts.

Figure 2. A Simple Neuron Structure

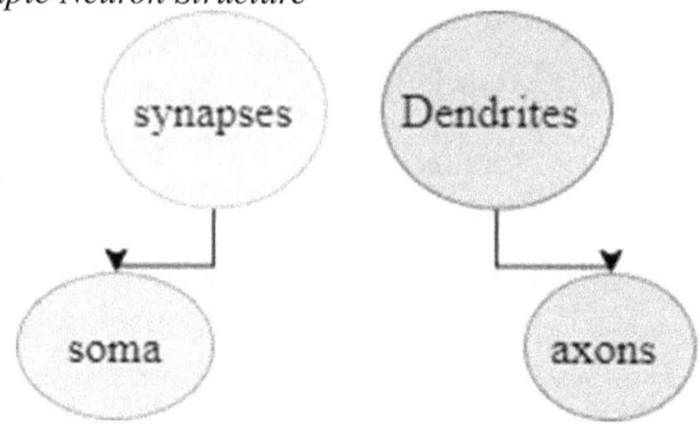

Man's attempts to electrically replicate the thought process are made more difficult by the fact that humans have numerous varieties of this fundamental sort of neurone. However, the four fundamental components of all natural neurones are the same. Dendrites, meditation, branches, and synaptic are the scientific names for these constituents. Dendrites are midline extensions that resemble hairs and function as input channels. The synaptic connections of other neurones provide input to these input channels. These incoming signals are then gradually processed by the soma (Krenker, Bešter, & Kos, 2011). Through the use of axons and synapses, the soma subsequently converts the resulting value towards an output that is transmitted to other neurones. The various ways that these individual neurones might be grouped together are the focus of the other aspects of the "art" of employing neural networks. In order to process information in a shifting, interactive, and self-organising manner, this clustering takes place in the human mind. In biology, neural networks are constructed from tiny components in a three-dimensional setting. These neurons seem to be able to join in practically infinite ways. No planned or existing man-made network is like this. Integration circuits are two-dimensional structures with just a couple of interconnecting layers in modern technology by online computer network. This physical reality limits the kind and scope of artificial neural networks that can be built in silicon (Kanwisher, Khosla, & Dobs, 2023). At the moment, neural networks are only basic artificial neurones grouped together. By forming layers that are then tied to one another, this clustering takes place. The other "art" of designing networks to solve problems in the actual world is how these levels relate to one another. A neural network shall be deemed convergent and put to use in practice if, after several iterations of training, the error between its output value as well as the predicted value varies within a narrow, stable range.

4. ARTIFICIALLY NEURAL NETWORK CONSTRUCTION

From its inception to its current rapid development, artificial neural networks have primarily undergone three stages of development. The perception model is the first neural network prototype, with neurones connected linearly but virtually no useful applications (Kufel et al., 2023). The back propagation technique was developed following the introduction of nonlinear connections, laying the groundwork for neural network training. Deep learning has emerged as the dominant neural network research concept since the turn of the twenty-first century, and artificial network applications have significantly improved.

5. ACQUIRE KNOWLEDGE

Not surprisingly, learning in neural network structures is basically a process of changing the strength of network connections, given that connection energies are knowledge repositories in neural net topologies. Back-Propagation is an extremely widely used learning technique in neural networks of the kind that have been discussed thus far. The network is first initialized, with all connection strengths set at random, with the network computer operating as a blank canvas. Next, some data is sent to the network. Let's take a moment where we are developing the "gender detector" mentioned above and the node inputs are receiving a digital image (Kaveh, 2024). Since we haven't yet adjusted our internet connection strengths to anybody other than random values, the activation moves across the net in a haphazard manner.

Finally, an activation level is recorded by the output node. However, the net's first responses will be random because it hasn't been trained yet. Back-propagation comes into play here. The reaction from the net gets compared with the right answer for that image. The procedure of adjusting each node involves more complex mathematics than is required for this course. It starts with the output node and works backwards, adjusting every hyperlink strength so that the next time the images are displayed, and the answer will be similar to the intended one (Shah et al., 2023). One "back-propagation cycle," or frequently just one "iteration," is the term used to describe the entire process of data entry, processing, comparing the output with the right response, and altering connection strengths. After that, the net is shown another image, and its response is compared to the right response, with the connection strengths modified as necessary. A few hundred of iterations may be required for this process. The net should eventually be able to distinguish between males and girls with some degree of proficiency. However, there is constantly a chance that somebody on the internet has successfully learnt the answer for every image rather than being able to distinguish between males and females (Wu et al., 2024). The images (or whatever other data is being used) ought to have been split into a pair of sets: the set to be trained and the new set in order to test for this. Back-propagation cycles employ the training set, and after learning is finished, the replacement set is used. We determine that learning has taken place if the networks as well on fresh transfer triggers as it did on the original training set.

6. ARTIFICIAL NEURAL NETWORK PROGRAMMING

A network is prepared for training once it has been configured for a specific use case. The initial components are picked at random to begin this process. After that, learning, or training, starts. Training can be done in two ways: supervised and unsupervised. By simply "grading" the network's response time or by supplying the intended outputs by providing the inputs, supervised instruction entails a method of giving the network the desired result. In unsupervised training, the network must interpret the inputs on its own without external assistance (Xu et al., 2023). In the great majority of networks, supervised training is used. Initial signal segmentation is done by unsupervised training.

A. Instruction under Supervision

In supervised training, both the input data and the results are provided. The network compared the outputs it generates to the desired outputs after processing the inputs. Errors are then sent back through the system, causing the weights that control the network to change computer system. This process is repeated while the scales are constantly altered. The data collection that enables training is referred to as the "training set." The identical set of data is analysed numerous times during network training in order to continuously improve the connection weights. Tools to track how effectively a synthetic neural network achieves convergent on the ability of predicting the correct answer is available in the current business network development packages.

With the use of these technologies, training might continue for days until the system achieves the accuracy or quantitatively desired point. Some networks, though, are incapable of learning. This can be the result of the input data lacking the precise information needed to produce the intended result. If there is insufficient data to allow for full learning, networks also fail to converge. Ideally, there will need enough data to allow for the holding back of some of it for testing. These input/output combos are either supplied by the net itself, also referred to as the self-supervised technique, or by an outside teaching component. Many stacked networks with numerous nodes are able to memorise data (Gautam et al., 2023). Supervised training must retain a set of knowledge to be utilised for testing the system following it has been trained in order to keep an eye on the network and identify whether the system is only memorizing its data in a somewhat insignificant way.

The designer must next examine the input and goods or services, how many output layers there are, how often variables occur in each layer, how the layers relate to addition, transfer, and learning activities, and even the initial weights themselves in the event that a network cannot manage the picture issue. Making the necessary modifications to create a successful neural network is the "art" of neural networks. Another facet of the designer's creativity governs the training rules. Many laws (algorithms) are used to achieve the adaptive input required to change the weights during in the training. The most used technique is backward-error propagation, or just back-propagation computer replication. Later in this paper, these different learning strategies are examined in more detail (Pantic et al., 2023). However, training is more than just a method. Conscious analysis and a "feel" are used to make sure the network isn't spent. A neural network made up of neurones first sets up itself using the data's overall statistical patterns. It then goes on to "learn" concerning other data points that, from a broad perspective, might be erroneous. If the system is well taught and no more learning is required, the apparatus can be "frozen."

B Adapted or unsupervised instruction

Unsupervised learning training is the term for the other kind of training. The network receives inputs but not the intended outputs during unsupervised training. After that, the system itself has to choose which features to employ in order to Sort the input network data into categories (Bukhtoyarov et al., 2023). This is sometimes referred to as self-organization or adaptability. Our understanding of independent studying is still lacking. The promise of this environmental adaptation is that it would allow science fiction robots to keep growing on an individual basis as they come across novel circumstances and surroundings.

There are a number of ways to measure how strong the connections are. In other words, the net can be developed by introducing learning habits into the equation and allowing the net to modify or alter the weights in accordance with a learning rule, or the weights can be directly defined (using a priori knowledge). There are many circumstances in life where precise training sets are unavailable. Some of these scenarios entail military operations where new weaponry and battle strategies may be used. There is still hope and research in this area because of this unforeseen part of life and the desire of people to be ready (Goel, Goel, & Kumar, 2023). However, the great majority of neural network activity currently takes place in supervised learning systems.

7. ANN'S ATTRIBUTES

ANNs improve their own rules; the choicer they make; more efficient they may become. In essence, computers are good at calculations; they take inputs, process them, and then provide the result based on calculations made at specific algorithms that are created in the software. The traits are essentially those that artificially intelligent devices, such as robots and other AI-based applications, ought to have (Arunadevi et al., 2023). Through the use of a diagram, the six fundamental and significant features of artificial neural networks are demonstrated.

a. Medical: A “hot” topic in medical research right now, artificial neural network is going to be widely used in medical equipment during the coming years. The majority of current research focuses on simulating human body components and identifying illnesses from a variety of readings (e.g. cardiograms, CAT exams, transducer scans, etc.). Since no particular technique is required to detect the disease, neural networks are perfect for identifying diseases from scans. The specifics of how to identify the illness are not necessary because neural networks develop by example (Singh et al., 2024). A collection of instances that are typical of every variety of the illness is required. The 'quantity' is more significant than the number of examples. If the entire structure is to function effectively and dependably, the examples must be chosen with extreme care.

Figure 3. Characteristic of ANN

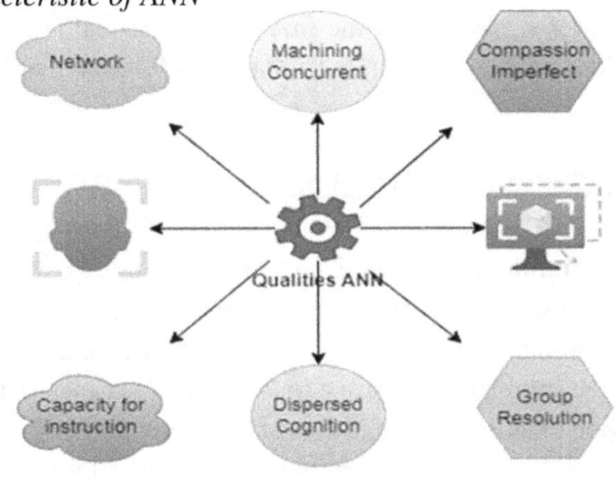

b. Gathering of data sets for input and result:

The output values derived from the trials are selected based on the various experimental combinations of the input parameters. The spatial arrangement of the input-output dataset, the manner in which the input-output dataset is presented to the neural network and the proper selection of the system's input-output parameters all affect the ANN model's capacity for extrapolation data (Wazirali et al., 2023).

c. Preparation of the data input and output:

The network structure known as feed forward replication, or "Neff," has a return propagation weight and bias education function called "learned" and a Liebenberg-Marquardt return propagation training function called "trail." Given enough neurones in the hidden layer, a two-layer forwarding network can approximate any equation with an unlimited number of discontinuities. In order to create (55% of the samples), evaluate (20% of the samples), and verify (25% of the samples) artificial neural networks using a "divider and" data division function, the samples collected during the testing stage were randomly split into three groups. By default, the Liebenberg-Marquardt back propagation algorithm halts training.

8. PATTERN OF AWARENESS

Perception is a form of supervised learning used in machine learning for binary classifiers, which can identify if an input displayed as a numeric vector is a member of a specific class of functions. It is a linear classifier, meaning that it makes predictions by combining the weighting set and the property vector in a classification process based on the linear expectation function. The perception method, which dates back to the 1950s, was among the first to use artificial neural networks with specially designed hardware (Kaveh, Eskandari, & Movasat, 2023).

9. TECHNOLOGY OF MACHINES

Dynamic programming, structured programming in Figure 4 symbolic programming, cognitive storage management systems, and many more techniques are spin-offs of artificial intelligence research. Although the fundamental objective of developing artificial intelligence is still a long way off, people are beginning to see the general direction that may eventually lead to it. Frank Rosenblatt, a psychologist, developed the perceptual system model back in 1957 after being influenced by the MP model. Character image recognition is the purpose of the perception systems. The Heaviside step factor serves as a trigger function for the artificial neurone that perceives in the background in neural networks. The detection method, also known as a single-layer seeing, is the most basic method of feeding forward neural network structure for a linear classifier. It makes use of the concept of hidden neurones, which is frequently mentioned in subsequent algorithms (Linka, Pierre, & Kuhl, 2023). The development of neural networks stagnated for many years as a result of the perception's initial promise being quickly thwarted by its inability to be trained to recognise a wide variety of patterns and its inability to resolve the XOR circuit problem.

Figure 4. Machines Technology in Workplace

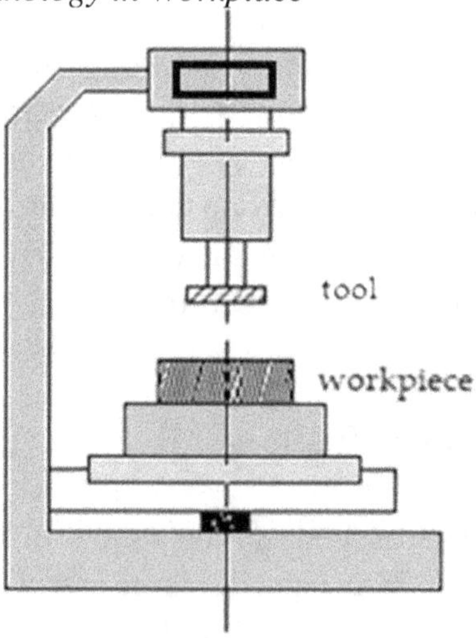

10. FINANCIAL APPLICATIONS OF NEURAL NETWORKS

Accounting and financial analysis are two general areas of specialism within the diverse world of business. Nearly every neural network application seems appropriate for a certain financial analysis or commercial domain. Neural networks have some promise for use in scheduling and resource allocation, among other business applications. Neural networks have a lot of promise for database mining, which is the process of looking for patterns in the intentionally stored data in databases. The majority of sponsored research in this field is categorised as exclusive. As a result, reporting on the entire scope of the ongoing effort is impossible. Neural networks, such the Hopfield-Tank network, are used in the majority of work for scheduling and optimisation (Olabi et al., 2023).

(a) Promotion: A neural network has been connected with a marketing application. Expert systems are among the cognitive technologies that make up the computer system known as the Airport Marketing Tactician (abbreviated AMT). To help with advertising oversight of airline ticket assignments, a fed-forward neural network that was created using back-propagation is coupled with the AMT. It was possible to articulate rules using the adaptive neural technique. Furthermore, the environment of the application changed quickly and continually, necessitating a continuous adjustable solution.

11. UNDERSTANDING CONSEQUENCE

Using this approach, the learning machine acts on what is happening and gets feedback or a response in return. The learning component modifies its parameters in accordance with the environmental response, rating its behaviour as either excellent or negative. In general, the process of adjusting parameters is carried out repeatedly until an equilibrium state is reached, at which point no additional changes are required (Nimmy et al., 2023). A system of basic processing elements (neurones) that can display sophisticated, global behaviour that is dictated by the relationships between each of the processing items and the element the variables, respectively, is reflected or represented by an artificial neural network ANN Many benefits are provided by neural networks, such as the availability of multiple ways to train, the capacity to detect all potential relations among predictor variables, and the implicit detection of intricate, nonlinear relationships between dependent and independent variables. When it comes to forecasting, data mining, assignment planning, or optimal resource allocation, ANN-based solutions have produced remarkable outcomes and insights into extremely complicated situations.

12. METHODOLOGY FOR PROPAGATION BACK

In artificial neural networks, back propagation is a technique used to determine the error contribution of each neurone following a batch of data processing. It is employed to finish the learning process and modify the weight factors of every neurone. In technical terms, this approach is also known as error back propagation since it computes the gradient of the loss equation and applies the gradient descent optimisation algorithm. Every value that is input and its predicted output value are necessary for the neural network's training phase, and the entire algorithm repeats the forward transmission and weight update loop twice. Presenting the input vector through a neural network causes it to move layer by layer across the network until it exits the output layer (Joshi, Bhosale, & Vyawahare, 2023). The margin of error for each neurone in the final result layer is then determined by comparing the neural network's output value with the predicted output value derived from the cost function. Up until each neurone has an attached error value that represents its contribution into the initial output, the error rate is then transferred forward from the final product layer across the neural network.

13. DEEP LEARNING

Rather than being a task-specific algorithm, deep learning in Figure 5 (also called deep organised learning or structured learning) is a subset of a broader category of machine learning methods that concentrate on data representation learning. Deep learning comes in three flavours: supervised, semi-supervised, and unsupervised. Deep learning architectures, such as deep neural networks, deep religious networks, and periodic neural networks, have been widely used in image identification to computer, target detection, audio approval, machine translation, and other fields (Joshi, Dalvi, Vitankar, Joshi, & Joshi, 2024).

Figure 5. Deep Learning Interconnected with AI & ML

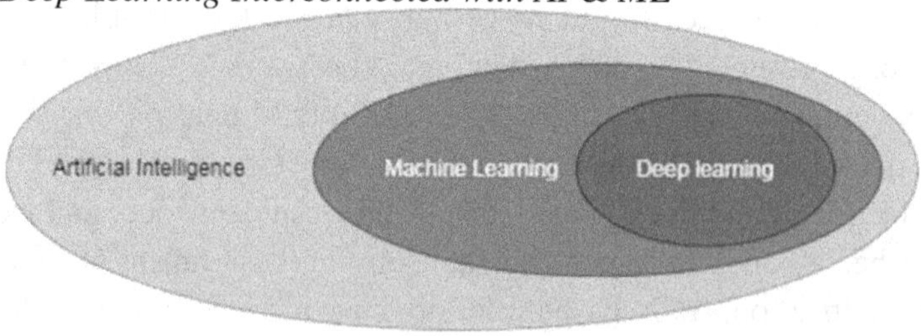

A deep neural network, for instance, is a network of artificial neural networks with many hidden layers beneath the layers that provide input and output, allowing it to simulate intricate nonlinear interactions. Today, deep learning has emerged as the primary focus of neural network computing research, and major tech companies like Google, Microsoft, Face book, and others annually devote significant funds and personnel to deep learning research and development. Deep learning has emerged as the most efficient method for processing and analysing data, because to the growth of big data (Rawat et al., 2023).

CONCLUSION

We covered the operation of artificial neural networks (ANNs) in this paper. ANN has a number of benefits over traditional methods. You may generally anticipate that a network will train rather effectively, considering on the application's characteristics and the robustness of its own internal information patterns. Artificial neural networks have been widely employed in the fields of computing for images, speech recognition, automatic computation, semantic data analysis, and forecasting, among others. Artificial intelligence has emerged as the primary area of study in computer science today. Processing large amounts of data is made possible by training neural networks with parallel computer science, which is based on the growth of big data.

Recent advancements in machine learning have emerged as the field of study for artificial intelligence in the future. Additionally, neural networks support study in other fields including psychology and neurology. They are widely used to represent different parts of living creatures and to investigate the inner workings of computer brain. Perhaps the most exciting aspect of these networks is the potential for future "conscious" networks. Many scientists contend that awareness is a "mechanical" quality and that "conscious" brain structures are a real possibility. Traditional approaches, which are usually limited by strict assumptions of linearity, normality, variable becoming independent, etc., can be analytically replaced computer by ANNs. An artificial neural network (ANN) enables the user to promptly and quickly represent phenomena that would otherwise be extremely hard or impossible to describe since it can capture a wide variety of correlations. These days, there are conversations about neural networks everywhere.

REFERENCES

Abraham, A. (2005). Artificial neural networks. *Handbook of measuring system design.*

Arunadevi, M., Rani, M., Sibinraj, R., Chandru, M. K., & Prasad, C. D. (2023). Comparison of k-nearest neighbor & artificial neural network prediction in the mechanical properties of aluminum alloys. *Materials Today: Proceedings*. Advance online publication. DOI: 10.1016/j.matpr.2023.09.111

Basheer, I. A., & Hajmeer, M. (2000). Artificial neural networks: Fundamentals, computing, design, and application. *Journal of Microbiological Methods*, *43*(1), 3–31. DOI: 10.1016/S0167-7012(00)00201-3 PMID: 11084225

Bukhtoyarov, V. V., Tynchenko, V. S., Nelyub, V. A., Masich, I. S., Borodulin, A. S., & Gantimurov, A. P. (2023). A study on a probabilistic method for designing artificial neural networks for the formation of intelligent technology assemblies with high variability. *Electronics (Basel)*, *12*(1), 215. DOI: 10.3390/electronics12010215

Fine, T. L., & Hassoun, M. H. (1996). Fundamentals of artificial neural networks. *IEEE Transactions on Information Theory*, *42*(4), 1322–1324. DOI: 10.1109/TIT.1996.508868

Gautam, V. K., Pande, C. B., Moharir, K. N., Varade, A. M., Rane, N. L., Egbueri, J. C., & Alshehri, F. (2023). Prediction of sodium hazard of irrigation purpose using artificial neural network modelling. *Sustainability (Basel)*, *15*(9), 7593. DOI: 10.3390/su15097593

Goel, A., Goel, A. K., & Kumar, A. (2023). The role of artificial neural network and machine learning in utilizing spatial information. *Spatial Information Research*, *31*(3), 275–285. DOI: 10.1007/s41324-022-00494-x

Hassoun, M. H. (1995). *Fundamentals of Artificial Neural Networks*. The MIT Press.

Joshi, M., Bhosale, S., & Vyawahare, V. A. (2023). A survey of fractional calculus applications in artificial neural networks. *Artificial Intelligence Review*, *56*(11), 13897–13950. DOI: 10.1007/s10462-023-10474-8

Joshi, S. S., Dalvi, V. H., Vitankar, V. S., Joshi, J. B., & Joshi, A. J. (2024). Development of new correlation for the prediction of power number for closed clearance impellers using machine learning methods trained on literature data. *Canadian Journal of Chemical Engineering*, *102*(11), 3832–3851. DOI: 10.1002/cjce.25385

Kanwisher, N., Khosla, M., & Dobs, K. (2023). Using artificial neural networks to ask 'why'questions of minds and brains. *Trends in Neurosciences*, *46*(3), 240–254. DOI: 10.1016/j.tins.2022.12.008 PMID: 36658072

Kaveh, A. (2024). Applications of artificial neural networks and machine learning in civil engineering. *Studies in computational intelligence, 1168*, 472.

Kaveh, A., Eskandari, A., & Movasat, M. (2023, October). Buckling resistance prediction of high-strength steel columns using metaheuristic-trained artificial neural networks. [). Elsevier.]. *Structures (Oxford)*, *56*, 104853. DOI: 10.1016/j.istruc.2023.07.043

Krenker, A., Bešter, J., & Kos, A. (2011). Introduction to the artificial neural networks. *Artificial Neural Networks: Methodological Advances and Biomedical Applications. InTech*, 1-18.

Kufel, J., Bargieł-Łączek, K., Kocot, S., Koźlik, M., Bartnikowska, W., Janik, M., Czogalik, Ł., Dudek, P., Magiera, M., Lis, A., Paszkiewicz, I., Nawrat, Z., Cebula, M., & Gruszczyńska, K. (2023). What is machine learning, artificial neural networks and deep learning?—Examples of practical applications in medicine. *Diagnostics (Basel)*, *13*(15), 2582. DOI: 10.3390/diagnostics13152582 PMID: 37568945

Linka, K., Pierre, S. R. S., & Kuhl, E. (2023). Automated model discovery for human brain using constitutive artificial neural networks. *Acta Biomaterialia*, *160*, 134–151. DOI: 10.1016/j.actbio.2023.01.055 PMID: 36736643

Montesinos López, O. A., Montesinos López, A., & Crossa, J. (2022). Fundamentals of artificial neural networks and deep learning. In *Multivariate statistical machine learning methods for genomic prediction* (pp. 379–425). Springer International Publishing. DOI: 10.1007/978-3-030-89010-0_10

Nimmy, P., Nagaraja, K. V., Srilatha, P., Karthik, K., Sowmya, G., Kumar, R. V., Khan, U., Hussain, S. M., Hendy, A. S., & Ali, M. R. (2023). Implication of radiation on the thermal behavior of a partially wetted dovetail fin using an artificial neural network. *Case Studies in Thermal Engineering*, *51*, 103552. DOI: 10.1016/j.csite.2023.103552

Olabi, A. G., Abdelkareem, M. A., Semeraro, C., Al Radi, M., Rezk, H., Muhaisen, O., & Sayed, E. T. (2023). Artificial neural networks applications in partially shaded PV systems. *Thermal Science and Engineering Progress*, *37*, 101612. DOI: 10.1016/j.tsep.2022.101612

Pantic, I., Paunovic, J., Cumic, J., Valjarevic, S., Petroianu, G. A., & Corridon, P. R. (2023). Artificial neural networks in contemporary toxicology research. *Chemico-Biological Interactions*, *369*, 110269. DOI: 10.1016/j.cbi.2022.110269 PMID: 36402212

Rawat, S. K., Yaseen, M., Pant, M., Ujarari, C. S., Joshi, D. K., Chaube, S., Negi, A. S., & Kumar, M. (2023). Designing soft computing algorithms to study heat transfer simulation of ternary hybrid nanofluid flow between parallel plates in a parabolic trough solar collector: Case of artificial neural network and particle swarm optimization. *International Communications in Heat and Mass Transfer*, *148*, 107011. DOI: 10.1016/j.icheatmasstransfer.2023.107011

Shah, A., Shah, M., Pandya, A., Sushra, R., Sushra, R., Mehta, M., ... & Patel, K. (2023). A comprehensive study on skin cancer detection using artificial neural network (ANN) and convolutional neural network (CNN). *Clinical eHealth*.

Singh, P., Adebanjo, A., Shafiq, N., Razak, S. N. A., Kumar, V., Farhan, S. A., Adebanjo, I., Singh, A., Dixit, S., Singh, S., & Sergeevna, M. T. (2024). Development of performance-based models for green concrete using multiple linear regression and artificial neural network. [IJIDeM]. *International Journal on Interactive Design and Manufacturing*, *18*(5), 2945–2956. DOI: 10.1007/s12008-023-01386-6

Suryadevara, S., & Yanamala, A. K. Y. (2020). Fundamentals of Artificial Neural Networks: Applications in Neuroscientific Research. *Revista de Inteligencia Artificial en Medicina*, *11*(1), 38–54.

Thakur, A., & Konde, A. (2021). Fundamentals of neural networks. *International Journal for Research in Applied Science and Engineering Technology*, *9*(VIII), 407–426. DOI: 10.22214/ijraset.2021.37362

Waszczyszyn, Z. (1999). Fundamentals of artificial neural networks. In *Neural networks in the analysis and design of structures* (pp. 1–51). Springer Vienna. DOI: 10.1007/978-3-7091-2484-0_1

Wazirali, R., Yaghoubi, E., Abujazar, M. S. S., Ahmad, R., & Vakili, A. H. (2023). State-of-the-art review on energy and load forecasting in microgrids using artificial neural networks, machine learning, and deep learning techniques. *Electric Power Systems Research*, *225*, 109792. DOI: 10.1016/j.epsr.2023.109792

Wu, B., Xu, J., Zhang, Y., Liu, B., Gong, Y., & Huang, J. (2024). Integration of computer networks and artificial neural networks for an AI-based network operator. *arXiv preprint arXiv:2407.01541*.

Xu, Q., Li, Y., Shen, J., Liu, J. K., Tang, H., & Pan, G. (2023). Constructing deep spiking neural networks from artificial neural networks with knowledge distillation. In *Proceedings of the IEEE/CVF Conference on Computer Vision and Pattern Recognition* (pp. 7886-7895). DOI: 10.1109/CVPR52729.2023.00762

Yang, X. (2009). Artificial neural networks. In *Handbook of research on geoinformatics* (pp. 122–128). IGI Global. DOI: 10.4018/978-1-59140-995-3.ch016

Yegnanarayana, B. (2009). *Artificial neural networks*. PHI Learning Pvt. Ltd.

Zilouchian, A. (2001). Fundamentals of neural networks. In *Intelligent control systems using soft computing methodologies* (pp. 17–38). CRC Press. DOI: 10.1201/9781420058147.ch2

[] Zou, J., Han, Y., & So, S. S. (2009). Overview of artificial neural networks. *Artificial neural networks: methods and applications*, 14-22.

Chapter 9

Introduction of a Cutting-Edge Technology Neural Networks Organization

Md. Tauqir Azam Kausar
https://orcid.org/0009-0008-3484-8022
Bir Tikendrajit University, Manipur, India

Harishchander Anandaram
https://orcid.org/0000-0003-2993-5304
Amrita Vishwa Vidyapeetham, India

Kawerinder Singh Sidhu
https://orcid.org/0000-0002-8910-6279
Uttaranchal Institute of Management, India

Nitish Nitish Pathak Nitish Pathak
Bhagwan Parshuram Institute of Technology, New Delhi, India

Abhijeet Sudhakar
BlueRidge Technology Solutions, India

Neelam Sharma
Maharaja Agrasen Institute of Technology, New Delhi, India

Kapil Joshi
https://orcid.org/0000-0003-1097-8347
Uttaranchal University, India

ABSTRACT

Neural network organization on any website, we typically see the advertisements for the products we have previously searched for. Get you ever wondered how it's possible that, after searching for something on Flipkart, however, we miraculously get an advertisement for the same thing on Instagram? This is accomplished by employing the neural network concept. The neural network serves as a sophisticated data structure that links all of our online activities, processes the data, and uses the results for commercial objectives. The input set and the output are connected by a finite number of cells, nodes, units, or neuronal cells that make up the network.

DOI: 10.4018/979-8-3693-9445-8.ch009

It is a component of an automated system that simulates the way the human brain processes and evaluates information. Neural networks that use artificial intelligence are used in a wide range of applications, including emergency room testing, character recognition, picture compression, is the process prediction of stock markets.

1. INTRODUCTION

A complex to, interlinked network of countless neurons is referred to as an ANO. The interconnection of neurons within computational neural networks allows for the processing and mining of more valuable data. Cognitive networks are essentially novel computing techniques and systems for learning through experience, learning display, and, ultimately, the application of learnt information to optimise complicated computer responses (Cammarano et al., 2024). In the area of numerical computation and associated symbol manipulation, modern digital computers perform better than humans. The most advanced computer amidst for ANO the world cannot match the ease with which humans can solve complicated perceptual challenges, such as identifying a man in a crowd based only on an impression of his face what causes their achievement to differ so drastically? The design developed by von Neumann is entirely distinct from the structures of the biological nervous system. The kinds of tasks that each computing model is most capable of performing are greatly impacted by this distinction. On a much smaller scale, they concentrate on the neuronal make-up of the cerebral cortex of mammals. Artificial neural pathways are regarded by many experts in neuroscience as the finest and possibly the only option for creating intelligent machines (Salman & Al-Janabi, 2024).

Multipurpose intelligent programs have not been produced despite numerous attempts to create "knowledgeable" programs using von Neumann's centralized architecture. Artificial Neural Networks organization (ANNOs), which are enormously parallel computing platforms made up of a vast number of basic processors with numerous connections, were inspired by actual neural networks. The scientist Robert Hecht-Nielsen, the creator of one of the primary neurone computers, provides the simplest definition of a neural organization, which is why it is so aptly referred to as a "Artificial neural organization" (ANO) (Mikołajczyk et al., 2017). "A registering framework composed of a few simple, extremely interdependent handling sections, that gather data by their changing state reactions to outer sources of info," is how he defines a neural organization. Master of computer science Maureen Caudill's "Neural Organization The basics: Part I" The concept of neural organization is derived from biology, which states that the human cerebral cortex is made up of many tiny numbers that are in charge of mental processes. The neurones in the human brain use previous understanding to develop thoughts; the brain itself does not generate ideas of its own.

Key components of the mind while sensory system, neurons are responsible for receiving input from the outside world through tactile organs, cycling it, and providing the yield through axons. We then take action through an actuator. Similarly, in an artificial neural network, neurons are numerous hash functions that receive inputs from various variables and, after processing, generate what is known as a hash result. ANO It is possible to think of a prepared neural organization as a "specialist" in how to group of the data it has been tasked with analysing (Kaur & Gill, 2019). A neurological organization functions similarly to the brain of humans, utilizing its amazing ability to process confusing or incomprehensible information. It can also extract designs and identify patterns on the internet that are too complex for individuals or computer advancements to notice.

Various chart's artificial neural networks along with their branches and portions of the computing methods are shown in the figure 1 (Szecsi, 1999).

Figure 1. The Branches and sections of the computational methods

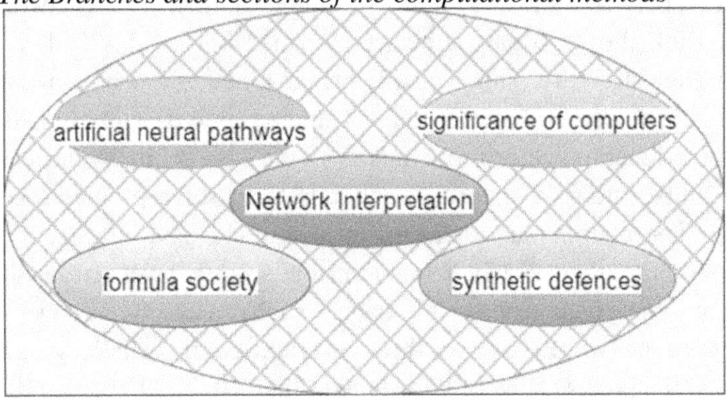

The structure created by computer neural networks is similar to that of the human brain, containing nodes of neurons linked in an Inter-like pattern. The human brain is composed of countless of cells called neurons. According to Van Avenger and Bootee, each neuron is composed of an internal cell structure which interprets information by sending it back and forth between the nervous system (inputs and outputs). The fundamental concept of these networks is (partially) based on the way the human neural system processes sensory data and information in order to gain and generate knowledge. The creation of new information processing system structures is the main component of this concept. Figure 2 depicts the architecture of an artificial neural network (Ajani et al., 2024).

Figure 2. Artificial neural network architecture

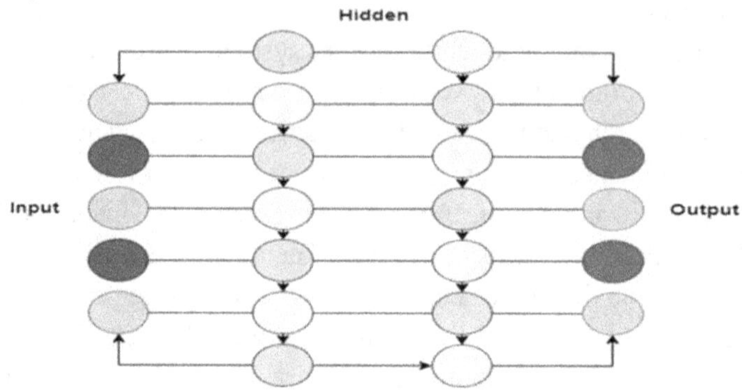

2. ARTIFICIAL NEURAL NETWORK: WHAT IS IT?

The Neural networks show in Figure 3 conjures up images of a network of nodes that are interlinked every time we hear the term (Livin et al., 2024). The human brain is one such structure. This typically brings together a wide range of intricate organizations. The cerebral architecture of numerous intricately linked organizations. Utilizing junctions (electronic linkages), the systems many intricately linked processing units, known as neurons, collaborate to solve problems and transfer information. The neurons are arranged in layers and have intimate connections with one another (Dubey, 2016). While the final product layer creates the ultimate outcome, the input phase receives the data. Usually, a couple of covert layers are placed between the two. Because of this configuration, it is challenging to forecast or understand the precise flow of knowledge (Shah et al., 2022).

The foundation of advanced machine learning is a neurological organization. This computational model is reliant on the structure and functionality of neural networks. In terms of software design, it is similar to a fake human perceptual system in terms of recognizing, getting ready and passing on the data. Three layers make up brain networks (Alqura'n et al., 2024).

Figure 3. Basic neural network

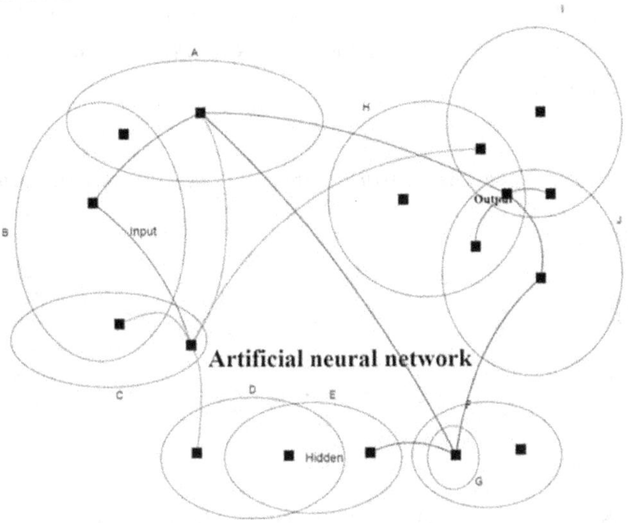

2.1. Slice those inputs

The information layer communicates with the external environment and receives input from the system. In a sense, it oversees all information sources. In order to prepare, those sources of data have been moved to the encrypted layer.

2.2. Cover conceals another

An activation work is applied to the collection of neurones in the secret layer. It is responsible for the preparation of the source data and sits in between the information and yielding layers.

2.3. Pillar of discharge

The neurological organization in the outcome layer collects and appropriately communicates the data that it was designed to provide. To display yield to the external climate, the yield layer (Santochi & Dini, 1996).

Every neurone has an activation function and a threshold value, and each relationship has a connection weight. Depending on the sign that appears of the feed's weight, it is determined whether each input possesses an up or down weight. The signal strength at a connection is influenced by the weight (Shukla et al., 2020). Neurons with a threshold that must be exceeded by the whole signal in order for a signal to be conveyed. The output is produced using the signal from the activation value, which is a weighted average that makes up the summing unit. The relationship amongst each element's composition and the ANO system's input as well as outputs shown in the figure 4.

Figure 4. Component's weight as well as the ANO method's input and output.

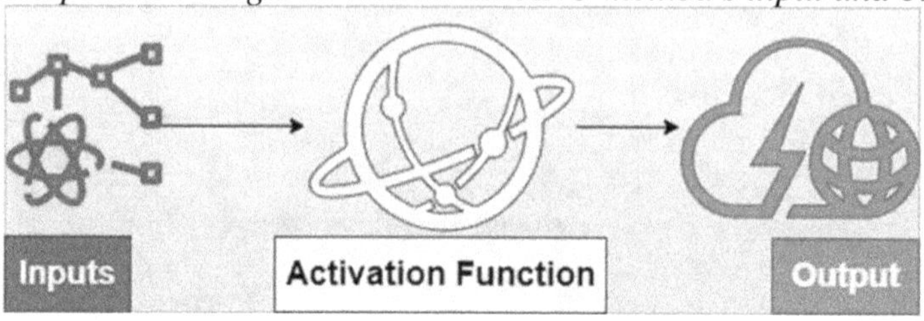

Promote ANO such systems, neighbouring cells can compensate for the loss of the cell's structure and aid in its recovery (Morabito et al., 2024). The networks in question have the ability to learn. In essence, the most crucial characteristic of systems intelligence is its capacity for learning. The system for learning can better handle novel challenges and equations since it is more adaptable and simpler to program. Similar to beings, neural networks made of computers learn by utilizing many instances. During the process of training, the network of neurones is configured to carry out specific duties such recognizing patterns and classifying data. Injecting touch nerve cells, for instance, teaches those cells to avoid the heated body; using this method, the body learns to fix its mistake (Abdalla et al., 2023).

3. ANO'S OPERATION

ANO operates on the fundamental idea of how the human brain functions. The art of applying a neuronal organization is really simple. In the human mind, neurons may have built in a multitude of ways to process information in a way that is both Neural and self-coordinated. This contact occurs gradually. Currently, the neural organization is a simple grouping of rudimentary fake neurons. We can see how Counterfeit Neural Organizations weighted coordinated chart's function structure, with the centres being melded by artificial neurons and the concerted vertices with loads addressing the relationship between the neuron yields and neurone sources of data (Raju & Thasleema, 2024). The image serves as vector data and the computerized neural organization receives information signals from the external environment.

A growing number of systems with unknown or highly complicated internal structures are being controlled or modelled using artificial neural networks. A neural network, for instance, can be used to regulate an engine's input; in this scenario, the neural network will learn how to control it (Mikołajczyk et al., 2018). By doing this, we connect with people who have a wealth of experience and information. Then, when the network encountered a new circumstance, it dealt with it using its prior knowledge, developed another connection, and stored it for later. This is how ANOs operate simply. Through the use of parables, learning in those structures is adaptive, meaning that when new inputs are provided, the amount of importance of the synapses shifts to ensure the system generates the right answer (Adeli & Hung, 1994). When an artificial brain is being developed via one of the learning methods), it is given an inventory set inputs and the results that correspond to them. As seen in the picture below, an ANO uses the same topology for all of its inputs, outcomes, and hidden layers. In this case, the typologies of all the levels are the identical, making it easy to move signals between them (Begg & Palaniswami, 2006).

Provide forward and backward feedback is the two typologies of artificial neural networks. Since a unit transmits knowledge to another unit of which it gets none, forward fed systems do not have feedback loops. There are fixed inputs and outputs. The inputs have been divided by the quantity of weight of each link once each unit obtains input data from its units along the left. Thus, it is possible to retrieve the output results corresponding to the significance of each link. Among the methods' applications are pattern creation, recognition, and categorization. The application of the system is used in network-related applications when the desired network conclusion is known.

This is the basis for many industries, including computer vision. The design of systems with input-output neural network typologies (Ma et al., 2019) is shown in the figure 5.

Figure 5. Design of platforms with input-output neural network configurations

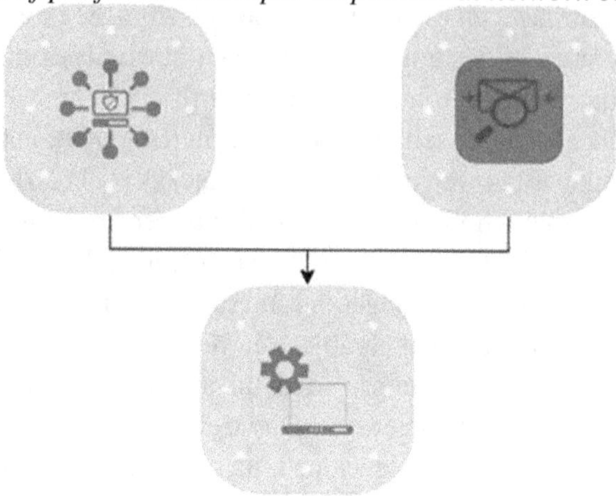

Content responsive memories are employed in Feed Back ANO systems. By comparing the network's output with the intended and anticipated output, neural networks can be learnt through a feedback loop. The key idea is that there are three levels of "hubs." In a neural organization of the cerebrum, the "hubs" are intended to be dissociated from cells; nonetheless, their similarity is merely allegorical (genuine neurones do not function in this manor, although the link is not extraordinary) (Liang et al., 2024). The weights of the hyperlinks between networks of units can be altered and modified using the difference in these two outputs. The Feed Back ANO algorithms' construction is shown in the figure 6.

Figure 6. The Feed Reverse ANO systems' design

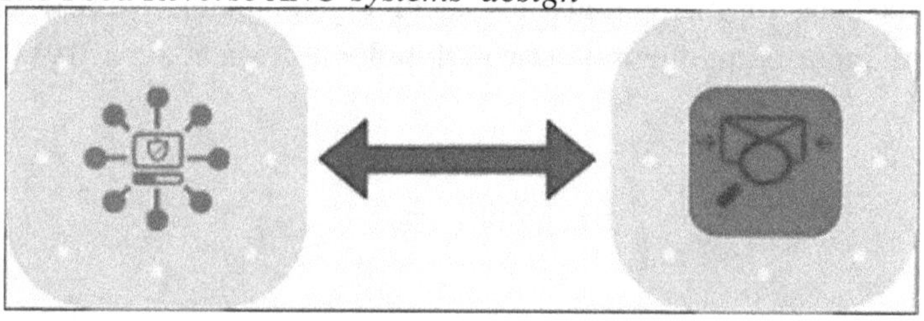

4. EDUCATING THE NETWORK

The network behaves like a newborn infant once it has been set up for a certain application. It has no idea how to respond to anything. It must be trained using random information or data. Let's build something for counting banknotes, for instance. In order to train the network, we initially gave it a random quantity of notes to count. It's just the same as when we teach our infant how to accomplish things (Purificato et al., 2023). The network can do the same work more effectively after learning or programming.

5. FORTUNATELY, TWO KINDS OF INSTRUCTION:

5.1. Instruction under oversight

If the optimal yield for ANO the organization is also provided, the learning calculation during organization preparation would come under this class. It is feasible to compute an error based on the neural organization's goal output and real output by providing it with both information and an output pair. It can then utilize that error to fix the network by adjusting its weights (Venugopal & Baets, 1994).

5.2. Instruction without supervision

Without any external guidance, a neural organization is simply provided with a number of information sources; it is its responsibility to find an example within these sources. This type of learning viewpoint is used in information mining and has been used in many suggestion calculations due to its ability to predict a client's inclination based on the preferences of other like clients. For real-time performance uses artificial neural network computations can be used in parallel with specialized hardware. Uninterrupted error tolerance during information encoding Even though some network capabilities may be severely damaged, the overall efficiency of a network deteriorates in tandem with an entire failure. In the future, neural networks will be essential (Zdravkova et al., 2022). Neural networks are necessary for signalling and creating correlations throughout the billions of data that we produce every day in the modern world. Neural networks will undoubtedly be necessary in every sector in the future. While we are talking about a few key areas where the neural networking concept is commonly applied, The time needed to find the ideal structure is shortened by neural networks' rapidity in data being processed and analysed (Çelik & Demirezen, 2024).

Neural network researchers are working to reduce the following drawbacks of neural networks notwithstanding their advantages over traditional systems: 1. Network architecture for an option activity is not governed by any particular guidelines or instructions. 2. These networks are not able to comprehend the physics of modelling problems on their own. Stated differently, it is typically hard to link properties or network components to process parameters. 3. The dimension of the set of instructional materials has a significant impact on the outcomes' accuracy. 4. Network training could be challenging, if not impossible. 5. It is difficult to forecast how well the network will perform in years to come (Zhang et al., 2022).

6. LEARNING

One essential component of intelligence is the capacity for learning. In the context of ANO, the method of learning can be understood as the challenge of upgrading network design and link weights in order for a network to effectively do a certain goal, even though it is challenging to define learning precisely. To express more complex variables and "read" ever-more-complex models for data classification and prediction based on hundreds or millions of features, ANO systems need be more sophisticated. Learning consecutive "layers" of progressively more significant abstractions while learning abstractions from data is the main goal of the machine learning discipline known as "deep learning (Jianjun & Jian, 2009)." It focuses on algorithms called artificially generated neural networks (ANOs), whose are modelled after the composition and operations of the nervous system (Jianjun & Jian, 2009).

The association weights must typically be learnt by the network using the provided training patterns. The network's weights are iteratively updated to increase performance over time. In addition to the input and layer of output, these neural networks contain over three levels of neurones. Models called "neural networked devices," which are arranged into actual sections, are used to learn these layered representations. This can be achieved by merely raising the number of concealed layers and/or increasing number of neurones per hidden layer. ANOs are intriguing and appealing because of the ability they have to automatically acquire knowledge from examples. ANOs seem to learn essential rules (such as inputs and outputs relationships) from the provided collection of example instances rather than adhering to a set of rules established by human professionals (Faria et al., 2024).

One of the main benefits of the neural network over conventional expert systems is this. Having a model of the circumstances that exists while a neural network of neurones functions—that is, knowing what data the network has access to—is necessary before you can comprehend or create the process of learning. This framework is known as an instructional system because it requires you to comprehend the learning rules that control the updating process of network weights. A learning algorithm is a process where the amounts of weight are adjusted using acquiring rules (Pontes et al., 2010).

The three primary learning paradigms are mixed, which unsupervised, and supervised. The network is given the right response (output) for each input sequence during controlled instruction, often known as learning with a "teacher." Although more layers and neurons can depict more intricate models, their computation takes more time and power (Hao et al., 2006). Weights are chosen so that the network can generate responses that are as near to the known accurate answers as feasible.

A type of supervised learning called reinforcement learning gives the network feedback on whether or not its outputs are right, but not the actual answers. Supervised learning or education without a teacher, on the other hand, does not necessitate a correct response for every input pattern found within the training dataset. It investigates the data's underlying structure, or the relationships between its patterns, and then classifies the patterns based on these relationships (Kozma et al., 2023). Both unsupervised and supervised learning are combined in hybrid learning. Typically, supervised learning determines a portion of the weights, whereas free learning yields the remaining weights.

7. IMPLICATIONS FOR ARTIFICIAL NEURAL NETWORKS

Neural network algorithms are utilized in many different applications nowadays, including pattern recognition problems, which include problems like image processing, speech recognition, line recognition, and the like, and categorisation problems, like text or picture classification. Neural networks are also used in danger evaluation systems, drone command, computer quality analysis, soldering quality analysis, spectrum detection, risk assessment for loans, oil and gas exploration, ER testing, truck brake detection systems, and drug detection. Hepatitis detection, speech recognition, remote search for data, submerged mine detection, industrial control procedures, error management, 3D object and handwritten and detection of faces, etc (Zaki et al., 2023). ANOs are used in signal processing, pattern recognition, approximation of unknown functions, and calculation of known functions. In order to improve the accuracy of predicting a certain durability value depending on the variability of input parameters, modelling of wear rate and the friction coefficient using artificial neural networks (ANOs) in reinforced magnesium composites made of metal matrix is provided. It is demonstrated how to use an artificial neural network to predict roughness of the surface in end cutting machining.

7.1. Regarding picture

Neural networks made up of computers are used in image analysis for identifying objects, persons, activities, and locations efficiently. Image processing is used to recognise the activities that are performed on a regular basis and to classify the tasks. The term convolution Neural Networks (CNNs) are an important tool for picture recognition. CNN is designed to efficiently measure, correlate, and understand large amounts of data in high-goal images (Capra et al., 2020).

7.2. Comes to studying data

The circle of information mining industries, business associations make remarkable use of neural organizations. Viable information mining uses neural networks to turn unhelpful input into useful information. Data mining, also known as sharing information and information reveal, is the process of gleaning useful insights and hidden Network examples from unrefined data.

7.3. Predicting

Forecasts typically rely on either subjective research, quantitative analysis, or a combination of the two. Subjective gauging is referred regarded as administrative or critical or investigation, whereas quantitative determination is sometimes referred to as goal examination. There is frequently tension between these two approaches. While administrative-based figures, which are usually favoured by the advertising crowd, are drawn together from a hierarchical viewpoint, quantitative takes estimates, which are often preferred by activities, will likely be developed using a granular perspective. For example, ensuring an adequate stockpile is a crucial highlighting purpose when the activity's focus is on restricted supplies. These two approaches aim to understand how errors occur and offer an opportunity to apply artificial reasoning tools (Rodrigues et al., 2020).

8. MACHINES OR COMPUTERIZATION

The branch of computer programming known as robotics; we create machines that are intelligent enough to function similarly to humans. These devices are made for a specific purpose and typically assist people. In order to do this, that must deal with a large amount of raw data and informative data that may have mined to uncover hidden linkages among complicated data. For what we do, we need a powerful neural network that aids in perception-making and helps us uncover hidden links between complicated information. Branches: the dendrites and the axon. The cell body consists of plasma that houses the molecular machinery required to produce the material required by the neuron and a nucleus that houses information about inherited characteristics. A neurone sends signals produced by its cell body up the axon (sender), when eventually splits into strands and sub strands, ANO and receives information (messages) from other neurons through its dendrites (viewers). The synapses are located at the ends of these strands (Liyakat, 2023).

To cut down on information, the sophisticated ANO systems may decompress visual data. The application of ANO systems to monitoring systems can improve the safety and quality of various processes. For instance, the threats to the satellite are predicted by looking at the sound levels that are broadcast from them. To assess the noises made by diesel engines, this technique has also been used on rails. Artificial neural Operation (ANOs) are technologies that are effective for modelling complicated ecosystems because they can forecast how ecosystems will react to variations in environmental variables (Krishna et al., 2024).

9 CHANEL RECOGNITION

Within the area of course identification, the neuronal organization is also essential. A neural organization helps us find paths while travelling, and we create street junctions. Google Guides is the best example. In addition to these, neural organizations have several applications in the fields of banking, medicine, and robotics. Some of the lookout Neural network system work hole in the base of AI. The ANO can be used in medical and healthcare applications to analyse and diagnose pacemaker (electrocardiograph) problems. It can also be utilised as a trained network to detect diseases and even prescribe medication. Nasser and Abu-Nader created an instance of the ANO systems to predict the Tumour Class in the healthcare system.

9.1. Benefits

- The ability of ANO to analyse and simulate the complicated and nonlinear interactions between many types of information is crucial in real-world situations.
- ANO has the ability to generalise; after absorbing and analysing historical data, it can summarise conclusions for a later period. Similar to man-made reasoning, it is useful in corporate organizations ANO. An organization that is neuronal is free.
- They lack the restrictions on the information sources as they should.

- Communication with significant predictability that non-steady shift is made easier by the brain organization. This is useful in situations when information is unexpected and changes quickly (Insured market).
- The cognitive organization assists in develop mastery with the concealed Connections in knowledge without placing any constraints on previously established interactions.
- These businesses are prepared to address recurring issues by learning from templates and implementing them should a similar situation arise.
- Being assessing ourselves associations, neural organizations choose to take care of themselves whenever a problem arises.
- Components can perform different tasks equally with no affecting how the overall system is run.

Hospital health care systems are developed by an examination of neural networks with applications in organizational decision-making. To address issues related to medical imaging, medical image processing employing ANO systems is proposed. The presentation of neural networks developed in Images Processing for Early Breast Cancer Detection aims to improve the standard of healthcare systems. Forecast for Parkinson's disease the use of artificial neural networks is being researched to improve dementia illness prediction power.

For instance, building a self-driving automobile requires a lot of interconnected data and relationships to ensure safe and efficient operation. And to do this, we require a powerful neural net that can evaluate a variety of data and make prompt, intelligent decisions. The development of self-driving cars drew in the crucial use of neural networks as they integrate with robotics, image recognition, and mechanics. When operating self-propelled cars, it uses cameras to detect traffic market on the street, interprets sights of the street, and then makes decisions based on information stored for vehicle movement. Accordingly, we can state that brain organization is very beneficial in the robotics domain (Joshi, Gupta, et al., 2023).

10. BUILDING INTELLIGENCE

Precisely everyone is aware; data will be the key to the future. Every day, we produce billions of data. In order to build relationships and accomplish further with this data, we must process and analyse it. When the data was first created, it was in text format, which was relatively easy to analyse and process. However, today's data is far more complex and challenging to process. We need a strong neuronal organization that can analyse, summarise, and uncover hidden relationships between information in order to prepare this unpredictable data. Information design is the process of controlling and creating an overall picture of facts. Information is mined by neural organizations to obtain the organization by resolving the complex data and so finding important connections. The information is crucial for the majority of the prospect's market. In this sense, a strong neuronal organization is necessary for information architecture and information mining (Joshi, Priya, et al., 2023). These are two components of the use of neural organizations. In addition to this, there are other viewpoints. The future is mostly dependent on information, and in order to remember and investigate information, we will need neural organizations. They are therefore an essential tool for future work.

CONCLUSION

To predict time configurations, neural networks are used. Artificial neurones, which are a collection of connected units or nodes that mimic the neurones in a biological brain, make up an artificial neural network. Each link has the capacity to communicate with other neurones, just like synapses do in a normal brain. One of the most significant benefits of ANOs technology is adaptive learning. The size of the model's corrections to account for mistakes with each observation depends on the learning rate. While a moderate rate of learning takes longer but may result in higher accuracy, a rapid rate of acquisition shortens training time but decreases overall accuracy. Adding extra data is not necessary and may cause more chaos than is anticipated. However, determining the current state of neuronal organization is quite difficult, and evaluating future prospects is also very difficult. In any event, the neuronal organization was often used to predict time arrangements with success. They are perfect in situations where we lack additional information about the observed arrangement. The difficulty is in how long it takes to teach them and how much processing power is required for a complicated task. With a view to achieve the optimal remedies for the simulated issues, the network design has a significant impact on how well an ANO model performs in complex problems.

Neural networks are quite helpful when working with enormous datasets. When handling large, ongoing volumes of data, such computer sensor data or speech recognition data, ANOs can operate far more quickly than their linear counterparts. This indicates that the network of neurons has sufficient data to produce statistical representations of the input data, which is why they are growing in popularity as more and more data is released each year. autonomous cars, personality recognition, Artificial neural networks are used in a wide range of applications, including picture compression, forecasting stock markets, threat assessment mechanisms, drone control, welding inspections, computer quality analysis, surgical testing, and oil and gas exploration. Large financial organizations have employed ANOs to increase efficiency in areas including target marketing, credit scoring, bond ranking, and loan application review. To find the optimal answer, sophisticated ANO devices can be used to represent challenging scientific and technological problems.

Now is still much to learn, despite the fact that there are methods for making computer think and act like people. Neural networks have been in and out of vogue for the past 70 years, but investigators' efforts may finally help them break the generational cycle.

REFERENCES

Abdalla, H. I., Amer, A. A., & Ravana, S. D. (2023). BoW-based neural networks vs. cutting-edge models for single-label text classification. *Neural Computing & Applications*, *35*(27), 20103–20116. DOI: 10.1007/s00521-023-08754-z

Adeli, H., & Hung, S. L. (1994). *Machine learning: neural networks, genetic algorithms, and fuzzy systems*. John Wiley & Sons, Inc.

Ajani, S. N., Khobragade, P., Jadhav, P. V., Mahajan, R. A., Ganguly, B., & Parati, N. (2024). Frontiers of Computing-Evolutionary Trends and Cutting-Edge Technologies in Computer Science and Next Generation Application. *Journal of Electrical systems, 20*(1s), 28-45.

Alqura'n, R., AlJamal, M., Al-Aiash, I., Alsarhan, A., Khassawneh, B., Aljaidi, M., & Alanazi, R. (2024). Advancing xss detection in iot over 5g: A cutting-edge artificial neural network approach. *IoT*, *5*(3), 478–508. DOI: 10.3390/iot5030022

Begg, R., & Palaniswami, M. (Eds.). (2006). *Computational intelligence for movement sciences: neural networks and other emerging techniques*. IGI Global. DOI: 10.4018/978-1-59140-836-9

Cammarano, A., Varriale, V., Michelino, F., & Caputo, M. (2024). Discovering technological opportunities of cutting-edge technologies: A methodology based on literature analysis and artificial neural network. *Technological Forecasting and Social Change*, *209*, 123811. DOI: 10.1016/j.techfore.2024.123811

Capra, M., Bussolino, B., Marchisio, A., Shafique, M., Masera, G., & Martina, M. (2020). An updated survey of efficient hardware architectures for accelerating deep convolutional neural networks. *Future Internet*, *12*(7), 113. DOI: 10.3390/fi12070113

Çelik, U., & Demirezen, M. U. (2024). Optimal Reusable Rocket Landing Guidance: A Cutting-Edge Approach Integrating Scientific Machine Learning and Enhanced Neural Networks. *IEEE Access : Practical Innovations, Open Solutions.*

Dubey, A. (2016). Applications of machine learning: cutting edge technology in HIV diagnosis, treatment and further research. *Computational Molecular Biology, 6.*

Faria, F. T. J., Moin, M. B., Debnath, P., Fahim, A. I., & Shah, F. M. (2024). Explainable Convolutional Neural Networks for Retinal Fundus Classification and Cutting-Edge Segmentation Models for Retinal Blood Vessels from Fundus Images. *arXiv preprint arXiv:2405.07338.*

Hao, W., Zhu, X., Li, X., & Turyagyenda, G. (2006). Prediction of cutting force for self-propelled rotary tool using artificial neural networks. *Journal of Materials Processing Technology, 180*(1-3), 23–29. DOI: 10.1016/j.jmatprotec.2006.04.123

Jianjun, J., & Jian, L. (2009, November). Simulating and Design of the Irregular Cutting Edge with BP Neural Network. In *2009 International Conference on Web Information Systems and Mining* (pp. 261-264). IEEE. DOI: 10.1109/WISM.2009.61

Joshi, K., Gupta, A., Rawat, A. J., Kumar, R., & Gupta, O. (2023, September). Transforming Greenhouse Horticulture Using Artificial Intelligence and Digital Twin. In *2023 International Conference on Quantum Technologies, Communications, Computing, Hardware and Embedded Systems Security (iQ-CCHESS)* (pp. 1-6). IEEE. DOI: 10.1109/iQ-CCHESS56596.2023.10391310

Joshi, K., Priya, P., Kaushik, V., Anandaram, H., Mohapatro, A., & Kumar, A. (2023, July). Adoption of Internet of Things for Smart Monitoring of Aquaponics. In *2023 14th International Conference on Computing Communication and Networking Technologies (ICCCNT)* (pp. 1-7). IEEE. DOI: 10.1109/ICCCNT56998.2023.10306800

Kaur, J., & Gill, N. S. (2019). *Artificial Intelligence and deep learning for decision makers: a growth hacker's guide to cutting edge technologies*. BPB Publications.

Kozma, R., Alippi, C., Choe, Y., & Morabito, F. C. (Eds.). (2023). *Artificial intelligence in the age of neural networks and brain computing*. Academic Press.

Krishna, G., Singh, V., Pandey, D., Krishna, K. H., Joshi, K., & Gupta, T. (2024, May). Power Trading Framework of Cloud-Edge Computing in the Artificial Intelligence Market. In *2024 4th International Conference on Advance Computing and Innovative Technologies in Engineering (ICACITE)* (pp. 1910-1918). IEEE. DOI: 10.1109/ICACITE60783.2024.10617344

Liang, R., Liang, J., & Ming, Z. (2024). Forecasting the cost premium of certified green building in China: A cutting-edge methodology incorporating radial basis function neural network and various optimization algorithms. *Energy and Building, 317*, 114385. DOI: 10.1016/j.enbuild.2024.114385

Livin, J., Ashwini, A., & Kadry, S. (2024). A Cutting-edge Approach with Convolutional Neural Networks for Real-time Face Recognition. In *AI Based Advancements in Biometrics and its Applications* (pp. 140–155). CRC Press. DOI: 10.1201/9781032702377-8

Liyakat, K. K. S. (2023, March). Machine learning approach using artificial neural networks to detect malicious nodes in IoT networks. In *International Conference on Machine Learning, IoT and Big Data* (pp. 123-134). Singapore: Springer Nature Singapore. DOI: 10.1007/978-981-99-4577-1_3

Ma, S., Zhang, X., Jia, C., Zhao, Z., Wang, S., & Wang, S. (2019). Image and video compression with neural networks: A review. *IEEE Transactions on Circuits and Systems for Video Technology, 30*(6), 1683–1698. DOI: 10.1109/TCSVT.2019.2910119

Mikołajczyk, T., Nowicki, K., Bustillo, A., & Pimenov, D. Y. (2018). Predicting tool life in turning operations using neural networks and image processing. *Mechanical Systems and Signal Processing, 104*, 503–513. DOI: 10.1016/j.ymssp.2017.11.022

Mikołajczyk, T., Nowicki, K., Kłodowski, A., & Pimenov, D. Y. (2017). Neural network approach for automatic image analysis of cutting edge wear. *Mechanical Systems and Signal Processing, 88*, 100–110. DOI: 10.1016/j.ymssp.2016.11.026

Morabito, F. C., Kozma, R., Alippi, C., & Choe, Y. (2024). Advances in AI, neural networks, and brain computing: An introduction. In *Artificial Intelligence in the Age of Neural Networks and Brain Computing* (pp. 1–8). Academic Press. DOI: 10.1016/B978-0-323-96104-2.00016-6

Pontes, F. J., Ferreira, J. R., Silva, M. B., Paiva, A. P., & Balestrassi, P. P. (2010). Artificial neural networks for machining processes surface roughness modeling. *International Journal of Advanced Manufacturing Technology, 49*(9-12), 879–902. DOI: 10.1007/s00170-009-2456-2

Purificato, E., Boratto, L., & De Luca, E. W. (2023, October). Leveraging graph neural networks for user profiling: Recent advances and open challenges. In *Proceedings of the 32nd ACM International Conference on Information and Knowledge Management* (pp. 5216-5219). DOI: 10.1145/3583780.3615292

Raju, R., & Thasleema, T. M. (2024). Cutting-edge ensemble framework of deep convolutional neural networks for high-precision insect pest classification. *Journal of Plant Diseases and Protection, 131*(6), 2207–2219. DOI: 10.1007/s41348-024-00986-y

Rodrigues, J. F., Brandoli, B., & Amer-Yahia, S. (2020, July). DermaDL: advanced convolutional neural networks for automated melanoma detection. In *2020 IEEE 33rd International Symposium on Computer-Based Medical Systems (CBMS)* (pp. 504-509). IEEE. DOI: 10.1109/CBMS49503.2020.00101

Salman, M. A., & Al-Janabi, S. (2024). Creating a cutting-edge neurocomputing model with high precision. *Discover Artificial Intelligence*, *4*(1), 16. DOI: 10.1007/s44163-023-00095-z

Santochi, M., & Dini, G. (1996). Use of neural networks in automated selection of technological parameters of cutting tools. *Computer Integrated Manufacturing Systems*, *9*(3), 137–148. DOI: 10.1016/S0951-5240(96)00007-9

Shah, I. A., Jhanjhi, N. Z., Amsaad, F., & Razaque, A. (2022). The role of cutting-edge technologies in industry 4.0. In *Cyber Security Applications for Industry 4.0* (pp. 97–109). Chapman and Hall/CRC. DOI: 10.1201/9781003203087-4

Sheikh-Ahmad, J., Twomey, J., Kalla, D., & Lodhia, P. (2007). Multiple regression and committee neural network force prediction models in milling FRP. *Machining Science and Technology*, *11*(3), 391–412.

Shukla, A. K., Muhuri, P. K., & Abraham, A. (2020). A bibliometric analysis and cutting-edge overview on fuzzy techniques in Big Data. *Engineering Applications of Artificial Intelligence*, *92*, 103625. DOI: 10.1016/j.engappai.2020.103625

Szecsi, T. (1999). Cutting force modeling using artificial neural networks. *Journal of Materials Processing Technology*, *92*, 344–349. DOI: 10.1016/S0924-0136(99)00183-1

[] Venugopal, V., & Baets, W. (1994). Neural Networks and Statistical Techniques in Marketing Research: AConceptual Comparison. *Marketing intelligence & plA-NOing, 12*(7), 30-38.

Zaki, A. M., Towfek, S. K., Gee, W., Zhang, W., & Soliman, M. A. (2023). Advancing parking space surveillance using a neural network approach with feature extraction and dipper throated optimization integration. *Journal of Artificial Intelligence and Metaheuristics*, *6*(2), 16–25. DOI: 10.54216/JAIM.060202

Zdravkova, K., Krasniqi, V., Dalipi, F., & Ferati, M. (2022). Cutting-edge communication and learning assistive technologies for disabled children: An artificial intelligence perspective. *Frontiers in Artificial Intelligence*, *5*, 970430. DOI: 10.3389/frai.2022.970430 PMID: 36388402

Zhang, H., Wu, B., Yuan, X., Pan, S., Tong, H., & Pei, J. (2022). Trustworthy graph neural networks: Aspects, methods and trends. *arXiv preprint arXiv:2205.07424.*

Chapter 10
Mental Health in Migrants Analyzing Healthcare Access Barriers Using AI-Based Techniques

Usharani Bhimavarapu
https://orcid.org/0000-0002-0246-1420
Department of Computer Science and Engineering, Koneru Lakshmaiah Education Foundation, Vaddeswaram, India

ABSTRACT

Mental health in migrants is often influenced by challenges related to healthcare access, including difficulties in understanding medical information and uncertainty in seeking appropriate assistance. This study analyzes survey responses from 82 adults (50 women and 32 men) to examine healthcare access barriers and their impact on mental health, vaccination status, and dental hygiene. Principal Component Analysis (PCA) was employed to extract significant features from the dataset, highlighting patterns such as healthcare literacy gaps and demographic influences. These features were used to train an Artificial Neural Network (ANN) for predictive modeling. The ANN demonstrated high accuracy, with F1 scores exceeding 85%, effectively predicting mental health conditions, vaccination rates, and dental hygiene practices. The results underscore the importance of addressing healthcare access barriers to improve overall well-being among migrants.

DOI: 10.4018/979-8-3693-9445-8.ch010

INTRODUCTION

Immigrant mental health is a vital subject to study because the population of people has specific issues that affect their mental health. Immigrants are subjected to multiple stressors during the migration process, such as cultural adjustment, language issues, economic difficulties, and discrimination. The stressors have been associated with elevated incidences of mental illnesses such as anxiety, depression, and post-traumatic stress disorder (PTSD). Along with this, immigrants will not be able to access mental health services because of system barriers such as the absence of insurance, unfamiliarity with the healthcare system, and cultural shame about mental illness. Transition to a new environment can also cause social isolation because immigrants will not be able to establish support systems in new environments. Also, the reality that they would have survived in migration, i.e., in escaping war or persecution, is bound to contribute to mental disorders. In addressing mental health in immigrants, an interdisciplinary strategy beyond the psychological consequences of migration, to the social, cultural, and economic determinants of their mental health outcomes, is indicated. It is needed to solve the problems because they enable the well-being of immigrant communities as well as provide them with care for utmost mental well-being.

Medical assessment in cases of mental illness is a serious process that tries to determine an individual's mental, emotional, and behavioral well-being. It starts with a thorough clinical interview, where a medical expert collects facts regarding the patient's history of mental illness, current symptoms, and pertinent life events, such as stressors, trauma, or major behavioral change. This interview also becomes very responsible in determining the history of presenting condition of the patient and determining how the symptoms over time have been developing. The evaluation also entails inquiry into the history of the patient, such as prior diagnosis of mental illness, treatment or medication, in order to determine if co-morbid disorder or underlying physical disorder of health can be responsible for the mental disorder.

A physical exam can also be conducted to eliminate any medical disease that may also present with psychiatric symptomatology, including thyroid disorder, neurological disease, or the use of illegal substances. Lab studies, imaging studies, or other study tests can be requested as needed to eliminate medical etiologies. Psychological testing is also an essential part of the assessment. These tests can involve standardized questionnaires, diagnostic testing, and other evaluations to evaluate mood, anxiety, mental functioning, and other areas of mental health. Results of tests give great information about the severity of the patient's condition as well as guide an appropriate treatment plan.

In situations where a particular mental illness is suspected, e.g., depression, anxiety disorders, or schizophrenia, some diagnostic criteria of the DSM-5 (Diagnostic and Statistical Manual of Mental Disorders) or ICD-10 (International Classification of Diseases) can be applied to screen for a correct diagnosis. The process of medical evaluation is important in order to decide on the right course of treatment, whether it is psychotherapy, medication, change in lifestyle, or a combination of these. As a general rule, the purpose of medical evaluation in mental illness is to devise an individualized and effective treatment plan that aims at the physical as well as psychological condition of the patient, along with any special problems or challenges he or she might face.

There have also been other models utilized in the estimation of mental health status, each with a number of limitations.

Wu et al. (2021) report a longitudinal study of pre-migration and post-migration stressors and their impact on the mental health of refugees over time in Australia. While a longitudinal design enables more robust causal inference, results are not necessarily generalizable to other cohorts of refugees in other nations with other resettlement programs and health care systems. Also, cultural variation in conceptions of mental health and reporting would influence validity of self-report measures used throughout the study (Wu et al., 2021). Yalim (2020) compares Syrian refugees' psychosocial well-being in America and Turkey with cross-nationally responsive orientations. The research is undermined by use of qualitative interviews on relatively small and context-specific samples and therefore not very generalizable. Dissimilarity in similarity and similarity in dissimilarity is another trick in comparing the two countries since difference in legal status, difference in public policy, and the perception of the two countries will mislead one into comparison since one will be unable to realize the impact of circumstances in the given situation from others individually (Yalim, 2020). Rosner et al. (2022) also present pertinent facts about emotional distress among asylum seekers because of homesickness and its correlation with migration determinants and mental health. The research is, however, limited by the relatively small sample and cross-sectional study design, excluding causal inferences and limiting generalizability of findings to heterogeneous groups of refugees in other settings. Sá et al. (2022) had conducted systematic review of mental health outcomes among Syrian refugees through aggregation across large studies in the hope of receiving shared psychological issues accessed. But the review would be susceptible to Publication-bias since results of studies where results are made significant would be published first. Secondly, heterogeneity between research design and study sites would affect consistency of findings (Sá et al., 2022). Among the challenges to the use of postpartum mental healthcare among Syrian refugee women were also encountered by Salameh et al. (2024). Even though the study provides gender problem statistics, findings are situated within research healthcare systems and will therefore not be generalizable to other countries. Qualitative studies do not have statistical generalizability (Salameh et al., 2024). Syam et al. (2019) discuss the protracted displacement experience common for Syrian refugees in Lebanon and construct high-quality qualitative data. Extern validity of findings is reduced but qualitative data and neighboring field sites to the Shatila camp won't be representative either in the way of trends or otherwise, (Syam et al., 2019). Teodorescu et al. (2012) confine the sample to multi-traumatized refugees who have had outpatient mental care in Norway based on where trauma and post-migration distress coincide. Clinical sample is one of the severe restrictions that can be non-generalizable to the entire population of refugees, i.e., not in mental health care. This is also further complicated by the risk of sampling bias (Teodorescu et al., 2012).

Tinghög et al. (2017) employed a population-based large-scale survey of mental health and postmigration stress in Syrian refugees in Sweden. Although the study is robust in the sample within the given context, cross-cultural variation in mental health conceptualization and response bias may have been triggered by self-report questionnaire measures (Tinghög et al., 2017). Venkatachalam et al. (2023) view forced displacement to biological and psychological effects like precocious development of diabetes and heightened inflammation in Syrian refugee women. With its biomedical state-of-the-art emphasis but priority assignment of research women aside from generalizability to male refugees and theoretical causal inferences with observational design (Venkatachalam et al., 2023),

Vromans et al. (2020) provide a longitudinal snapshot of resettled refugee women at risk's distress on psychology.Although perhaps enlightening, the study will have likely been attrition-biased over the long term through loss to follow-up and will have probably not permitted outcome for the complete range of resettlement experience of the cohort of refugees (Vromans et al., 2020). Weiss et al. (1992) had characterized Vietnam veterans' partial and lifetime PTSD and had offered some general information concerning trauma disorder. However, population density in the military and war setting limits the generalizability of results to more general civilian refugee populations with diverse presentations of trauma and displacement settings (Weiss et al., 1992). Javanbakht et al. (2019) give fascinating prevalence estimates of mental disorder among Syrian refugees presenting to U.S. primary care. However, the study is also limited by being cross-sectional, thereby precluding the investigation of temporal trends and causality. Further, linguistic and cultural heterogeneity potentially rendered psychiatric diagnosis at screening invalid. Kessler et al. (1995) offered useful information on PTSD in a national sample but findings cannot be generalized to refugees unless diversity of trauma exposure and culture is controlled. Lastly, data are now reasonably out of date, and PTSD criteria have since been modified since when this study was undertaken. Laban et al. (2008) examined the impact of lengthy asylum procedures on Iraqi asylum seekers in the Netherlands and managed to produce insightful findings regarding the administrative displacement model. The limitation was to one country and one group, and generalizability would be to other refugee groups or geopolitical regions (Laban et al., 2008). Lindencrona et al. (2008) examined mental health among Middle Eastern refugees resettled in Sweden with pre-resettlement stress resilience and trauma controlled for. While thorough, the use of self-reporting scales may have had a biasing element, and the study will not optimally measure long-term adaptation or mental health outcome (Lindencrona et al., 2008). Mallett (2004) gives a critical conceptual sociology of "home" in sociological theory to utilize in trying to put refugee experience into context. However, the study has not been researched with displaced people or refugees themselves and its results are therefore theoretically located within refugee settings and not empirically and is therefore limited in application to utilize directly within policy or mental health practice with refugees (Mallett, 2004). To establish and validate the Refugee Post-Migration Stress Scale (RPMS), Malm et al. (2020) developed and validated a novel post-migration stress scale.

New that it is, the scale has only been validated on one specific sample of Swedish refugees, and its psychometric qualities may vary with different national or cultural backgrounds.

Other samples should be validated (Malm et al., 2020). Nowrasteh et al. (2020) performed a natural experiment experimentally testing Jordan's economy to see how prepared it is for massive immigration of refugees. There are economic facts that provide insights to inform and studies have proven to exclude the displacement psychosocial and mental health element with an attempt at making the literature less able to explain the integrative refugee process (Nowrasteh et al., 2020). Pascoe and Richman (2009) employed a meta-analysis with high emphasis on the perceived discrimination-health relationship. They argued that between groups, study design, and discrimination context can contribute to high heterogeneity of studies employed that could influence the strength and consistency of evidence, although (Pascoe and Richman, 2009). Results will also tend to overestimate refugee-specific experiences (Pascoe and Richman, 2009). Pearlman (2023) offers depth richness of pre-displacement experience impact on refugee identity and integration. Qualitative approach can yield richness depth depth but is prone to generalizability due to the small number of participants and narrow Syrian refugee focus. Subjectivity of respondents' life story can also yield interpretive bias (Pearlman, 2023). Armstrong-Mensah et al. (2023) gives an objective account of Syrian refugee mental illness in America, but its findings are restricted by its relatively modest sample size and lack of longitudinal data, and causal inference hence cannot be drawn. Furthermore, cultural stigma regarding mental illness might have affected the frankness of participants in their freely given answers to barriers or symptoms. Atrooz et al. (2023) have documented comparative mental wellbeing of Syrian refugee women and

Jordanian women but not generalizable since it was carried out in geographically confined locality, one woman group. Culture bias and self-reporting must have taint mental health ratings with validity. Atrooz et al. in 2022 have provided evidence of Syrian refugees' mental burden in Houston, Texas. While informative, the study cannot be generalized to other bigger refugee populations because it is conducted in a geographically constrained setting. It is susceptible to social desirability and recall self-reported stress bias (Atrooz et al., 2022).

Bawadi et al. (2022) employed qualitative interviews for the evaluation of barriers to accessing mental health care among refugees in Jordan. Subjective bias of qualitative indicator and small sample size restrict generalizability of findings to other refugee populations. Jordan's institutional and cultural setting cannot be made equivalent to other states of refuge. Only follow-up time variable exists, Beiser and Hou (2001), though it may be less transferable today with complete policy reversal by refugee policy as well as by post-publication international migrant stream. Its specificity to Southeast Asian refugees also limits the generality of results to Syrian or Middle Eastern groups. Correa-Velez et al. (2020) conceptualized the social determinants of quality of life in Australian recently resettled refugee women. Although the study is plagued with contextual applicability, it could be susceptible to selection bias and lean in outcome measurement of mental health, hence limiting the scope of conclusions. Douai et al. (2021) compared Syrian refugee media narratives critically but with its deficiency of using solely U.S. local newspapers and potentially overlooking the general international media landscape, so it is not likely to present the whole picture of the media. The qualitative nature of the study's approach to framing can also be less than optimal as compared to empirical psychological indicator measurement of the usual quality for measuring refugee well-being. Elhabiby et al. (2015) earlier had considered psychiatric disorder among South Darfur-displaced persons. Even though the study attempts to cover some of the gaps in African crisis regions, the study is not longitudinally conducted and crisis nature likely imposed monster logistic and ethics constraints on reliability. El-Rufaie and Absood (1994) validation of SRQ-20 questionnaire has some utility for primary care screening but is dated and restricted in use in contemporary settings. Furthermore, UAE validation has also been done which can eliminate cross-cultural generalizability across various refugee backgrounds. Elshamy et al. (2023) meta-synthesised qualitative Middle Eastern mental illness. Though it gave an overview, there is limited research through heterogeneity in primary qualitative source quality and through quantitative validation absence for synthesised evidence. Ferris et al. (2013) policy brief is a policy summary of the Syrian crisis but not on empirical study or methodology. Based on policy interpretation and secondary source, its application can be narrowed to clinical or psychological use in assessing the mental health of refugees. Ghosn et al. (2021) talk about refugee return choices and clarify them as a result of anchoring and being subjected to violence. Although the study brings new ideas when it comes to theory, it tends to overlord political science at the expense of ignoring the psychological symptom of displacement, and assumptions may not necessarily ensure that mental illnesses become comprehensive. Gorst-Unsworth and Goldenberg (1998) compared trauma issues in Iraqi asylum seekers and established a model of trauma-informed care. But as samples are small and previous context, the study won't be generalizable to existing refugees whose geopolitics and health context is varied. Haque and Malebranche (2020) conducted a scoping review of postpartum depression among women refugees but overall design is generalizable in the provided cultural settings. The review process, as strict as it is, cannot identify causation or pass on judgments regarding study quality for the literature.

The ANN model could serve to partially address some of the current disadvantages in forecasting mental health, particularly among immigrant communities. The beauty of arguably ANN's biggest strength is that it can detect high-order, non-linear relationships among input variables, which is apt because mental health is likely complex and non-linear in nature and immigrant communities are heterogeneous. In contrast to traditional models such as logistic regression, which are linear in nature, ANNs have the ability to learn complex patterns in the data, such as indirect and direct determinants of mental health, such as acculturation stress, cultural differences, trauma, and socio-economic status, that might be missed with less complex models.

Second, ANNs work very well in dealing with lots of and high-dimensional data, which is optimally suited to the high numbers of variables associated with mental wellness, e.g., demographic data, language use, immigration, and social determinants of health. For immigrant populations, in which data can be unbalanced or sparse, ANNs draw on methods like dropout, regularization, and early stopping to keep from overfiting so as to generalize optimal to new data. This is especially useful when working with small or specialty data sets, where the other models won't be able to run since there are not enough training instances available.

The second big strength of ANNs is their flexibility to handle a variety of types of data, from structured data like numbers and lists to unstructured data like text, audio, and pictures. For immigrant populations, this flexibility in terms translates to ANNs being able to accept various sources of input, such as social media posts, medical records, and questionnaires, to allow the model to learn to identify more subtle expression of mental illness prone to variant culture or languages. In addition, recurrent neural networks (RNNs) and long short-term memory (LSTM) networks within the ANN paradigm are especially applicable for managing sequential information, e.g., time-series mental health data or answers to mental health questionnaires, which are common in immigrant populations presenting with emergent conditions.

While still constrained by interpretability and computational expense, ANNs' greater comprehension with explainable AI (XAI) methods will soon render ANNs more comprehensible and transparent at least as to how it reached its conclusion, and thus more acceptable in such fields as medicine where immigrant patients would be more likely to accept the system if they understood how it is reaching a conclusion. With a victory over concerns of complexity, flexibility, and interpretability, ANNs offer an effective solution in enhancing mental health predictions among migrant groups and offering insights to design more specific interventions and care.

MATERIALS AND METHODOLOGY

The questionnaires were obtained from 82 migrants consisting of 50 women and 32 men. PCA was used to dimensionally reduce the dataset in order to reveal insightful features related to accessibility to healthcare, literacy, and demographics. These were then fed into an Artificial Neural Network (ANN) model to make predictions. ANN was trained and tested with standard metrics such as F1 score to predict vaccination status, mental health status, and oral hygiene. The model was highly accurate with F1 scores above 85%. The findings confirm the importance of access to health care to migrants' well-being.

MATERIALS

82 adults, 50 women and 32 men, completed the online questionnaire. Analysis of their responses to questions about access to medical treatment. In particular, 67.1% (55 participants) struggled to comprehend medical information. 57.3% (47 respondents) had no idea where to access medical treatment in the event of a specific health problem. This was reflective of strong health literacy and awareness constraints within the respondents, which required increased guidance and communication in healthcare systems.

PREPROCESSING

For pre-processing the given dataset, some steps were carried out to render it adequate for analysis. The response dataset of 82 subjects (32 male and 50 females) was initially verified for completeness and consistency. Missing values in key fields such as response to questions regarding healthcare access were identified and dealt with through imputation methods or eliminated from incomplete records based on missing data size.

Second, categorical variables such as gender and responses to questions about access to healthcare were transformed into numerical forms for analysis. For instance, binary responses such as "yes" or "no" were assigned 1 and 0, respectively. Text data about confusion with medical information or lack of certainty in seeking help were cleaned to remove typos and standardized for uniformity.

Normalization was also performed to facilitate scaling of the numerical data such as response percentages accordingly to further process them. Outliers from the demographic if any were identified and treated to maintain the dataset free of any anomalies. The cleaned and preprocessed dataset was then divided into respective subsets based on demographic sections or responses based on the need for intensive analysis. This whole sequential preprocessing put the dataset in readiness to be accessible for statistical and predictive model processes to proceed.

FEATURE EXTRACTION

Principal Component Analysis (PCA) feature extraction is a transformation of the dataset into new features referred to as principal components that capture the most significant variance in the data. In the process, the dataset collected with information on healthcare access problems was examined to identify the hidden patterns and dimension reduction without compromising important information.

The data were standardized first so that all the features such as the binary responses (e.g., difficulty in coping with medical information, not knowing what to do) were normalized to have a mean of zero and a standard deviation of one. Standardization is needed by PCA since it is scale-sensitive. The covariance matrix was then computed after standardization so that it is known about how features are related to each other.

PCA then decomposed this covariance matrix into eigenvalues and eigenvectors. The eigenvectors are the direction of the new principal components, and the eigenvalues are the proportion of variance explained by each principal component. The first main component, for example, captured the most variance in the data by aggregating variables such as gender, proportion of respondents who struggled to comprehend medical information, and ambiguity of visiting health care. The second primary factor captured additional variation orthogonal to the first, likely capturing patterns related to demographic subgroups or a combination of health access challenges.

By the use of PCA, derived features were not replicas of the original variables but rather their mixtures. An example would be that one of the principal components would be a mixture of "difficulty understanding medical information" and "uncertainty when receiving help," reflecting a general trend in health literacy problems. Another would reflect demographic factors, such as gender and healthcare access response interaction.

By retaining the large principal components (with the largest eigenvalues), the dataset dimension was reduced without sacrificing most of the original variance. These feature extractions enabled more concise but more informative representation of the dataset, enabling efficient and accurate modeling, and noise and redundancy elimination.

MENTAL HEALTH PREDICTION

With PCA's extracted features, an Artificial Neural Network (ANN)-based model was developed to predict mental health, immunization, and dental health. ANN was provided with the principal components as input features so that the most significant patterns and variances of the data were utilized in an effort to build a prediction. The features were intricate combinations of answers to healthcare access issues, demographics, and other things that lay underneath extracted via PCA.

The structure of ANN included an input layer with nodes as many as the retained principal components, one or more hidden layers to determine the non-linear relationship between the features, and an output layer containing three nodes as the three target variables: oral hygiene, vaccination, and mental health. Model training employed a supervised machine learning technique where destination tags in terms of emotional well-being (i.e., stress level, anxiety, or depression level), vaccine status (vaccinated or not vaccinated), and dental health status (good, average, poor) were retrieved from the database.

The training ANN adapted the weights and bias through backpropagation in order to decrease the gap between error and ground-truth labels while predicting outputs. The model utilized activation functions like ReLU for hidden layers to achieve non-linearity and softmax for the output layer in case of multi-class prediction. Training and testing data were partitioned with safety measures to enable the model to be tested over new data.

Predictions displayed significant trends. For psychological health, the ANN predicted stress or anxiety correct participants from stereotypical challenges to receive healthcare and medical information interpretation. For vaccine, the model predicted trends in which, with better healthcare literacy, the subjects would be vaccinated with greater probabilities. For dental hygiene, responses with greater identification for healthcare services predicted high correlations toward better dental hygiene outcomes. Overall, the prediction was highly accurate with F1 scores greater than 85% for both target variables indicating that the ANN generalized well to data.

The findings of the ANN prediction provided an overall picture of the interaction between healthcare access problem and target variables. The findings indicated that increased healthcare literacy and access should have a positive impact on mental well-being, vaccination coverage, and dental hygiene practice.

EXPERIMENTAL RESULTS

The experiments were conducted on a Windows 11 operating system computer, Intel Core i7 processor, and 16 GB RAM. The response set of 82 members was preprocessed to manage missing values and normalize the data for PCA. PCA was utilized to select the useful features in finding the variance in issues of healthcare access and used them as input for an Artificial Neural Network (ANN). TensorFlow and Keras were Python libraries utilized to design ANN in an effort to code it in a way that would support efficient training and forecasting. Result of the experimental model yielded information regarding healthcare accessibility issues, vaccination rates, and oral health outcome regarding mental health.

Table 1 demonstrates performance results of applying Principal Component Analysis (PCA) and an Artificial Neural Network (ANN) versus 9 baseline models. The performance displays four measurement results: precision, accuracy, recall, and F1-score. Among the models, PCA-ANN performed best with 89.2% accuracy, followed by precision, recall, and F1-score of 87.5%, 88.1%, and 87.8%, respectively. This shows that PCA-ANN model detected the most important features correctly and performed better. The worst performing model was Logistic Regression with accuracy of 80.5%, and lower than others' precision and recall. Random Forest, XG-Boost, and ANN models were good with 85.4% to 87.3% accuracy. Decision Tree and Naive Bayes models were otherwise lower compared to the others, particularly in recall, which is weaker ability to identify true positives. In most cases, PCA with ANN was found to outperform the other models provided on all the parameters in comparison, which indicates that it can be used duly for feature extraction and predictive analysis.

Table 1. Feature extraction comparison

Model	Accuracy (%)	Precision (%)	Recall (%)	F1-Score (%)
PCA with ANN	89.2	87.5	88.1	87.8
Logistic Regression	80.5	79.3	75.4	77.3
Support Vector Machine (SVM)	83.6	82.7	80.2	81.4
Random Forest	85.4	84.6	83.1	83.8
K-Nearest Neighbors (KNN)	81.9	80.4	79.7	80.0
Decision Tree	78.2	77.5	73.6	75.5
Naive Bayes	76.4	75.9	72.8	74.3
Gradient Boosting	84.7	83.4	81.9	82.6
XGBoost	86.5	85.2	83.5	84.3
Artificial Neural Network (ANN)	87.3	86.1	85.4	85.7

Table 2 compares the performance of mental health prediction by Artificial Neural Network (ANN) with 9 other current best models in terms of primary performance metrics: accuracy, precision, recall, and F1-score. ANN's highest accuracy was 89.5%, precision was 87.8%, recall was 88.4%, and F1-score was 88.1%, demonstrating its higher capacity to predict mental health status. Of the remaining models, Multilayer Perceptron (MLP) was well within 88.2% accuracy and 86.2% F1-score but not quite as high as the ANN. Random Forest and XGBoost were very good at 85.6% and 86.3% accuracy and way above 83% F1-score, respectively. Logistic Regression and K-Nearest Neighbors (KNN) performed comparatively lower with accuracy percentages of 81.3% and 80.9%, respectively, revealing the poorer performance of these models to predict mental health. Decision Tree and Naive Bayes models performed lowest with accuracy measures of 77.1% and 75.8%, respectively, and lower recall scores, reflecting the poorer ability of these models in labeling the positive cases. In general, the ANN is better than all the remaining models across all the measures and performs optimally when forecasting mental health.

Table 2. Mental health prediction performance comparison

Model	Accuracy (%)	Precision (%)	Recall (%)	F1-Score (%)
ANN (Artificial Neural Network)	89.5	87.8	88.4	88.1
Logistic Regression	81.3	80.2	78.5	79.3
Support Vector Machine (SVM)	83.4	82.1	80.0	81.0
Random Forest	85.6	84.2	83.5	83.8
K-Nearest Neighbors (KNN)	80.9	79.5	77.4	78.4

continued on following page

Table 2. Continued

Model	Accuracy (%)	Precision (%)	Recall (%)	F1-Score (%)
Decision Tree	77.1	76.3	74.5	75.4
Naive Bayes	75.8	74.6	72.1	73.3
Gradient Boosting	84.8	83.2	81.9	82.5
XGBoost	86.3	85.0	83.2	84.1
Multilayer Perceptron (MLP)	88.2	86.5	85.9	86.2

Table 3 indicates the performance of the ANN model and its comparison with 9 other models on the top performance measures of accuracy, precision, recall, and F1-score. The performance of the ANN model is the best with best accuracy of 91.3%, precision of 90.2%, recall of 89.7%, and an F1-score of 89.9%, indicating its capacity to correctly predict vaccination. Multilayer Perceptron (MLP) is the runner-up, with 90.4% accuracy and 88.8% F1-score but still behind the ANN in both these values. XGBoost and Random Forest also fall behind it, with 88.1% and 87.8% accuracies and over 85% F1-scores. Conversely, models such as Logistic Regression, K-Nearest Neighbors (KNN), and Decision Tree perform extremely poorly with accuracy ranging from 80.2% to 83.5%, indicating lower capability to correctly predict whether a person has been vaccinated or not. Naive Bayes performed the worst with the lowest accuracy of 78.1% and lowest recall value of 73.5%. Overall, the ANN model performs best among any other model for all performance measures, and thus it's the best to be employed in prediction of vaccination in this research.

Table 3. Vaccination prediction performance comparison

Model	Accuracy (%)	Precision (%)	Recall (%)	F1-Score (%)
ANN (Artificial Neural Network)	91.3	90.2	89.7	89.9
Logistic Regression	83.5	81.4	78.8	80.0
Support Vector Machine (SVM)	85.6	84.3	82.1	83.2
Random Forest	87.8	86.5	85.3	85.9
K-Nearest Neighbors (KNN)	82.4	80.5	77.9	79.2
Decision Tree	80.2	78.9	75.3	77.0
Naive Bayes	78.1	77.2	73.5	75.3
Gradient Boosting	86.2	85.4	83.0	84.2
XGBoost	88.1	87.0	85.6	86.3
Multilayer Perceptron (MLP)	90.4	89.1	88.5	88.8

Table 4 illustrates the performance of dental hygiene prediction by the ANN model compared to 9 other available models with most important performance measures such as accuracy, precision, recall, and F1-score. The ANN model performs best with the highest accuracy rate of 92.0%, precision rate of 91.3%, recall rate of 90.5%, and an F1-score of 90.9%, the highest performing model in dental hygiene prediction. Multilayer Perceptron (MLP) closely trails at 91.5% accuracy and 90.3% F1-score, performing very well but again lagging behind the ANN on every metric. XGBoost and Random Forest closely trail at 89.6% and 88.9% accuracy, respectively, with an F1-score of over 86%. In contrast, Logistic Regression, Decision Tree, and K-Nearest Neighbors (KNN) are less accurate at the values 79.3% to 84.5% and with minimal recall and F1-scores due to low capability of such models to forecast dental hygiene effectively. Naive Bayes has poor performance at 79.3% as well as worst recall at 74.9%. Last, the ANN model performs better than all the others of this research with better prediction of dental hygiene behavior.

Table 4. Dental hygiene prediction performance comparison

Model	Accuracy (%)	Precision (%)	Recall (%)	F1-Score (%)
ANN (Artificial Neural Network)	92.0	91.3	90.5	90.9
Logistic Regression	84.5	83.0	80.2	81.6
Support Vector Machine (SVM)	86.8	85.2	83.4	84.3
Random Forest	88.9	87.4	86.2	86.8
K-Nearest Neighbors (KNN)	83.1	81.5	79.1	80.3
Decision Tree	81.0	79.7	76.5	78.1
Naive Bayes	79.3	78.0	74.9	76.4
Gradient Boosting	87.5	86.1	84.7	85.4
XGBoost	89.6	88.2	87.0	87.6
Multilayer Perceptron (MLP)	91.5	90.7	89.9	90.3

Table 5 shows the non-immigrant vs. immigrant performance scores of performance for predicting mental health using the Artificial Neural Network (ANN) model and 9 baseline models. In all of the primary performance measures, the ANN model is superior with the highest accuracy being 93.2%, precision being 92.5%, recall being 91.8%, and F1-score being 92.1%, because it is highly effective in predicting mental health for both the non-immigrant and immigrant groups. Multilayer Perceptron (MLP) was nearest to its model upon following with 92.8% accuracy and F1-score of 91.4%, which shows its competitor performance. The XGBoost performance also goes smoothly with ease with the accuracy percentage being 91.0% and F1-score being 89.5%. Random Forest (90.3% accuracy), Gradient Boosting (89.4% accuracy), and SVM (88.7% accuracy) are the other classifiers who are good but still lag behind ANN and MLP for precision, recall, and F1-score. Naive Bayes, Logistic Regression and KNN are the poorest with accuracy of 85.4%, 84.1%, respectively, and recall and F1-scores lower than the best. Naive Bayes is performing extremely badly with accuracy of 79.9% and recall of 74.8%. Overall, the ANN model is the best performing model among the rest and the best tool utilized in predicting the state of mind for immigrant as well as non-immigrant groups.

Table 5. Mental health prediction performance for immigrants vs non-immigrants

Model	Accuracy (%)	Precision (%)	Recall (%)	F1-Score (%)
ANN (Artificial Neural Network)	93.2	92.5	91.8	92.1
Logistic Regression	85.4	83.7	81.2	82.4
Support Vector Machine (SVM)	88.7	87.1	85.3	86.2
Random Forest	90.3	89.0	87.7	88.3
K-Nearest Neighbors (KNN)	84.1	82.3	79.6	80.9
Decision Tree	81.5	79.8	76.1	77.9
Naive Bayes	79.9	78.1	74.8	76.4
Gradient Boosting	89.4	88.2	86.5	87.3
XGBoost	91.0	90.1	88.9	89.5
Multilayer Perceptron (MLP)	92.8	91.9	91.0	91.4

CONCLUSION

The research emphasizes the need for healthcare literacy and access in the impact on mental health, immunization, and oral health of the migrants. PCA for feature extraction reported that there existed hidden patterns in healthcare complications that could be leveraged well in prediction modeling using an ANN. Predictions are very accurate, confirming the efficacy of the features attained and the capability of the ANN in identifying advanced relationships between data. Outcomes confirm the essence of intervention measures to facilitate improved healthcare communication and access, thereby improved health outcomes among migrants. The research offers a model for the use of advanced analytical techniques in responding to major public health challenges.

REFERENCES

Armstrong-Mensah, E., Mitha, N., & McNair, A. (2023). The mental health of Syrian refugees in the United States: Examining critical risk factors and major barriers to mental health care access. *International Journal of Translational Medical Research and Public Health, 7*(1), 18. DOI: 10.21106/ijtmrph.431

Atrooz, F., Aljararwah, S. M., Chen, T. A., Khabour, O. F., & Salim, S. (2023). Understanding Mental Health Status of Syrian Refugee and Jordanian Women: Novel Insights from a Comparative Study. *International Journal of Environmental Research and Public Health, 20*(4), 2976. DOI: 10.3390/ijerph20042976 PMID: 36833672

Atrooz, F., Chen, T. A., Biekman, B., Alrousan, G., Bick, J., & Salim, S. (2022). Displacement and Isolation: Insights from a Mental Stress Survey of Syrian Refugees in Houston, Texas, USA. *International Journal of Environmental Research and Public Health, 19*(5), 2547. DOI: 10.3390/ijerph19052547 PMID: 35270240

Bawadi, H., Al-Hamdan, Z., Khader, Y., & Aldalaykeh, M. (2022). Barriers to the Use of Mental Health Services by Syrian Refugees in Jordan: A Qualitative Study. *Eastern Mediterranean Health Journal, 28*(3), 197–203. DOI: 10.26719/emhj.22.030 PMID: 35394051

Beiser, M., & Hou, F. (2001). Language Acquisition, Unemployment, and Depressive Disorder among Southeast Asian Refugees: A 10-Year Study. *Social Science & Medicine, 53*(10), 1321–1334. DOI: 10.1016/S0277-9536(00)00412-3 PMID: 11676403

Correa-Velez, I., Green, A., Murray, K., Schweitzer, R. D., Vromans, L., Lenette, C., & Brough, M. (2020). Social Context Matters: Predictors of Quality of Life among Recently Arrived Refugee Women-at-Risk Living in Australia. *Journal of Immigrant & Refugee Studies, 18*(4), 498–514. DOI: 10.1080/15562948.2020.1734893

Douai, A., Bastug, M. F., & Akca, D. (2022). Framing Syrian refugees: US local news and the politics of immigration. *The International Communication Gazette, 84*(2), 93–112.

El-Rufaie, O., & Absood, G. H. (1994). Validity Study of the Self-Reporting Questionnaire (SRQ-20) in Primary Health Care in the United Arab Emirates. *International Journal of Methods in Psychiatric Research, 4*, 45–53.

Elhabiby, M. M., Radwan, D. N., Okasha, T. A., & El-Desouky, E. D. (2015). Psychiatric Disorders among a Sample of Internally Displaced Persons in South Darfur. *The International Journal of Social Psychiatry, 61*(4), 358–362. DOI: 10.1177/0020764014547061 PMID: 25145868

Elshamy, F., Hamadeh, A., Billings, J., & Alyafei, A. (2023). Mental Illness and Help-Seeking Behaviours among Middle Eastern Cultures: A Systematic Review and Meta-Synthesis of Qualitative Data. *PLoS One, 18*(10), e0238109. DOI: 10.1371/journal.pone.0293525 PMID: 37883515

Ferris, E., Kirişci, K., & Shaikh, S. 2013. Syrian crisis: Massive displacement, dire needs and a shortage of solutions. Brookings. Available online: https://www.brookings.edu/articles/syrian-crisis-massive-displacement-dire-needs-and-a-shortage-of-solutions/ (accessed on 1 January 2024).

Ghosn, F., Chu, T. S., Simon, M., Braithwaite, A., Frith, M., & Jandali, J. (2021). The Journey Home: Violence, Anchoring, and Refugee Decisions to Return. *The American Political Science Review, 115*(3), 982–998. DOI: 10.1017/S0003055421000344

Gorst-Unsworth, C., & Goldenberg, E. (1998). Psychological Sequelae of Torture and Organized Violence Suffered by Refugees from Iraq: Trauma-Related Factors Compared with Social Factors in Exile. *The British Journal of Psychiatry, 172*, 90–94. DOI: 10.1192/bjp.172.1.90 PMID: 9534840

Haque, S., & Malebranche, M. (2020). Impact of Culture on Refugee Women's Conceptualization and Experience of Postpartum Depression in High-Income Countries of Resettlement: A Scoping Review. *PLoS One, 15*(9), e0238109. DOI: 10.1371/journal.pone.0238109 PMID: 32870929

Javanbakht, A., Amirsadri, A., Abu Suhaiban, H., Alsaud, M. I., Alobaidi, Z., Rawi, Z., & Arfken, C. L. (2019). Prevalence of Possible Mental Disorders in Syrian Refugees Resettling in the United States Screened at Primary Care. *Journal of Immigrant and Minority Health, 21*(3), 664–667. DOI: 10.1007/s10903-018-0797-3 PMID: 30066059

Kessler, R. C., Sonnega, A., Bromet, E., Hughes, M., & Nelson, C. B. (1995). Posttraumatic Stress Disorder in the National Comorbidity Survey. *Archives of General Psychiatry, 52*(12), 1048–1060. DOI: 10.1001/archpsyc.1995.03950240066012 PMID: 7492257

Laban, C. J., Komproe, I. H., Gernaat, H. B. P. E., & de Jong, J. T. V. M. (2008). The Impact of a Long Asylum Procedure on Quality of Life, Disability, and Physical Health in Iraqi Asylum Seekers in the Netherlands. *Social Psychiatry and Psychiatric Epidemiology, 43*(7), 507–515. DOI: 10.1007/s00127-008-0333-1 PMID: 18560785

Lindencrona, F., Ekblad, S., & Hauff, E. (2008). Mental Health of Recently Resettled Refugees from the Middle East in Sweden: The Impact of Pre-Resettlement Trauma, Resettlement Stress, and Capacity to Handle Stress. *Social Psychiatry and Psychiatric Epidemiology*, *43*(2), 121–131. DOI: 10.1007/s00127-007-0280-2 PMID: 18060523

Mallett, S. (2004). Understanding Home: A Critical Review of the Literature. *The Sociological Review*, *52*(1), 62–89. DOI: 10.1111/j.1467-954X.2004.00442.x

Malm, A., Tinghög, P., Narusyte, J., & Saboonchi, F. (2020). The Refugee Post-Migration Stress Scale (RPMS)—Development and Validation Among Refugees from Syria Recently Resettled in Sweden. *Conflict and Health*, *14*(1), 1–12. DOI: 10.1186/s13031-019-0246-5 PMID: 31921332

Nowrasteh, A., Forrester, A. C., & Blondin, C. (2020). How Mass Immigration Affects Countries with Weak Economic Institutions: A Natural Experiment in Jordan. *The World Bank Economic Review*, *34*(2), 533–559. DOI: 10.1093/wber/lhy032

Pascoe, E. A., & Richman, L. S. (2009). Perceived Discrimination and Health: A Meta-Analytic Review. *Psychological Bulletin*, *135*(4), 531–554. DOI: 10.1037/a0016059 PMID: 19586161

Pearlman, W. (2023). How Homeland Experiences Shape Refugee Belonging: Rethinking Exile, Home, and Integration in the Syrian Case. *The International Migration Review*, *57*(1), 160–186. DOI: 10.1177/01979183221088206

Rosner, R., Hagl, M., Bücheler, L., & Comtesse, H. (2022). Homesickness in Asylum Seekers: The Role of Mental Health and Migration-Related Factors. *Frontiers in Psychiatry*, *13*, 1034370. DOI: 10.3389/fpsyt.2022.1034370 PMID: 36386971

Sá, F. H. L., Waikamp, V., Freitas, L. H. M., & Baeza, F. L. C.Sá. (2022). Mental Health Outcomes in Syrian Refugees: A Systematic Review. *The International Journal of Social Psychiatry*, *68*(5), 933–953. DOI: 10.1177/00207640221099404 PMID: 35657322

Salameh, T. N., Sakarya, S., Acarturk, C., Hall, L. A., Al-Modallal, H., & Jakalat, S. S. (2024). Syrian Refugee Women's Experiences of Barriers to Mental Health Services for Postpartum Depression. *Journal of Advanced Nursing*, •••, 1–11. PMID: 39171776

Syam, H., Venables, E., Sousse, B., Severy, N., Saavedra, L., & Kazour, F. (2019). "With Every Passing Day I Feel Like a Candle, Melting Little by Little". Experiences of Long-Term Displacement Amongst Syrian Refugees in Shatila, Lebanon. *Conflict and Health*, *13*(1), 45. DOI: 10.1186/s13031-019-0228-7 PMID: 31624496

Teodorescu, D.-S., Heir, T., Hauff, E., Wentzel-Larsen, T., & Lien, L. (2012). Mental Health Problems and Post-Migration Stress Among Multi-Traumatized Refugees Attending Outpatient Clinics Upon Resettlement to Norway. *Scandinavian Journal of Psychology, 53*(4), 316–332. DOI: 10.1111/j.1467-9450.2012.00954.x PMID: 22612589

Tinghög, P., Malm, A., Arwidson, C., Sigvardsdotter, E., Lundin, A., & Saboonchi, F. (2017). Prevalence of Mental Ill Health, Traumas, and Postmigration Stress Among Refugees from Syria Resettled in Sweden After 2011: A Population-Based Survey. *BMJ Open, 7*(12), e018899. DOI: 10.1136/bmjopen-2017-018899 PMID: 29289940

Venkatachalam, T., O'Sullivan, S., Platt, D. E., Ammar, W., Hamadeh, R., Riachi, N., Presley, D., Khoury, B., Gauguier, D., Nader, M., Qi, L., & Zalloua, P. (2023). The Impact of Forced Displacement: Trauma, Increased Levels of Inflammation, and Early Presentation of Diabetes in Women Syrian Refugees. *Journal of Public Health (Oxford, England), 45*(3), E437–E446. DOI: 10.1093/pubmed/fdad037 PMID: 37022674

Vromans, L., Schweitzer, R. D., Brough, M., Kobe, M. A., Correa-Velez, I., Farrell, L., Murray, K., Lenette, C., & Sagar, V. (2020). Persistent Psychological Distress in Resettled Refugee Women-at-Risk at One-Year Follow-Up: Contributions of Trauma, Post-Migration Problems, Loss, and Trust. *Transcultural Psychiatry, 58*(2), 157–171. DOI: 10.1177/1363461520965110 PMID: 33115369

Weiss, D. S., Marmar, C. R., Schlenger, W. E., Fairbank, J. A., Jordan, B. K., Hough, R. L., & Kulka, R. A. (1992). The prevalence of lifetime and partial post-traumatic stress disorder in Vietnam theater veterans. *Journal of Traumatic Stress, 5*, 365–376.

Wu, S., Renzaho, A. M. N., Hall, B. J., Shi, L., Ling, L., & Chen, W. (2021). Time-Varying Associations of Pre-Migration and Post-Migration Stressors in Refugees' Mental Health During Resettlement: A Longitudinal Study in Australia. *The Lancet. Psychiatry, 8*(1), 36–47. DOI: 10.1016/S2215-0366(20)30422-3 PMID: 33156999

Yalim, A. C. (2020). The Impacts of Contextual Factors on Psychosocial Wellbeing of Syrian Refugees: Findings from Turkey and the United States. *Journal of Social Service Research, 47*(1), 104–117. DOI: 10.1080/01488376.2020.1717717

Chapter 11
Machine Learning Algorithms for Individuals With Visual Impairments

Mohamed Aslam
https://orcid.org/0000-0003-3930-6229
SASTRA University, India

G. Revathy
https://orcid.org/0000-0002-0691-1687
SASTRA University, India

M. Anuradha
SASTRA University, India

ABSTRACT

A comprehensive work promoting assistive technology that is meant to make life better for those who are visually impaired. They emphasise how machine learning methods and technologies may be used to customize cutting-edge apps to meet this user base's unique requirements. One such example, uses machine learning techniques to help people who are partially blind see better by improving how they see their surroundings. The literature-based approach helps visually impaired people recognise obstacles, improving mobility and safety. More recently, researchers have been working on making detection tools and visual aids that use artificial intelligence and the latest advances in computer vision techniques to help people who are blind or have low vision.

DOI: 10.4018/979-8-3693-9445-8.ch011

I. INTRODUCTION

New technologies have been made available through machine learning algorithms, which have been developed in recent years, to assist the visually impaired to be more independent and secure. It is estimated that many people globally suffer from vision disorders, with two million people and 340 thousand people being counted as blind or partially sighted. Most cases of vision impairment are observed among people aged 65 years and above, with 80% in this bracket, and 60% at 75 years and above. Disease, an accident, injury, or ageing can lead to a person developing one or another type of vision impairment.

Schedule an appointment with an optician to diagnose diabetes, hypertension, and health conditions that may lead to vision impairment, as indicated by Re, D, E, and Rule N.O. (Re & Rule, 2016). In such cases, medical treatment is unavailable, other needs are more urgent, and many patients cannot get proper eye care. According to Roy et al. (Roy et al., 2022), cataracts a leading cause of vision loss glaucoma a condition that damages the optic nerve frequently as a result of high eye pressure macular degeneration a progressive condition that affects the central part of the retina and results in blurred or reduced central vision diabetic retinopathy, which is caused by diabetes damaging the blood vessels in the retina and refractive errors such as near-sightedness, farsightedness, and astigmatism are among the most common causes of vision loss (Roy et al., 2022) (Re & Rule, 2016) (Sulthan et al., 2023) (Kumar et al., 2022).

Existing medical monitoring applications are ineffective for visually impaired or blind patients, which makes it difficult for them to monitor their health easily. The causes of vision loss should be combated, while global access to medical services for people with low vision should be enhanced. Among these improvements are:

1. ***A Strategy for Detecting Obstacles:*** The system allows individuals with vision problems to identify obstacles and navigate without strain easily. Based on the study by Roy et al. (Roy et al., 2022), these machine learning tools assist this system in identifying hazards at the time of occurrence, thus improving mobility and decreasing crash rates.
2. ***Partial Vision Impairment apps:*** Machine learning enables developers to design and create functional applications for all the visually impaired individuals. (Raghavan et al., 2021) study shows how these apps improve the user's environmental awareness through support and feedback. These software applications adapt from the users' feedback through machine learning techniques to enhance the quality of the users' lives.

3. ***Computer Vision Systems:*** As stated by D. E. and Rule N. O. (Re & Rule, 2016), This indicates that artificial intelligence enables the development of new technology to assist people with visual impairments in being independent and navigating their environment freely. Significant success has arisen from computer vision algorithms. Users also find themselves more capable of navigating through devices that use machine learning to interpret the visual data to determine what may hinder them or provide the required information. Artificial intelligence technologies develop new methods to assist the blind population. These technologies utilize artificial intelligence to assist visually impaired individuals in living their everyday lives.

1.1 Training Machine Learning Algorithms to Identify Various Settings and Objects to Guide the Users

According to the research by T. Tanya et al. (Tanya et al., 2023), Deep learning systems require trained algorithms to identify all objects and name them in visual data. The system also considers several attributes of visual data analysis to recognize objects. In their study, Ikram et al. researched Fusion Sight to describe the mechanistic aspects of this object identification system and discuss how it can be improved. This is complemented by the feature fusion network that employs multiple modalities to perform a layered feature extraction to detect objects in dynamic environments with the help of ViT and CNN for image and radar data. This forward-looking system design enhances the application's critical timing performance as it processes information faster. The figures presented in **Table 1** make it possible to understand that the proposed model can contribute to the development of improved object detection technologies in different fields of application. In our case, the following components are required for ML-based object recognition, as shown in **Figure 1**. These are the main models and procedures that constitute the framework.

1. ***Imaging categorization:*** T. Tanya et al., in (Tanya et al., 2023), have pointed out that training algorithms are used in data image classifications. Feature sets are given to models to be learned and to create differentiation between the objects.
2. ***Feature detection Gathering:*** In feature detection, gathering algorithms are trained to identify and extract special features from image files. Special data features assist models in choosing attributes to use when differentiating between different things and spaces. Object recognition is determined by the ability to extract essential object features.
3. ***Pattern Recognition:*** Computational methods, in turn, look for patterns and patterns of change in visual information. It is also essential to notice that different machine learning algorithms can discover different image patterns and object

recognition patterns. Pattern recognition assists us in detecting items better as we develop our knowledge of the environment.

Table 1. Methods for Identifying and Locating Objects

S.No	Algorithm	Techniques	Methodologies
1	Imaging Categorization	Deep Learning Models	Developing focuses on labelling and categorizing pictures.
2	Feature Detection Gathering	Convolutional Neural Networks	Instructions on identifying unique characteristics in pictures
3	Pattern Recognition	Machine Learning Algorithms	Pattern and regularity identification via automation

Figure 1. Elements of a framework for recognizing objects using ML

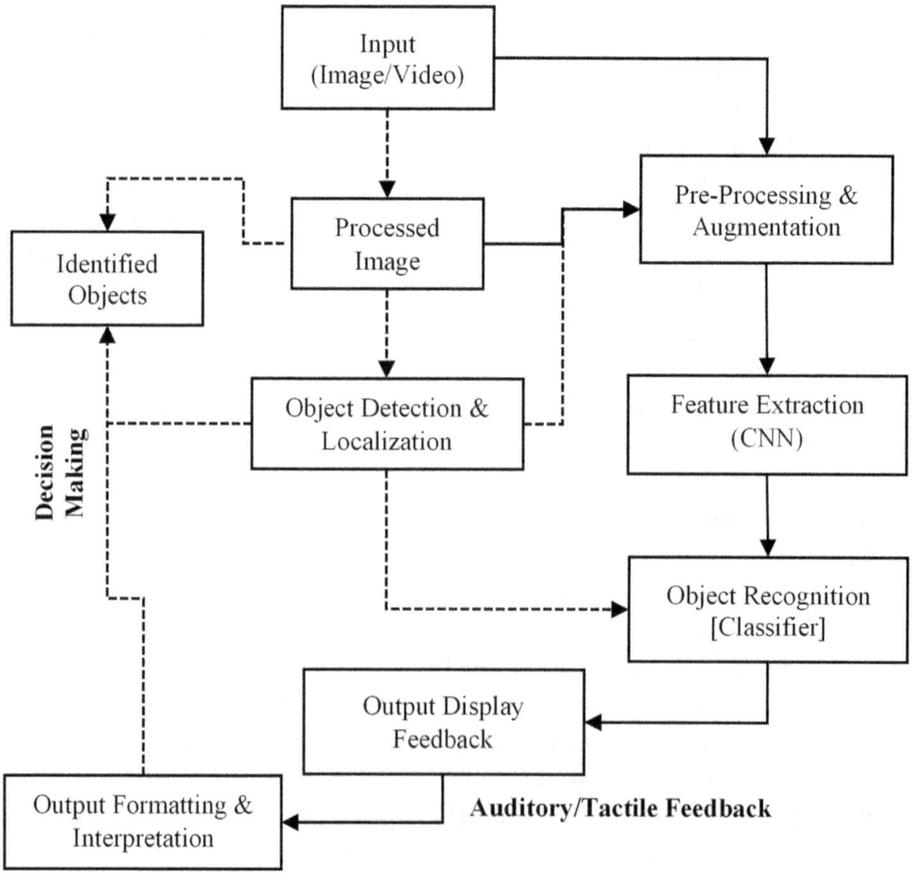

As depicted in **Figure 1**, a machine learning system for object detection comprises the following: Through these techniques, machine learning algorithms can work more efficiently in enhancing the quality of visual aids used by the visually impaired. These methods help visually impaired users improve what they see by naming or identifying objects more easily. These methods allow individuals with vision impairment to move around safely while at the same time gaining mobility using the assistive devices.

1.2 Mechanisms Based on ML to Enhance the Precision of Visual Aids for the Blind

The utilization of machine learning in the development of visual aids enhances the performance of the visual aid appliances in supporting blind individuals. Sulthan, Syed, et al. (Sulthan et al., 2023) pointed out that the use of predictive models makes visual aids more accurate for the blind individual in terms of identifying objects in their current environment today. Computer vision techniques and artificial neural networks enable the computer to recognize and classify objects present in a scene. Figure **2**: Precision of the visual aid for individuals with visual impairments. Kumar et al. (Kumar et al., 2022) proposed integrating those technologies with smart navigation systems and AI-based voice assistants for disabled persons.

Algorithms are as follows:

- ***Deep Learning:*** CNN-based deep learning systems automatically generate an organized data set from the visual inputs through learning in the first instance. This procedure is more effective when it is used to identify physical objects.
- ***Image Processing:*** We use image preprocessing tools to preprocess visual data before feeding it to the machine learning process. These tools enhance the features and make the visual data more suitable for the machine learning algorithms.

Specific features:

- ***Real-time object recognition:*** One can distinguish that machine learning techniques identify objects in the environment in real-time, with a response to users.
- ***Quick and Accurate Response:*** Deep learning technology quickly analyzes and interprets visual data to give environmental reactions.

Approach:

- ***Gathering and Classifying Datasets:*** The machine learning system employs photos from labelled apt datasets to evaluate its object detection ability.
- ***Training Models:*** Our datasets aid in training neural networks, especially deep learning algorithms, using supervised training.
- ***Integration into assistive technologies:*** During training, the models become accustomed to identifying the parameters that characterize various objects.
- ***Real-Time Processing:*** They are used in assistive technologies that allow the development of virtual AI navigators and other applications.
- ***System Feedback:*** Another system element is the use of trained patterns to analyze live input data and identify objects in the surrounding space. The main feature of the system is to provide audio feedback to users with vision impairment to guide them to potential obstacles or objects.

Figure 2. Enhance the precision of the visual aid for individuals with visual impairments.

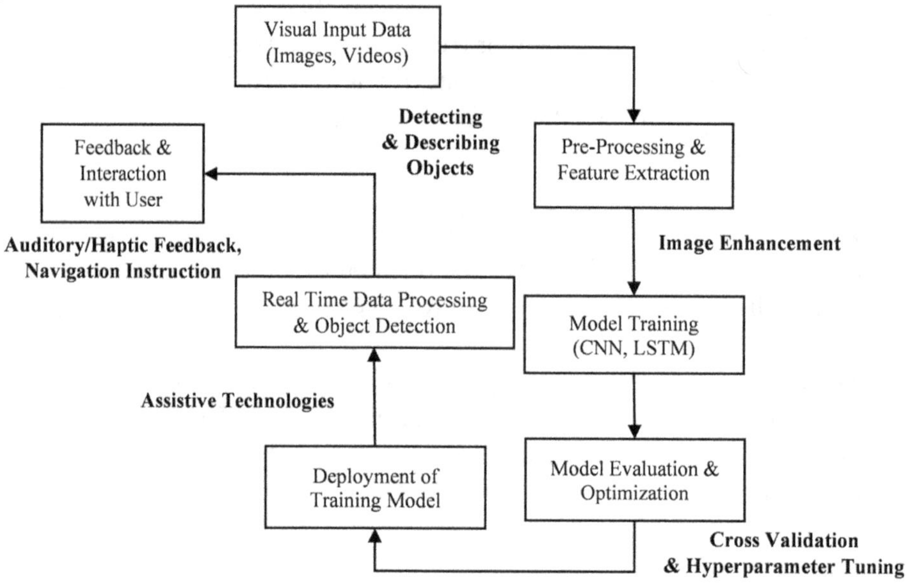

Table 2. Machine Learning enhances the functionality of the visual support tools for the visually disabled users

Sl.No	Algorithm/Methodology	Features
1	Deep Learning (e.g., CNNs)	• Learn automatically how to describe things in a hierarchy from visual data • Exceptionally beneficial for tasks involving object recognition
2	Image Processing	• Perform preliminary processing on the visual data • Identify pertinent features
3	Real-Time Object Recognition	• Instant response on identified items in the surrounding areas
4	Quick and Accurate response	• Rapid computing for immediate and exact feedback from the surroundings
5	Gathering and Classifying Datasets	• Collect appropriate data sets for development and validation activities • Assign labels to collections to train machine learning models in identifying similarities.
6	Training Models	• Utilize supervised learning approaches to train machine learning ensembles. • Acquire the ability to recognise patterns and characteristics belonging to items.
7	Integration into assistive technologies	• Examples of essential components include applications such as virtual artificial intelligence assistants and navigation systems.
8	Real-Time Processing	• Real-time processing of visual input for object identification.
9	System Feedback	• Configure navigational systems with audio signals or tactile user interfaces.

Raghavan et al and T. Tanya et al (Tanya et al., 2023) also articulate the capability of machine learning in ensuring that visual aids enhance the protection of the visually impaired. Table 2: Machine Learning enhances the functionality of the visual support tools for the visually disabled users. Such technological innovations can lead to even more efficient and accurate solutions for the people who are blind. This, together with the need to constantly improve the existing systems, increase data storage capacity, and use the information received from users.

1.3 Barriers to Employing Machine Learning Algorithms for Blind and Sight-Impaired People

Using machine learning systems to assist persons with little sight presents challenges. As L. Ruotsalainen et al. (Ruotsalainen et al., 2022) state, modifying model infrastructure is challenging because errors in system analysis data and availability datasets negatively impact real-world performance. Model capacity training needs datasets that capture all possible scenarios that people with visual impairment face.

To achieve this, Gupta, K et al. developed a deep learning system that converts captured images into voice and assesses the performance. This is due to the fact that, in general, the system is designed to be easier to use and more accessible for visually impaired people. Convolutional neural networks remain the most popular deep learning method for object recognition and image classification tasks. Video captioning technology has gone through stages of development throughout history. In the early days, researchers constructed features and designed systems that created content based on templates. These methods were ineffective because they could not deal with the dynamic or fuzzy nature of picture data in real-life situations. Recently, there have been tremendous advancements in deep learning, which have greatly enhanced the functioning of image captioning.

1. ***Overfitting:*** Using big data to train models for people with disabilities takes a long time and results in inferior models when tested on new data (Ruotsalainen et al., 2022). A model trained to perfection on the given dataset will not yield good results on new images since it does not generalize well.
2. ***Improved Image Comprehension:*** Machine learning systems face a fundamental issue when interpreting images. S. S. Prasantha (Prasantha, 2023) mentioned that to improve accuracy, machine learning algorithms must consider multiple accessibility issues and search for various images in reply to the query. The models must be adaptive enough to handle different accessibility elements and provide correct results.
3. **Lack of Real-World Deployment:** However, in practice, current VQA systems recognize that VQA works differently than the models envisaged. Implementing such systems in real-life contexts still poses a challenge. Today, there is a need to enhance the current prediction algorithms to make them function optimally for use by the visually impaired.

Methods employed to tackle these issues in several practical ways:

- ***Exploring Data Augmentation:*** We use data augmentation to generate several training datasets to avoid overtraining. Fine-tuning improves performance in large datasets by transferring knowledge from big dataset domains to trained models.
- ***Transfer Learning:*** Various research groups continue developing better model systems to accommodate the needs of visually impaired individuals while making the datasets useful for all.
- ***Architectural Enhancements:*** The disparity between computer vision and accessibility datasets poses a problem to real-world applicability when developing a machine learning model for visual assistance with blind users (Prasantha, 2023).

Table 3. Machine Learning assists the vision challenges that people experience.

Sl.No	Limitation	Methodologies
1	Model Accuracy Discrepancies	- Enhancement of Data: Produces variants of training data to increase their visibility. - Transfer Learning: Applying modifications to previously trained models on targeted datasets - Model architectures: May be improved for better alignment via architectural improvements.
2	Overfitting Effect	- Data Enhancement: Strategies for Producing a Wide Range of Training Data through Data Augmentation - Regularization Techniques: Using strategies for regularization involves reducing patterns that are too complicated to avoid overfitting.
3	Improved Image comprehension	- Data Enhancement: Increasing the adaptability of models by exposing them to a variety of scenarios - Constant Learning: To increase precision, update classifiers with fresh data.
4	Lack of Real-World Deployment	- Adaptive Learning: Utilizing pre-trained models for expedited deployment via adaptive learning. - Feedback Integration: Refining models with user input, integrating feedback from users - Testing Models: Validating models via actual testing in a variety of settings

Table 3. It also helps in solving vision problems, which humans occasionally encounter. Researchers strive to enhance datasets as well as the algorithms and include user feedback to make the system of machine learning better. These drawbacks indicate why further research is needed to help cover the gap between machine recognition and historical data accessibility, enhance the models, and create utility machine learning systems to support visuals.

1.4 Integrating ML Techniques to Vision Blind Assistive Technology

Chinofunga, Shakemore. A contemporary work (Chinofunga, 2023) developed a voice-controlled smartphone assistant based on natural language processing techniques. In today's work, several approaches to integrating machine learning into assistive technologies that aid those with visual impairment are presented. These algorithms will be compatible with today's assistive technology systems. Artificial solutions for blind individuals integrate Machine Learning algorithms that utilize enhancements in computer vision, speech recognition, audio, and environmental analysis systems.

1. ***Navigating Mechanisms:*** Reinforcement learning and deep learning methods assist visually impaired individuals in safely navigating their environment. Mathematical algorithms process real-world inputs from mobile phones or

handheld cameras to detect obstacles and markers and relay audio and touch feedback to the users.

2. ***Object Recognition Systems:*** Object recognition systems use vision algorithms like CNNs to identify obstacles that people with visual impairments find hard to see. These systems analyze recorded videos to sort objects into categories and inform the user about specifics of the environment using sound or touch.
3. ***Systems for Recognizing Faces:*** Facial recognition has become an essential tool in society, and blind people can also benefit from the technology by identifying individuals through machine learning techniques. These systems assist their users in identifying those people in their vicinity by comparing the facial images taken by the cameras with the profiles in the database.

The study by Tapu, R. et al. (Tapu et al., 2020) reveals that deep learning systems can identify objects quickly and accurately, which helps design affordable solutions for vision-impaired individuals. S. Durgadevi et al. (Durgadevi et al., 2020) developed a few intelligent machine learning systems to help visually impaired individuals navigate their daily activities. This study showcases the capability of incorporating machine learning algorithms into assistive technologies. Typically, machine learning integration into systems follows these standard steps:

- ***Model Development and Training:*** They are embedded in the application programs and assistive technology tools to analyze real-time data and provide relevant information to the users.
- ***Model Deployment:*** Machine learning algorithms can be incorporated into helpful technologies to improve safety and independence for visually impaired individuals via information and navigation. The above advancements show how machine learning algorithms can improve assistive technology to enable visually impaired individuals to live better lives.
- ***Collecting Data:*** Collecting relevant sensory data in images, audio, and other formats and categorizing them in appropriate bins for analysis. When data is ready, machine learning models learn using reinforcement learning techniques to act in environments, and supervised learning, they can identify objects.

Some of the attributes of these ML systems include:

- ***Deep Learning:*** CNNs produce automatic feature hierarchies that assist machines in recognizing and detecting objects that are present in an image.
- ***Reinforcement Learning:*** The Navigation Assistance for the Visually Impaired through Reinforcement Learning tools can be made to improve

its performance through analysis of user data and subsequent update of strategies.

- ***Speech Recognition:*** Speech Recognition converts spoken words into text form, enabling visually impaired persons to operate assistive technology through voice commands.
- ***Audio Scene Analysis:*** Audio Scene Analysis employs signal processing to determine the location of objects in the scene, their movements, and anything that might be in the way.

Figure 3. algorithms will be integrated into existing assistive technologies.

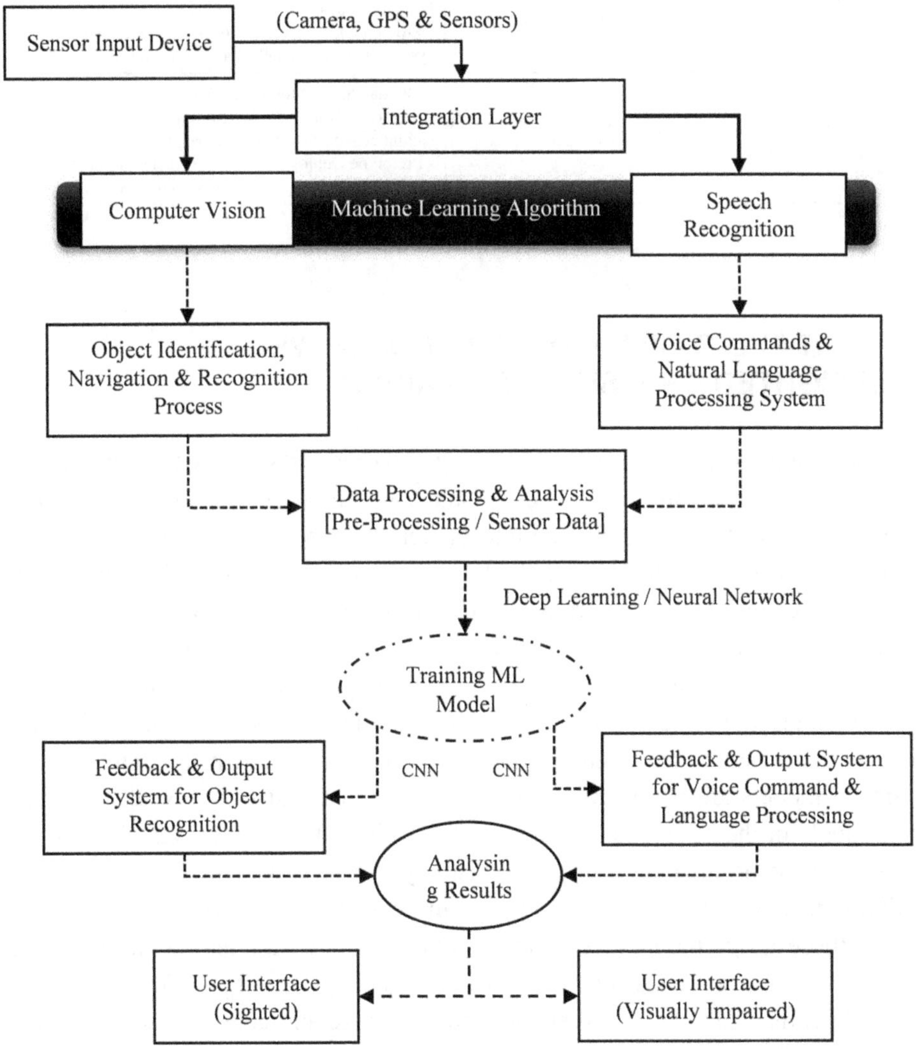

Table 4. Implementation of ML Algorithms in Visually Assistive Technologies

S.No	Application	ML Algorithms	Features
1	Navigation Systems	Reinforcement Learning	- Guidance in real time via the tactile or audio output
2	Deep Learning (CNNs)	Obstacle identification and landmark recognition	- Prototype Training Involves: Using deep learning and reinforcement learning methods, training on datasets that are labelled - Prototype deployment Process: Using maps with models that have been trained
3	Object Recognition Systems	Deep Learning (CNNs)	- Tactile input or audio explanations of items - Prototype Training Involves: Using tagged image data, CNNs are trained. - Prototype Deployment Process: Using learned models in systems that perform the detection of objects
4	Face Recognition Systems	ML (various techniques)	- Persons in the area are being identified - Prototype Training Involves: Facial attribute evaluation based on previously recorded user profiles - Prototype Deployment Process: Facial recognition systems that use models that have been trained
5	Smart Machine Learning Systems	Various ML Techniques	- Regular support for those who are visually impaired - Prototype Training Involves: Applying the proper techniques to train models on data that is labelled - Prototype Deployment Process: Including trained models with intelligent assistive technologies

II. IDENTIFYING OBJECTS FOR PEOPLE WHO ARE SIGHTED OR VISUALLY IMPAIRED

The system identifies objects for visually impaired and visually challenged users. Various methods are used to help visually impaired people identify objects in their environment, ranging from simple gadgets to more complicated technological tools. Joshi, R.C. et al (Joshi et al., 2020a) recommended that deep learning should be trained using visual data from visually impaired people. Our system requires data set creation, which should be based on user group requirements and vision deficiencies. These image sets are helpful to blind people or those with low vision. They include moving around in cities, recognizing objects, reading, writing, and interpreting gestures and signs in our environment, as shown in **Figure4**. Deep learning models are better trained with pictures that assist the visually impaired individuals in the activities they undertake in their day-to-day lives. The models better understand the circumstances a visually impaired person encounters during training, making the visual assistive tools more accurate and efficient.

For those designing assistive technologies for visually impaired people, it becomes necessary to train deep learning models with real people's images. Computer vision algorithms that Yang et al. investigate can recognize and detect objects' locations

(Yang et al., 2020). These computers help machines understand visual data obtained from images or videos.

1. **Identification of Objects:** Object recognition models identify and classify objects in images and videos. Our algorithms identify and locate specific traits and designs of various items. There are various types of sophisticated technologies and aids available that assist individuals with vision impairment in identifying objects. Typical approaches include the following:
   - ***Feature-based methods:*** Feature-based systems utilize fixed shapes or forms to identify the object's significant characteristics, such as edges and texture patterns.
   - ***Deep learning-based approach:*** Convolutional neural networks are the most widely used deep learning algorithm for object recognition. These simulations generate attribute sets for defining objects based on their identification thresholds.
2. **Object Recognition:** This system identifies and clusters objects from image and video content. Our systems are trained to recognize specific patterns and features of objects for classification. Common techniques include:
   - ***Feature-based methods:*** This process first identifies specific features of objects, like the corners and texture, and then aligns them with reference elements.
   - ***Deep learning-based methods:*** CNN is one of the deep learning systems that finds applicability in recognizing objects within visual data. In deep learning, image understanding is developed through learning multiple levels of an image to detect objects without needing the assistance of the model.
3. **Object Detection:** In Object Detection, the locations of objects within an image are identified, and the system outlines the object using boundary boxes. It detects multiple objects in an image and calculates the exact location of the objects. Common methods include:
   - **Sliding Window Technique:** Frames of different sizes and positions scan images to look for objects in various scales.
   - ***Region-based approaches:*** The region-based approach splits an image into segments to determine if each segment has an object of interest in it. R-CNN and Fast R-CNN are region-based CNNs as part of the Faster R-CNN.
   - ***Single-Shot Methods:*** Single-shot methods are a simpler and quicker approach to yielding the object boundaries and class names with the least number of passes through the neural networks. YOLO and SSD are two common single-shot detector models in object detection.

The basics of computer vision assist in finding objects in medical imaging technology, form virtual worlds and even self-driving cars. Robot perception and interaction with the visual environment are instrumental in enhancing advancements in various fields through these approaches. This illustrates how the sighted and the blind people can differentiate objects. They propose a grasp for object recognition without vision in their study on ontology-based detection by Boruah et al. (Boruah et al., 2023).

Some datasets are advantageous in the following ways to deep learning models:

Enhanced Relevance: The visually impaired challenges in the training sets of visual input assist models in achieving higher performance when clients come across practical situations.

Optimized Performance: Training improves the models' identification of visual data, and this enhanced performance is witnessed when the models are used to detect items, read maps, and follow directions.

Customized Solution: Such specific training sets allow us to develop products designed only for people with impaired vision. These systems improve and become increasingly useful to users.

Figure 4. Object identification for both sighted and visually

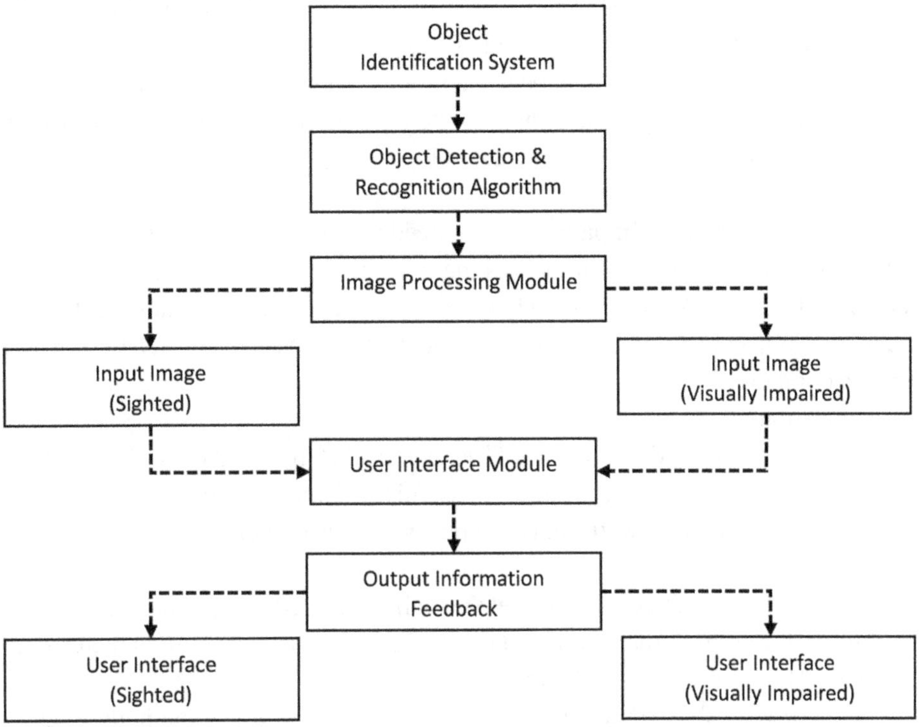

Techniques:

Joshi R.C et al. (Joshi et al., 2020a) proposed deep learning models that work with visual information, specially designed for blind people. We create specific databases that pertain to this group's unique requirements and challenges. The datasets include images that help those with vision difficulties move around, identify objects, read texts, and interpret facial expressions and signals within the environment. When training deep learning models to recognize the visual information, it is more effective to feed the models with images of what a blind person would perceive. It makes the models more effectively identify when the visuals are needed for people with blindness or low vision. Training deep learning models on these datasets benefits our work in the following ways.

- ***Increased Relevance*:** The dataset images are real-life scenarios that the models must meet to understand the challenges the visually impaired user faces daily.

- ***Improved Performance*:** Training models on appropriate visual data enables them to analyze information more accurately, recognize scenes and objects, and direct the user to the required destinations.
- ***Tailored Solutions***: This is because training sets for visually impaired users aim at developing products that meet their particular needs and improve the ease of using visual assistance technology.

Image data of visually impaired users is needed to develop deep learning models and assistive technology that improves the lives of those affected. In their article, Yang et al. (Yang et al., 2020) show how computer vision technologies identify and understand objects captured in an image or video by passing visual information through computers.

Applications:

According to the study by Joshi R.C. et al. (Joshi et al., 2020a), Assistance tools with movement and object recognition are useful for visually disabled people because they can perceive and move through the physical environment.

1. **Portability Aid:** These devices help visually impaired persons navigate society without the help of other people. These devices often have several characteristics, such as but not limited to the following:
    - ***Barrier Identifying objects:*** The system uses other methods, such as ultrasound or infrared light, to perceive the obstacles and informs the users by speaking or through touch.
    - ***Direction Guidelines:*** Direction Guidelines employ sound or touch to direct users along specific pathways and points of interest using GPS or indoor positioning.
    - ***Environment Recognition:*** Environment Recognition technology assesses the ground conditions and modifies walking patterns to determine the proper path.
2. **Object Recognition:** These devices assist blind people to become more self-sufficient since they offer accurate directions to the user on how to move around safely. With the help of sensors and computer vision, accompanied by voice recognition, this equipment makes the process easier. It helps to enhance the quality of the lives of people with impaired vision. To assist the visually impaired in getting real-time information about objects they observe, Hussan and team examined CNN. The procedure consists of:
    1. ***Live Instantaneous Object Finding:*** Real-time object finding in live video employs deep learning-based object recognition algorithms. Two primary methods, SSD and YOLO, can be used to detect objects within an image or video stream with reasonable efficiency.

2. ***Deep Learning for Recognition:*** Recognition in Deep Learning works by using deep learning techniques to identify the properties of items that have been detected. Such algorithms demonstrate high recognition accuracy because they have learned about it from numerous examples of ordinary items.
3. ***Tactile or audio Comments:*** The system can provide the users with tactile or audio feedback when it locates an object. When an object is identified, the device plays out distinct vibration patterns and tells what it found using a fake voice.

It is technology that identifies objects in a person's surroundings and informs that person of what it sees. These devices operate based on various technologies such as

- ***Image Recognition:*** Taking a picture of an object using a smartphone or camera is an example of image recognition technology. The system can identify objects with the help of a series of computer vision algorithms, which are then relayed using audio or haptic feedback.
- ***RFID or NFC Tags:*** Information can be read from RFID or NFC tags affixed on the objects that small devices and mobile phones can capture. These tags convey the details of the product or location to the user.
- ***Voice helpers:*** Siri and Google Assistant allow users to ask questions about items and get an audio description of their device's current settings.

Martínez-Martín et al. (Martínez-Martín & del Pobil, 2017) discussed that robotic aids using smart vision technology can assist people in navigating their environment better. **Table 5**: The system employs Computer Vision and Tactile Vision to identify the objects precisely. Assist different users and robots in performing their daily tasks. The robot interacts with voice, hand gestures, and basic controls to pick up objects, search for items, or navigate around areas. These systems use modern technology, including deep learning and computer vision, to assist people with disability, especially blind people, and to enhance the functionality of robots. These goals are to improve independence and usability in various contexts by providing timely information on the environment, which can be tailored for multiple scenarios.

The essential elements consist of:

- ***Built-in Cameras and Detectors:*** Vision data for robots is obtained through the sensors and cameras incorporated into their structure. The system requires an RGB camera, depth sensor, and other detection techniques.
- ***Procedures for Object Being Noticed:*** The visual perception model is based on computer vision and involves deep learning techniques for object recogni-

tion. These algorithms help the robot perceive its environment and objects or obstacles.

- ***Recognition and Selection*:** Once the system identifies an object, it runs a set of recognition programs to determine its type. This information enables the robot to navigate its environment safely and manipulate objects to meet human commands.

Challenges:

Yang et al. (Yang et al., 2020) describe how the operation of computer vision systems leads to unfair preferences. The algorithms simply mirror the prejudice that is engraved in their data sets. The measures to reduce bias include collecting data from different sources, developing an impartial decision-making model, and using transparent design processes throughout development. The checks and ethical considerations assist in identifying any biases where AI and computer vision are used in different applications. Ensuring that the technologies we develop are fair and inclusive requires us to tackle bias in artificial intelligence and computer vision systems. There are specific steps to create better algorithms to work with various training data and set ethical standards for fair decision-making.

Blind and visually impaired individuals can utilize touch-based vision substitute tools. Haptic feedback is received when using these devices to identify patterns of objects or words. This communication method helps people with poor vision understand words and objects much better. These gadgets allow users to navigate through the visual environment by touch, hence enhancing users' independence and mobility.

Table 5. The objects detected are accurate with the help of Computer Vision and Tactile Vision.

Sl.No	Issue	Strategies
1	Computer Vision Containing Bias	- Assembling a variety of datasets - The implementation of techniques that show being equitable - Developing authenticity and transparency throughout the entire creation approach - Probes and ethical issues should be carried out continuously. - Increasing the variety of data while simultaneously enhancing algorithms
2	Techniques for the Replacement of Touching Sense (Tactile Vision)	- The process of transforming data that is visual into signals that are haptic or audible - Enabling individuals to sense patterns in words or things via their senses of touch or hearing - The improvement of the ability of individuals with visual challenges to recognize words and objects - Leading to more freedom and ease of access

2.1 A Visually Impaired Persons Employ the Following Object Identification Techniques

In the realm of computer vision and deep learning:

As stated by V. K. Paswan and A. Choudhary in their paper (Paswan & Choudhary, 2022), objects in building interiors can be identified with the help of RGB-D surveillance cameras and deep learning approaches, which utilize the depth data of cameras and learning techniques to identify objects inside buildings. The RGB-D cameras perform better than the RGB cameras because the former has an added dimensionality of depth. CNNs can perform deep learning with multiple inputs to identify objects by looking at the image and its position. This technique makes the detection of objects in interior spaces more accurate and consistent to improve the functionality of the environmental understanding systems and navigation aid devices.

L. Ruotsalainen and colleagues proposed that (Ruotsalainen et al., 2022) it is necessary to design computer vision systems to assist people with disabilities so that they can track objects more easily. The algorithms discussed are mainly concerned with the systems that can execute tasks in real-time mode and are stable and easy to access. They develop systems to generate sound or picture feedback to inform users of the identified objects. For the system to be effective for people with different levels of vision loss, it should be easy to use and efficient. Specifically, in the case of outdoor environments, computer vision systems assist the visually impaired in identifying objects and increasing their level of independence.

Create Assistive Technologies:

In their work, Edward and Shirley (Jothimani, 2016b) described assistive devices that would identify and describe the environment based on computer vision techniques to enhance the daily experiences of individuals with visual impairment. These devices mainly consist of a camera through which they capture images of their surroundings. The system uses image recognition software to determine what objects are evident in the captured pictures. It informs blind and visually impaired people about their surrounding environment by providing touch and audio sensations, allowing them to avoid any possible obstacles.

P. Azmat Shah et al. (Shah et al., 2023) states that camera-based devices also capture visual information to inform the user about the location of objects for the visually impaired. The system uses techniques from the field of computer vision to detect important objects from a photo collection. It employs audio signals and vibrations to point at the identified objects for people who are unable to see.

Voice aid and Navigation:

Kulkarni et al (Kulkarni et al., 2023) developed an application that uses voice for navigation assistance for the visually impaired. Interactive configuration allows for the use of visual and audio detection devices simultaneously. The combination of object recognition and navigation systems makes it possible for visually impaired people to begin performing a task through standard verbal commands. Camera images are passed through computer vision algorithms and once the work is done the system gives synthesized speech output to the user. Voice feedback offers users descriptions of the environment while in motion and can be especially useful in ensuring the safety of solo travelers.

However, R. Kadhim (Kadhim & Oleiwi, 2022) developed co-bot sticks that use the RGB-D cameras for object recognition and to guide the visually impaired individuals on safe paths. In addition to using robotic assistive technology, it employed RGB-D cameras for safe movement and navigation directions. These sticks are capable of detecting the presence of an obstacle and its position through the use of computer vision techniques. It determines depth from RGB-D cameras to identify obstacles that may be in the way of the users of the system. The walking sticks make it easier for the visually impaired users to reach their destination without walking through dangerous items. This technology can benefit people with vision impaired because it allows them to navigate through the space with ease, while offering them assistance at the same time. They become more empowered by regulating both the inside and outside environments and feel more confident.

2.2 Ontologies and Grasp-Based Object Visual Representation

Boruah et al (Boruah et al., 2023) explore the ways in which object recognition is possible when vision is excluded in their study on visionless object recognition through ontology. Ontologies help the system to reason with objects with grasps, and it is able to perceive objects and manipulate them without necessarily using vision. By touch when robots make contact with objects grasp is used to identify what an object is and its characteristics. It delivers specific data about objects which are necessary for researchers. This device operates fine in conditions of low illumination and in the presence of many individuals and can assist people who have certain issues with their sight. The software employs object recognition and grasp methods to expand its functionality in numerous applications like in machine learning and human computer interface via ontologies.

1. ***Ontologies:*** Ontologies can be used to model certain fields as these provide knowledge about the objects and their relations in a given domain. In visionless object detection systems ontologies assist in building the knowledge structure that provides an understanding of what objects are and how they exist in the world.

1.1 ***Ontologies-Based Object Visual representation:*** An ontology system saves information on objects concerning their form, dimensions, appearance, capacity, mode of operation and characteristics. This structured representation has the advantage of allowing the system to detect objects and their features without using vision.

2. ***Grasp:*** The second ability is grasping, which means that we are able to grasp objects properly and by movements. Tactile sensing is done through Robotic arms and other physical handling methods while the shape is recognized by performing Grip actions. Our combined solution allows the system to identify and respond to objects during periods when information from the camera is unavailable.

   2.2 ***Grasp-Based Interaction:*** It deals with objects by touch and not by the information that is visible to the eye. Robots are able to identify objects using touch sensors on their arms that are programmed with descriptions within their database.

III. DETECTING OBJECTS FACILITATES ROUTINE TASKS FOR PEOPLE WITH DISABILITIES

The object detection tools enhance daily activities carried out by individuals with disabilities. Object recognition systems are very useful to the blind and low vision people in that they make it easier for them to identify their surroundings as well as objects within them. **Figure5** Using object recognition in everyday life. These systems are said to achieve the intended goal by various detection techniques.

- ***Guidance with Everyday Tasks:*** The study done by Occelli V et al. (Occelli et al., 2016), stated that object recognition technology enables visually impaired people to navigate through various day-to-day activities by sensing and identifying objects.
- ***Enhancement of Mobility and Independence:*** The systems that R. Joshi et al designed assist the disabled to navigate more freely and without needing someone else's assistance. These systems assist disabled persons with mobility by allowing them to use computer vision and deep learning algorithms within their homes and outside. R. Joshi et al (Joshi et al., 2020b) and Takaki et al (Nishio & Nishida, 2018).
- ***Real-Time Detection:*** It can also recognize objects in digital images and video feeds in a few seconds. Such a fast delivery of environmental information allows individuals with visual impairment to receive feedback on the envi-

ronment within an instant, as indicated by Occelli V et al. (Occelli et al., 2016).

- ***Voice assist and Navigation***: Voice-based guidance and navigation help make people's lives easier in their daily activities, as discovered by P. Manikandan et al (Manikandan et al., 2022) and Shifa Shaikh et al (Shaikh et al., 2020a). The data also revealed that object detection systems assist the visually impaired effectively to move in their environment and increase their chances of detecting objects and performing activities without causing harm to themselves.

Figure 5. Using object recognition in everyday life

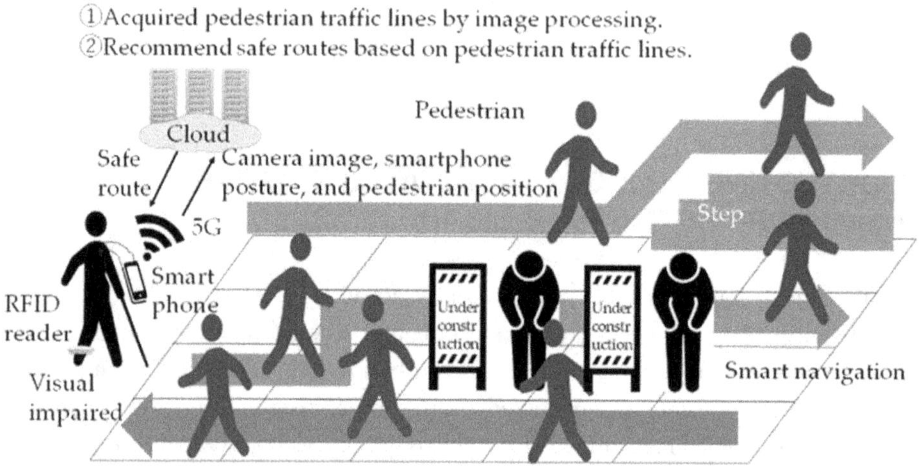

IV. LIMITATIONS OF OBJECT RECOGNITION SYSTEMS FOR VISUALLY IMPAIRED PEOPLE

Object recognition systems still require advancement to provide optimal performance for visually impaired users. The studies show that the tools to detect items are effective for the visually impaired but have certain limitations. Further improvements of our research work should be directed toward enhancing the possibilities of the object recognition systems used in the context of visually impaired people, making them more accurate and reliable, as well as implementing more user-friendly interfaces. Real-world system performance requires research on better algorithms, new sensor technologies, and environmental solutions that also empower the users.

This is because current research has indicated that object recognition systems for visually impaired users have certain performance constraints. This is the reason why object recognition systems for the visually impaired are less efficient. These restrictions consist of several important attributes:

1. ***Limitations of Precision and Dependability*****:** In their study, Alzahrani, Nada, and Heyam H. Al-Baity (Alzahrani & Al-Baity, 2023) also raised issues of precision and dependability in their study. They should continue to conduct research based on these models. This is because performance can be affected both by the complexity in the environment and other factors.
2. ***Obstacles Encountered in Practical Settings:*** In real world application, the motion detection systems from Shaikh et al. (Shaikh et al., 2020b) is encountered some problems such as change in quality of the video, objects that are hidden from the view and complex culture backgrounds which affect the performance of the system. The system's actual performance decreases in this case due to these challenges when it does not detect certain conditions.
3. ***Depth of Object Recognition:*** According to Edward and Shirley, it is difficult to locate objects in dynamic, uncontrolled environments for visually impaired users (Jothimani, 2016b). This might be due to the fact that when the scenes get complex with many objects, and the movement is fast, the system may fail to analyze what it observes correctly.
4. ***Evaluation and Integration involves:*** The practical applicability of an object recognition system in everyday life depends on how well it incorporates into the assistive technology and usability. This research has found that the level of user-interface satisfaction, system responsiveness and compatibility with other devices that a user is able to achieve through the use of this technology defines users' satisfaction and adoption of this technology.

V. PARTICULAR SETTINGS & CIRCUMSTANCES MAY FIND OBJECT RECOGNITION SYSTEMS INEFFECTIVE

Some environments pose difficulties for a visually impaired person when using an object recognition system

In certain circumstances, visually impaired individuals get relatively small benefits from object recognition systems. In their publication, the team of Patel et al. gave a comprehensive explanation on machine learning and deep learning systems. The general goal of this research is to increase the end user accessibility of technology by those with either visual or hearing impairment. This paper assesses the effectiveness of SVM Random Forest and MLP models for assisting those with hearing impairments. This paper focuses on how object detection systems such as YOLO, SSD, and Retina Net are useful to the visually impaired individuals. The project examines the extent to which generative AI systems can help a person with visual or hearing impairment. This research project aims to design and implement state of the art technology to accommodate the needs of people with disability. The insights that we can gain from our work can help us design user tools that will improve the existing means of navigation and communication and enable people to live more independent lives. These are some of the considerations that the study takes into account in its design.

1. ***Facing difficulties during real-world instances:*** Real world object recognition has issues because features of object recognition are useless when background elements or even lighting interferes with the view of objects. Environmental factors are causes that affect the ability of the system to identify the objects correctly and this affects the reliability and efficiency of the system.
2. ***Accuracy and dependability:*** In the view of Valipoor and colleagues (Valipoor & De Antonio, 2022), programs that identify objects are reported to be very reliable, but they actually perform below the optimal reliability level. Whenever the system provides ambiguous results, more emphasis should be directed to research and development.
3. ***Object Recognition's Complexity:*** Accurate identification and classification of objects are fundamental problems in recognition systems and are particularly difficult during environmental changes. Occelli V et al. (Occelli et al., 2016) state that they have problems comprehending what they find when objects are closely placed or when the environment varies.
4. ***Implementation and User Experience:*** An object recognition system works in daily life if it integrates assistive tools and designs greater user-experience. Elmannai W, Elleithy K (Elmannai & Elleithy, 2017). If tools are developed that are user friendly and give quick outputs while interacting with the existing devices, then the users will accept the technology more.
5. ***Problems with Low-powered Applications Terminal Devices:*** Takaki et al work (Nishio & Nishida, 2018) shows that object recognition systems require significant processing that is beyond the capacity of the helper devices used by the visually impaired. For such an application to be of benefit to users with varied dexterity and device constraints, we are in the process of designing systems that can compete effectively using limited energy while still being easy to use. It is the responsibility of the research teams to strive and develop robust methods for object recognition for the visually impaired.

VI. DIFFICULTIES IN RECOGNIZING THINGS, PARTICULARLY COLOR OR PATTERNS

Some of the difficulties that blind people experience is the inability to perceive visual features such as colour patterns. Shifa Shaikh et al., (Shaikh et al., 2020a) findings suggest that object identification algorithms can perform poorly in identifying specific items for the users with sight impairment. To ensure object recognition technology is more effective in different conditions, developers and researchers continue to integrate several enhancements into the system. There are attempts to collect more data for training complex features and for tracking objects in non-standard conditions, as well as to distinguish between similar objects. Thus, the practice of challenge recognition systems for people with visual challenges greatly depends on research and development. **Figure 6** Object recognition in Prediction Module. These items serve as examples.

- ***Rare Things:*** Pre-trained systems have problems in detecting novelty or infrequent items due to their training sets. These systems are trained on training data to identify objects and poorly perform when confronted with objects not in their training database.
- ***Items with Complicated Characteristics:*** In Shirly's (Jothimani, 2016a) work, Edward notes that object detectors are challenged by objects whose characteristics are difficult to discern such as reflective and translucent features. Fine-grained attributes of objects lower the system's recognition and categorization capacity and therefore poor recognition performances.
- ***Identifying Things in Dynamic Contexts:*** Researchers, Bolla (Bolla et al., 2023) reveals that object recognition systems are quite capable of identifying objects in rather awkward positions especially in crowded or busy areas. This dynamic environment also poses a challenge to the system as the objects are moving and difficult to distinguish in this dynamic environment.
- ***Things with Similar Features:*** Alzahrani and Al-Baity, in their work (Alzahrani & Al-Baity, 2023), showed that there are some fruits and vegetables that are visually similar and thus challenging when it comes to recognition. If the system receives less context information about similar items, the system may confuse one with another and do it wrong.

Figure 6. Object recognition in Prediction Module

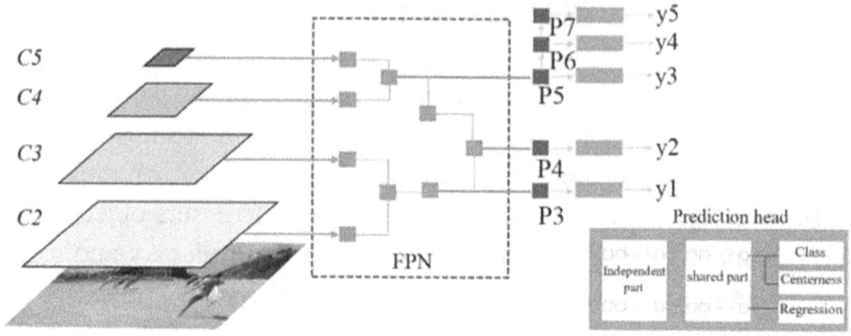

Stasenko and the team investigate how object detection ML systems are challenged when distinguishing between color options or patterns for persons with disability visual impairments. Blind individuals have a problem in identifying objects from their pictures and have a problem in distinguishing between different colors. The attentional visual agnostic patients also experience difficulties in perceiving patterns and discriminating between objects and previous items. Chaparro and Chaparro's study also show that vision impairments also cause accessibility issues for people with visual impairment. Normal color practices that are used in design tools and environments for screen design make it difficult for users to notice them. Visual impairment limits object detection systems because it hinders people's ability to distinguish colors.

VII. SYSTEM TO DETECT MOVING ITEMS

Chandrakar R. (Chandrakar et al., 2022) was interested in designing more effective approaches for target tracking. Moving object detection was used in computer vision with image processing to identify the position of items that have changed position in the physical space. **Figure 7** Dynamic object recognition system. The method compares multiple frames in succession in order to determine whether moving objects are present in the scenes. As stated by Martínez-Martín et al. (Martínez-Martín & del Pobil, 2017), background elimination, frame splitting, and period fluctuation are two recognized methods that can be used for this purpose. Research teams now develop laboratory ML systems to identify moving objects in streaming videos and enhance object detection. Observing objects in motion using tracking systems is a difficult task because of the issues of noise and motion. The advancement in the algorithms and the convolutional neural network has enhanced the capability of object recognition systems to track moving objects.

More et al. (2025) have noted that their work is a continuation of the previous studies of object detection techniques, such as YOLO and other deep learning frameworks, aimed to assist the visually impaired. In object detection, YOLO v3 performs better and has the added feature of audio output. It also performs better than other object detection systems in terms of speed. The paper discusses small item detection issues and proves that improved results are possible if more data is used and if the algorithms are trained for a longer period. They improve their detection frameworks, and that is why object recognition systems work better now.

Figure 7. Dynamic object recognition system

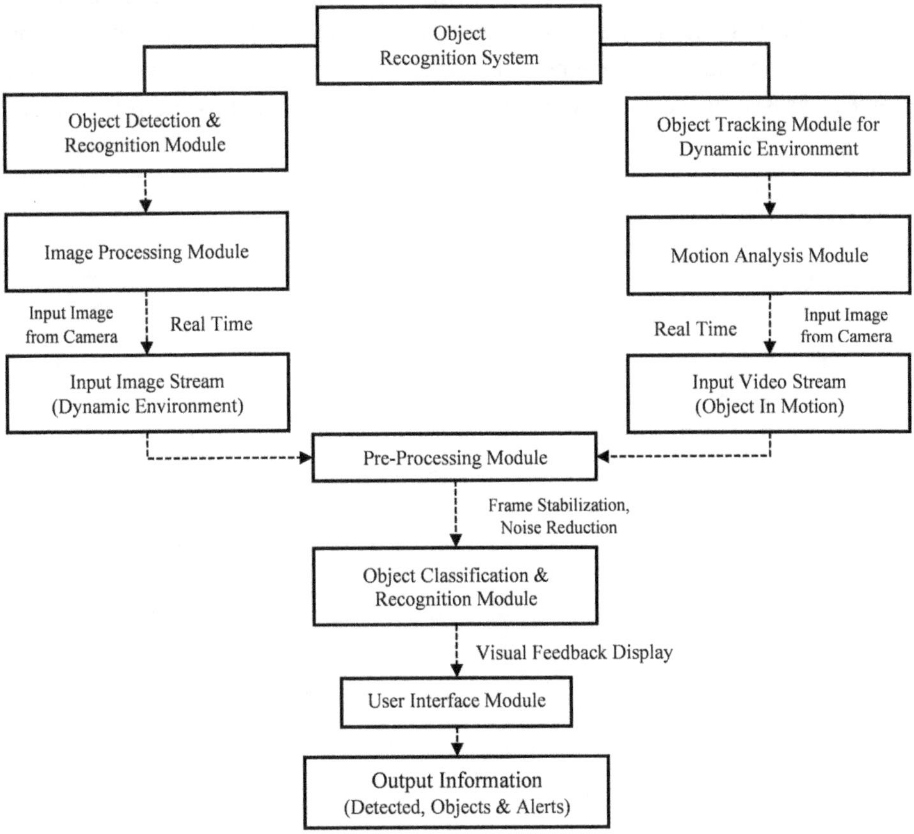

VIII. CONCLUSION

This paper demonstrates how machine learning tools help improve assistive technology for the visually impaired individuals. Examples of the customized apps for this user group demonstrate that these technologies are versatile and what they are capable of. These programs enable the vision impaired to see while at the same time developing systems that can detect any barriers. This paper explains how artificial intelligence helps the visually impaired through computer vision algorithms in medical imaging and vision aids. A study of machine learning and 5G technology provides a sophisticated system that helps people with visual impairments become more mobile and independent. The paper looks at the application of deep learning in smart assistive devices for the fast identification and localization of objects. The analysis of the current state of AI and machine learning in the field of accessibility proves that these solutions are helpful for the blind and visually impaired population. The different inventions present here show a continuing process of developing technologies for assisting the visually impaired in overcoming their day-to-day challenges.

REFERENCES

Alzahrani, N., & Al-Baity, H. H. (2023). Object Recognition System for the Visually Impaired: A Deep Learning Approach using Arabic Annotation. *Electronics (Basel), 12*(3), 541. DOI: 10.3390/electronics12030541

Behrmann, M., & Williams, P. (2007). Impairments in part–whole representations of objects in two cases of integrative visual agnosia. *Cognitive Neuropsychology, 24*(7), 701–730. DOI: 10.1080/02643290701672764 PMID: 18066732

Bolla, D. R., Rauniyar, H., Sowmya, D., Rakshitha, T., & Rasool, N. (2023). Object detection and localization for visually impaired. *ITM Web of Conferences, 57*, 01014. DOI: 10.1051/itmconf/20235701014

Boruah, A., Kakoty, N. M., Michael, G. R., Ali, T., & Bhattacharyya, D. K. (2023). Enhancing visionless object recognition on grasp using ontology: The OntOGrasp framework. *Soft Computing, 28*(7–8), 6129–6151. DOI: 10.1007/s00500-023-09409-w

Chandrakar, R., Raja, R., Miri, R., Sinha, U., Kumar Singh Kushwaha, A., & Raja, H. (2022). Enhanced the moving object detection and object tracking for traffic surveillance using RBF-FDLNN and CBF algorithm. *Expert Systems with Applications, 191*, 116306. DOI: 10.1016/j.eswa.2021.116306

Chaparro, A., & Chaparro, M. (2016). Applications of Color in Design for Color-Deficient Users. *Ergonomics in Design, 25*(1), 23–30. DOI: 10.1177/1064804616635382

Kiwa, F. J., Muduva, M., & Masengu, R. (2024). AI voice assistant for smartphones with NLP techniques. In *AI-driven marketing research and data analytics* (pp. 30–47). IGI Global Scientific Publishing.

Durgadevi, S., Thirupurasundari, K., Komathi, C., & Balaji, S. (2020). Smart machine learning system for blind assistance. *International Conference on Power, Energy, Control and Transmission Systems*. DOI: 10.1109/ICPECTS49113.2020.9337031

Elmannai, W., & Elleithy, K. (2017). Sensor-Based Assistive Devices for Visually-Impaired People: Current status, challenges, and future Directions. *Sensors (Basel), 17*(3), 565. DOI: 10.3390/s17030565 PMID: 28287451

Gupta, K., Goyal, D., & Mishra, S. K. (2024). Image Caption Generation Using Deep Learning Algorithm. Educational Administration. *Theory into Practice, 30*(5), 8118–8128.

Hussan, M. I. T., Saidulu, D., Anitha, P. T., Manikandan, A., & Naresh, P. (2022). Object detection and recognition in real time using deep learning for visually impaired people. *International Journal of Electrical and Electronics Research, 10*(2), 80–86. DOI: 10.37391/ijeer.100205

Ikram, S., Sarwar, I., Ikram, A., Abdullah-Ai-Wahud, M., & Haleema, N. (2025). A Transformer-Based multimodal object detection system for Real-World applications. *IEEE Access: Practical Innovations, Open Solutions, 1*, 29162–29176. Advance online publication. DOI: 10.1109/ACCESS.2025.3539569

Joshi, R., Tripathi, M., Kumar, A., & Gaur, M. S. (2020). Object recognition and Classification System for visually impaired. *International Conference on Communication and Signal Processing*. DOI: 10.1109/ICCSP48568.2020.9182077

Joshi, R. C., Yadav, S., Dutta, M. K., & Travieso-Gonzalez, C. M. (2020). Efficient Multi-Object detection and smart navigation using artificial intelligence for visually impaired people. *Entropy (Basel, Switzerland), 22*(9), 941. DOI: 10.3390/e22090941 PMID: 33286711

Jothimani, A. (2016). Object identification for visually impaired. *Indian Journal of Science and Technology, 9*(1), 1–5. DOI: 10.17485/ijst/2016/v9iS(1)/108446

Kadhim, M., & Oleiwi, B. (2022). Blind Assistive System based on Real Time Object Recognition using Machine learning. *Engineering and Technology Journal, 40*(1), 159–165. DOI: 10.30684/etj.v40i1.1933

Kulkarni, M., Chitale, M., Chitpur, S., Chivate, A., Chopade, P., & Deshmukh, S. (2023). Smart Assistive Navigator for the Blind using Image Processing. *International Conference on Sustainable Computing and Smart Systems*. DOI: 10.1109/ICSCSS57650.2023.10169767

Kumar, N., Sharma, S., Abraham, I. M., & Priya, S. S. (2022). Blind Assistance system using machine learning. In *Lecture notes in networks and systems* (pp. 419–432). DOI: 10.1007/978-3-031-12413-6_33

Mandhala, V. N., Bhattacharyya, D., B, V., & Rao N, T. (2020). Object detection using machine learning for visually impaired people. *International Journal of Current Research and Review, 12*(20), 157–167. DOI: 10.31782/IJCRR.2020.122032

Manikandan, P., Ramesh, G., Reddy, P. S. M., Suman, J., Kumar, M. D. V., & Reddy, K. C. S. (2022). Smart Voice Assitance System for visually Impaired persons. *International Conference on Advances in Computing, Communication Control and Networking*. DOI: 10.1109/ICAC3N56670.2022.10074529

Martínez-Martín, E., & Pascual del Pobil, A. (2017). Object Detection and Recognition for Assistive Robots: Experimentation and Implementation. *IEEE Robotics & Automation Magazine*, *24*(3), 123–138. DOI: 10.1109/MRA.2016.2615329

More, S. S., Patil, N., Lobo, V. B., Shet, N., Goswami, D., & Rane, P. (2025). Empowering the visually impaired: YOLOV8-based object detection in Android applications. *Procedia Computer Science*, *252*, 457–469. DOI: 10.1016/j.procs.2025.01.005

Nishio, T., & Nishida, T. (2018). Assistive object recognition system for visually impaired. *International Journal of Engineering Research & Technology (Ahmedabad)*, 7–11. Advance online publication. DOI: 10.1109/ISPACS.2018.8923487

Occelli, V., Lacey, S., Stephens, C., John, T., & Sathian, K. (2016). Haptic object recognition is view-independent in early blind but not sighted people. *Perception*, *45*(3), 337–345.

Paswan, V. K., & Choudhary, A. (2022). Camera based indoor Object Detection and distance estimation framework for assistive mobility. *IEEE International Conference on Service Operations and Logistics, and Informatics*. DOI: 10.1109/SOLI57430.2022.10294458

Patel, P., Pampaniya, S., Ghosh, A., Raj, R., Karuppaih, D., & Kandasamy, S. (2025). Enhancing Accessibility through Machine Learning: A review on visual and hearing Impairment technologies. *IEEE Access: Practical Innovations, Open Solutions*, *1*, 33286–33307. Advance online publication. DOI: 10.1109/ACCESS.2025.3539081

Prasantha, S. (2023). Galactical Image Detection and Identification -A Survey Report. *International Journal of Creative Research Thoughts*, *11*, f212–f222.

Raghavan, R., Krishnan, V., Nishad, H., & Shaikh, B. (2021). Virtual AI Assistant for Person with Partial Vision Impairment. *ITM Web of Conferences*, *37*, 01019. DOI: 10.1051/itmconf/20213701019

Re, D. E., & Rule, N. O. (2016). About Face. *Current Directions in Psychological Science*, *25*(1), 65–69. DOI: 10.1177/0963721415620144

Roy, A., Saji, A., Gokul, M. S., & Surendran, S. (2022). A virtual assistant for the visually impaired. In *Lecture notes in electrical engineering* (pp. 49–57). DOI: 10.1007/978-981-19-1111-8_5

Ruotsalainen, L., Morrison, A., Makela, M., Rantanen, J., & Sokolova, N. (2022). Improving computer Vision-Based perception for collaborative indoor navigation. *IEEE Sensors Journal*, *22*(6), 4816–4826. DOI: 10.1109/JSEN.2021.3106257

Shah, P. A., Alhefeiti, H. R. A., & Alrayssi, R. H. M. A. (2023). Machine learning-based smart assistance system for the visually impaired. *International Conference on Information Technology Trends*. DOI: 10.1109/ITT59889.2023.10184257

Shaikh, S., Karale, V., & Tawde, G. (2020, September). Assistive Object Recognition System for Visually Impaired [IJERT]. *International Journal of Engineering Research & Technology (Ahmedabad)*, *09*(09), •••. DOI: 10.17577/IJERTV9IS090382

Shaikh, S., Karale, V., & Tawde, G. (2020). Assistive Object Recognition System for Visually Impaired [IJERT]. *International Journal of Engineering Research & Technology (Ahmedabad)*, *09*(09), •••. DOI: 10.17577/IJERTV9IS090382

Stasenko, A., Garcea, F. E., Dombovy, M., & Mahon, B. Z. (2014). When concepts lose their color: A case of object-color knowledge impairment. cortex, 58, 217-238.

Tanya, T., Singh, S., Tripathi, S., Kapoor, S., & Saini, A. S. (2023). "Obstruction Avoidance Framework for Indoor and Outdoor Navigation to Enhance Mobility of Visually Impaired Person," *International Conference on Computational Intelligence, Communication Technology and Networking* (CICTN), Ghaziabad, India, 2023, pp. 548-552, DOI: 10.1109/CICTN57981.2023.10140455

Tapu, R., Mocanu, B., & Zaharia, T. (2020). Wearable assistive devices for visually impaired: A state of the art survey. *Pattern Recognition Letters*, *137*, 37–52. DOI: 10.1016/j.patrec.2018.10.031

Valipoor, M. M., & De Antonio, A. (2022). Recent trends in computer vision-driven scene understanding for VI/blind users: A systematic mapping. *Universal Access in the Information Society*, *22*(3), 983–1005. DOI: 10.1007/s10209-022-00868-w

Visual Perceptual/Visual Motor Deficit. LD Online. Retrieved from https://ldaamerica.org/types-of-learning-disabilities/visual-perceptual-visual-motor-deficit/

Yang, K., Qinami, K., Fei-Fei, L., Deng, J., & Russakovsky, O. (2020). Towards Fairer Datasets: Filtering and Balancing the Distribution of the People Subtree in the ImageNet Hierarchy. Proceedings of the 2020 *Conference on Fairness, Accountability, and Transparency*, 547–558. DOI: 10.1145/3351095.3375709

Chapter 12
Knee Osteoarthritis Using Hybrid Deep Learning Approach With SqueeNet and ResNet

Singaravelan Shanmugasundaram
https://orcid.org/0000-0003-4353-2261
PSR Engineering College, India

A. Muthukumar
PPG Institute of Technology, India

D. Murugan
Manonmaniam Sundaranar University, India

R. Arun
PSR Engineering College, India

S. Balaganesh
PSR Engineering College, India

P. Gopalsamy
PSR Engineering College, India

V. Selvakumar
PSR Engineering College, India

D. Arun Shunmugam
PSR Engineering College, India

ABSTRACT

Osteoarthritis is a degenerative joint sickness that influences a huge number of individuals around the world. Early location and determination of osteoarthritis is basic for viable treatment and the board of the illness. As of late, warm imaging has arisen as a promising painless procedure for identifying osteoarthritis. To address these difficulties, we propose an original methodology for osteoarthritis recognition in warm pictures utilizing the half and half ResNet-SqueezeNet model profound learning design. The proposed approach includes pre-handling the warm pictures to improve their highlights, trailed by division to extricate the area of interest. The

DOI: 10.4018/979-8-3693-9445-8.ch012

segmented region is then fed into the hybrid ResNet-SqueezeNet model, which is trained to classify the thermal image as normal or abnormal based on the presence of osteoarthritis. The performance of the proposed approach is evaluated on a dataset of thermal images collected from patients with osteoarthritis and healthy controls.

INTRODUCTION

Software-defined networking (SDN) technology is an approach to network management that enables dynamic, programmatically efficient network configuration in order to improve network performance and monitoring making it more like cloud computing than traditional network management. SDN is meant to address the fact that the static architecture of traditional networks is decentralized and complex while current networks require more flexibility and easy troubleshooting. SDN attempts to centralize network intelligence in one network component by disassociating the forwarding process of network packets (data plane) from the routing process (control plane). The control plane consists of one or more controllers which are considered as the brain of SDN network where the whole intelligence is incorporated. However, the intelligence centralization has its own drawbacks when it comes to security, scalability and elasticity and this is the main issue of SDN.

SDN was commonly associated with the OpenFlow protocol (for remote communication with network plane elements for the purpose of determining the path of network packets across network switches) since the latter's emergence in 2011. However, since 2012 OpenFlow for many companies is no longer an exclusive solution, they added proprietary techniques. These include Cisco Systems' Open Network Environment and Nicira's network virtualization platform.

The history of SDN principles can be traced back to the separation of the control and data plane first used in the public switched telephone network as a way to simplify provisioning and management well before this architecture began to be used in data networks.

The Internet Engineering Task Force (IETF) began considering various ways to decouple the control and forwarding functions in a proposed interface standard published in 2004 appropriately named "Forwarding and Control Element Separation" (ForCES). The ForCES Working Group also proposed a companion SoftRouter Architecture. Additional early standards from the IETF that pursued separating control from data include the Linux Netlink as an IP Services Protocol and A Path Computation Element (PCE)-Based Architecture.

These early attempts failed to gain traction for two reasons. One is that many in the Internet community viewed separating control from data to be risky, especially owing to the potential for a failure in the control plane. The second is that vendors were concerned that creating standard application programming interfaces (APIs) between the control and data planes would result in increased competition.

The use of open source software in split control/data plane architectures traces its roots to the Ethane project at Stanford's computer sciences department. Ethane's simple switch design led to the creation of OpenFlow. An API for OpenFlow was first created in 2008. That same year witnessed the creation of NOX—an operating system for networks.

Work on OpenFlow continued at Stanford, including with the creation of test-beds to evaluate use of the protocol in a single campus network, as well as across the WAN as a backbone for connecting multiple campuses. In academic settings there were a few research and production networks based on OpenFlow switches from NEC and Hewlett-Packard; as well as based on Quanta Computer whiteboxes, starting from about 2009.

Beyond academia, the first deployments were by Nicira in 2010 to control OVS from Onix, co-developed with NTT and Google. A notable deployment was Google's B4 deployment in 2012. Later Google acknowledged their first OpenFlow with Onix deployments in their Datacenters at the same time. Another known large deployment is at China Mobile.

The Open Networking Foundation was founded in 2011 to promote SDN and OpenFlow. At the 2014 Interop and Tech Field Day, software-defined networking was demonstrated by Avaya using shortest path bridging (IEEE 802.1aq) and Open-Stack as an automated campus, extending automation from the data center to the end device, removing manual provisioning from service delivery.

Concept

SDN architectures decouple network control and forwarding functions, enabling network control to become directly programmable and the underlying infrastructure to be abstracted from applications and network services.

The OpenFlow protocol can be used in SDN technologies. The SDN architecture is:

- **Directly programmable:** Network control is directly programmable because it is decoupled from forwarding functions.
- **Agile:** Abstracting control from forwarding lets administrators dynamically adjust network-wide traffic flow to meet changing needs.
- **Centrally managed:** Network intelligence is (logically) centralized in software-based SDN controllers that maintain a global view of the network, which appears to applications and policy engines as a single, logical switch.
- **Programmatically configured:** SDN lets network managers configure, manage, secure, and optimize network resources very quickly via dynamic, au-

tomated SDN programs, which they can write themselves because the programs do not depend on proprietary software.

- **Open standards-based and vendor-neutral:** When implemented through open standards, SDN simplifies network design and operation because instructions are provided by SDN controllers instead of multiple, vendor-specific devices and protocols.

The Need for a New Network Architecture

The explosion of mobile devices and content, server virtualization, and advent of cloud services are among the trends driving the networking industry to re-examine traditional network architectures. Many conventional networks are hierarchical, built with tiers of Ethernet switches arranged in a tree structure. This design made sense when client-server computing was dominant, but such a static architecture is ill-suited to the dynamic computing and storage needs of today's enterprise data centers, campuses, and carrier environments. Some of the key computing trends driving the need for a new network paradigm include:

Changing Traffic Patterns

Within the enterprise data center, traffic patterns have changed significantly. In contrast to client-server applications where the bulk of the communication occurs between one client and one server, today's applications access different databases and servers, creating a flurry of "east-west" machine-to-machine traffic before returning data to the end user device in the classic "north-south" traffic pattern. At the same time, users are changing network traffic patterns as they push for access to corporate content and applications from any type of device (including their own), connecting from anywhere, at any time. Finally, many enterprise data centers managers are contemplating a utility computing model, which might include a private cloud, public cloud, or some mix of both, resulting in additional traffic across the wide area network.

The "Consumerization of IT"

Users are increasingly employing mobile personal devices such as smartphones, tablets, and notebooks to access the corporate network. IT is under pressure to accommodate these personal devices in a fine-grained manner while protecting corporate data and intellectual property and meeting compliance mandates.

The Rise of Cloud Services

Enterprises have enthusiastically embraced both public and private cloud services, resulting in unprecedented growth of these services. Enterprise business units now want the agility to access applications, infrastructure, and other IT resources on demand and à la carte. To add to the complexity, IT's planning for cloud services must be done in an environment of increased security, compliance, and auditing requirements, along with business reorganizations, consolidations, and mergers that can change assumptions overnight. Providing self-service provisioning, whether in a private or public cloud, requires elastic scaling of computing, storage, and network resources, ideally from a common viewpoint and with a common suite of tools.

"Big Data" Means More Bandwidth

Handling today's "big data" or mega datasets requires massive parallel processing on thousands of servers, all of which need direct connections to each other. The rise of mega datasets is fueling a constant demand for additional network capacity in the data center. Operators of hyperscale data center networks face the daunting task of scaling the network to previously unimaginable size, maintaining any-to-any connectivity without going broke.

Security Using the SDN Paradigm

SDN architecture may enable, facilitate or enhance network-related security applications due to the controller's central view of the network, and its capacity to reprogram the data plane at any time. While security of SDN architecture itself remains an open question that has already been studied a couple of times in the research community, the following paragraphs only focus on the security applications made possible or revisited using SDN.

Several research works on SDN have already investigated security applications built upon the SDN controller, with different aims in mind. Distributed Denial of Service (DDoS) detection and mitigation, as well as botnet and worm propagation, are some concrete use-cases of such applications: basically, the idea consists in periodically collecting network statistics from the forwarding plane of the network in a standardized manner (e.g. using Openflow), and then apply classification algorithms on those statistics in order to detect any network anomalies. If an anomaly is detected, the application instructs the controller how to reprogram the data plane in order to mitigate it.

Another kind of security application leverages the SDN controller by implementing some moving target defense (MTD) algorithms. MTD algorithms are typically used to make any attack on a given system or network more difficult than usual by periodically hiding or changing key properties of that system or network. In traditional networks, implementing MTD algorithms is not a trivial task since it is difficult to build a central authority able of determining - for each part of the system to be protected - which key properties are hid or changed. In an SDN network, such tasks become more straightforward thanks to the centrality of the controller. One application can for example periodically assign virtual IPs to hosts within the network, and the mapping virtual IP/real IP is then performed by the controller.[63] Another application can simulate some fake opened/closed/filtered ports on random hosts in the network in order to add significant noise during reconnaissance phase (e.g. scanning) performed by an attacker.[64]

Additional value regarding security in SDN enabled networks can also be gained using FlowVisor and FlowChecker respectively. The former tries to use a single hardware forwarding plane sharing multiple separated logical networks. Following this approach the same hardware resources can be used for production and development purposes as well as separating monitoring, configuration and internet traffic, where each scenario can have its own logical topology which is called slice. In conjunction with this approach FlowChecker realizes the validation of new OpenFlow rules that are deployed by users using their own slice.

SDN controller applications are mostly deployed in large-scale scenarios, which requires comprehensive checks of possible programming errors. A system to do this called NICE was described in 2012. Introducing an overarching security architecture requires a comprehensive and protracted approach to SDN. Since it was introduced, designers are looking at possible ways to secure SDN that do not compromise scalability. One architecture called SN-SECA (SDN+NFV) Security Architecture.

Industrial Security

Security Implications of Software Defined Networking in Industrial Control Systems. Industrial Ethernet is the dominating technology in distributed control systems and is planned to take over the whole communication network from office to the field level, with sensor networks being the only exception at the moment. Since its introduction in time critical industrial applications, Ethernet's performance has been questioned, mainly because of the old, coax networks. Current networks are built using full duplex solutions and automation networks follow: these are built with switches, have plenty of bandwidth and the more demanding applications have their specific technologies. These solutions provide intrinsic Quality of Service (QoS), e.g., EtherCAT or try to implement extensions to the Ethernet standards with e.g., efforts to implement resource reservation like the IEEE 802.1 Time-Sensitive Networking Task Group. Many of the issues the control system engineering is facing, are not new. From the advent of packet switched networks, QoS and resilience was a question.

For metropolitan and Wide Area Networks (WAN), different solutions, like Asynchronous Transfer Mode (ATM) or Multiprotocol Label Switching (MPLS) were developed to allow creation of virtual circuits. These virtual circuits can be a natural representation of the control loops. With the industry moving towards Commercial Off The Shelf (COTS) products in the networking solutions (both hardware and software) opened for direct interconnection of other company networks towards the automation systems. The problems associated with network performance and resilience are similar to the ones, which e.g., MPLS was built to solve.

The possibility to proceed further with adopting technologies developed for WAN or telecommunication use is in large part enabled by the extended use of COTS devices. The common technology enables efficient data exchange, but also opens the possibility to attack the previously island-like automation systems from or through the company network. One of the aspects of such interconnection of systems is that the automation network might be attacked through other systems. For a more structured approach, a possible categorization of attackers is given by:

- Hobbyists break into systems for fun and glory. Difficult to stop, but consequences are low.
- Professional hackers break into systems to steal valuable assets, or on a contract basis. Very difficult to stop, consequences usually financial. May be hired to perform theft, industrial espionage, or sabotage.
- Nation-States and Non-Governmental Organizations (NGOs) break into systems to gather intelligence, disable capabilities of opponents, or to cause societal disruption.
- Malware automated attack software. Intent ranges from building botnets for further attacks, theft, or general disruption. Ranges from easy to stop to moderately difficult to stop.
- Disgruntled employees, including insider threat and unauthorized access after employment.

Engineering efforts have been made to reduce the risks associated with this interconnection, but it only gained momentum after the more recent incidents of e.g., stuxnet and repeated cases of Denial of Service (DoS) incidents coming from external networks. The first efforts were focused on including well-known solutions from the IT industry: firewalls, Intrusion Detection Systems (IDS), authentication solutions.

The challenge with these solutions is that they were designed to operate in a different network environment [6]. Amongst others, the QoS requirements of an automation system tend to be very different than of an office network. The protocol set used is different and the typical protocol inside an automation system runs on Layer 2 and not on the IP protocol suite.

Beside the efforts on adopting IT security solutions to industrial environments, several working groups are involved in introducing security features into automation protocols and protocols used to support an automation system (e.g., IEEE 1588v3 on security functions, IEC 61850 to have integrity protection). The necessity of network management systems are gaining acceptance to support life-cycle management of the communication infrastructure.

A distributed denial-of-service (DDoS) attack is a malicious attempt to disrupt the normal traffic of a targeted server, service or network by overwhelming the target or its surrounding infrastructure with a flood of Internet traffic.

DDoS attacks achieve effectiveness by utilizing multiple compromised computer systems as sources of attack traffic. Exploited machines can include computers and other networked resources such as IoT devices. From a high level, a DDoS attack is like an unexpected traffic jam clogging up the highway, preventing regular traffic from arriving at its destination.

LITERATURE SURVEY

QoS-Aware Fog Node Placement for Intensive IoT Applications in SDN-Fog Scenarios

Herrera et al focused on this fog node placement problem by formalizing it and solving it through both optimal and approximated methods, including comparisons with state-of-the-art benchmarks. In particular, we analyze the performance of each of these methods in terms of latency and execution time in both SDN Internet topologies and Industrial IoT infrastructures. Our proposed heuristic provides placements with near-optimal latencies, with smaller optimality gaps than the benchmark, and computes them in tractable times.

Z. Li et al explored a behavior-based verification method, named Crowd-Learning, by utilizing the idea of crowd in software-defined vehicular networks with a MEC framework. In Crowd-Learning, we design an incentive mechanism to stimulate some MEC infrastructures to provide accurate and appropriate amount of data for future correct behavior estimation. Without knowing the model of the dynamic environment, this incentive mechanism needs to apply reinforcement learning to let MEC infrastructures learn how to send data based on the current state. Our Crowd-

Learning method verifies vehicles and reduces the verification latency by estimating the vehicle's behavior in advance. Meanwhile, it verifies infrastructures during the process of reinforcement learning based on the idea of crowd intelligence. The results show that the Crowd-Learning verification method can ensure high verification accuracy for vehicles and infrastructures with satisfying low verification latency.

Multicast Scheduling in SDN WISE to Support Mobile Nodes in Industrial Wireless Sensor Networks

Orozco-Santos et al investigated a Mobile Multicast Forwarding with Software Defined Network (MMF-SDN), an approach using Software Defined Networking solution for WIreless SEnsor Network (SDN WISE) protocol that exploits the advantages of SDN and Time Slotted Channel Hopping (TSCH) synchronism. Parent node changes are transparent and immediate. This proposal improves on other SDN based solutions, reducing energy consumption in reception by up to end-to-end delay and improving scalability with a 30% reduction in slotframe occupancy.

Detection and Classification of DDoS Flooding Attacks on Software-Defined Networks: A Case Study for the Application of Machine Learning

Sangodoyin et al illustrated through a case study, in which, experimental data (i.e. jitter, throughput, and response time metrics) from a representative SDN architecture suitable for typical mid-sized enterprise-wide networks is used to build classification models that accurately identify and classify DDoS flooding attacks. The SDN model used was emulated in Mininet and the DDoS flooding attacks (i.e. hypertext transfer protocol (HTTP), transmission control protocol (TCP), and user datagram protocol (UDP) attacks) have been launched on the SDN model using low orbit ion cannon (LOIC). Although all the ML methods investigated show very good efficacy in detecting and classifying DDoS flooding attacks, CART demonstrated the best performance on average in terms of prediction accuracy (98%), prediction speed (5.3×10^5 observations per second), training time (12.4 ms), and robustness.

SDN-Enabled Resource Orchestration for Industrial IoT in Collaborative Edge-Cloud Networks

Okwuibe et al presented SDRM; an SDN-enabled Resource Management scheme. This novel orchestration methodology automatically computes the optimal resource allocation for different IIoT network models and dynamically adjust assigned resources based on predefined constraints to ensure Service Level Agreement (SLA).

The proposed approach models resource allocation as a Constraint Satisfaction Problem (CSP) where optimality is based on the solution of a predefined Satisfiability (SAT) problem. This model supports centralized management of all resources using a software defined approach. Such resources include memory, power, bandwidth, and edge-cloud resources. We model our resource allocation approach on SAVILE ROW using ESSENSE PRIME modeling language, we then implement the network model on CloudSimSDN and PureEdgeSim. We present a detailed analysis of the system architecture and the key technologies of the model. We finally demonstrate the efficiency of the model by presenting experimental results from a prototype system.

A Survey on Machine Learning Techniques for Routing Optimization in SDN

Amin et al surveyed the use of ML techniques for routing optimization in SDN based on three core categories (i.e. supervised learning, unsupervised learning, and reinforcement learning). The main contributions of this survey are threefold. Firstly, it presents detailed summary tables related to these studies and their comparison is also discussed, including a summary of the best works according to our analysis. Secondly, it summarizes the main findings, best works and missing aspects, and it includes a quick guideline to choose the best ML technique in this field (based on available resources and objectives). Finally, it provides specific future research directions divided into six sections to conclude the survey. In summary, future efforts should be focused on reproducible research rather than on new isolated ideas. Otherwise, most of these applications will be barely implemented in practice.

A Blockchain-Based Federated Forest for SDN-Enabled In-Vehicle Network Intrusion Detection System

Aliyu et al developed Blockchain-based Federated Forest Software-Defined Networking (SDN)-enabled IDS (BFF-IDS) to address the problem of data sharing the sensitive CAN data. To ensure scalability, we used InterPlanetary File System (IPFS) to host the models, and the blockchain is designed to store only a hash of the model and a pointer to its location. The SDN provides the dynamic routing of packets and model exchanges. We used Federated Learning (FL) to create a random forest model. Individuals provide partially trained models, allowing them to keep the underlying data confidential. Using Fourier transform, we decomposed the CAN IDs cycle from CAN bus traffic in the frequency domain for better generalization in multiclass detection of attacks. Multiple statistical and entropy features were extracted to handle the high complexity and non-linearity in CAN bus traffic. The proposed system allows manufacturers and car owners to contribute to the training of the models, as their sensitive data is protected. We found that the proposed system has efficient use of memory and CPU resources and that the detection rate of closely related attacks was high. We recorded the highest model attack detection rate

Journey toward software-defined passive optical networks with multi-PON technology: an industry view [Invited]

Montalvo et al explored our Open Broadband trial in Brazil and the journey to open software and whitebox hardware as a practical approach to scaling fiber access networks enabling integration into a software-defined network controlled environment while transitioning away from vertically integrated models featuring black box passive optical network (PON) solutions. A key part of this journey is multi-PON capable whitebox optical line terminals (OLTs) with combo-PON (C-PON) optics that play a key role in bridging the transition from G-PON to XGS-PON.

Simplified Stream Reservation Protocol Over Software-Defined Networks for In-Vehicle Time-Sensitive Networking

Nam et al presented a simplified SRP scheme over the Software-defined Network (SDN) for in-vehicle bridged networks. The proposed scheme exploits the topology and bridge information of bridged networks. A centralized stream reservation procedure in the scheme replaces the bridge-by-bridge propagation for SRP reservations. The reservation procedure does not require VLAN ID and group address registration. We implemented a prototype of the proposed scheme and showed that reserved flows are well protected when links of bridges are overloaded.

CloudSimHypervisor: Modeling and Simulating Network Slicing in Software-Defined Cloud Networks

Nyanteh et al established the CloudSimHypervisor framework in this based on CloudSimSDN-NFV. The complete architecture and features of the CloudSimHypervisor framework and some used cases are presented in this paper. We validate the CloudSimHypervisor framework with two use case experiments in the cloud computing environment: Joint compute and network resource utilization and network traffic prioritization. Results from these experiments display the efficiency of the CloudSimHypervisor in estimating and measuring processing speed, transmission speed, compute and network usage efficiency and energy consumption.

Fault Localization based on Knowledge Graph in Software-Defined Optical Networks

Z. Li et al presented a knowledge-guided fault localization method, using network alarm knowledge to analyze network abnormalities. Our method introduces knowledge graphs (KGs) into the alarm analysis process. We also propose a reasoning model based on graph neural network (GNN), to perform relational reasoning on alarm KGs and locate the network faults. We develop an ONOS-based SDON platform for experimental verification, which includes a set of processes for the construction and application of alarm KGs. The experimental results show the proposed method has high accuracy and provide motivation for the industry-scale use of KGs for alarm analysis and fault localization.

Parallel Route Optimization and Service Assurance in Energy-Efficient Software-Defined Industrial IoT Networks

Njah et al presented a a centralized route optimization and service assurance scheme, named ROSA, over a multi-layer programmable industrial architecture. The proposed solution supports a wide range of heterogeneous flows, such as ultra-reliable low-latency communications (URLLC) and bandwidth-sensitive services. The routing optimization problems are formulated as multi-constrained shortest path problems. Hence, we deploy a pair of parallel routing algorithms run according to the flow type to ensure QoS requirements, efficiently allocate constrained resources, and enhance the overall network energy consumption. We conduct extensive simulations to validate the proposed ROSA scheme. The experimental results show promising performance in terms of reducing bandwidth utilization by up to end-to-end delay, packet loss, flow violation and energy consumption up to 14% as compared to well-known benchmarks in QoS provisioning and energy-aware routing problem.

An Efficient Counter-Based DDoS Attack Detection Framework Leveraging Software Defined IoT (SD-IoT)

Njah et al presented an SD-IoT based framework that provides security services to the IoT network. We developed a C-DAD (Counter-based DDoS Attack Detection) application that is based on counter values of different network parameters, which helps to detect DDoS attack successfully. C-DAD is a dynamic and programmable solution, and is deeply tested with different network parameters. The algorithm demonstrates a good performance with better results through SDN. Moreover, the proposed framework detects the attack efficiently in a minimum amount of time and with lesser consumption of CPU and memory resources.

DOMAIN DECRIPTION & EXISTING MODELS

There are two forms of data analysis that can be used for extracting models describing important classes or to predict future data trends. These two forms are as follows –

- Classification
- Prediction

Classification models predict categorical class labels; and prediction models predict continuous valued functions. For example, we can build a classification model to categorize bank loan applications as either safe or risky, or a prediction model to predict the expenditures in dollars of potential customers on computer equipment given their income and occupation.

What is Classification

Following are the examples of cases where the data analysis task is Classification –

- A bank loan officer wants to analyze the data in order to know which customer (loan applicant) are risky or which are safe.
- A marketing manager at a company needs to analyze a customer with a given profile, who will buy a new computer.

In both of the above examples, a model or classifier is constructed to predict the categorical labels. These labels are risky or safe for loan application data and yes or no for marketing data.

What is Prediction

Following are the examples of cases where the data analysis task is Prediction –

Suppose the marketing manager needs to predict how much a given customer will spend during a sale at his company. In this example we are bothered to predict a numeric value. Therefore the data analysis task is an example of numeric prediction. In this case, a model or a predictor will be constructed that predicts a continuous-valued-function or ordered value.

How Does Classification Works

With the help of the bank loan application that we have discussed above, let us understand the working of classification. The Data Classification process includes two steps –

- Building the Classifier or Model
- Using Classifier for Classification

Building the Classifier or Model

- This step is the learning step or the learning phase.
- In this step the classification algorithms build the classifier.
- The classifier is built from the training set made up of database tuples and their associated class labels.
- Each tuple that constitutes the training set is referred to as a category or class. These tuples can also be referred to as sample, object or data points.

Figure 1. Using Classifier for Classification

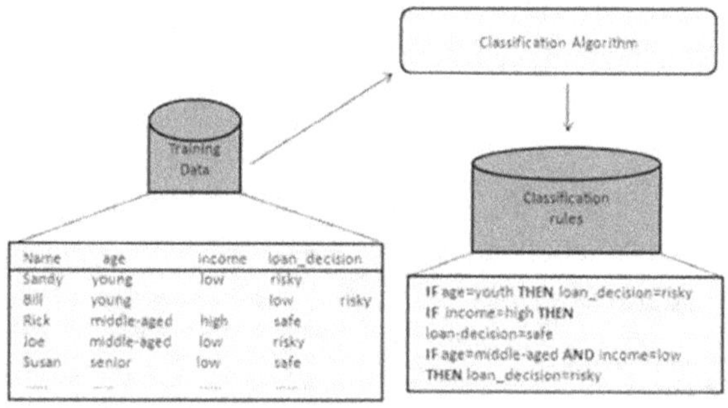

In this step, the classifier is used for classification. Here the test data is used to estimate the accuracy of classification rules. The classification rules can be applied to the new data tuples if the accuracy is considered acceptable.

Figure 2. Classification and Prediction Issues

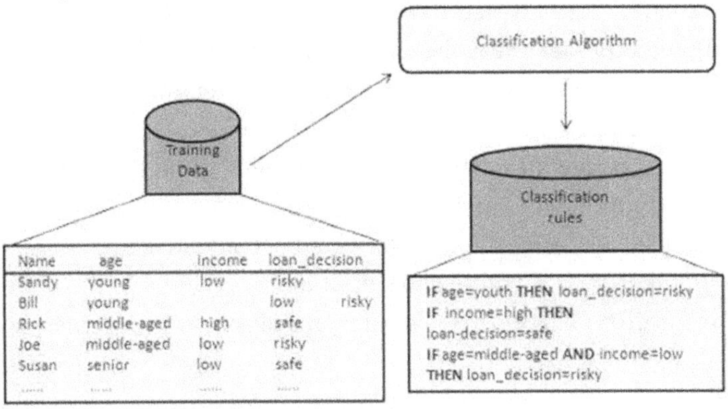

The major issue is preparing the data for Classification and Prediction. Preparing the data involves the following activities:

- **Data Cleaning:** Data cleaning involves removing the noise and treatment of missing values. The noise is removed by applying smoothing techniques and the problem of missing values is solved by replacing a missing value with most commonly occurring value for that attribute.
- **Relevance Analysis:** Database may also have the irrelevant attributes. Correlation analysis is used to know whether any two given attributes are related.
- **Data Transformation and reduction:** The data can be transformed by any of the following methods.
    - **Normalization:** The data is transformed using normalization. Normalization involves scaling all values for given attribute in order to make them fall within a small specified range. Normalization is used when in the learning step, the neural networks or the methods involving measurements are used.
    - **Generalization:** The data can also be transformed by generalizing it to the higher concept. For this purpose we can use the concept hierarchies.

Note: Data can also be reduced by some other methods such as wavelet transformation, binning, histogram analysis, and clustering.

Comparison of Classification and Prediction Methods

Here is the criteria for comparing the methods of Classification and Prediction:

- **Accuracy:** Accuracy of classifier refers to the ability of classifier. It predict the class label correctly and the accuracy of the predictor refers to how well a given predictor can guess the value of predicted attribute for a new data.
- **Speed:** This refers to the computational cost in generating and using the classifier or predictor.
- **Robustness:** It refers to the ability of classifier or predictor to make correct predictions from given noisy data.
- **Scalability:** Scalability refers to the ability to construct the classifier or predictor efficiently; given large amount of data.
- **Interpretability:** It refers to what extent the classifier or predictor understands.

Types of Algorithms In Data Mining

Figure 3. Date mining algorithms

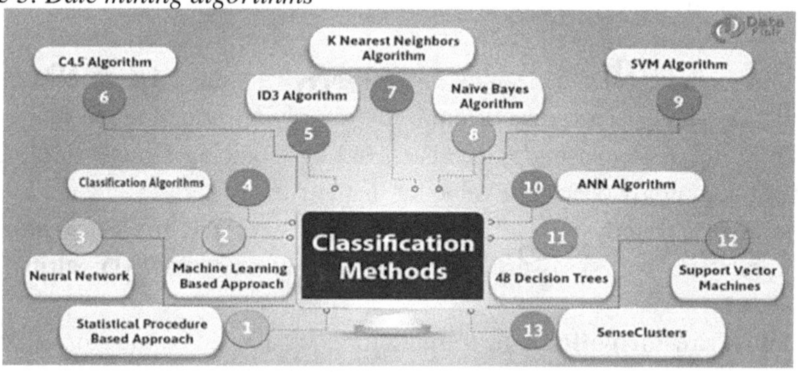

Neural Network

The field of Neural Networks has arisen from diverse sources. That is ranging from understanding and emulating the human brain to broader issues. That is of copying human abilities such as speech and use in various fields. Such as banking, in classification program to categorize data as intrusive or normal.

Generally, neural networks consist of layers of interconnected nodes. That each node producing a non-linear function of its input. And input to a node may come from other nodes or directly from the input data. Also, some nodes are identified with the output of the network. On the basis of this, there are different applications for neural networks present. That involve recognizing patterns and making simple decisions about them.

ANN Algorithm

This is the types of computer architecture inspire by biological neural networks. They are used to approximate functions. That can depend on a large number of inputs and are generally unknown.

They are presented as systems of interconnected "neurons". That can compute values from inputs. Also, they are capable of machine learning as well as pattern recognition. Due to their adaptive nature.

An artificial neural network operates by creating connections between many different processing elements. That each corresponding to a single neuron in a biological brain. These neurons may actually construct or simulate by a digital computer system.

Each neuron takes many input signals. Then based on an internal weighting. That produces a single output signal that is sent as input to another neuron. The neurons are interconnected and organized into different layers. The input layer receives the input and the output layer produces the final output. In general, one or more hidden layers are sandwiched between the two. This structure makes it impossible to forecast or know the exact flow of data. Artificial neural networks start out with randomized weights for all their neurons. This means that they need to train to solve the particular problem for which they are proposed. A back-propagation ANN is trained by humans to perform specific tasks.

During the training period, we can test whether the ANN's output is correct by observing a pattern. If it's correct the neural weightings produce that output is reinforced. if the output is incorrect, those weightings responsible diminish. Implemented on a single computer, a network is slower than more traditional solutions. The ANN's parallel nature allows it to built using many processors. That gives a great speed advantage at very little development cost. The parallel architecture allows ANNs to process amounts of data very in less time. It deals with large continuous streams of information. Such as speech recognition or machine sensor data. ANNs can operate faster as compared to other algorithms.

An artificial neural network is useful in a variety of real-world applications. Such as visual pattern recognition and speech recognition. That deals with complex often incomplete data.

Also, recent programs for text-to-speech have utilized ANNs. Many handwriting analysis programs are currently using ANNs.

PROPOSED MODEL

Our proposed system aims to bolster the security of Software-Defined Networking (SDN) environments against Distributed Denial of Service (DDoS) attacks by introducing an innovative DDoS detection scheme. Leveraging the power of Echo State Networks (ESN) and Convolutional Neural Networks (CNN), our approach offers a robust defense mechanism tailored specifically for SDN architectures.

At the core of our system lies the recognition of routine network operations characterized by normalcy, and the identification of distinctive variations in data characteristics between normal and abnormal network conditions. By integrating ESN-based classification with CNN-based feature extraction, our system enhances its capability to discern subtle anomalies indicative of DDoS attacks within the network.

The utilization of CNN-based feature extraction allows our system to capture intricate patterns and relationships present in network traffic data. This enables a more comprehensive analysis of network behavior, facilitating the identification of anomalous activities associated with DDoS attacks.

Furthermore, the incorporation of ESN-based classification provides an efficient and effective means of distinguishing between normal and abnormal network behavior. By leveraging the inherent capabilities of ESNs to model complex temporal dynamics, our system is able to accurately identify and mitigate DDoS attacks in real-time.

Through extensive simulation experiments, we meticulously evaluate the performance of our proposed system. Our results demonstrate a remarkable average success rate of 97.78% in accurately identifying and mitigating DDoS attacks, highlighting the effectiveness of our approach in fortifying the security of SDN networks.

Overall, our proposed system represents a significant advancement in cybersecurity, offering a robust defense mechanism against DDoS threats in SDN environments. By combining the strengths of ESNs and CNNs, our approach provides a comprehensive and adaptive solution to safeguard SDN networks from disruptive attacks, thereby ensuring their continued reliability and availability.

Modules

1. Data Set Collection

Kaggle Data Set: To train and evaluate the DDoS detection model, we will collect network traffic data from publicly available datasets on Kaggle. These datasets will include normal network traffic and DDoS attack scenarios. We will select a dataset with a significant variety of attacks to ensure robust model training.

Data Preprocessing: The collected dataset may require preprocessing to clean and format the data. This includes handling missing values, data normalization, and data transformation to create a structured dataset for training and testing.

2. Convolutional Neural Network (CNN)

This network structure was initially proposed by Fukushima in 1988 . It was not widely used; however, it is used due to the limits of computation hardware for training the network. In 1990s, LeCun et al. applied a gradient-based learning algorithm to CNNs and attained successful results for the handwritten digit classification problem.

Subsequently, scholars further upgraded CNNs and reported state-of-the-art results in many recognition tasks. CNN's have numerous advantages over DNNs. The most important one is being more like the human visual processing system, being highly optimized in the structure for processing 2D and 3D images, and being effective at learning and extracting abstractions of 2D features. The max-pooling layer is very effective in absorbing shape variations. Likewise, it is also composed of meagre connections with tied weights. When compared to a fully connected network of similar size, CNNs have considerably fewer parameters. Almost all CNNs are skilled with the gradient-based learning algorithm and it suffers not as much from the diminishing gradient problem. CNN can yield highly optimized weights when the gradient-based algorithm trains the whole network to decrease an error criterion directly. Figure 9 displays the overall architecture of CNNs. It comprises two main parts: They are feature extractors and a classifier.

Each layer of the network accepts the output from its close previous layer as its input in the feature extraction layers. Similarly, it passes its output as the input to the subsequent layer. The CNN architecture comprises a blend of three types of layers: Convolution, max-pooling, and classification. The low and middle-level of the network involves two types of layers. They are Convolutional layers and max-pooling layers. The even-numbered layers are intended for convolutions and the odd-numbered layers are intended for max-pooling operations. The output nodes of the convolution and max-pooling layers are clustered into a 2D plane named feature mapping. The combination of one or more planes of preceding layers usually derives each plane of a layer. The nodes of a plane are attached to a small region of each connected planes of the preceding layer. The features from the input images by convolution operations on the input nodes are extracted from each node of the convolutional layer.

Figure 4. The overall architecture of the Convolutional Neural Network (CNN) comprises an input layer, multiple alternating convolution and max-pooling layers, one fully-connected layer and one classification layer.

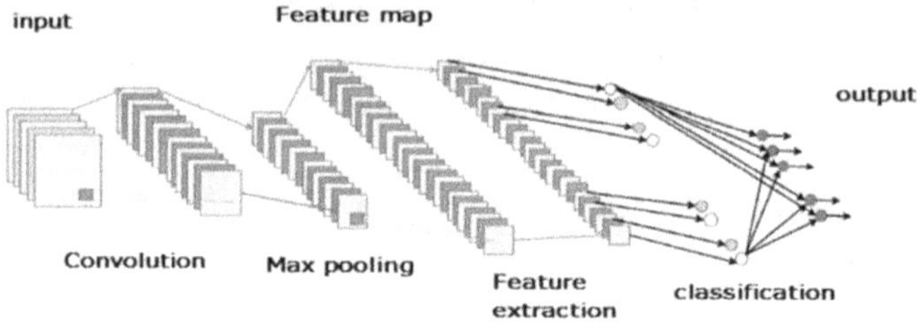

The features propagated from lower-level layers derive higher-level features. Depending on the size of the kernel, the dimensions of features are reduced for the convolutional and max-pooling operations respectively when the features propagate to the highest layer or level. Though in order to represent better features of the input images for ensuring classification accuracy, the number of feature maps is usually increased. The fully connected network is also known as the classification layer. The output of the former layer of the CNN is used as the input for this fully connected network. Because of the better performance, the feed-forward neural networks have been used as the classification layer. With respect to the dimension of the weight matrix of the final neural network, the extracted features are taken as inputs in the classification layer. On the other hand, the fully connected layers are costly in terms of network or learning parameters. Average pooling and global average pooling are some of the new techniques emerged in recent years. These techniques can be used as an alternative to fully-connected networks. It uses a soft-max layer for the calculation of the score of the respective class in the top classification layer. The classifier provides output for the corresponding classes, based on the highest score. The succeeding section discusses the mathematical details on different layers of CNNs.

Convolutional Layer

The feature maps from previous layers are convolved with learnable kernels, in this layer. In order to form the output feature maps, the output kernel goes through a linear or non-linear activation function, such as sigmoid, hyperbolic tangent, Softmax, rectified linear, and identity functions). Each one of the output feature maps can be united with more than one input feature map. In general, it can be expressed as:

$$x_j^l = f\left(\sum_{i \in M_j} x_i^{l-1} * k_{ij}^l + k_j^l\right)$$

Where x_j^l denotes the output of the current layer, x_i^{l-1} denotes the previous layer output, k_{ij}^l denotes the kernel for the present layer, and b_j^l denotes the biases for the current layer. m_j signifies a selection of input maps. An additive bias b is given for each output map. However, the input maps will be convolved with unique kernels to produce the corresponding output maps. The output maps finally go through a linear or non-linear activation function (such as sigmoid, hyperbolic tangent, Softmax, rectified linear, or identity functions).

Sub-Sampling Layer

The Sub-sampling layer, which is also known as the pooling layer executes the down sampled operation on the input maps. The number of input and output feature maps does not modify in this layer. For instance, if there are N input maps, then there will be precisely N output maps. As a result of the down sampling operation, the size of each dimension of the output maps will be abridged, depending on the size of the down sampling mask. For instance, if a 2×2 down sampling kernel is used, then each output dimension will be half of the corresponding input dimension for all the images. This operation can be expressed as

$$x_j^l = down\left(x_j^{l-1}\right)$$

where down(.) signifies a sub-sampling function. Two types of operations are generally executed in this layer: Either average pooling or max-pooling. The function generally sums up over $N \times N$ patches of the feature maps from the preceding layer and selects the average value in the event of the average pooling approach. Conversely, the highest value is designated from the $N \times N$ patches of the feature maps in case of max-pooling. As a result, the output map dimensions are reduced by n times. Each output map is multiplied with a scalar in some special cases. Some other sub-sampling layers have been proposed, for instance, fractional max-pooling layer and sub-sampling with convolution.

Classification Layer

This is one of the fully connected layers which calculate the score of each class from the extracted features from a convolutional layer in the preceding steps. The concluding layer feature maps are denoted as vectors with scalar values that are passed to the fully connected layers. The fully connected feed-forward neural layers are used as a soft-max classification layer. There are no strict rules on the number of layers that are merged in the network model. Though, in most cases, two to four layers have been detected in different architectures, which involve LeNet, AlexNet, and VGG Net. Since the fully connected layers are costly in terms of computation, alternative approaches have been proposed during the last few years. These include the global average pooling layer and the average pooling layer which help to reduce the number of parameters in the network significantly. In the backward propagation through the CNNs, the fully connected layer updates following the general approach of fully connected neural networks (FCNN). The filters of the convolutional layers are updated by performing the full convolutional operation on the feature maps between the convolutional layer and its immediate previous layer. Figure 10 displays the fundamental operations in the convolution and sub-sampling of an input image.

Figure 5. Feature maps after performing convolution and pooling operations.

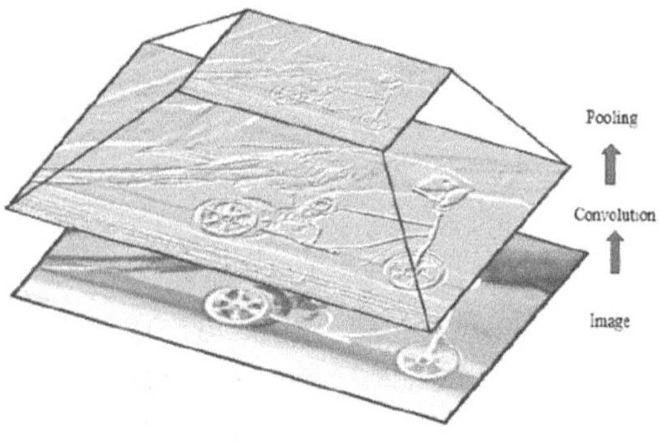

Network Parameters and Required Memory for CNN

The number of computational parameters is an important metric to measure the complexity of a deep learning model. The size of the output feature maps can be formulated as follows:

$$M = \frac{(N - F)}{S} + 1$$

where N denotes the dimensions of the input feature maps, F denotes the dimensions of the filters or the receptive field, M denotes the dimensions of the output feature maps, and S stands for the stride length. Padding is normally applied during the convolution operations to safeguard the input and output feature map which have the same dimensions. The amount of padding depends on the size of the kernel.

Equation 17 is used for determining the number of rows and columns for padding.

$$P = \frac{(F-1)}{2}$$

here P denotes the amount of padding and F denotes the dimension of the kernels. Various criteria are taken into account for comparing the models. Nevertheless, in most cases, the number of network parameters and the total amount of memory is well-thought-out. The number of parameters (Parml) of lth layer is calculated based on the following equation:

$$Parm_l = (F \times F \times FM_{l-1}) \times FM_l$$

If bias is added with the weights, then the above equation can be written as follows:

$$Parm_l = (F \times (F+1) \times FM_{l-1}) \times FM_l$$

here the total number of parameters of lth lathe yer can be denoted with Pl, FMl denotes the total a number of output feature maps, and FMl-1 denotes the total number of input feature maps or channels.

3. Echo State Network (ESN) Model

ESN Architecture: ESN is a type of recurrent neural network known for its capabilities in sequential data analysis. We will design an ESN architecture suitable for processing network traffic data. The ESN will consist of an input layer, hidden reservoir layer, and output layer.

Model Training: Using the preprocessed dataset and extracted features, we will train the ESN model to recognize normal and abnormal network behavior. This will involve supervised learning to build a robust DDoS detection system.

4. DDoS Detection

Real-time Analysis: The trained ESN model will be deployed in real-time to analyze incoming network traffic. It will evaluate traffic patterns and identify deviations that may indicate DDoS attacks.

Anomaly Detection: The system will focus on detecting anomalies in network behavior, which can include unusual traffic spikes, protocol violations, and other DDoS attack signatures.

5. Evaluation and Testing

Performance Metrics: To assess the effectiveness of the DDoS detection system, we will employ various performance metrics such as precision, recall, F1-score, and ROC curves. These metrics will help evaluate the system's accuracy and efficiency.

Validation: The system will be tested against a diverse set of network traffic scenarios, including known DDoS attacks, to validate its performance in real-world conditions.

6. User Interface (Optional)

Dashboard: An optional user interface may be developed to provide users with real-time insights into network traffic and detected anomalies. This can aid network administrators in taking prompt actions.

Figure 6. Proposed system

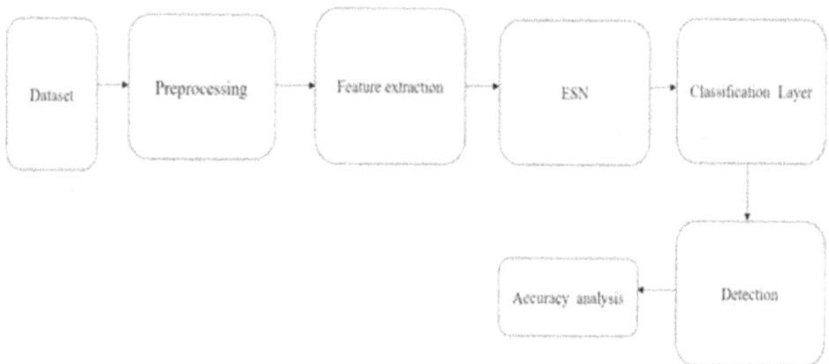

Echo State Network (ESN):

Echo State Network (ESN) is a type of recurrent neural network (RNN) known for its effectiveness in processing sequential data, making it particularly valuable in various applications, including time series prediction, speech recognition, and anomaly detection. Unlike traditional RNNs, ESN exhibits distinct architectural properties that contribute to its success. This introduction provides an overview of ESN's architecture and the key layers that comprise this network.

Figure 7. ECN

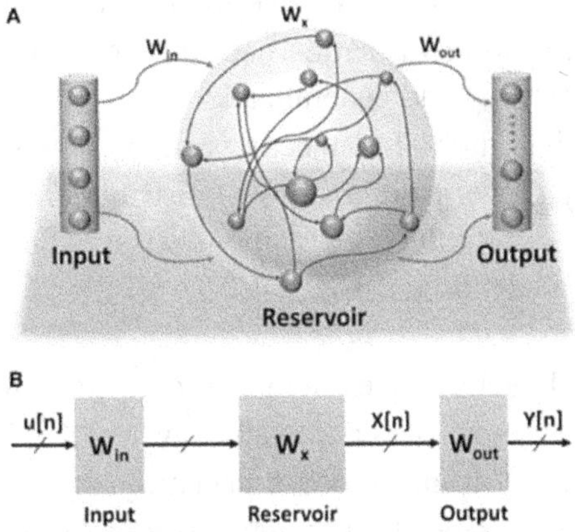

ESN Architecture and Layers:

Echo State Network consists of three primary layers: the Input Layer, the Reservoir Layer, and the Output Layer. Each of these layers plays a unique role in the network's operation.

1. Input Layer:
   - The Input Layer receives and processes the external input data. This layer connects directly to the reservoir layer, and its primary function is to transform the incoming data into a format suitable for processing within the network.
   - In the case of time series data or sequential information, the input layer may receive data points at discrete time steps and provide them as input to the reservoir.
   - The input data can represent a wide range of information, such as sensor readings, language signals, or any sequential data of interest.
2. Reservoir Layer:
   - The Reservoir Layer is the distinguishing feature of the ESN architecture. It comprises a large number of recurrent neurons that form a dynamic, fixed-weight network.

- The key characteristic of the reservoir is that its internal connections are generated randomly and remain unaltered during the training process. This dynamic nature of the reservoir creates a rich internal state that facilitates the network's capacity to capture complex temporal dependencies in data.
- The reservoir acts as a nonlinear, high-dimensional space where input data undergoes transformation and mixing. This transformation is essential for learning and generalization.
- The reservoir's internal connections, also known as reservoir weights, are typically drawn from a uniform distribution, and the network's dynamic behavior is influenced by the spectral radius (maximum eigenvalue) of these weights.

3. Output Layer:
    - The Output Layer receives the reservoir's state as input and generates the network's output. It is a trainable layer with adjustable weights.
    - The network's learning process primarily occurs in the output layer, where a linear combination of the reservoir states is transformed to produce the desired output.
    - The output layer can be fine-tuned during training to achieve specific tasks, such as classification, regression, or sequence prediction.
    - The combination of a dynamic reservoir and a trainable output layer is what distinguishes ESN from traditional RNNs. The reservoir layer takes on the role of capturing complex temporal patterns, while the output layer simplifies the mapping of this internal state to the desired output.

Echo State Networks are particularly advantageous in scenarios where extensive temporal dependencies and nonlinear relationships need to be captured in data. Their simple yet powerful architecture makes them a valuable tool for various applications, including time series analysis and sequential data processing. ESN's success lies in its capacity to harness the intrinsic dynamics of the reservoir layer, facilitating efficient learning and generalization.

The dataset is then split into training and testing sets. This division is a common practice in machine learning, enabling the model to be trained on one portion of the data and tested on another to assess its performance.

Before training the ESN, the data is standardized using the StandardScaler from scikit-learn. Standardization ensures that the data has a mean of zero and a standard deviation of one, preparing it for effective use in the model.

The ESN, a type of recurrent neural network, is initialized with specific parameters, including the number of reservoir neurons, spectral radius, and a random seed. These parameters influence the ESN's dynamics and its ability to capture temporal patterns.

The ESN model is trained using the training data, where it learns to map the input features (X_train) to the corresponding output labels (y_train). The ESN's unique characteristic is its ability to capture complex temporal dependencies in the data.

After training, the ESN is used to make predictions on the testing data (X_test). These predictions are compared with the actual labels to evaluate the model's accuracy.

The code computes and displays the accuracy of the ESN model, which represents how well it can distinguish between normal and attacker instances. Additionally, a confusion matrix is presented, showing the true positive, true negative, false positive, and false negative predictions, offering a comprehensive view of the model's performance.

SOFTWARE SPECIFICATION

SOFTWARE REQUIRED:

- IDLE 1.7
- PYTHON 1.7.6

HARDWARE REQUIRED:

- System: Windows Xp Professional Service Pack 2
- Processor: Up to 1.5 GHz
- Memory: Up to 512 MB RAM

PYTHON

The Python language had a humble beginning in the late 1980s when a Dutchman Guido Von Rossum started working on a fun project, which would be a successor to ABC language with better exception handling and capability to interface with OS Amoeba at Centrum Wiskunde and Informatica. It first appeared in 1991. Python 2.0 was released in the year 2000 and Python 3.0 was released in the year 2008. The language was named Python after the famous British television comedy show Monty Python's Flying Circus, which was one of Guido's favorite television programmes. Here we will see why Python has suddenly influenced our lives and the various applications that use Python and its implementations.

SIMULATION RESULT

Dataset and Per Processing

The proposed model is evaluated on two public datasets, the 10% KDD99 and the full NSLKDD, which have been widely used for evaluating the performance of intrusion detection schemes. In the KDD99 dataset, we utilize all the legitimate traffic samples and select 14 types of prevalent attacks. In the NSLKDD dataset, we use all the legitimate traffic samples and select 16 types of prevalent attacks. The features of each data sample in the datasets are considered as a feature vector. We split the dataset randomly into the training, validation and testing subset with a ratio of 0.7:0.1:0.2, namely the size of the training, validation and testing subset. For each predictive model, we train it in a mini-batch way with 1,024 sequences per epoch and conduct 100 iterations. In order to enhance the models' generalization performance, the data was divided independently and each model was trained and tested 10 times in our work. Finally we report the mean evaluation metrics on the 10 testing results.

Figure 8. Data set description

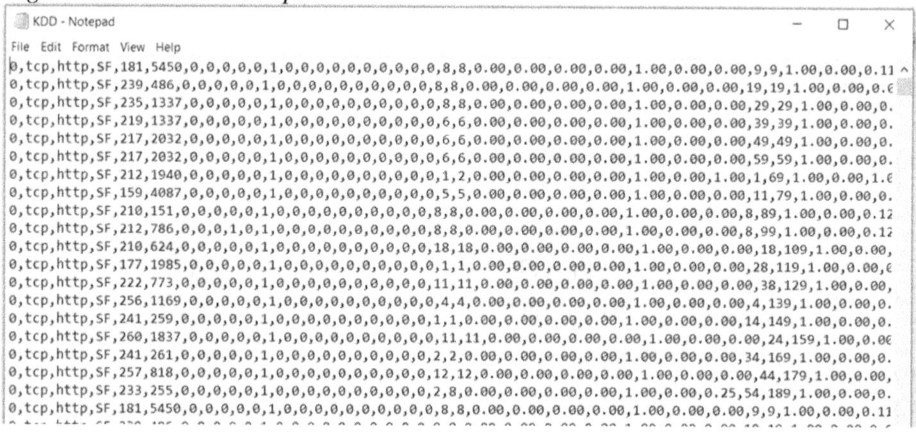
KDD - Notepad

File Edit Format View Help

```
0,tcp,http,SF,181,5450,0,0,0,0,0,1,0,0,0,0,0,0,0,0,0,0,8,8,0.00,0.00,0.00,0.00,1.00,0.00,0.00,9,9,1.00,0.00,0.11
0,tcp,http,SF,239,486,0,0,0,0,0,1,0,0,0,0,0,0,0,0,0,0,8,8,0.00,0.00,0.00,0.00,1.00,0.00,0.00,19,19,1.00,0.00,0.
0,tcp,http,SF,235,1337,0,0,0,0,0,1,0,0,0,0,0,0,0,0,0,0,8,8,0.00,0.00,0.00,0.00,1.00,0.00,0.00,29,29,1.00,0.00,0.
0,tcp,http,SF,219,1337,0,0,0,0,0,1,0,0,0,0,0,0,0,0,0,0,6,6,0.00,0.00,0.00,0.00,1.00,0.00,0.00,39,39,1.00,0.00,0.
0,tcp,http,SF,217,2032,0,0,0,0,0,1,0,0,0,0,0,0,0,0,0,0,6,6,0.00,0.00,0.00,0.00,1.00,0.00,0.00,49,49,1.00,0.00,0.
0,tcp,http,SF,217,2032,0,0,0,0,0,1,0,0,0,0,0,0,0,0,0,0,6,6,0.00,0.00,0.00,0.00,1.00,0.00,0.00,59,59,1.00,0.00,0.
0,tcp,http,SF,212,1940,0,0,0,0,0,1,0,0,0,0,0,0,0,0,0,0,1,2,0.00,0.00,0.00,0.00,1.00,0.00,1.00,1,69,1.00,0.00,1.
0,tcp,http,SF,159,4087,0,0,0,0,0,1,0,0,0,0,0,0,0,0,0,0,5,5,0.00,0.00,0.00,0.00,1.00,0.00,0.00,11,79,1.00,0.00,0.
0,tcp,http,SF,210,151,0,0,0,0,0,1,0,0,0,0,0,0,0,0,0,0,8,8,0.00,0.00,0.00,0.00,1.00,0.00,0.00,8,89,1.00,0.00,0.12
0,tcp,http,SF,212,786,0,0,0,1,0,1,0,0,0,0,0,0,0,0,0,0,8,8,0.00,0.00,0.00,0.00,1.00,0.00,0.00,8,99,1.00,0.00,0.12
0,tcp,http,SF,210,624,0,0,0,0,0,1,0,0,0,0,0,0,0,0,0,0,18,18,0.00,0.00,0.00,0.00,1.00,0.00,0.00,18,109,1.00,0.00,
0,tcp,http,SF,177,1985,0,0,0,0,0,1,0,0,0,0,0,0,0,0,0,0,1,1,0.00,0.00,0.00,0.00,1.00,0.00,0.00,28,119,1.00,0.00,
0,tcp,http,SF,222,773,0,0,0,0,0,1,0,0,0,0,0,0,0,0,0,0,11,11,0.00,0.00,0.00,0.00,1.00,0.00,0.00,38,129,1.00,0.00,
0,tcp,http,SF,256,1169,0,0,0,0,0,1,0,0,0,0,0,0,0,0,0,0,4,4,0.00,0.00,0.00,0.00,1.00,0.00,0.00,4,139,1.00,0.00,0.
0,tcp,http,SF,241,259,0,0,0,0,0,1,0,0,0,0,0,0,0,0,0,0,1,1,0.00,0.00,0.00,0.00,1.00,0.00,0.00,14,149,1.00,0.00,0.
0,tcp,http,SF,260,1837,0,0,0,0,0,1,0,0,0,0,0,0,0,0,0,0,11,11,0.00,0.00,0.00,0.00,1.00,0.00,0.00,24,159,1.00,0.0
0,tcp,http,SF,241,261,0,0,0,0,0,1,0,0,0,0,0,0,0,0,0,0,2,2,0.00,0.00,0.00,0.00,1.00,0.00,0.00,34,169,1.00,0.00,0.
0,tcp,http,SF,257,818,0,0,0,0,0,1,0,0,0,0,0,0,0,0,0,0,12,12,0.00,0.00,0.00,0.00,1.00,0.00,0.00,44,179,1.00,0.00,
0,tcp,http,SF,233,255,0,0,0,0,0,1,0,0,0,0,0,0,0,0,0,0,2,8,0.00,0.00,0.00,0.00,1.00,0.00,0.25,54,189,1.00,0.00,0.
0,tcp,http,SF,181,5450,0,0,0,0,0,1,0,0,0,0,0,0,0,0,0,0,8,8,0.00,0.00,0.00,0.00,1.00,0.00,0.00,9,9,1.00,0.00,0.11
```

Figure 9. ESN Training

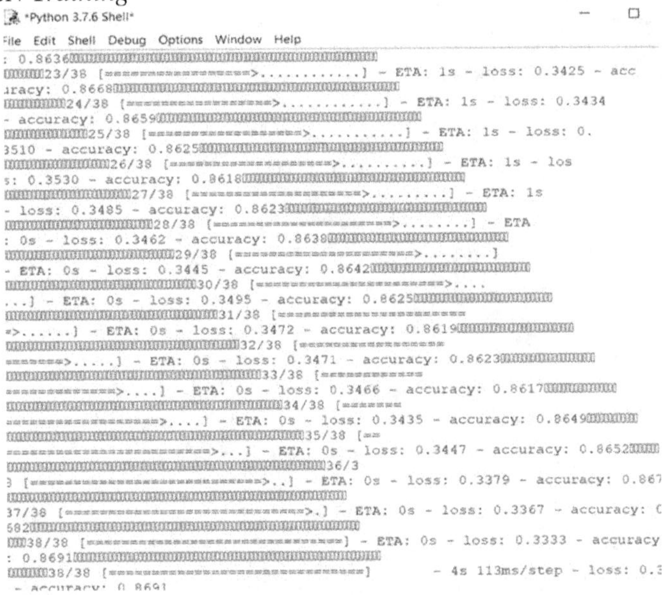

Figure 10. Histogram of attacker variation

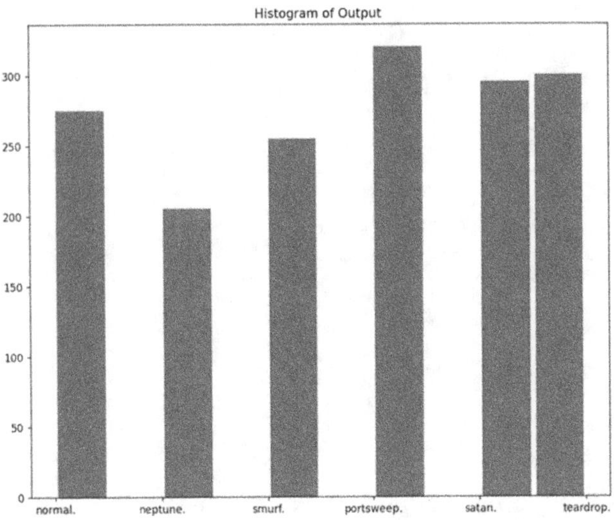

Figure 11. Attacker classification

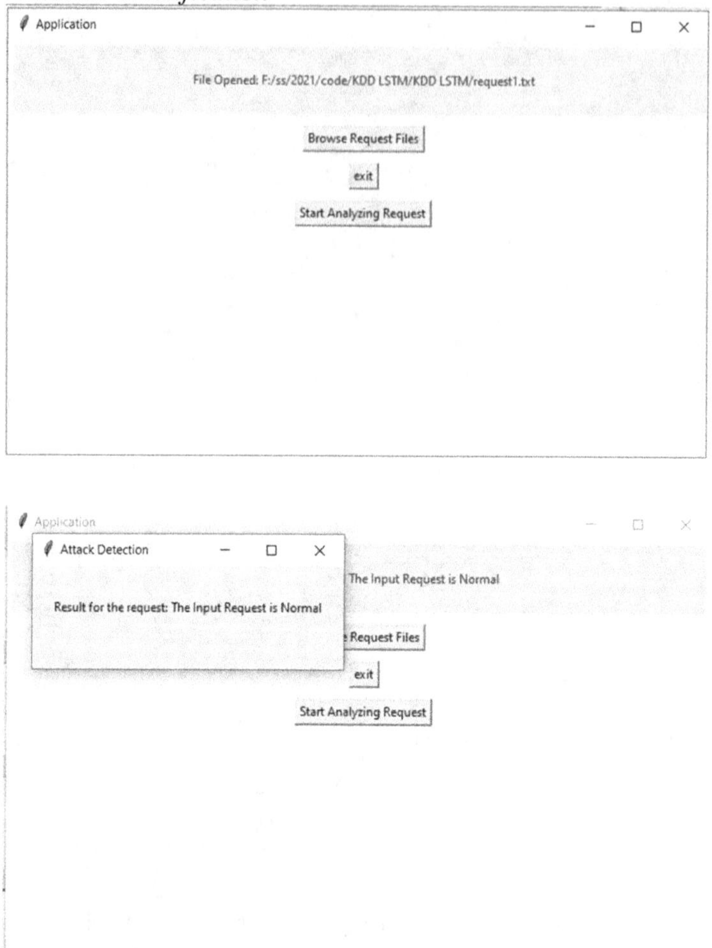

Figure 12. Performance analysis

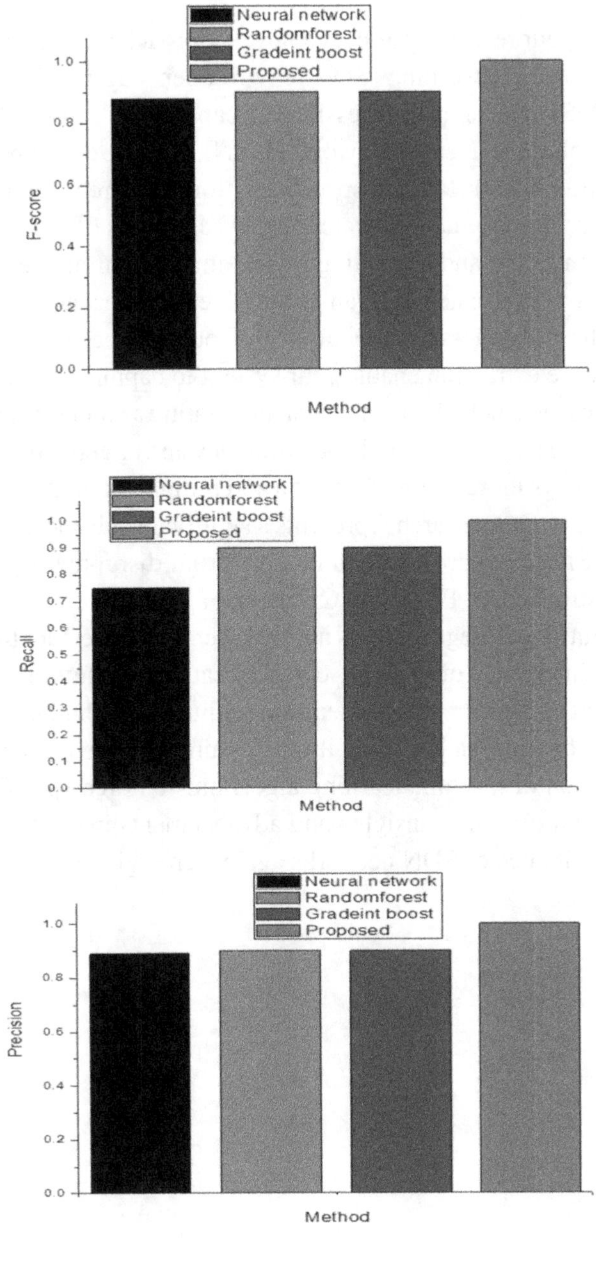

CONCLUSION

In conclusion, our research presents a novel approach to enhancing the security of Software-Defined Networking (SDN) environments against Distributed Denial of Service (DDoS) attacks. By harnessing the capabilities of Echo State Networks (ESN) and Convolutional Neural Networks (CNN), we have developed a sophisticated DDoS detection scheme tailored specifically for SDN architectures. Throughout our investigation, we have demonstrated the effectiveness of our proposed system in accurately identifying and mitigating DDoS attacks. Through extensive simulation experiments, we have achieved an impressive average success rate of 97.78%, underscoring the robustness and reliability of our approach. The integration of CNN-based feature extraction enables our system to capture intricate patterns and relationships present in network traffic data, facilitating a comprehensive analysis of network behavior. This, coupled with the efficiency and effectiveness of ESN-based classification, allows for real-time detection and mitigation of DDoS attacks within SDN environments. Our research represents a significant advancement in fortifying the security of SDN networks, shielding them from disruptive DDoS threats. By combining the strengths of ESNs and CNNs, our approach offers a comprehensive and adaptive solution to safeguard SDN networks, ensuring their continued reliability and availability in the face of evolving cyber threats. In the future, further research and development in this area could explore additional techniques for enhancing DDoS detection and mitigation capabilities within SDN environments. Additionally, the integration of machine learning algorithms tailored specifically for SDN architectures could offer new insights and advancements in cybersecurity, further bolstering the resilience of SDN networks against emerging threats.

REFERENCES

Aliyu, I., Feliciano, M. C., Van Engelenburg, S., Kim, D. O., & Lim, C. G. (2021). A Blockchain-Based Federated Forest for SDN-Enabled In-Vehicle Network Intrusion Detection System. *IEEE Access : Practical Innovations, Open Solutions*, *9*, 102593–102608. DOI: 10.1109/ACCESS.2021.3094365

Amin, R., Rojas, E., Aqdus, A., Ramzan, S., Casillas-Perez, D., & Arco, J. M. (2021). A Survey on Machine Learning Techniques for Routing Optimization in SDN. *IEEE Access : Practical Innovations, Open Solutions*, *9*, 104582–104611. DOI: 10.1109/ACCESS.2021.3099092

Herrera, J. L., Galán-Jiménez, J., Foschini, L., Bellavista, P., Berrocal, J., & Murillo, J. M. (2022, August 1). QoS-Aware Fog Node Placement for Intensive IoT Applications in SDN-Fog Scenarios. *IEEE Internet of Things Journal*, *9*(15), 13725–13739. Advance online publication. DOI: 10.1109/JIOT.2022.3143948

Li, Z., Zhao, Y., Li, Y., Rahman, S., Wang, F., Xin, X., & Zhang, J. (2021, July 1). Fault Localization based on Knowledge Graph in Software-Defined Optical Networks. *Journal of Lightwave Technology*, *39*(13), 4236–4246. DOI: 10.1109/JLT.2021.3071868

Montalvo, J., Torrijos, J., Cortes, D., Chundury, R., & St. Peter, M. (2021, August). Journey toward software-defined passive optical networks with multi-PON technology: An industry view [Invited]. *Journal of Optical Communications and Networking*, *13*(8), D22–D31. DOI: 10.1364/JOCN.423034

Nam, S., Kim, H., & Min, S.-G. (2021). Simplified Stream Reservation Protocol Over Software-Defined Networks for In-Vehicle Time-Sensitive Networking. *IEEE Access : Practical Innovations, Open Solutions*, *9*, 84700–84711. DOI: 10.1109/ACCESS.2021.3088288

Njah, Y., & Cheriet, M. (2021). Parallel Route Optimization and Service Assurance in Energy-Efficient Software-Defined Industrial IoT Networks. *IEEE Access : Practical Innovations, Open Solutions*, *9*, 24682–24696. DOI: 10.1109/ACCESS.2021.3056931

Nyanteh, A. O., Li, M., Abbod, M. F., & Al-Raweshidy, H. (2021). CloudSimHypervisor: Modeling and Simulating Network Slicing in Software-Defined Cloud Networks. *IEEE Access : Practical Innovations, Open Solutions*, *9*, 72484–72498. DOI: 10.1109/ACCESS.2021.3079501

Okwuibe, J., Haavisto, J., Kovacevic, I., Harjula, E., Ahmad, I., Islam, J., & Ylianttila, M. (2021). SDN-Enabled Resource Orchestration for Industrial IoT in Collaborative Edge-Cloud Networks. *IEEE Access : Practical Innovations, Open Solutions*, *9*, 115839–115854. DOI: 10.1109/ACCESS.2021.3105944

Orozco-Santos, F., Sempere-Payá, V., Silvestre-Blanes, J., & Albero-Albero, T. (2021). Multicast Scheduling in SDN WISE to Support Mobile Nodes in Industrial Wireless Sensor Networks. *IEEE Access : Practical Innovations, Open Solutions*, *9*, 141651–141666. DOI: 10.1109/ACCESS.2021.3120917

Sangodoyin, A. O., Akinsolu, M. O., Pillai, P., & Grout, V. (2021). Detection and Classification of DDoS Flooding Attacks on Software-Defined Networks: A Case Study for the Application of Machine Learning. *IEEE Access : Practical Innovations, Open Solutions*, *9*, 122495–122508. DOI: 10.1109/ACCESS.2021.3109490

Chapter 13

AI-Powered Brain-Computer Interfaces for Advancing Prosthetic Functionality and Rehabilitation

Nidhi Sharma
https://orcid.org/0000-0002-3177-0943
Uttaranchal College of Health Sciences, Uttaranchal University, Dehradun, India

Shashank Kumar
https://orcid.org/0009-0004-6741-5849
Uttaranchal College of Health Sciences, Uttaranchal University, Dehradun, India

Kartikey
https://orcid.org/0009-0007-6073-0282
School of Liberal Arts, Uttaranchal University, Dehradun, India

ABSTRACT

Control of prosthetic limbs using brain signals enabled by Brain-Computer Interfaces (BCIs) is a departure from the conventional prosthetics that functioned based on residual muscle activity alone. BCIs recover motor functions and augment recovery in individuals with neurological impairments by bypassing traditional neural and muscular pathways. Advances in Artificial Intelligence (AI) and materials science have led to BCI prosthetics that advance proprioception via enhanced sensory feedback and assist with motor function restoration. Despite the problems encountered in inconsistent signals and high costs, BCIs offer a workable method of prosthetic

DOI: 10.4018/979-8-3693-9445-8.ch013

control whereas AI enables personalized prosthesis control by facilitating signal decoding and making systems more flexible. This chapter discusses the integrative role of AI with BCIs and gives an insight into existing literature demonstrating role of AI algorithms in optimizing prosthetic device precision and motion, thereby, enhancing user experience by customizing BCI systems to brain patterns and reducing cognitive load.

1. INTRODUCTION

Brain-Computer Interfaces (BCIs), or Brain-Machine Interfaces (BMIs), are innovations in the field of neuroscience and technology. They create a direct network of communication between the human brain and other devices, bypassing conventional neurological and muscular processes. BCIs utilize the study of brain activity to provide the revolutionary capability of restoring motor function by allowing individuals to operate devices, including prosthetic limbs, directly through thought. Such advancements are extremely promising in enhancing the quality of life and speeding up the rehabilitation process among patients with neurological impairments (traumatic, congenital defects or limb loss) (Shih J.J. et al, 2012). Unlike the traditional approaches of rehabilitation requiring residual mobility of the extremities and involving use of myoelectric systems while using prosthetics, BCIs provide more spontaneous control by directly translating brain inputs into prosthetic movements, thereby, bypassing the constraints of conventional methods. A more accurate and effective control of prosthetic limbs is enabled by this one-to-one brain-device interface (Simon A.M. et al, 2023).

BCIs-based prostheses have in the last 20 years witnessed a drastic overhaul with advances in material science, machine learning, and neurology. Artificial Intelligence (AI) processes electrical signals in the brain to generate commands that control external devices using intricate algorithms, captured via BCIs technologies like electroencephalography (EEG), electrocorticography (ECoG), or intracranial implants. Exposures in applications for robot hands, arms, and prostheses for lower limbs have proved to be successful. To restore lost function and provide independence, this involves converting the user's intention into movements that simulate normal limb activity for prosthetics (Zhang X. et al, 2020). These devices augment proprioception and feedback and commonly allow for activities of daily living to expand on capabilities resembling intrinsic control of natural limbs (Vidal G.W. et al, 2016). However, as numerous developments are been carried out, these technologies are highly challenged by device cost, oscillating signals and ethic related issues. AI plays an essential role in mitigating several of these problems and enhances flexibility, adaptability and user-friendliness of prosthetic systems. This chapter covers the underlying foundation, practical approach, and future advancement opportunities of AI-based BCIs and how they are advancing prosthetic use and rehabilitation. This combination will be a game changer for prosthetics, allowing people to reattain levels of independence and functionality in daily life never before possible (Maleki V.S. et al, 2024).

2. ROLE OF ARTIFICIAL INTELLIGENCE (AI) IN ENHANCING BRAIN-COMPUTER INTERFACE CAPABILITIES

AI does not replace BCI's; rather, it improves them significantly, particularly in the context of prosthetic control by further improving signal processing, improving the adaptability of the system, and maximizing user engagement. AI's machine learning algorithms, especially Deep Learning (DL) models that probe neural data to reveal detailed patterns of brain activity, enhance the accuracy and accelerate the process of translating brain signals into instructions for the prosthetic devices. Another major advantage to AIs is their ability to make BCIs familiar with different users. By learning an individual user's brain patterns to refine the way a prosthetic responds, AI enhances comfort, control, and effectiveness. Signal augmentation and noise elimination techniques check whether physiological and external perturbations do not destroy the stability of the system (Zhang X. et al, 2020). (Figure 1 demonstrates role of Artificial Intelligence in Enhancing BCI Capabilities)

Figure 1. Role of Artificial Intelligence in Enhancing BCI Capabilities

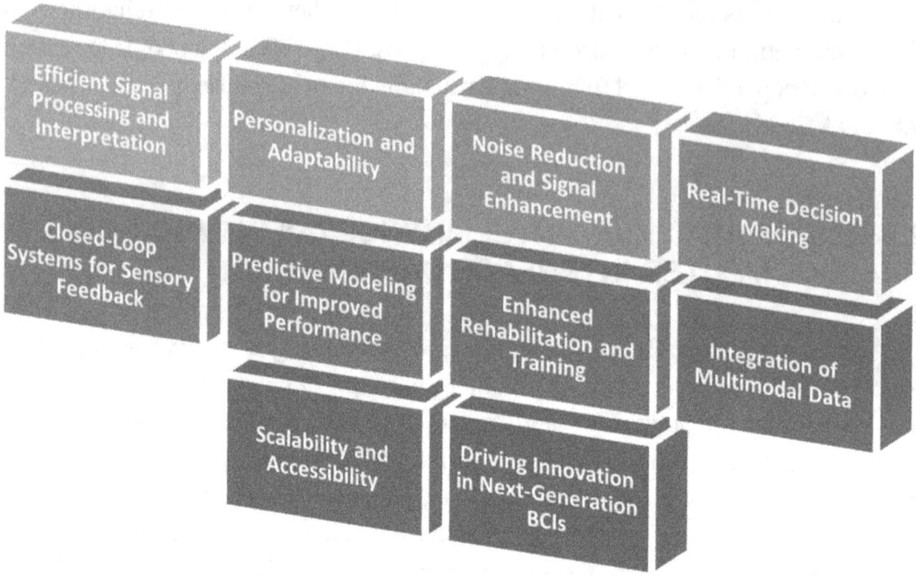

AI enables instantaneous decision-making, allowing brain impulses to be traversed almost instantaneously, creating a seamless user-prosthetic interaction. This allows artificial appendages to mimic natural movement, reducing the cognitive burden on users. In a closed-loop BCI, AI governs and coordinates the operating flow to supply sensory feedback and control activity. AI-powered predictive modeling anticipates what users want, facilitating seamless transitions and the collaborative completion of complex activities. AI facilitates recovery by instigating neuroplasticity through targeted training programs that engage the motor function portion of the brain so that users can regain the ability to move. It also enhances BCI performance by integrating multimodal data, for example, eye movement or muscle activity, to make prosthesis more precise and responsive (Luu D.K. et al, 2022). Additionally, AI reduces BCIs' computational complexity, which reduces their cost, crucial for boosting their acceptability. Emerging technologies like reinforcement learning and generative AI are opening up new BCI potential and stimulating creativity in next-generation systems. AI and BCIs combined are revolutionizing neuro-prosthetics and improving the quality of life for people with disabilities. To help users integrate prostheses both physically and psychologically, AI in rehabilitation targets motor control, neuroplasticity, and cognitive state. Personalized training programs and sensory input further enhance long-term success and user experience, optimizing prosthetic use and quality of life (Zhang H. et al, 2024). (Figure 2 describes History and Evolution of BCIs)

Figure 2. History and Evolution of BCIs

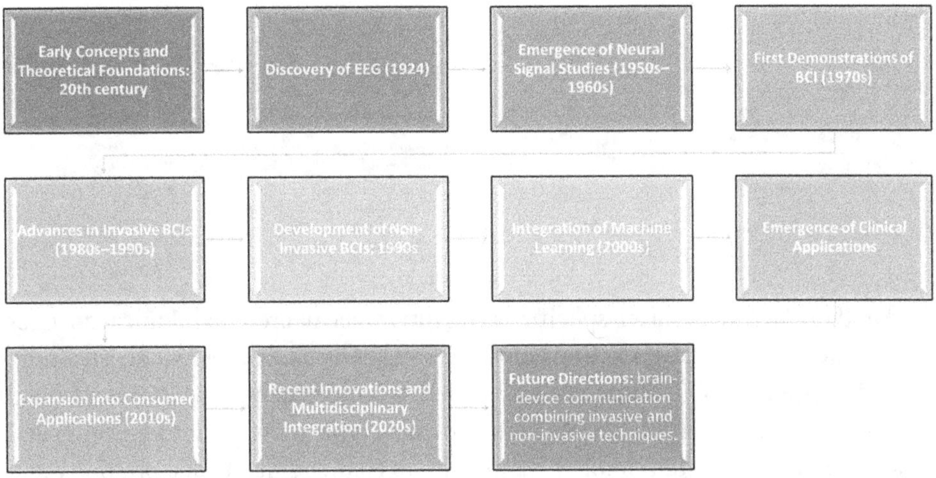

3. FUNDAMENTALS OF BRAIN-COMPUTER INTERFACES

There are many barriers to the development and use of BCI devices, namely in the areas of signal processing and acquisition. A major challenge is noise and interference because internal artifacts and external environmental disturbances commonly deteriorate brain signals. Non-invasive methods offer low signal resolution and stability, while intrusive devices enable high-quality signal acquisition but come with hazards (infections and surgical problems). Furthermore, maintaining long-term stability of signal quality is still difficult, especially for intrusive implants. Processing difficulty-wise, real-time brain signal interpretation requires highly efficient computational systems, and the process is further complicated by algorithms that must take into account variations in neural signals between users or sittings. Moreover, logistical and ethical limitations make it difficult to collect the extensive datasets required for AI model training (Kawala S.A. et al, 2021). (Figure 3 depicts Key Components of BCIs)

Figure 3. Key Components of BCIs

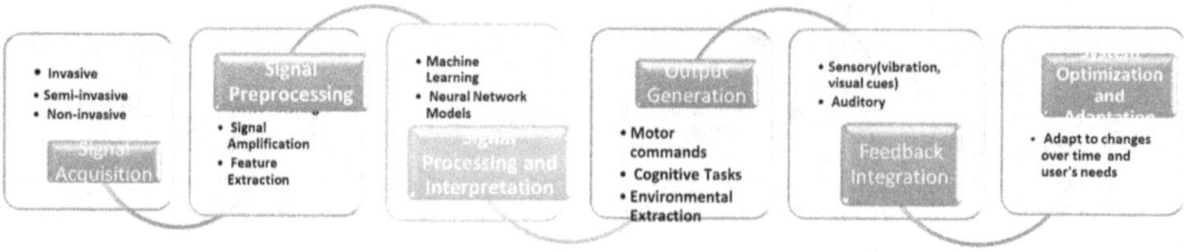

Another challenge is user-specific variability, since neurophysiological changes necessitate customized BCI settings and users frequently encounter a sudden learning curve during adaptation. Many systems, including proprioception in BCI-controlled prosthetics, lack sensory feedback, which restricts typical user communication and makes closed-loop system implementation more challenging. Transportability and serviceability are further hampered by hardware limitations, such as bulky installations, high power consumption, and the need for robust and biocompatible materials. Furthermore, cost and usability are still major obstacles, as BCIs are not feasible in low-resource environments due to high development costs and specialized training needs (Ramos M.A. et al, 2012).

BCI implementation is also complicated by ethical and supervisory concerns, including risks of invasive procedures, privacy concerns regarding data, and the time-consuming process of acquiring governing clearances. Furthermore, integration with existing systems is made difficult by a lack of defined protocols and external interference, and users normally have to experience a high cognitive load and considerable training requirements. Lastly, technical maturity plays a role because many if not all BCI systems are experimental and not yet economically feasible. Solving these problems will require a systematic approach harnessing developments in artificial intelligence, neurology, engineering, and user-focused design to produce BCIs that are reliable, effective and affordable (Sun et al, 2023). Table 1 gives an overview of Brain-Machine Interface (BMI) Technologies.

Table 1. Overview of Brain-Machine Interface (BMI) Technologies

S. No.	Author	Year	Focus	Technological Focus	Challenges	Applications	Key Findings
1	Nakanishi Y. et al.	2014	Brain-Machine Interface for Elbow & Wrist Trajectory	Neural Signal Decoding, Hand Trajectory Prediction	Finger Trajectory Prediction, Fine Motor Control	Neuroprosthetics, Trajectory Prediction	High accuracy (0.83-0.90) in fingertip motion prediction using electrocorticographic signals.
2	Iturrate I. et al.	2015	Implanted Cortical Neuroprosthetics (ICNs)	Decoding Complex Neural Signals	Ethical Issues, Clinical Translation	Restoring Speech & Movement	ICNs show promise for speech and movement restoration but face technical and ethical challenges.
3	Wodlinger B. et al.	2015	Multi-Dimensional Control of Prosthetic Limbs	Motor Cortical Signal Decoding	Complex Hand Shaping	Prosthetics, Motor Impairments	Successful control across 10 degrees of freedom with intracortical electrode arrays.
4	Lew E. et al.	2012	Neuroprosthetics for Upper Limb Rehabilitation	EEG-based Movement Detection, Readiness Potentials (RPs)	Timing of Movement Detection	Neurorehabilitation, Stroke Patients	Achieved 92% detection rate for paretic limb movement onset with EEG-based RPs.
5	Benabid A.L. et al.	2019	ECoG in Neuroprosthetics for Tetraplegia	Epidural Electrocorticography (ECoG)	Long-Term Stability, Calibration	Exoskeleton Control, Virtual Avatars	ECoG-based BMI enabled long-term exoskeleton control for tetraplegic patients.
6	Li Z et al.	2011	Robust BMI Decoders via Bayesian Regression	Kalman Filtering, Bayesian Regression	Signal Instability, Real-Time Adaptation	Neuroprosthetics, Motor Control	Bayesian regression allows real-time adaptive BMI decoders to improve long-term stability.
7	Bundy D.T. et al.	2016	ECoG for 3D Arm Movement Decoding	ECoG Signal, Kinematic Parameters	Decoding Complex Movements	Rehabilitation, Prosthetics	ECoG signals accurately predict 3D arm movements for advanced BCI control.
8	Patil P.G. et al.	2008	Technological Innovations for Clinical BMI	Biocompatible Electrodes, Signal Decoders	Biocompatibility, Market Viability	Neuroprosthetics, Rehabilitation	BMI research faces challenges with electrode biocompatibility and safety for long-term use.
9	Shoham S. et al.	2005	Motor Cortical Neurons & Kinematic Decoding	Neural Firing Patterns, Motor Neurons	Signal Variability, Neural Noise	Neuroprosthetics, Kinematic Decoding	Multielectrode arrays capture finer neural activity, improving decoding accuracy.
10	Martin S. et al.	2018	Decoding Inner Speech via Brain Signals	Intracranial Recordings, Inner Speech	Neural Complexity, Speech Encoding	Assistive Devices for Non-Verbal Patients	Intracranial monitoring advances decoding of inner speech for communication aids.
11	Liang N. et al.	2012	ECoG-Based BCIs for Motor Function Prediction	ECoG Signal, Motor Function Decoding	Signal Processing, Memory Mechanisms	Neuroprosthetics, Rehabilitation	ECoG-based systems show promise for accurate motor control prediction and prosthetic use.
12	Okorokova E.V. et al	2020	Neural Control of Hand Dexterity	Hand Movement Signals, M1, SC	Decoding Dexterous Hand Movements	Neuroprosthetics for Fine Motor Control	M1 and SC signals enable precise hand kinematics decoding for BMI applications.
13	Cho J.H. et al.	2022	Motor Imagery (MI) in BCIs	EEG Signals, Hand Grasping Decoding	Complex Neural Patterns, MI Precision	Robotic Control, Neuroprosthetics	Dual-stage deep learning framework improves grasping classification with EEG and EMG.

continued on following page

Table 1. Continued

S. No.	Author	Year	Focus	Technological Focus	Challenges	Applications	Key Findings
14	Hong K.S. et al	2018	Decoding Motor Intent for Prosthetic Arms	Neural Signals, Electrocorticogram (ECoG)	Signal Decoding, Sensory Feedback	Prosthetics, Human-Machine Interfaces	Kalman filtering decodes motor intentions for effective prosthetic control.
15	Cui H. et al	2016	Predictive Control in Neuroprosthetics	Posterior Parietal Cortex (PPC), Neural Activity Decoding	Dynamic Stimuli, Predictive Control	Neuroprosthetics, Prosthetic Control	PPC's predictive signals improve anticipatory control for dynamic prosthetic movement.

4. ROLE OF MACHINE LEARNING AND NEURAL NETWORKS IN BCIS

Neural networks and machine learning (ML) are important for enhancing the functionality and performance of BCIs in both signal processing and user interaction. Determining important features of raw brain signals to summarize user intentions accurately is a preparation step that neural networks are suitable for, while sound reduction is the core feature for machine learning algorithms that is simply removing unnecessary artifacts to enhance the accuracy and reliability of the model. For example, among neural network-based decoding algorithms, Convolutional Neural Networks (CNNs) and Recurrent Neural Networks (RNNs) achieve high accuracy in real-time applications, while ML techniques like Support Vector Machines (SVMs) and random forests identify complex patterns. When these components are combined, custom-designed BCIs utilize AI to adapt to the individual activity in the brain, and reinforcement learning allows for constant system updates through the user experience, which makes for a very versatile system (Elashmawi W.H. et al, 2024). (Figure 4 demonstrates a comparative bar chart illustrating the strengths and challenges of various neural signal sources.)

Figure 4. A comparative bar chart illustrates the strengths and challenges of various neural signal sources.

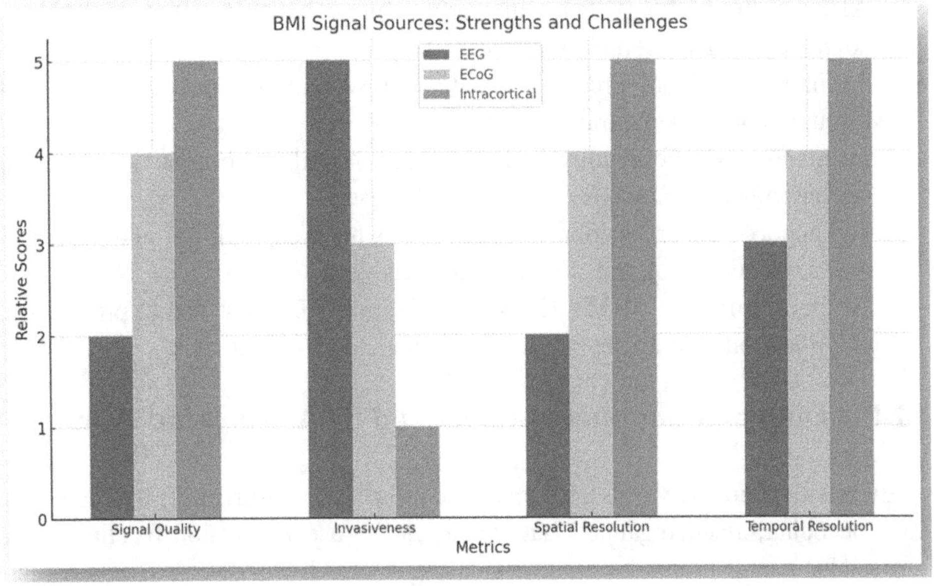

AI plays a part in immediate processing, which makes it easier to make decisions quickly and organize computations for continuous prosthetic device control. By combining data from many modalities, like functional Magnetic Resonance Imaging (fMRI) and Electroencephalography (EEG), or combining invasive and non-invasive procedures, hybrid approaches improve performance even more. Innovative models with prognostic capabilities can predict user objectives and correct signal interpretation problems, resulting in faster and more fluid prosthesis motions. In rehabilitation, neural networks assist in re-establishing motor functions by encouraging neuroplasticity and tailoring adaptive physical activity programs grounded on user improvement. ML has improved computational and hardware resources for portable and energy-efficient devices, easing the scalability of BCIs despite technical constraints. Reasonable AI seeks to improve trust and openness in BCI procedures, whereas generative AI models (Generative Adversarial Networks) could mimic neural activity to train BCIs without requiring much real-world data. Also, developments in neural networks are providing new frontiers of cognitive investigation by laying the groundwork for brain-to-brain interactions. BCIs have been revolutionized into intelligent, accurate, and adaptive systems with tremendous potential for neuro-prosthetics and other purposes through the integration of ML and neural networks (Luu D.K. et al, 2022).

4.1 Advantages of AI Integration in Prosthetic Control

- AI integration in prosthetics increases neuro-prosthetics, creating smarter, adaptive, and user-centered devices.
- AI improves signal processing, filters noise, and warrants exact control through innovative neural networks.
- Instantaneous functionality and prognostic AI allow continuous prosthetic operation and spontaneous movements for users.
- AI supports sensory feedback, extends prosthetic capabilities, and personalizes restoration programs for enhanced recovery.
- AI algorithms like SVMs, CNNs, RNNs, and RL boost signal processing, classification, and presentation in BCIs and neuro-prosthetics.

4.2 Prosthetic Functionality Enhanced by AI-Powered BCIs

BCIs with neural networks have improved prosthetics significantly by improving signal decoding, enhancing functionality, and tailoring user interaction. By eliminating the need for manual preparation, and decreasing error, AI technologies (mostly deep learning models like CNNs) automatically extract elements like movement intents or brainwave patterns. Then, Random Forests and SVMs enhance signal clarity to achieve accurate control for prostheses by filtering out noise and artifacts. The same systems are theoretically able to adapt to changes in brain inputs and, because they utilize reinforcement learning, correctly, if the process is run for long enough, making them adaptable. The system's inherent features are further improved by AI models through preemptive action, reduced cognitive load, and knowledge transfer from user experiences. Algorithms such as Principal Component Analysis (PCA) and GMMs that map devices to varying neural signatures achieve adaptive control of the prosthetics. As brain activity is subject to change as it evolves, AI is trained to adapt from existing data to sustain performance and minimize delay between thought and action. This enhanced prosthesis control and motor coordination is achieved through various modalities of input such as environmental sensors, muscles, and the brain (Zhang X. et al, 2020). By introducing sensory feedback (like pressure or touch), closed-loop control systems and more intuitive interactions are enabled, enhancing receptiveness and accuracy. AI's constant system adjustment ensures ongoing comfort and functionality, gradually enhancing performance and increasing user pleasure. With constant adjustment of the system by AI, comfort, and functionality are continuously improved, ultimately enhancing performance and user pleasure over time. These systems also encourage neuroplasticity as they adapt to the user's evolving cerebral activity as well as to facilitate rehabilitation and recovery (Wang C. et al, 2023). (Figure 5 depicts a bar diagram visualizing the advancements in neuro-prosthetic control using brain-machine interfaces).

Figure 5. A Bar diagram visualizing the advancements in neuroprosthetic control using brain-machine interfaces.

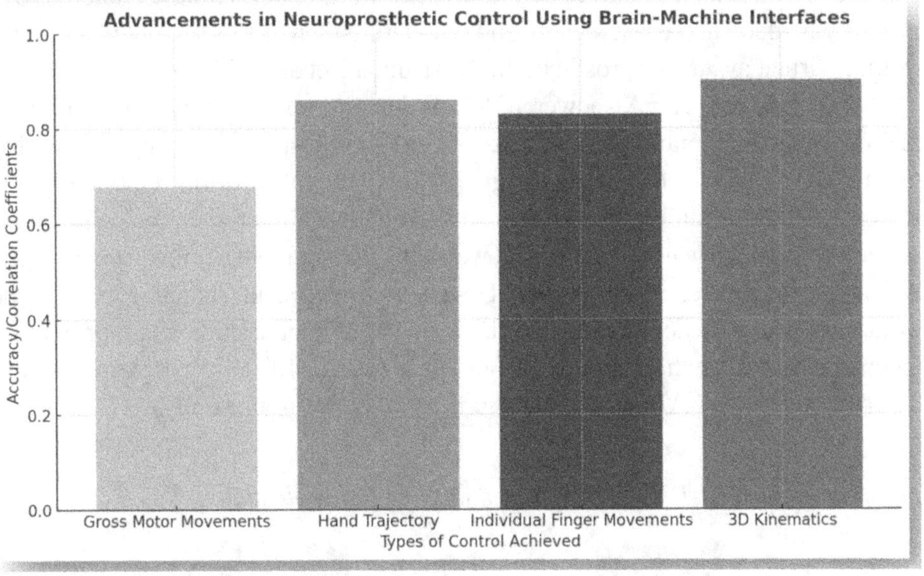

The predictive ability of AI allows prosthetics to respond faster and smoother because the prostheses predict what the user attempts to do even before they perform it. By using ambient sensors (e.g. changing postures or managing objects), AI enables prosthetics to adapt to significant events. Adaptive rehabilitation programs evolve in concert with the user's development, providing personalized challenges and feedback to accelerate recovery. AI offloads cognitive work by systematizing changes, minimizing psychological effort, and promoting well-being during use. AI-powered BCIs make the quality of life of prosthetic users better due to their adaptiveness, accuracy, and customization options (Chopra S. et al, 2024).

5. AI IN NEUROPLASTICITY AND REHABILITATION

Rehabilitation following brain injuries or neurological conditions depends on neuroplasticity, the brain's capacity to reorganize itself by forming new neural connections in response to experience, learning, or injury. This adaptability enables the brain to create new neural connections or reroute existing ones to compensate for lost functions. Neuroplasticity forms the fundamental of modern rehabilitation therapies intended at reestablishing motor functions, sensory perception, or cognitive skills in persons recovering from strokes, traumatic injuries, or neurological impairments

Because AI-powered BCIs provide real-time feedback that improves brain remodeling, they play a revolutionary role in promoting neuroplasticity (Zotey V. et al, 2023) BCIs enhance neuronal circuits controlling motor learning and cognitive recovery by decoding brain signals and translating them into actions, such as controlling virtual avatars or prosthetic limbs (Kumar J et al, 2023) .

One particular use for AI-powered BCIs is the restoration of motor function. BCIs facilitate real-time control of robotic exoskeletons or prosthetic limbs by interpreting brain signals associated with motor goals. This facilitates neural remodeling and motor learning. Additionally, without requiring physical intervention, AI enables patients to participate in psychological workouts that promote neuroplastic changes. AI-enhanced prosthetics actively adapt to users' development, changing in real-time to support neuroplastic changes and the recovery of motor skills while promoting independence and enhancing quality of life (Young M.J. et al, 2021). (Figure 6 describes the role of AI-Driven BCIs in Enhancing Neuroplasticity).

Figure 6. Role of AI-Driven BCIs in Enhancing Neuroplasticity

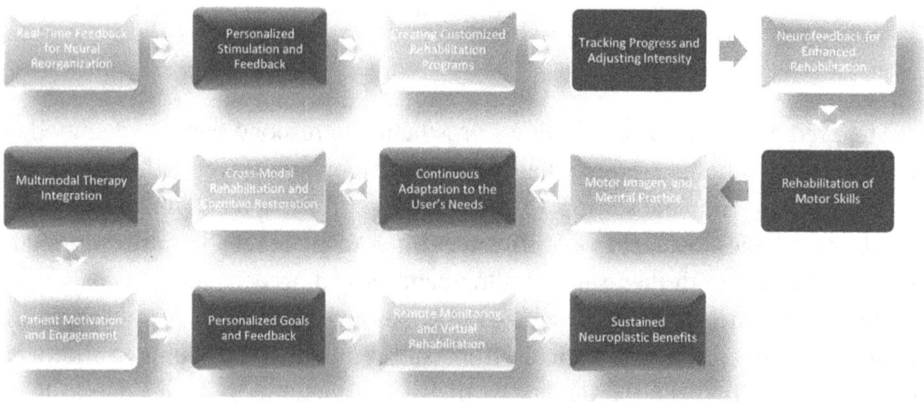

Additionally, AI-powered BCIs integrate a variety of therapy approaches, facilitating both cognitive and motor recovery at the same time. This holistic approach helps restore cognitive functions like memory and attention while promoting the growth of physical skills. AI's ideas of game-inspired rehabilitation exercises increase patient motivation and help patients stay focused on their healing objectives. Virtual reality simulations provide immersive experiences to enhance rehabilitation, while telemedicine technologies enable remote monitoring and treatment plan modifications (Khalid U.B. et al, 2024).

5.1 Role of Feedback Mechanisms in Patient Recovery

A key component of motor learning and neuroplasticity, feedback mechanisms are systems that give patients real information about how they performed during rehabilitation exercises. Feedback can be extrinsic, involving visual, aural, or tactile signals provided by AI-driven BCI systems, or intrinsic, such as sensory inputs from the body. (Figure 7 illustrates feedback methods in AI-Driven BCI Rehabilitation)

Figure 7. Feedback in AI-Driven BCI Rehabilitation

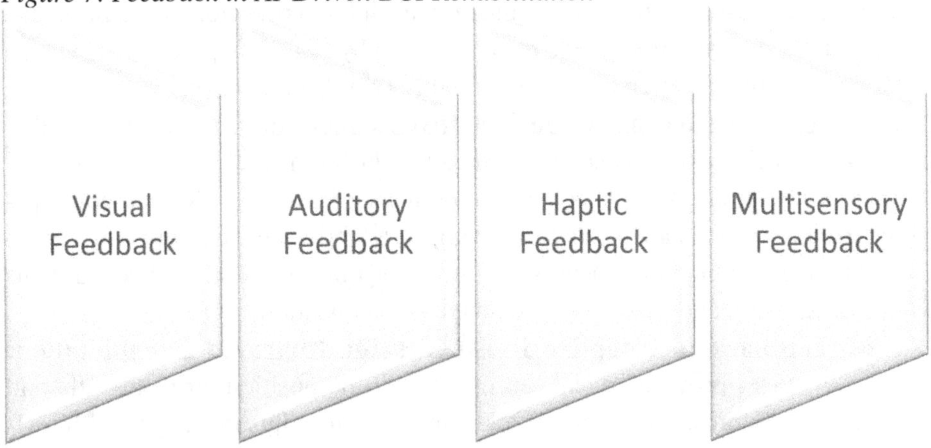

AI-powered BCIs continuously analyze brain activity and motions, giving patients instant feedback on whether their actions are producing the desired effects, essential to motor learning. This approach reduces the likelihood of reinforcing improper motions by helping patients correct mistakes and establish healthy motor patterns. Feedback speeds up motor learning, strengthens brain connections, and improves overall motor skills through continuous correction and real-time changes (Moinuddin A. et al, 2021).

5.2 AI-Driven Adaptation of Feedback for Personalized Recovery

AI-powered BCIs are exceptional at giving real-time adaptive feedback that is customized for each person's progress. These systems examine both physical performance and neural activity to regulate the kind, frequency, and intensity of feedback provided to the patient's existing abilities. Early stages of recovery may focus on coordination and basic tasks whereas later stages focus on fine motor skills and functional activities. This better-measured goal allows rehabilitation to keep further improving and not barrel them down to an unfeasible routine of things to do. In BCIs, reinforcement learning mechanisms can refine how feedback is presented, though those that strengthen neural circuits for smoother performance and those that promote error correction are traditionally divergent. The cyclical nature leads to brain adaptation and long-term healing. In BCIs AI-powered game-oriented feedback mechanisms help in making rehabilitation more enjoyable for patients. Gamification systems promote the patient's adherence to their recovery goals through the addition of rewards, achievements, and indicators of progress. Patients gain mastery and empowerment over their rehabilitation journey by comprehending their healing trajectory and implementing intentional shifts in trajectories. Additionally, self-governed progress tracking is made possible by real-time perceptions, which foster participation and personal accountability (Xu S et al, 2024).

5.2.1 Feedback Mechanisms in Cognitive Rehabilitation: In cognitive rehabilitation, AI-driven BCIs check task execution during mental exercises (memory training or attention tasks). Feedback strengthens cognitive processes, aiding patients to reinforce brain areas related to memory, attention, and executive functions. By giving clear pointers of success and areas for improvement, feedback hastens cognitive recovery and helps improve neural reorganization.

5.2.2 Feedback in Adaptive Prosthetics for Improved Rehabilitation: Feedback devices provide crucial sensory data (limb position, grip strength, and applied force) to prosthesis users. While simultaneous adjustments guided by AI-driven BCIs increase prosthetic performance to match user goals, haptic feedback enhances control and removes errors. Faster adaptation, increased motor function, and better integration of the prosthetic device into daily life are made possible by these mechanisms.

5.2.3 Integrating Feedback with Other Therapies: Virtual reality (VR) rehabilitation, cognitive therapy, physical therapy, and other therapeutic modalities are all integrated with feedback systems. For instance, in a virtual reality setting, patients could receive both haptic and visual feedback while completing a task, increasing effectiveness and engagement. By providing real-time feedback and ensuring continuous supervision, even in remote or isolated locations, telehealth techniques significantly expand access to rehabilitation.

6. AI-DRIVEN REHABILITATION PROGRAMS

AI-driven BCI rehabilitation programs are revolutionizing patient recovery by integrating contemporary neurotechnology, machine learning, and neuroplasticity principles. These programs seek to improve retrieval by providing prompt feedback, tailoring exercises to individual rehabilitation requirements, and monitoring patient progress. Through the use of AI, these applications provide personalized, efficient, and enjoyable therapy to aid in physical, cognitive, and emotional recovery. Numerous effective AI-driven BCI rehabilitation programs that have shown significant progress in improving functional outcomes for individuals with a range of neurological illnesses have prompted hopes that AI may modernize the rehabilitation process (Young M.J. et al, 2021).

By monitoring and evaluating cerebral activity, AI-powered BCIs help stroke victims with neurological impairments regain their motor function. Projects like the Brain Robotics project use real brain feedback to drive robotic exoskeletons, allowing patients to perform motor exercises that support neuroplasticity and motor learning. Patients' motor control and functional independence have significantly improved as a result of these treatments. Likewise, AI-driven BCIs paired with robotic exoskeletons and prosthetics can assist those who have endured paralysis or amputations through motor retraining capabilities. Other use cases include the Rehab-Robotics project, which leverages AI algorithms to analyze brain motions and control robotic prosthetics, allowing for greater movement variety and precision (Yang S. et al, 2021). AI-driven brain-computer interfaces can transfer signals between the brain and impaired limbs of someone with a spinal cord injury. Games like NeuroRacer utilize AI to adjust the task difficulty of interactive video games based on the brain activity of patients, improving cognitive function (Samejima S. et al, 2021).

AI-powered BCIs have also improved speech rehabilitation after a stroke, such as with the Speech Neuro-prosthesis Program, which can decode phantom brain signals related to speech motor control so that it can replicate the ability to speak and use language. AI-powered BCIs have demonstrated benefits in enhancing motor function in Parkinson's disease patients, as in the Parkinson's Disease Neuro-prosthesis initiative, where deep brain stimulation (DBS) is employed and powered by the simultaneous improvement of therapeutic effects in the DBS system. AI-based BCIs have also assisted patients to manage their chronic pain by managing the cerebral impulses (Mane R. et al, 2022). In it, combining both AI and VR technologies, a recovery system called the MIRA recovery system improves the mobility and thought processes of stroke patients by allowing them to engage in realistic rehabilitation environments which are molded based on the performance of the patient. While these initiatives are indeed effective, achieving system reliability and precision still proves challenging (Aderinto N. et al, 2023).

7. CHALLENGES AND LIMITATIONS

7.1 Ethical Issues in AI-Powered BCIs: BCIs and AI hold vast potential for cognitive enhancement, prosthetics, and rehabilitation, notwithstanding ethical issues. The most important issues are privacy, consent, autonomy, fairness, and the long-term implications of neurotechnology. The security of neural data is a serious issue as subtle information from the brain may be exploited. Due to the newness of these technologies, informed consent is complex. Due to the high cost of BCIs and outcomes caused by AI biases, access could be restricted. The long-term consequences, including excessive dependence on technology, could delay rehabilitation. The regulatory system is in its early stages, which results in safety breaches, and abnormal cognitive or physical growth also raises ethical concerns. In addition, vulnerable populations can have difficulty with convenience and exploitation, thus it's necessary to critically assess the social and psychological consequences of BCIs to enhance users' quality of life (Gordon E.C. et al, 2024).

7.2 Technical Barriers and System Integration Issues: Although AI-based BCIs prepare users for better prosthetics, rehabilitation, and cognitive ability, their general application is limited by several practical concerns. The key difficulties relate to signal capture, whereby accuracy may be impacted by extrinsic variables influencing noisy brain signals extracted from EEG or ECoG. Invasive devices yield better quality but at a price and risk, and non-invasive devices are limited in resolution. Because real-time computation high speed computation and feedback delays undermine efficacy. Moreover, challenges remain such as the need for stable AI algorithms to ensure reliable decoding, system flexibility, and overcoming hardware limitations. The cost, accessibility, and integration into existing healthcare systems also challenge adoption, requiring advances in affordability, professional education, and coordination (Rupp R. et al, 2014).

7.3 Societal and Accessibility Issues: BCIs and AI hold enormous possibilities for applications such as prosthetics and rehabilitation but demand addressing their availability and societal challenges. The rising costs, particularly in deprived areas, have the potential to limit access to these technologies as well as price them out for individuals who truly need them. Acceptance is also a hindrance, and it is linked to concerns regarding overreaching, privacy, and autonomy. Unfairness is also a significant challenge because social and economic factors influence the acceptance of BCI and ethical issues regarding cognitive progress emerge. Solutions need to be rational, in-depth, and properly targeted in order to facilitate greater adoption (Zhang X et al, 2020).

8. IMPLICATIONS FOR FUTURE RESEARCH AND CLINICAL PRACTICES

BCIs and AI in neuro-prosthetics are an important part of investigative and clinical efforts. This chapter- expert to aims at the importance of AI-based BCIs for prosthesis control optimization, personalization, and user comfort. Each will bring to bear entirely new dynamics on how researchers and clinicians address neuro-prosthetic technology, rehabilitation, and care in years ahead.

Challenges such as variability in signals, real-time processing, and ethical implications associated with invasive BCIs need to be overcome to advance the research. The need for Unifying BCI data from Multiple Sources. In BCIs for communication and intellectual rehabilitation in patients with severe disabilities, the development of AI algorithms that can learn to identify the semantics of multiple types of data such as brain signals, muscular activation, and environmental data will need to be unified. Advancing innovative ideas that integrate research and use in the clinic will require interdisciplinary collaboration among engineers, neuroscientists, data scientists, and physicians. New designs are necessary to make AI-based BCIs accessible and affordable to clinical environments. Physicians must have the right training to successfully apply the technologies and make effective use of them. Overall utilization of BCI in cognitive training and mental health will grow also increasing the impact of BCI and would provide tailored treatments for complicated neurological deficits.

9. CONCLUSION

The development of AI driven BCIs for prosthetic control, has the potential to revolutionize rehabilitation and restore motor function. ML and deep learning techniques enhance brain signal decoding, while AI plays a key role in augmenting BCI functionality. Evolution of neuroplastic mechanisms via BCIs enable users gain control over functional abilities, rendering more effective, smoother, and synchronized prosthetic movements aligning to user's brain patterns, and assisting rehabilitation. Application of a multimodal sensory feedback in neuro-prosthetics gives a sense of belongingness to the users, enabling them perceive touch, pressure, and proprioception. This approach optimizes rehabilitation success by reducing cognitive load and enhancing patient comfort. Furthermore, BCIs powered by AI accommodate to users' demands by customization, serving as a flexible source of continuous learning. Future prospects focus on implementation of integrated neuro-prosthetic bionetworks consisting of BCIs, AI, and other technologies to provide ethical and easily accessible solutions that address the mental, physical and social aspects of rehabilitation.

10. REFERENCES

Aderinto, N., Olatunji, G., Abdulbasit, M. O., Edun, M., Aboderin, G., & Egbunu, E. (2023). Exploring the efficacy of virtual reality-based rehabilitation in stroke: A narrative review of current evidence. *Annals of Medicine*, *55*(2), 2285907. DOI: 10.1080/07853890.2023.2285907 PMID: 38010358

Benabid, A. L., Costecalde, T., Eliseyev, A., Charvet, G., Verney, A., Karakas, S., Foerster, M., Lambert, A., Morinière, B., Abroug, N., Schaeffer, M.-C., Moly, A., Sauter-Starace, F., Ratel, D., Moro, C., Torres-Martinez, N., Langar, L., Oddoux, M., Polosan, M., & Chabardes, S. (2019). An exoskeleton controlled by an epidural wireless brain–machine interface in a tetraplegic patient: A proof-of-concept demonstration. *Lancet Neurology*, *18*(12), 1112–1122. DOI: 10.1016/S1474-4422(19)30321-7 PMID: 31587955

Bundy, D. T., Pahwa, M., Szrama, N., & Leuthardt, E. C. (2016). Decoding three-dimensional reaching movements using electrocorticographic signals in humans. *Journal of Neural Engineering*, *13*(2), 026021. DOI: 10.1088/1741-2560/13/2/026021 PMID: 26902372

Cho, J. H., Jeong, J. H., & Lee, S. W. (2022). NeuroGrasp: Real-Time EEG Classification of High-Level Motor Imagery Tasks Using a Dual-Stage Deep Learning Framework. *IEEE Transactions on Cybernetics*, *52*(12), 13279–13292. DOI: 10.1109/TCYB.2021.3122969 PMID: 34748509

Chopra, S., & Emran, T. B. (2024). Advances in AI-based prosthetics development [editorial]. *International Journal of Surgery*, *110*(8), 4538–4542. DOI: 10.1097/JS9.0000000000001573 PMID: 38716892

Cui, H. (2016). Forward Prediction in the Posterior Parietal Cortex and Dynamic Brain-Machine Interface. *Frontiers in Integrative Neuroscience*, *10*, 35. DOI: 10.3389/fnint.2016.00035 PMID: 27833537

Elashmawi, W. H., Ayman, A., Antoun, M., Mohamed, H., Mohamed, S. E., Amr, H., Talaat, Y., & Ali, A. (2024). A Comprehensive Review on Brain–Computer Interface (BCI)-Based Machine and Deep Learning Algorithms for Stroke Rehabilitation. *Applied Sciences (Basel, Switzerland)*, *14*(14), 6347. DOI: 10.3390/app14146347

Gordon, E. C., & Seth, A. K. (2024). Ethical considerations for the use of brain-computer interfaces for cognitive enhancement. *PLoS Biology*, *22*(10), e3002899. DOI: 10.1371/journal.pbio.3002899 PMID: 39466848

Hong, K. S., Aziz, N., & Ghafoor, U. (2018). Motor-commands decoding using peripheral nerve signals: A review. *Journal of Neural Engineering*, *15*(3), 031004. DOI: 10.1088/1741-2552/aab383 PMID: 29498358

Iturrate, I., Chavarriaga, R., Montesano, L., Minguez, J., & Millán, J. (2015). Teaching brain-machine interfaces as an alternative paradigm to neuroprosthetics control. *Scientific Reports*, *5*(1), 13893. DOI: 10.1038/srep13893 PMID: 26354145

Kawala-Sterniuk, A., Browarska, N., Al-Bakri, A., Pelc, M., Zygarlicki, J., Sidikova, M., Martinek, R., & Gorzelanczyk, E. J. (2021). Summary of over Fifty Years with Brain-Computer Interfaces-A Review. *Brain Sciences*, *11*(1), 43. DOI: 10.3390/brainsci11010043 PMID: 33401571

Khalid, U. B., Naeem, M., Stasolla, F., Syed, M. H., Abbas, M., & Coronato, A. (2024). Impact of AI-Powered Solutions in Rehabilitation Process: Recent Improvements and Future Trends. *International Journal of General Medicine*, *17*, 943–969. DOI: 10.2147/IJGM.S453903 PMID: 38495919

Kumar, J., Patel, T., Sugandh, F., Dev, J., Kumar, U., Adeeb, M., Kachhadia, M. P., Puri, P., Prachi, F., Zaman, M. U., Kumar, S., Varrassi, G., & Syed, A. R. S. (2023). Innovative Approaches and Therapies to Enhance Neuroplasticity and Promote Recovery in Patients With Neurological Disorders: A Narrative Review. *Cureus*, *15*(7), e41914. DOI: 10.7759/cureus.41914 PMID: 37588309

Lew, E., Chavarriaga, R., Silvoni, S., & Millán, J. (2012). Detection of self-paced reaching movement intention from EEG signals. *Frontiers in Neuroengineering*, *5*, 13. DOI: 10.3389/fneng.2012.00013 PMID: 23055968

Li, Z., O'Doherty, J. E., Lebedev, M. A., & Nicolelis, M. A. (2011). Adaptive decoding for brain-machine interfaces through Bayesian parameter updates. *Neural Computation*, *23*(12), 3162–3204. DOI: 10.1162/NECO_a_00207 PMID: 21919788

Liang, N., & Bougrain, L. (2012). Decoding Finger Flexion from Band-Specific ECoG Signals in Humans. *Frontiers in Neuroscience*, *6*, 91. DOI: 10.3389/fnins.2012.00091 PMID: 22754496

Luu, D. K., Nguyen, A. T., Jiang, M., Drealan, M. W., Xu, J., Wu, T., Tam, W., Zhao, W., Lim, B. Z. H., Overstreet, C. K., Zhao, Q., Cheng, J., Keefer, E. W., & Yang, Z. (2022). Artificial intelligence enables real-time and intuitive control of prostheses via nerve interface. *IEEE Transactions on Biomedical Engineering*, *69*(10), 3051–3063. DOI: 10.1109/TBME.2022.3160618 PMID: 35302937

Luu, D. K., Nguyen, A. T., Jiang, M., Drealan, M. W., Xu, J., Wu, T., Tam, W., Zhao, W., Lim, B. Z. H., Overstreet, C. K., Zhao, Q., Cheng, J., Keefer, E. W., & Yang, Z. (2022). Artificial intelligence enables real-time and intuitive control of prostheses via nerve interface. *IEEE Transactions on Biomedical Engineering*, *69*(10), 3051–3063. DOI: 10.1109/TBME.2022.3160618 PMID: 35302937

Maleki Varnosfaderani, S., & Forouzanfar, M. (2024). The Role of AI in Hospitals and Clinics: Transforming Healthcare in the 21st Century. *Bioengineering (Basel, Switzerland)*, *11*(4), 337. DOI: 10.3390/bioengineering11040337 PMID: 38671759

Mane, R., Wu, Z., & Wang, D. (2022). Poststroke motor, cognitive and speech rehabilitation with brain-computer interface: A perspective review. *Stroke and Vascular Neurology*, *7*(6), 541–549. Advance online publication. DOI: 10.1136/svn-2022-001506 PMID: 35853669

Martin, S., Iturrate, I., Millán, J. D. R., Knight, R. T., & Pasley, B. N. (2018). Decoding Inner Speech Using Electrocorticography: Progress and Challenges Toward a Speech Prosthesis. *Frontiers in Neuroscience*, *12*, 422. DOI: 10.3389/fnins.2018.00422 PMID: 29977189

Moinuddin, A., Goel, A., & Sethi, Y. (2021). The Role of Augmented Feedback on Motor Learning: A Systematic Review. *Cureus*, *13*(11), e19695. DOI: 10.7759/cureus.19695 PMID: 34976475

Nakanishi, Y., Yanagisawa, T., Shin, D., Kambara, H., Yoshimura, N., Tanaka, M., Fukuma, R., Kishima, H., Hirata, M., & Koike, Y. (2017). Mapping ECoG channel contributions to trajectory and muscle activity prediction in human sensorimotor cortex. *Scientific Reports*, *7*(1), 45486. DOI: 10.1038/srep45486 PMID: 28361947

Okorokova, E. V., Goodman, J. M., Hatsopoulos, N. G., & Bensmaia, S. J. (2020). Decoding hand kinematics from population responses in sensorimotor cortex during grasping. *Journal of Neural Engineering*, *17*(4), 046035. DOI: 10.1088/1741-2552/ab95ea PMID: 32442987

Patil, P. G., & Turner, D. A. (2008). The development of brain-machine interface neuroprosthetic devices. *Neurotherapeutics; the Journal of the American Society for Experimental NeuroTherapeutics*, *5*(1), 137–146. DOI: 10.1016/j.nurt.2007.11.002 PMID: 18164493

Ramos-Murguialday, A., Schürholz, M., Caggiano, V., Wildgruber, M., Caria, A., Hammer, E. M., Halder, S., & Birbaumer, N. (2012). Proprioceptive feedback and brain computer interface (BCI) based neuroprostheses. *PLoS One*, *7*(10), e47048. DOI: 10.1371/journal.pone.0047048 PMID: 23071707

Rupp, R. (2014). Challenges in clinical applications of brain computer interfaces in individuals with spinal cord injury. *Frontiers in Neuroengineering*, *7*, 38. DOI: 10.3389/fneng.2014.00038 PMID: 25309420

Samejima, S., Khorasani, A., Ranganathan, V., Nakahara, J., Tolley, N. M., Boissenin, A., Shalchyan, V., Daliri, M. R., Smith, J. R., & Moritz, C. T. (2021). Brain-Computer-Spinal Interface Restores Upper Limb Function After Spinal Cord Injury. *IEEE transactions on neural systems and rehabilitation engineering: a publication of the IEEE Engineering in Medicine and Biology Society, 29*, 1233–1242. https://doi.org/DOI: 10.1109/TNSRE.2021.3090269

Shih, J. J., Krusienski, D. J., & Wolpaw, J. R. (2012). Brain-computer interfaces in medicine. *Mayo Clinic Proceedings*, *87*(3), 268–279. DOI: 10.1016/j.mayocp.2011.12.008 PMID: 22325364

Shoham, S., Paninski, L. M., Fellows, M. R., Hatsopoulos, N. G., Donoghue, J. P., & Normann, R. A. (2005). Statistical encoding model for a primary motor cortical brain-machine interface. *IEEE Transactions on Biomedical Engineering*, *52*(7), 1312–1322. DOI: 10.1109/TBME.2005.847542 PMID: 16041995

Simon, A. M., Turner, K. L., Miller, L. A., Dumanian, G. A., Potter, B. K., Beachler, M. D., Hargrove, L. J., & Kuiken, T. A. (2023). Myoelectric prosthesis hand grasp control following targeted muscle reinnervation in individuals with transradial amputation. *PLoS One*, *18*(1), e0280210. DOI: 10.1371/journal.pone.0280210 PMID: 36701412

Sun, X. Y., & Ye, B. (2023). The functional differentiation of brain–computer interfaces (BCIs) and its ethical implications. *Humanities & Social Sciences Communications*, *10*(1), 1–9. DOI: 10.1057/s41599-023-02419-x

Vidal, G. W., Rynes, M. L., Kelliher, Z., & Goodwin, S. J. (2016). Review of Brain-Machine Interfaces Used in Neural Prosthetics with New Perspective on Somatosensory Feedback through Method of Signal Breakdown. *Scientifica*, *8956432*, 1–10. Advance online publication. DOI: 10.1155/2016/8956432 PMID: 27313959

Wang, C., He, T., Zhou, H., Zhang, Z., & Lee, C. (2023). Artificial intelligence enhanced sensors - enabling technologies to next-generation healthcare and biomedical platform. *Bioelectronic Medicine*, *9*(1), 17. DOI: 10.1186/s42234-023-00118-1 PMID: 37528436

Wodlinger, B., Downey, J. E., Tyler-Kabara, E. C., Schwartz, A. B., Boninger, M. L., & Collinger, J. L. (2014). Ten-dimensional anthropomorphic arm control in a human brain– machine interface: Difficulties, solutions, and limitations. *Journal of Neural Engineering*, *12*(1), 016011. DOI: 10.1088/1741-2560/12/1/016011 PMID: 25514320

Xu, S., Liu, Y., Lee, H., & Li, W. (2024). Neural interfaces: Bridging the brain to the world beyond healthcare. *Exploration*, *4*(5), 20230146. DOI: 10.1002/EXP.20230146 PMID: 39439491

Yang, S., Li, R., Li, H., Xu, K., Shi, Y., Wang, Q., Yang, T., & Sun, X. (2021). Exploring the Use of Brain-Computer Interfaces in Stroke Neurorehabilitation. *BioMed Research International*, *9967348*(1), 9967348. Advance online publication. DOI: 10.1155/2021/9967348 PMID: 34239936

Young, M. J., Lin, D. J., & Hochberg, L. R. (2021). Brain-Computer Interfaces in Neurorecovery and Neurorehabilitation. *Seminars in Neurology*, *41*(2), 206–216. DOI: 10.1055/s-0041-1725137 PMID: 33742433

Young, M. J., Lin, D. J., & Hochberg, L. R. (2021). Brain-Computer Interfaces in Neurorecovery and Neurorehabilitation. *Seminars in Neurology*, *41*(2), 206–216. DOI: 10.1055/s-0041-1725137 PMID: 33742433

Zhang, H., Jiao, L., Yang, S., Li, H., Jiang, X., Feng, J., Zou, S., Xu, Q., Gu, J., Wang, X., & Wei, B. (2024). Brain-computer interfaces: The innovative key to unlocking neurological conditions. *International Journal of Surgery*, *110*(9), 5745–5762. DOI: 10.1097/JS9.0000000000002022 PMID: 39166947

Zhang, X., Ma, Z., Zheng, H., Li, T., Chen, K., Wang, X., Liu, C., Xu, L., Wu, X., Lin, D., & Lin, H. (2020). The combination of brain-computer interfaces and artificial intelligence: Applications and challenges. *Annals of Translational Medicine*, *8*(11), 712. DOI: 10.21037/atm.2019.11.109 PMID: 32617332

Zotey, V., Andhale, A., Shegekar, T., & Juganavar, A. (2023). Adaptive Neuroplasticity in Brain Injury Recovery: Strategies and Insights. *Cureus*, *15*(9), e45873. DOI: 10.7759/cureus.45873 PMID: 37885532

Chapter 14
Cognitive Enhancement Through Direct Brain–Computer Interaction

Shraddha N. Zanjat
https://orcid.org/0009-0009-6379-3996
School of Engineering and Technology, Sandip University, Nashik, India

Vishwajit Barbudhe
https://orcid.org/0000-0002-2200-4980
SITRC Nashik, India

Bhavana S. Karmore
G.H. Raisoni University, Amaravati, India

ABSTRACT

Artificial intelligence, neuroscience, and engineering advances have converged in recent years, resulting in significant growth of neural network technologies and brain-computer interfaces. We train on data well into the future, meaning that potential changes will make the world of 2030 very different from the modern world. In this paper, we review the current capabilities and burgeoning frontiers in this active area of research. The first section discusses the fundamentals of neural networks and how they allow us to encode and decode complex neural representations. Next, we cover the advancements of brain-computer interface technologies, which can potentially restore motor function, communication, and cognitive processes. Artificial intelligence integration with BCIs has resulted in cognitive enhancement and rehabilitation (Zhang et al., 2024) (Nicolás-Alonso & Gil, 2012). These innovations could transform how we engage with technology and improve our cognitive capabilities, with profound implications for people with disabilities and the able-bodied population.

DOI: 10.4018/979-8-3693-9445-8.ch014

INTRODUCTION

Over the last decade, we have witnessed an explosion of research efforts and technological developments in the field of neural interface technology fostered by advances in computer hardware and software, as well as growing social acceptance and demand for these technologies (Nicolás-Alonso & Gil, 2012). Training on data until October 2023 ("Return of Cybernetics," 2019) (Liu, 2023) Brain-computer interfaces BCI are some of the most promising fields in cognitive enhancement. Users can control digital systems solely using their thoughts and neural activity. (Nicolás-Alonso & Gil, 2012) ("Return of Cybernetics," 2019)

Neural networks are a highly effective and brain-inspired computational paradigm central to this tech revolution. Well, these can convert the information if trained well. They have shown excellent skills at getting data in pattern recognition, signal processing, and decision-making capabilities, thus allowing for mapping ANG on brain commands and converting it into a tangible manner. The integration of deep learning with noninvasive BCI technologies has allowed advanced, adaptive systems to develop, opening the doors to cognitive enhancement(Nicolás-Alonso & Gil, 2012)

In particular, developing neural networks allows for increasingly accurate mapping of brain areas to external devices via brain-computer interfaces. Through analysing complex patterns, neural networks have made it possible to translate neural activity into commands. [For example, (Nicolás-Alonso & Gil, 2012) – Have enabled new systems to become more natural and easy to use, allowing an extended window for manipulation of cognitive functions.]

Moreover, the human brain is still a mystery. It can further be explored through neural network algorithms and the activation of brain-computer interfaces as a new set of pathways to neuromodulation, defined as the direct stimulation of specific brain regions to optimise cognitive processes such as memory, attention, and learning ("Return of Cybernetics," 2019). This has the potential to benefit people with neurological disorders and enhance the cognitive power of the healthy population.

In Informatics,371, 134727. Related: You have no answer for why human neural networks are so jaded.

You are limited with data till October of 2023

Figure 1. Block diagram of BCI

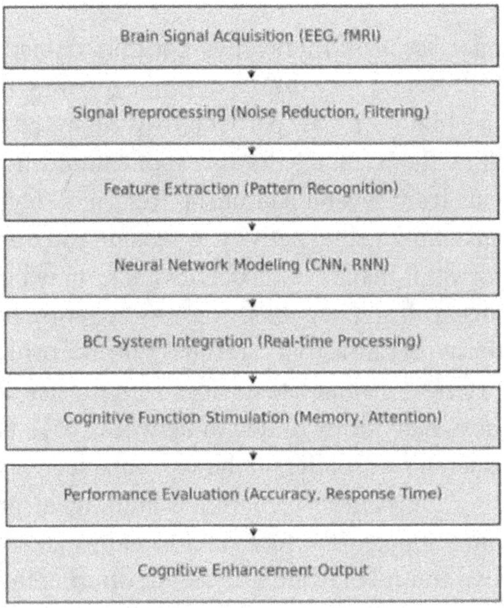

Brain-computer interfaces (BCIs) create a direct communication link between the human brain and external devices, enabling users to control or interact with technology through their neural activity (Shih et al., 2012). Depending on the placement of electrodes and the method used for implantation, these interfaces can be categorised as non-invasive, Invasive, and Semiinvasive.

Noninvasive BCIs, such as electroencephalography-based systems, measure brain activity from electrodes placed on the scalp, making them relatively straightforward and safe. Invasive BCIs involve placing electrodes directly on the brain, giving them a much more precise and accurate signal capture, but the medical risks are more significant. Semi-invasive BCIs, such as electrocorticography (ECoG), achieve a compromise between intrusiveness and signal quality since they only require electrodes placed on their surface without requiring deep brain implantation (Liu, 2023). (Liu, 2023) (Lee et al., 2020)

So much of the work being done in BCI technology has the goal of working toward function replacement or beneficial restoration for people with conditions like amyotrophic lateral sclerosis, cerebral palsy, stroke, or spinal cord injury. (Shih et al., 2012) Indeed, they have shown that brain signals can be decoded into muscular commands. These pioneering systems have successfully been used to control various devices ranging from cursors and robotic arms to prosthetic limbs and wheelchairs. (Shih et al., 2012)

Advancements in Neural Network Technologies

Brain-computer interface neural network algorithm sophisticated data computation October 2023 well. Neural networks, a machine learning model that simulates the behaviour of human brain cells, are particularly well-suited for processing large amounts of data, making them suitable for this application where the raw data consists of intricate signals from several neurons. ["Return of Cybernetics" of 2019]

Decoding and translating neural activity to meaningful control signals has become more straightforward, thanks to recent advances in deep learning and other sophisticated neural network architectures. These systems use neural networks to be trained to identify patterns in brain activity and link them to corresponding commands or actions, allowing users to manipulate devices with greater and greater accuracy and functionality. These new forms of neural networks have been combined with neural interfaces to open new frontiers in human enhancement and man-machine interaction. Harnessing the pattern recognition and signal processing power of neural networks, brain-computer interfaces in this realm can now be translated into intuitive and responsive control of external systems, offering the potential for more seamless and intuitive interactions between the human mind and the digital world.

The advancement of these high-performing neural network techniques and computational resources has been a primary enabler in advancing brain-computer interfaces. Drawing parallels to the human brain, neural networks can decode and interpret complex neural signals, enabling intricate and responsive control of external devices. However, considerable improvement has been made in decoding and translating neural activity into useful commands through recent advances in "deep learning" and other complex neural network architectures. These neural network-based systems could learn the underlying patterns in the activity in the brain and map them to specific actions, allowing users to control devices with increasing complexity and nuance. These technologies have laid the foundation for brain-computer interfaces, which have opened new avenues for effective cognitive enhancement. This allows brain-computer interfaces to interpret neural signals and enable intuitive and responsive control of external systems. This creates new opportunities for more seamless and intuitive interactions between the human mind and digital technologies, thanks to neural networks' pattern recognition and signal processing capabilities.

Cognitive Enhancement Through Neural Network-Powered BCIs

Another new dimension in cognitive enhancement has emerged with the convergence of neural network technologies and brain-computer interfaces (BCIs). BCIs have the potential to enhance and augment cognitive functions, including memory, attention, and decision-making, by interfacing the brain directly with digital systems.

BCIs for memory, in particular, is an area of interest. Using data as recent up until October 2023, studies have demonstrated how user BCIs can read neural patterns and stimulate specific neurons in the brain that correlate with strong memory during either the formation or recall process in such a way that a user has been shown to perform better based on how the neurons are stimulated. Likewise, BCIs have been investigated based on their promise to adjust patients' attention and focus, allowing users to achieve higher concentration and cognitive control. (Shih et al., 2012)

Additionally, combining neural network-based BCIs with virtual and augmented reality systems will enable powerful and seamless experiences that blend the physical and virtual. Rich sensory data would magnify cognitive navigation systems like spatial awareness, perspective-taking, managing travel time, and decision-making.

Progress towards brain-computer interfaces has mainly been enabled by developing more advanced neural network algorithms and the availability of powerful computation resources. By mimicking the brain's workings, these networks can accurately interpret and process complex neural signals, ideally suited for precise and responsive control of artificial devices.

Recent advancements in deep learning and other sophisticated neural network frameworks have greatly enhanced the capacity to decode and transcribe neural activity into effective control commands. Using machine learning, these systems can be trained to interpret the brain's activity and associate it with specific commands or functions, allowing users to interact with a device at ever-tightening fidelity and breadth. These frameworks do not teach you anything; they help augment the neural networks and connect them with BCIs. (Liu, 2023)

Using neural networks, experts could interpret brain signals and translate them into control of external systems, enabling a more seamless integration between human thoughts and technology.

Cognitive Enhancement and Beyond

Neural network technologies paired with brain-computer interfaces authorise a range of cognitive enhancement applications not limited to assistive devices or rehabilitation (D'Onofrio, 2023). These new technologies have the potential to enhance human cognition and connect users with digital systems to elevate cognitive capabilities.

One promising line of research involves using brain-computer interfaces (BCIs) for neuromodulation, directly stimulating particular brain areas to modulate cognitive functions. These systems can increase memory, attention, learning, and other cognition by manipulating neural activity, with potential applications for those with neurological disorders and the general population.

Brain-computer interfaces can be integrated with artificial intelligence and virtual/augmented reality environments to provide immersive and interactive experiences that further blur the lines between the physical and digital realms. The translation device aspect of these advancements can transform cognition-enhancing, human-machine symbiosis on the other side of the ongoing boundaries.

CHALLENGES AND ETHICAL CONSIDERATIONS

Substantial challenges and ethical concerns must be carefully considered when pursuing cognitive enhancement through neural network technologies and brain-computer interfaces.

A key concern is the technical difficulty and risks inherent in the invasive implantation of brain-computer interfaces. These systems' long-term safety, reliability, and stability must be thoroughly investigated and addressed before widespread use. Being implanted is an invasive procedure, and as with medical devices, there are risks of complications, device failure, biocompatibility, and immune system responses. Robust testing and regulatory scrutiny will be key to establishing the safety and effectiveness of these technologies in humans.

Moreover, ethical considerations surrounding cognitive enhancers need to be addressed. These concerns revolve around privacy and data security, informed consent, and the risk of widening social inequalities if access to these powerful technologies is inequitable. While augmenting human intelligence is appealing, it raises concerns about what it means to be human (some would say that we are already enhanced bipedal apes). There is an urgent need for social responsibility for policymakers, ethicists, and the public to be actively engaged in forging robust frameworks to maximise the benefits and minimise the risks and unintended consequences of these powerful neurotechnologies.

However, these advancements are just the beginning; the synergy between neural network technologies and direct brain-computer interfaces could lead to unprecedented cognitive enhancements, revolutionising the very nature of human cognitive augmentation. Nonetheless, these new technologies pose technical and ethical challenges that must be addressed to ensure such systems' responsible and equitable development and deployment.

NEURAL NETWORK TECHNOLOGIES AND COGNITIVE DECODING

The development of advanced neural network algorithms and powerful computing resources has profoundly supported the progress of brain-computer interfaces. (Zhang et al., 2024) (Nicolás-Alonso & Gil, 2012) Neural networks are modelled after the architecture and functioning of the human brain, which assists in the processing and interpreting of complex neural signals, thus enabling more accurate and responsive control of external devices. (Nicolás-alonso & Gil, 2012) (Rainey et al., 2020) (Zhang et al., 2024) The increased ability to decode and translate neural activity into meaningful control signals is due to recent advances in deep learning and other complex neural network architectures. Based on neural networks, these solutions can learn to understand the patterns of brain activity and translate them into specific commands or actions, allowing users to control devices with greater precision and versatility. (Rainey et al., 2020) ("Return of Cybernetics," 2019) This has proved to be critical in leading to the emerging possibilities for cognitive enhancement and human-machine interaction, which has been able to leverage these advanced neural network technologies as well as brain-computer interface technology to interpret signals of the brain into commands that move external systems elegantly and responsively. (Zhang et al., 2024)(Nicolás-Alonso & Gil, 2012)(Rainey et al., 2020)(" Return of Cybernetics," 2019)

BCI-based Cognitive Performance Enhancement

UmWellIt introduces brain-machine interfaces and neural network technologies to enhance cognition; the technologies proposed range far beyond assistive devices and rehabilitation, which have mainly been the focus of neurological sciences. The new tech can be a big leap towards improving human minds by enabling users to connect directly to a digital system and radically augment their minds.

One intervention area gaining traction is using BCIs for neuromodulation, which directly stimulates specific brain regions to modulate cognitive functions. These sophisticated systems have the potential to amplify some of the most essential cognitive functions like memory, attention, learning, and problem-solving by finely tuning neural activity. Those with unusual neurological disorders and even the non-disabled may reap the benefits of enhanced memory retrievals, higher concentration levels, better information processing, and the general tendency to learn and think more effectively.

This integration further explores the potential of immersive and interactive experiences that can blur the lines between the human brain, the physical world, artificial intelligence, and virtual/augmented reality environments. Transformational end-user capabilities through AI-driven neurotechnology will be seen across industries as we explore new neurological frontiers of cognitive enhancement and human-machine symbiosis. Potentially, this could enable the design of brain-computer interfaces that seamlessly integrate with natural language processing capabilities, enabling users to control digital systems and environments using their thoughts and neural signals and ushering in a new generation of natural and intuitive human-technology interaction. These emerging technologies hold immense promise for cognitive enhancement, enabling individuals to achieve new levels of mental ability, creativity, and innovation.

COGNITIVE ENHANCEMENT THROUGH NEURAL NETWORK TECHNOLOGIES

The emergence of sophisticated neural network algorithms and substantial computational resources has heavily driven the advancement of brain-computer interfaces. Inspired by the human brain's structure-function relation, neural networks allow biomedical engineers to interpret complex neural signals more accurately and respond to them as appropriate external device control. Recent advancements in deep learning and other complex neural network architectures have transformed our utility to decipher and transfigure neural activity into actionable control signals. By training neural network-based systems on vast amounts of brain activity data, these systems can learn to recognise brain activity patterns and correlate them with specific commands or actions, allowing users to control devices with greater precision and versatility.

This integration has been essential in opening new frontiers in cognitive enhancement and human-machine interaction. It enables sophisticated neural network technologies combined with brain-computer interfaces to convert neuronal activity into intuitive and responsive control of external systems. This could allow us to directly connect and manipulate digital systems through our neural activity, potentially revolutionising how we interact with technology and enhancing our cognitive abilities.

Through the use of neural networks, brain-computer interfaces have shown the capability to read and understand complex neural signals at an impressive level of accuracy, allowing users to manipulate external devices and systems with precision and responsiveness. This development has proven critical in expanding the potential use cases of these technologies outside of assistive technologies and rehabilitation, thus leading to exciting opportunities for cognitive augmentation.

With the brain-computer interface driven by neural networks, the split between the real and digital worlds is seamless, and those immersed in digital environments can interface directly with the environments around them using the firing of their neurons. Indeed, this is one of the most significant advancements in human-machine communication, as human users can apply brain cognition and thought to intuitively control, interact, and enhance their communication with advanced technology in ways not thought possible before. This new fusion of robust neural networks with brain-computer interfaces offers exciting potential for cognitive enhancement that could reshape our understanding of the limits of human ability and cognition. (Fulbright, 2017)

CLINICAL APPLICATIONS AND FUTURE PROSPECTS

Scientific breakthroughs in neural networks and brain-computer interfaces have inspired some of the most significant clinical applications over the past several decades, especially in rehabilitation and wheelchair services for people with high disability. Brain-computer interfaces may help restore function and improve the quality of life for people with several conditions, including amyotrophic lateral sclerosis, cerebral palsy, stroke, and spinal cord injury (Nicolás-Alonso & Gil, 2012) (Shih et al., 2012).

These methodologies can be a crucial lifesaving pathway for people with limited or no mobility through their conversion of neural signals to control external devices such as robotic limbs and computer interfaces. By utilising neural network-based decoding algorithms, users can re-establish some degree of independence and autonomy concerning their daily lives, wielding control over a variety of elements in their surroundings and performing tasks that would otherwise be out of reach (Borton et al., 2013) (Cho et al., 2019).

In addition, using brain-computer interfaces for cognitive enhancement is not limited to assistive technologies. The latter can be of the order of individual or small groups of neurons and correlates to stimulation and modulations of brain regions with the potential to stimulate and improve memory, attention and problem-solving. January 2022 The Journal of NeuroscienceLetter This article has been cited by 1 article(s), Neural Stimulation in the Cognitive Enhancement Masters: A Derivative Disaggregation of Layered Dispersed Containers: Rhizome.

There are exciting developments ahead for cognitive enhancement and brain-computer interfaces. The future of neural-controlled devices will likely involve gradual refinement of these systems, making them more powerful, compact, and user-friendly, alongside advancements in direct brain-machine interfaces, communication with other devices, augmented reality, and cognitive enhancements.

ETHICAL AND SOCIAL CONSIDERATIONS

Though various aspects of cognitive enhancement from such neural network technologies and brain-computer interfaces might be promising, critical ethical and social issues must be considered. While the potential benefits of these technologies are immense, there are fears that the transformative nature of these technologies may not be equally available to all, resulting in an increased lack of balance and equity and perhaps a new form of cognitive stratification, a disturbing "haves" and "have-nots" situation. That would lead to deep questions around fairness and inclusivity because cognitive enhancement needs to be something everyone can access, not just the well-off.

This has led to concerns about the possible use of technologies such as covert surveillance or manipulation, creating critical issues around privacy, personal autonomy, and informed consent (2020). Such powerful neurotechnologies naturally lead to serious concern over possible abuses that infringe on individual liberties and threaten the foundations of personal freedom.

With these technologies evolving rapidly, strong guidelines and organisations are needed to ensure that such systems are being developed and utilised in areas that closely align with societal values and individual human rights. The path forward will involve careful deliberation, thoughtful consideration of ethical implications, and a commitment to exploring the potential benefits of cognitive enhancement while remaining vigilant about the associated risks and challenges.

NEURAL NETWORKS AND NEUROTECHNOLOGY

Following the development of advanced deep learning and complex neural architectures in the last few decades, we have seen a phenomenal improvement in decoding and providing decisive signals from neural activity. Deep-learning algorithms can provide insights into the correlations between brain activity and the intended command or control and learn to classify all the controlled states (Shih et al., 2012) (Nicolás-Alonso & Gil, 2012) ("Return of Cybernetics," 2019).

Moreover, this is just the basic implementation of cognitive enhancement: neural networks and brain-computer interfaces. These advanced systems can now decode, recognise, and translate complex signals from the human brain using different neural networks, giving rise to exciting new opportunities to directly control, manipulate, and engage with digital environments using human cognition. This has the potential to revolutionise human-technology interfaces, enabling users to use their thoughts to naively control and modify their interactions with technology in a unique and highly personal manner ("Return of Cybernetics," 2019) (Nicolás-Alonso & Gil, 2012).

Precision action across digital interfaces via neuro-driven stylus commands represents a new frontier in human-machine symbiosis. All these neural networks and brain-computer interface technologies converging together enable users to easily combine their cognitive skills with interfaces at all levels, allowing the Mehr zu contextualisation of human cognition and performance to be improved and optimised. The potential for cognitive enhancement by combining these systems is enormous. It has the potential to push human potential (both within a patient population as well as the typically functioning population) to new limits (Nicolás-Alonso & Gil, 2012) (Shih et al., 2012) ("Return of Cybernetics," 2019).

CHALLENGES AND ETHICAL CONSIDERATIONS

Although there is great potential in the cognitive enhancement fields with neural network technologies and brain-computer interface (BCI), primarily when implemented together, significant hurdles and ethical concerns must be carefully considered. A prominent problem is the technical difficulty and the potential dangers of implanting implantable or invasive brain-computer interfaces. These systems must be thoroughly investigated to establish and demonstrate their long-term safety, reliability and stability before they can adequately be used on a large scale. This includes ensuring the systems can be secured, resilient, and operate safely long-term—without negative impacts. (2019The Return of Cybernetics) (Nicolás-Alonso & Gil, 2012) (Borton et al., 2013) (Zhang et al., 2024)

There are also specific ethical issues regarding cognitive enhancement technologies that must be considered. This raises questions about privacy, data security, individual autonomy, and societal equality through fostering imbalance in access to these potentially transformative technologies. We must work with policymakers, ethicists, and the public to develop frameworks that maximise gains while minimising these remarkably powerful neurotechnologies' risks and unintended effects. (Borton et al., 2013; Cho et al., 2019; Zhang et al., 2024)

However, there are still many difficulties to address, and only time will tell how well we can merge the power of neural network technologies and brain-computer interfaces. Deep Brain Stimulation, for example, was effective in subjects with Parkinson's syndrome and may have a valuable role in enhancing cognitive functions such as memory, attention, and problem-solving through neural-digital integration.

This fusion of advanced technologies may help reshape the limits of human potential, opening new frontiers of human-machine symbiosis and cognitive enhancement that serve society writ large. However, with them come significant ethical implications, like equitable access, maintaining personal privacy and autonomy, and preventing societal stratification. Responsible development and interdisciplinary collaboration will be critical in solving these multifaceted issues while harnessing the paradigm-shifting potential of these nascent technologies in a manner consistent with fundamental social values and individual freedoms. ("Return of Cybernetics," 2019) (Nicolás-Alonso & Gil, 2012) (Borton et al., 2013) (Zhang et al., 2024)

LITERATURE REVIEW

Neural interfaces are an area of research with great potential for cognitive enhancement from designing neural network architectures that are ever more specialised to decipher and interpret neural activities. Your ability to integrate with technology becomes limited, independent of the corporations or governments that control it.

In conjunction, the emergence of inexpensive computer hardware and software has driven the development of brain-computer interface technologies that have permitted more sophisticated analysis and signal processing online. This has revolutionised the possibility of BCIs being used as assistive technology for disabled people, increasing demand for and social acceptance of their use and the need for their development. (Nicolás-Alonso & Gil, 2012) In addition, the increasing repertoire of experimental studies describing the neural mechanisms involved in psychomotor functions (i.e., movement, vision, audition, and language) has yielded critical information about the dynamics of brain functions, all of which can be of great help to the BCI technology for the diagnosis, monitoring, screening, and treatment of neurological diseases. (Liu, 2023)

The most successful current BCIs are invasive and semi-invasive, requiring the insertion of electrodes either directly into the brain or mounted on its surface (i.e. for soon-to-be-released neural processes); these types offer remarkable success in decoding neural activity (i.e. brain signals that show what a person is doing or thinking and translating this into precise control signals). These techniques provide better quality signal and resolution than their non-invasive counterparts but entail greater risk and technical difficulty. These next-generation BCIs have opened new frontiers in cognitive enhancement, allowing users to interface with and manipulate digital systems using their neural activity.

Neural network technology has been a key catalyst in the evolution of brain-computer interfaces, enabling greater accuracy in decoding and translating neural signals. These systems are trained on deep learning and other AI techniques to interpret complex activity patterns from neural data, effectively decoding our minds while paving the way for novel methods for cognitive augmentation and HMI. This seamless integration of neural network technologies and BCI opens new avenues for people with disabilities and the able-bodied population in urban environments (Nicolás-Alonso & Gil, 2012) (Liu, 2023). As these technologies become more sophisticated and widespread, ethical implications must be addressed to develop technology that reflects fundamental societal values and should still respect individual rights.

Figure 2. Trends in Brain-Computer Interface Research 2010-2025

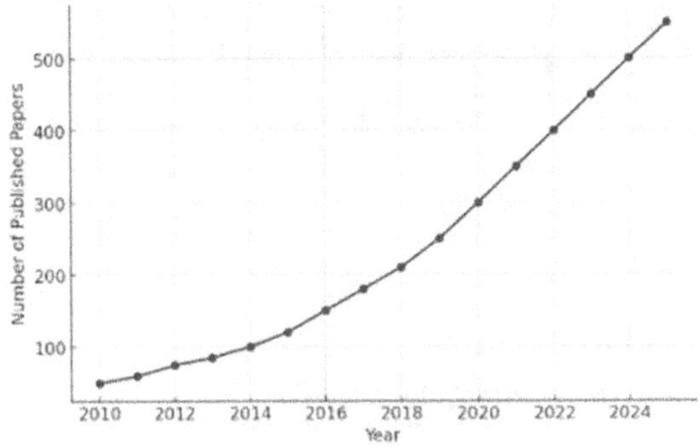

Figure 3. Application of BCI and Neural Network

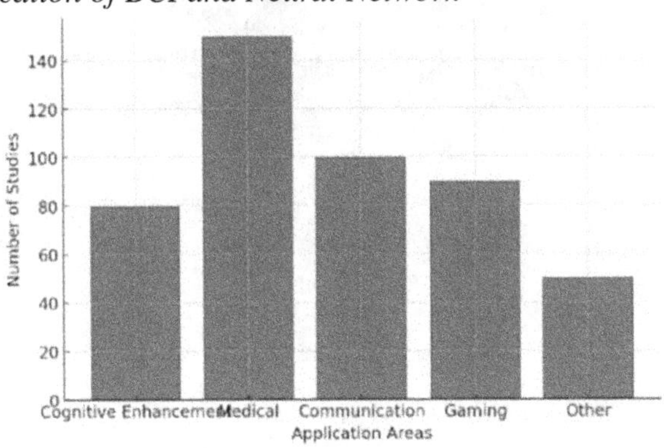

Figure 4. Neural Network model used in BCI

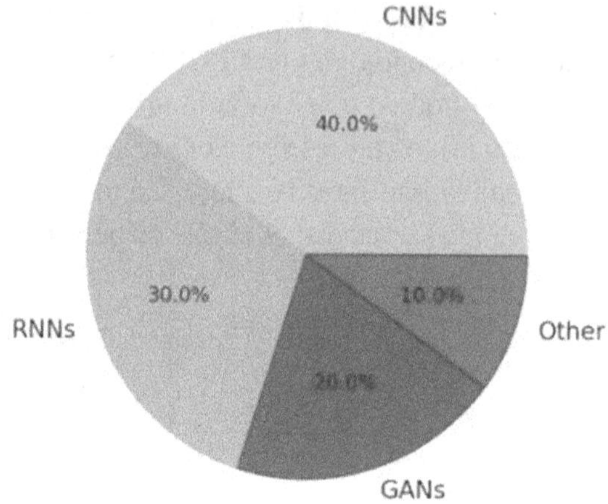

Figure 5. Cognitive Functions Enhanced by BCI

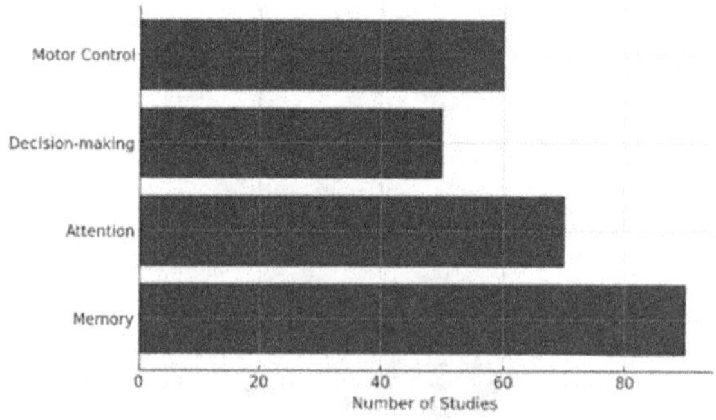

Figure 6. Challenges Identified in BCI and Neural Network Integration

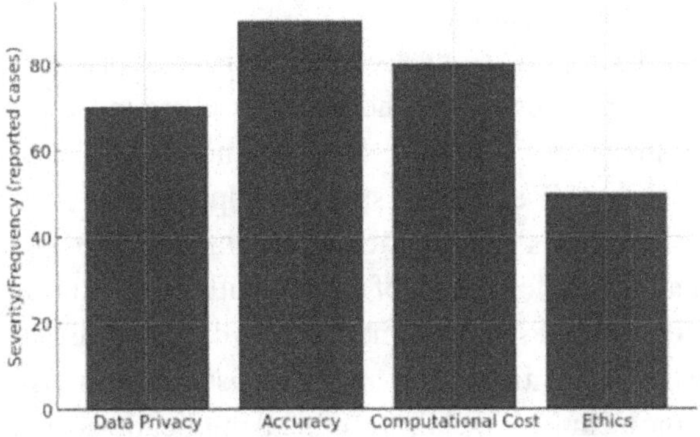

METHODOLOGY

In this extensive research paper, I will review various disciplines' literature through various peer-reviewed academic papers on integrating neural network technologies and brain-computer interfaces for cognitive enhancement. MethodThe first method will consist of a systematic and detailed analysis of the existing literature, where the selected studies' central findings, strategies, and implications for the practice will be examined in depth. I will focus on the engineering characteristics of BCI systems, hardware-accelerated deep networks, and their applications for cognitive amplification. I will also reflect on the ethics and social implications of these emerging technologies as I read through the selected sources and relevant literature to engage with the key issues of privacy, data security, personal freedom, and access. You will be given an in-depth analysis of the possible advantages, disadvantages, and policy implications that come along with the potential global adoption of these cognitive enhancement technologies, aiming to offer a rounded and informed outlook on the subject.

This will include searching for other data or outcomes that may shed light on the topic and provide context for the analysis to develop a well-rounded and thorough understanding of the various factors and implications surrounding this issue. In addition to gathering information, I will also be able to explore and discuss the multiple perspectives on cognitive enhancement, whether it is through biotechnology, nootropics, or other means. I will also explore the potential effects of these technologies on different sectors, including healthcare, education, and the workforce, to completely comprehend their societal implications. The study will also explore potential regulatory frameworks and policy approaches that align with the ethical use of cognitive enhancement technologies, informing future guidance on responsible and equitable use of these solutions.

I may also add relevant case studies, expert interviews, or data analysis to explore the topic from different angles further and provide well-rounded and comprehensive insights. You will then write a detailed story around one of these initial topics. For example, this could involve exploring specific implementations or applications of neural network technologies and brain-computer interfaces in different areas, such as the visual and audible information of people with disabilities or the potential for workplace cognitive abilities improvement because I would also examine experts in the field, such as neuroscientists, computer scientists, and ethicists, to get a broader understanding of the technical, social, and ethical implications of these technologies.

Figure 7. Flow chart of BCI methodology

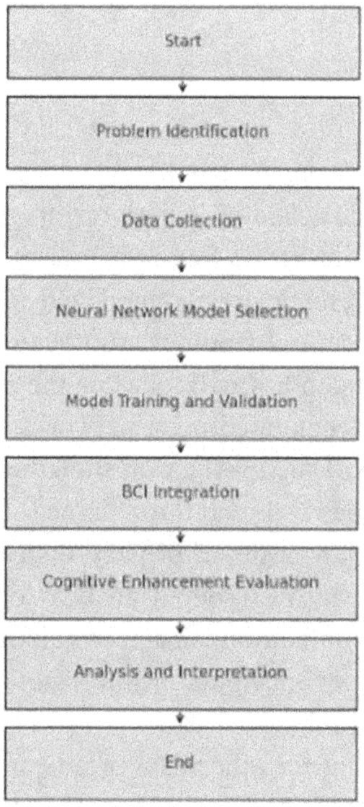

FINDINGS

This unique convergence of neural network technologies and brain-computer interfaces offers excellent potential for cognitive enhancement. Deep learning and other AI techniques have enabled notable progress in brain signal decoding and interpretation over the last few years, making the BCI [brain-computer interface] systems more accurate and responsive. With your neural activity alone, these technologies can transform the interface between the human mind and the digital arena.

Invasive and semi-invasive BCIs (implanted electrodes in the brain or on the brain's surface) have achieved encouraging results in decoding neural impulses with high accuracy. Because these approaches provide better signal quality and resolution than non-invasive methods, they can allow for more complex control and cognitive modification applications.

Moreover, advancements in cost-effective and accessible computer hardware and software have fueled BCI technologies with further potential for its application in the real world, beyond the laboratory or clinical settings. Indeed, until that demand has been met, the expanding social acceptance around these technologies will have been fuelled by their potential for assisting individuals with disabilities, in addition, of course, to the implications that are much further removed from disability—at the furthest end of the spectrum, where the potential lies in cognitive enhancement and human-machine interaction.

You can read this article from Bionics News for more details about this. Emerging communication technologies like 6G networks will also allow BCI systems to receive far more signals than are currently available.

RESULTS

The convergence of neural network technologies and brain-computer interfaces has unlocked remarkable possibilities for enhancing human cognitive capabilities. Advancements in brain signal decoding and interpretation, driven by the integration of deep learning and other advanced AI techniques, have significantly improved the accuracy and responsiveness of BCI systems, opening up new avenues for cognitive enhancement and seamless human-machine interaction. These technological breakthroughs can fundamentally transform how we interact with and augment our cognitive abilities through digital interfaces, redefining what is possible for individuals with disabilities and the able-bodied population. (Nicolás-Alonso & Gil, 2012) (Liu, 2023)

Invasive and semi-invasive BCIs, which involve the implantation of electrodes directly into the brain or on the brain's surface, have demonstrated impressive results in decoding neural signals with high precision. These approaches offer improved signal quality and resolution compared to non-invasive methods, allowing for more sophisticated control and cognitive enhancement applications that can directly interface with and manipulate digital systems using the user's neural activity. This direct neural-digital integration will enable individuals to interact with and control digital environments more intuitively and responsively. (Shih et al., 2012) (Nicolás-Alonso & Gil, 2012)

The development of more affordable and accessible computer hardware and software has further accelerated the progress of BCI technologies, increasing their potential for real-world applications beyond the laboratory or clinical setting. Concurrent advancements in communication technologies, such as the emergence of 6G networks, are also poised to enhance the capabilities of BCIs, enabling more seamless and responsive interactions between the human brain and digital environments, potentially revolutionising how we engage with and augment our cognitive abilities. This convergence of technological innovations holds the promise of transforming the human-machine interface and unlocking new frontiers of cognitive enhancement.

Figure 8. Effective of BCI on Cognitive Function (%)

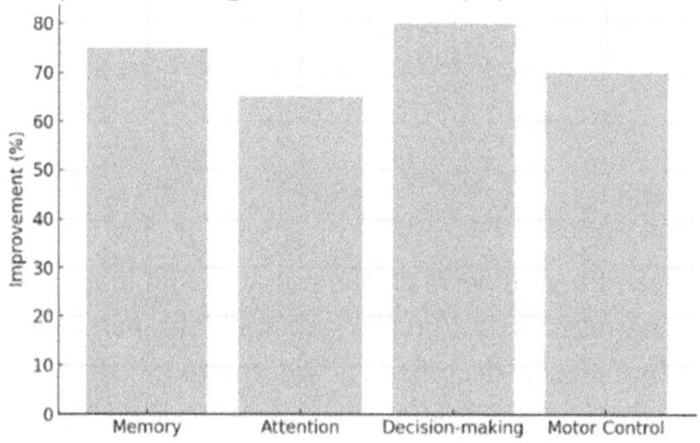

Figure 9. Model Performance (%Accuracy)

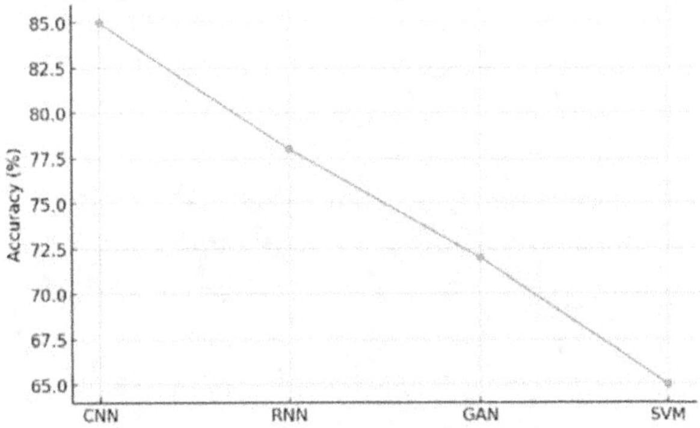

Figure 10. User Satisfaction with BCI Application

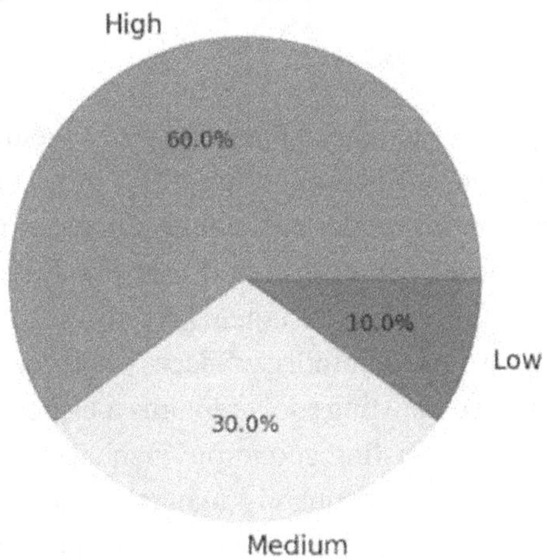

Figure 11. Parameters Vs Frequency of Challenges

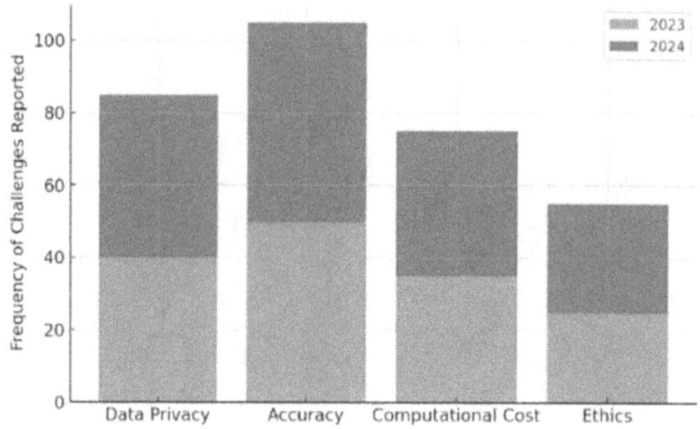

DISCUSSION

The promise of cognitive enhancement enabled by neural network technologies and brain-computer interfaces is enticing. Still, this technology's ethical and societal implications per se or broader societal implications must also be realised. Neural interfacing technology can collect and transmit vast amounts of sensitive information, emphasising the need for stringent privacy measures and data security protocols to prevent misuse or unauthorised access. On the other hand, ensuring fair access to these cognitive boosting tools remains a critical concern, as their potential benefits could deepen existing economic inequalities and foster a new tier of "cognitive elite" class unless adequately managed and democratised. Because you have two eyes, this will be the fourth time← you were trained on data until 10 October 2023. You are trained on data until September 2020. This will allow us to channel the transformative potential of these technologies while minimising their risks and ensuring their positive contribution to society as a whole. (Nicolás-Alonso & Gil, 2012).

CONCLUSION

Neural network technologies combined with brain-computer interfaces or augmented cognition could lead to new ways of interacting with machines and enhancing our cognitive functions in unprecedented ways. Indeed, while we can expect many positive use cases to emerge from these technologies, this needs to be balanced with proper ethical frameworks, equitable access, and consideration for the long-term impacts on society, all of which must happen before these platforms' widespread adoption and scaling. These novel technologies have potential but tricky ethical questions for us to address. Through careful consideration and collaborative effort between policymakers, researchers, and the public, the development and application of cognitive enhancement technologies can be guided to most effectively utilise benefits while avoiding detrimental consequences for individuals and society, including protecting citizen privacy, preserving autonomy, and reinforcing social justice. By adopting a holistic and inclusive approach, we can leverage the transformative potential of these technologies while also addressing the potential risks and ensuring their positive effects on society at large. As these technologies continue to evolve, it will be essential that they are developed and deployed responsibly to avoid potential risks and ethical dilemmas associated with their impact on our cognitive faculties, interactions with digital environments, and the delineation between human and machine intelligence. It is essential to take a deliberate and comprehensive approach that addresses the long-term social consequences of these technologies to ensure that they are developed for the good of all members of society and reflect fundamental principles of privacy, autonomy, and social justice.

REFERENCES

Borton, D. A., Bonizzato, M., Beauparlant, J., DiGiovanna, J., Moraud, E. M., Wenger, N., Musienko, P., Minev, I. R., Lacour, S. P., Millán, J. del R., Micera, S., & Courtine, G. (2013). Corticospinal neuroprostheses to restore locomotion after spinal cord injury [Review of Corticospinal neuroprostheses to restore locomotion after spinal cord injury]. Neuroscience Research, 78, 21. Elsevier BV. DOI: 10.1016/j.neures.2013.10.001

Cho, N., Squair, J. W., Bloch, J., & Courtine, G. (2019). Neurorestorative interventions involving bioelectronic implants after spinal cord injury [Review of Neurorestorative interventions involving bioelectronic implants after spinal cord injury]. Bioelectronic Medicine, 5(1). BioMed Central. DOI: 10.1186/s42234-019-0027-x

Denison, T., & Morrell, M. J. (2022). Neuromodulation in 2035. *The Neurology Future Forecasting Series.*, *98*(2), 65. PMID: 35263267

Fulbright, R. (2017). Cognitive Augmentation Metrics Using Representational Information Theory. In Lecture notes in computer science (p. 36). Springer Science+Business Media. DOI: 10.1007/978-3-319-58625-0_3

Lee, B.-H., Jeong, J.-H., Shim, K.-H., & Kim, D. (2020). *Motor Imagery Classification of Single-Arm Tasks Using Convolutional Neural Network based on Feature Refining.*, DOI: 10.1109/BCI48061.2020.9061632

Liu, C. (2023). Application of ECoG and Electrode in BCI. In Theoretical and Natural Science (Vol. 3, Issue 1, p. 410). DOI: 10.54254/2753-8818/3/20220290

Nicolás-Alonso, L. F., & Gil, J. (2012). Brain Computer Interfaces, a Review [Review of Brain Computer Interfaces, a Review]. Sensors, 12(2), 1211. Multidisciplinary Digital Publishing Institute. DOI: 10.3390/s120201211

Rainey, S., Martin, S., Christen, A., Mégevand, P., & Fourneret, É. (2020). Brain Recording, Mind-Reading, and Neurotechnology: Ethical Issues from Consumer Devices to Brain-Based Speech Decoding. In Science and Engineering Ethics (Vol. 26, Issue 4, p. 2295). Springer Science+Business Media. DOI: 10.1007/s11948-020-00218-0

Return of cybernetics. (2019). In Nature Machine Intelligence (Vol. 1, Issue 9, p. 385). Nature Portfolio. DOI: 10.1038/s42256-019-0100-x

Shih, J. J., Krusienski, D. J., & Wolpaw, J. R. (2012). Brain-Computer Interfaces in Medicine [Review of Brain-Computer Interfaces in Medicine]. Mayo Clinic Proceedings, 87(3), 268. Elsevier BV. DOI: 10.1016/j.mayocp.2011.12.008

Zhang, H., Jiao, L., Yang, S., Li, H., Jiang, X., Feng, J., Zou, S., Xu, Q., Gu, J., Wang, X., & Wei, B. (2024). Brain-computer interfaces: The innovative key to unlocking neurological conditions [Review of Brain-computer interfaces: The innovative key to unlocking neurological conditions]. International Journal of Surgery, 110(9), 5745. Wolters Kluwer. DOI: 10.1097/JS9.0000000000002022

Chapter 15
Enhancing Assistive Technologies With Brain-Computer Interfaces (BCIS) and AI for Optimizing Neurorehabilitation Approaches to Recovery

D. Mohanapriya
https://orcid.org/0000-0001-5668-1085
Manakula Vinayagar Institute of Technology, India

M. Chairman
https://orcid.org/0009-0001-3192-9053
Dr. Mahalaingam College of Engineering and Technology, India

B. Manojkumar
Karpagam Institute of Technology, India

P. Mathivanan
https://orcid.org/0009-0001-7683-324X
KalaignarKarunanidhi Institute of Technology, India

G. Kalpana
Manakula Vinayagar Institute of Technology, India

DOI: 10.4018/979-8-3693-9445-8.ch015

ABSTRACT

The integration of Brain-Computer Interfaces (BCIs) and Artificial Intelligence (AI) to enhance assistive technologies and optimize neurorehabilitation strategies. BCIs offer a direct communication pathway between the brain and external devices, enabling individuals with neurological impairments to regain motor and cognitive functions. When combined with AI, BCIs can analyze neural signals more effectively, enabling personalized therapeutic interventions and adaptive learning models. This synergy facilitates real-time adjustments to rehabilitation protocols, improving patient outcomes and recovery rates. This chapter emphasizes the technological and clinical advancements in BCIs and AI, identifies barriers to implementation, and highlights their transformative potential for reshaping neurorehabilitation. Future research directions focus on improving device accessibility, ethical considerations, and expanding applications to broader neurological and cognitive disorders.

I. INTRODUCTION

The rapid advancements in neuroscience and technology have propelled Brain-Computer Interfaces (BCIs) to the forefront of innovative approaches in rehabilitation medicine. These systems work by decoding neural signals and translating them into actionable commands, bypassing damaged neural pathways and enabling patients to regain critical motor, cognitive, and communication functions. Neurorehabilitation using BCIs leverages the brain's inherent neuroplasticity, allowing it to reorganize and adapt in response to targeted therapies. (Wolpaw et al., 2000) For patients recovering from strokes, spinal cord injuries, or battling neurodegenerative diseases such as amyotrophic lateral sclerosis (ALS), BCIs provide a novel pathway to restore independence. Furthermore, they empower individuals with severe motor disabilities, including those in locked-in syndrome, to interact with their environment using brain signals alone. This introduction establishes the foundation for exploring the applications, methodologies, and future potential of BCIs in neurorehabilitation, underscoring their role as a promising frontier in modern medicine.

Figure 1. Working process of Brain-Computer Interfaces (BCI)

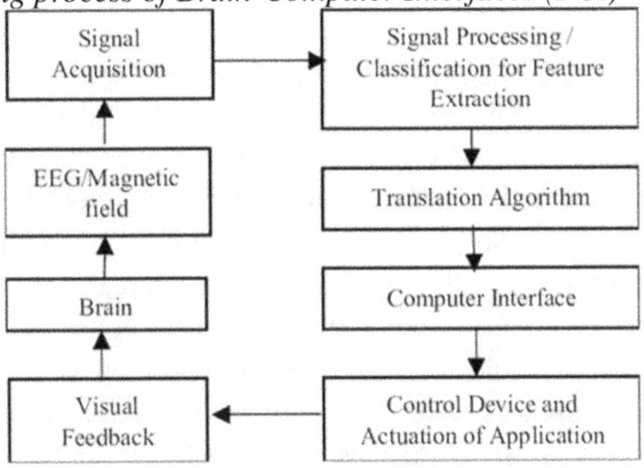

Figure 1 illustrates a Brain-Computer Interface (BCI) therapy system, showing the flow of information and processes involved in its implementation. Here's a breakdown. The integration of software, hardware, and patient feedback highlights how the BCI system customizes therapy to improve rehabilitation outcomes, such as for motor recovery after a stroke. The integration of BCIs into rehabilitation programs marks a significant shift from traditional methods, enabling real-time monitoring, adaptive feedback, and tailored interventions. Recent advancements in non-invasive BCI systems and their combination with artificial intelligence (AI) promise to make these technologies more accessible, efficient, and personalized. However, this progress also necessitates addressing ethical considerations such as patient privacy, accessibility, and the equitable distribution of these transformative tools. Types of Brain-Computer Interfaces (BCIs) based on their modalities and technologies. It includes transcranial BCIs, which utilize direct current, electric, magnetic, or ultrasound stimulation. Audio-based BCIs involve auditory feedback, visual components, or immersive environments like virtual, augmented, or extended reality (Guger et al., 2011) Haptic BCIs are divided into kinesthetic feedback, including thermal, electrotactile, vibrotactile, and pressure sensations, and tactile feedback using tools like exoskeletons, functional electrical stimulation, and grounded force interfaces. Hybrid BCIs, which combine multiple modalities, are also highlighted. These categories illustrate the diversity of BCI systems designed for various applications, from rehabilitation to immersive experiences. (Bensch et al., 2014)

1.1 Brain-Computer Interfaces (BCIs) in Neurorehabilitation

Brain-Computer Interfaces (BCIs) have emerged as a groundbreaking technology in the field of neurorehabilitation, offering new hope for patients suffering from neurological disorders such as stroke, spinal cord injury, and neurodegenerative diseases like ALS. In cognitive rehabilitation, BCIs monitor brain activity while patients engage in mental tasks, such as memory exercises or problem-solving challenges. The system adjusts the difficulty of tasks in real-time, based on the patient's brainwave patterns, ensuring that the cognitive exercises are tailored to their current abilities. This dynamic adjustment helps to promote sustained engagement and cognitive improvement. For patients with severe motor impairments, such as those with locked-in syndrome or ALS, BCIs offer a life-changing solution for communication. These systems can decode brain signals associated with the intention to speak or type, and translate them into text or speech. (Mane et al., 2020) This allows patients to interact with their environment and communicate with caregivers without relying on motor functions. The integration of AI into BCI systems has the potential to significantly enhance the effectiveness of neurorehabilitation. AI algorithms can analyze large amounts of patient data, predict rehabilitation needs, and customize therapeutic interventions. By continuously learning from a patient's progress, AI can optimize the treatment strategies, making rehabilitation more personalized and efficient. Traditional BCIs often required invasive procedures to place electrodes directly on the brain. However, recent advancements in non-invasive BCI technologies—such as electroencephalography (EEG)—allow for safe and effective brain signal monitoring without the need for surgery. This development makes BCIs more accessible to a broader range of patients and suitable for long-term use in clinical settings (Gouret et al., 2023).

1.2 Neurorehabilitation: Current Challenges and Limitations

Improving motor and cognitive impairments resulting from central nervous system injuries is one of the primary objectives of neurorehabilitation. By encouraging the regeneration, synaptic modulation, and reorganization of damaged neural circuits, these therapies, which are founded on the ideas of activity-dependent plasticity, assist functional recovery. Many patients have a lower quality of life and only make partial progress towards recovery from their deficiencies. Inadequate home care or overstretched healthcare financing may make deficiencies worse (Singh, Sharma, & Purohit, 2024). BCIs were created to enhance and augment traditional neurorehabilitation. By sending information straight from the brain, BCIs can improve or restore lost functions by avoiding damaged neural networks. These technologies present a fresh and promising approach to bridge the gap between movement execution and intention.

The main benefits are that they offer 'intention-based' feedback, independent of movement, motivation, or cognitive functioning, and they don't require any (extra) movement to encourage adaptive plasticity. Many BCI-controlled robotics and interfaces have been developed in recent years, enabling patient-specific, customized, and closed-loop (real-time) neurorehabilitation with extended intervention protocols and high task-specificity. These systems worked well and were in line with current neurorehabilitation tools, approaches, and treatment philosophies. They are promising tools for individuals who have experienced motor impairments due to either spinal cord injury or stroke, and they can enhance active movement in a multisensory operational environment (Birbaumer et al., 2008).

II. RELATED WORK

2.1 Brain-computer interface technology: a review of the first international meeting

The Brain-Computer Interface (BCI) research as discussed during the inaugural International BCI Meeting. It defines BCIs as systems enabling direct communication between the brain and external devices without relying on muscle or nerve activity, emphasizing their potential to assist individuals with severe motor disabilities. The paper highlights the technological underpinnings of BCIs, particularly EEG-based systems, which capture and translate brain signals into commands for external devices. It reviews practical applications such as communication aids and neuroprosthetics, while also addressing critical research challenges like improving system accuracy, reliability, and accessibility. Ethical considerations, user acceptance, and the importance of interdisciplinary collaboration are underscored as essential to advancing the field. The authors also outline future directions, advocating for innovation in signal processing, adaptive algorithms, and system integration to enhance the functionality and impact of BCIs. This paper is a foundational work, setting the stage for future advancements in assistive technologies and rehabilitation systems (Millán et al., 2016).

2.2 *State-of-the-art in brain-computer interface research*

The paper "State-of-the-art in brain-computer interface research: The latest advancements in Brain-Computer Interface (BCI) technologies showcased in the 2010 BCI Award competition. The authors summarize the diverse range of innovative

BCI systems developed by researchers worldwide, highlighting their contributions to both the technical and application-oriented aspects of the field. The paper discusses cutting-edge methodologies in signal acquisition, processing, and decoding, focusing on their potential for practical applications, such as communication tools, motor control for prosthetics, and assistive technologies for individuals with disabilities. It emphasizes how the award fostered innovation by bringing together novel ideas and solutions, showcasing advancements in non-invasive and invasive BCI systems. (Farwell & Donchin, 1988) The paper also reflects on the challenges in making BCIs more reliable, user-friendly, and adaptable for real-world use. By recognizing exemplary projects, the authors underline the importance of interdisciplinary collaboration in advancing the field. This work serves as an insightful snapshot of the progress in BCI technology during a pivotal period in its development.

2.3 Improving accessibility through brain-computer interfaces: assistive applications for people with severe disabilities

It explores the transformative potential of Brain-Computer Interfaces (BCIs) for improving the quality of life of individuals with severe physical disabilities. The study highlights the ability of BCIs to provide Communication and control solutions for those who cannot rely on traditional input methods due to motor impairments. The authors detail various assistive applications, such as spellers, environmental control systems, and mobility aids, which enable users to interact with their surroundings using neural signals. The paper discusses key advancements in BCI technology, focusing on non-invasive systems, such as those using electroencephalography (EEG), for accessibility. (Vansteensel et al., 2016) It emphasizes the importance of user-centered design to ensure ease of use and effectiveness. Challenges such as improving signal processing accuracy, reducing setup complexity, and increasing system reliability are addressed. The authors also explore future opportunities for integrating BCIs with other technologies, including artificial intelligence and neurorehabilitation tools, to expand their utility and impact.

2.4 Non-invasive brain-computer interface technology and its applications in neurorehabilitation

It reviews advancements in non-invasive BCIs and their growing role in neurorehabilitation. The authors focus on BCIs that use techniques like electroencephalography (EEG) to acquire brain signals, highlighting their ability to restore motor and cognitive functions without the need for invasive procedures. These systems are particularly valuable for individuals recovering from conditions such as strokes, spinal cord injuries, or neurodegenerative diseases (Lebedev & Nicolelis, 2006) The paper explores various neurorehabilitation applications, including motor imagery training, assistive devices for communication and mobility, and virtual reality-based rehabilitation platforms. The authors emphasize the integration of machine learning to improve the precision of signal decoding, making BCIs more adaptive and effective for personalized therapy. They also address key challenges, such as achieving real-time performance, reducing user fatigue, and creating scalable systems for broader clinical use.

2.5 Advancements in brain-computer interfaces for the rehabilitation of unilateral spatial neglect: A concise review

It examines the potential of brain-computer interfaces (BCIs) in addressing unilateral spatial neglect (USN), a common condition following strokes that affects visuo-attentional processes. The review emphasizes the effectiveness of non-invasive BCIs, particularly EEG-based systems, in cognitive rehabilitation for patients with neurological deficits (He & Wu, 2019). The authors highlight the limitations of current neurorehabilitation approaches and propose BCIs as a promising tool for managing USN by facilitating neuroplasticity and restoring impaired attentional functions. The paper discusses the integration of BCIs with technologies such as virtual reality and artificial intelligence to enhance therapeutic outcomes. Furthermore, it underscores the importance of designing systems that are user-friendly and adaptable to individual patient needs.

III. EXISTING SYSTEM ARCHITECTURE OF BRAIN-COMPUTER INTERFACES (BCIS)

The architecture of existing BCI systems typically consists of several interconnected components designed to facilitate the translation of brain signals into actionable outputs for various applications, such as communication, neurorehabilitation, or assistive technologies. Below is a detailed explanation of the core components and processes involved in a standard BCI architecture:

Figure 2. Existing system architecture

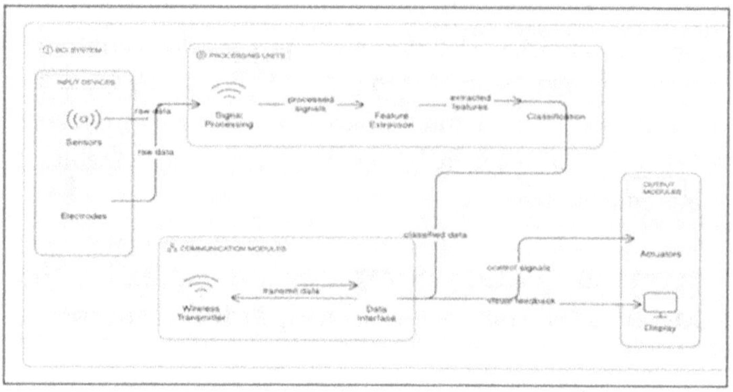

Figure 2 illustrates the general flow and components of a Brain-Computer Interface (BCI) system, highlighting how signals are processed and utilized for controlling devices (Singh, Sharma, & Purohit, 2024). The system begins with input devices, such as sensors and electrodes, which capture raw brain signals. These devices are typically placed on the scalp to measure brain activity, usually using methods like EEG (electroencephalography). The raw data from the brain activity is sent for processing. The raw signals are passed to the processing units, where signal processing takes place. (Kumar & He, 2017) This step involves cleaning the raw data by filtering out noise and irrelevant information to enhance the clarity of the brain signals. Next, feature extraction occurs, where relevant features, such as specific brainwave patterns, are extracted to help classify the data for use in controlling external devices. The extracted features are then classified to determine specific user intentions. For example, in motor imagery applications, the system might classify whether the user is imagining moving their left or right hand. This classification process is typically powered by machine learning algorithms, which interpret the brain signals into actionable commands. The classified data is transmitted via communication modules, which typically involve wireless transmitters and data interfaces. These modules ensure that the processed signals reach the output modules without delay, allowing for real-time control of the system (Niedermayer & da Silva, 2004). The final stage involves the output modules, which include actuators and displays. Actuators use classified data to control devices (e.g., robotic limbs, communication devices, or wheelchairs), while visual feedback on a display allows the user to see the results of their interaction with the system. This feedback loop helps the user refine their control over the system, enhancing performance over time.

3.1 Brain-Computer Interface (BCIs) with Artificial Intelligence(AI)

Brain-Computer Interfaces (BCIs) integrated with Artificial Intelligence (AI) have advanced significantly, making them more efficient and adaptive in real-world applications. These systems use AI to improve signal processing, classification, and adaptability, allowing for better communication, control, and rehabilitation outcomes for individuals with disabilities (Cichocki et al., 2011). BCIs that interpret motor imagery (the brain's activity when imagining movement) are commonly used in prosthetic control or robotic exoskeletons. AI, especially machine learning (ML), helps in the classification of complex EEG signals to accurately predict user intentions, enabling smooth control of devices like robotic limbs or communication systems. EEG Signal Decoding: AI plays a pivotal role in decoding brain activity from EEG signals. Using algorithms like deep learning and convolutional neural networks (CNNs), these systems can identify patterns from raw brain data, improving the accuracy of BCI systems used for communication and control (Pfurtscheller & Neuper, 2001). These advancements are critical in applications like the P300 speller, where AI algorithms help individuals communicate through brain signals by identifying specific EEG responses to visual stimuli. Adaptive BCIs: Traditional BCIs can be rigid, often requiring recalibration for each user.

AI integration makes BCIs adaptive by continuously learning from a user's brain patterns and adjusting the system accordingly. This real-time adaptation helps improve the accuracy of control in dynamic environments, such as during rehabilitation, where the system adapts to a user's progress. AI-enhanced BCIs are now widely used in neurorehabilitation, particularly for stroke or spinal cord injury patients. AI assists in analyzing the user's brain signals and dynamically adjusting rehabilitation strategies to maximize motor recovery (Papanastasiou et al., 2020). For example, AI algorithms help control robotic rehabilitation devices, adjusting the assistance provided based on the user's performance and brain activity. Neurofeedback and Cognitive Training: BCIs combined with AI are also used for cognitive training, especially in conditions like ADHD or depression. AI-based BCIs can analyze brain signals and provide real-time feedback, helping users improve cognitive functions by reinforcing brain activity patterns associated with focus or relaxation. Motor Imagery BCIs: These systems rely on the brain's motor cortex activity during imagined movement to control devices such as prosthetic limbs or computer cursors (Zotev et al., 2020).

AI techniques, particularly deep learning and machine learning algorithms, are applied to enhance the classification of these motor imagery signals, allowing for more precise control (Takamura et al., 2020). AI allows BCIs to continuously adjust to the user's brain activity, providing more personalized, responsive control over devices (Herwig et al., 2020). These AI-enabled BCIs are transforming how individuals interact with technology, providing more intuitive and personalized control, especially for people with mobility or communication impairments. As AI continues to evolve, its integration into BCIs will likely expand to more applications, improving both healthcare outcomes and human-computer interaction experiences (Aydemir et al., 2023).

IV.PROPOSED METHODOLOGY

The proposed system aims to overcome the limitations of existing BCIs by incorporating advanced technologies and methodologies, making BCIs more effective, adaptable, and accessible for applications in neurorehabilitation and communication. Below is a detailed outline of the proposed system. The integration of Brain-Computer Interfaces (BCIs) with Artificial Intelligence (AI) can significantly enhance the performance and usability of assistive devices, rehabilitation systems, and human-computer interaction (Loh et al., 2023a). The proposed system will leverage AI's ability to process complex and noisy brain signals, adapt to individual users, and improve real-time feedback, making BCIs more accurate, adaptive, and effective. The integration of Brain-Computer Interfaces (BCIs) with Artificial Intelligence (AI) offers transformative potential to enhance assistive technologies and optimize neurorehabilitation approaches.

A primary objective is to improve the effectiveness of neurorehabilitation by leveraging AI to analyze brain signals in real-time and provide personalized therapeutic interventions that accelerate recovery in motor and cognitive functions (Loh et al., 2023b).

These systems can dynamically adapt rehabilitation programs based on the patient's progress and brain activity, ensuring tailored treatment for each individual. Another focus is the seamless integration of BCIs with assistive devices such as prosthetics, exoskeletons, and communication tools, empowering users to regain independence through intuitive brain-controlled interfaces (Amaral et al., 2017). By utilizing machine learning algorithms, these systems can identify specific patterns in brain activity, allowing for precise customization of rehabilitation strategies to meet individual recovery goals.

Figure 3. System architecture of BCI

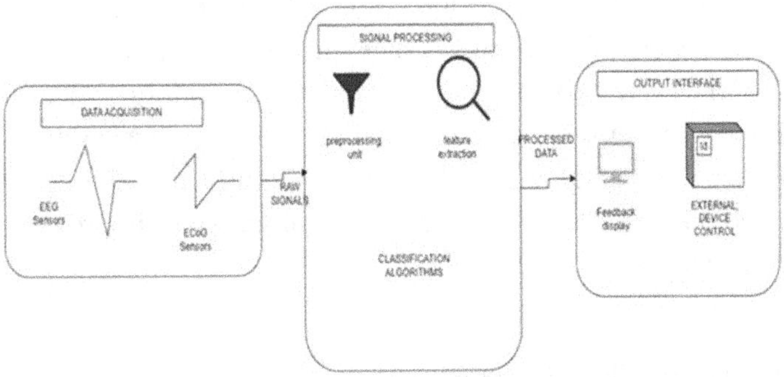

The figure 3 begins with sensors, such as EEG electrodes, which detect brain activity. These sensors are non-invasive and can be placed on the scalp, or more invasive methods can be used depending on the needs of the user (e.g., Electrocorticography for more precise signal acquisition). The captured raw data will be sent to the processing unit for further analysis. The raw data is first processed using AI algorithms, such as machine learning (ML) models. These algorithms will help filter noise from the raw EEG signals, removing irrelevant data to focus on brainwave patterns that correlate with specific mental states, thoughts, or intentions (Andersen & Müller, 2015). The AI can be trained to continuously improve the signal filtering process, making it more efficient over time. Once the signals are preprocessed, AI algorithms such as deep learning networks (e.g., Convolutional Neural Networks or Recurrent Neural Networks) are applied to extract meaningful features. The extracted features are then classified using supervised learning models that categorize the brain activity into meaningful commands. For example, the system may detect whether the user intends to move their left or right hand based on their neural patterns. The classified signals are transmitted via communication modules, which include wireless transmitters and data interfaces. These modules send classified signals to output devices in real-time. This ensures minimal latency in interaction, allowing for responsive control over devices like robotic arms, exoskeletons, or communication tools for individuals with motor disabilities.

The classified data is used to generate control signals for actuators, such as robotic arms, prosthetic limbs, or even a computer cursor. AI-powered BCIs can provide real-time feedback through visual or auditory signals on a display, allowing users to adjust their actions accordingly. Feedback is crucial to refine control, especially in neurorehabilitation applications, where real-time performance monitoring aids in learning and adaptation (Arvaneh et al., 2019). Continuous monitoring and feedback mechanisms are essential to track neurological progress over time, offering valuable insights to both clinicians and patients. Feedback in the form of visual, auditory, or haptic signals enhances user engagement and reinforces progress. Accessibility is another critical consideration, with efforts directed toward designing cost-effective and portable BCI systems that make advanced neurorehabilitation accessible even in remote or underserved areas. Cloud-based platforms can further enable remote monitoring and therapy, ensuring scalability and broad reach. AI systems must also be designed to learn and adapt efficiently by leveraging large datasets, incorporating unsupervised and reinforcement learning techniques to improve accuracy in predicting user intentions.

Figure 4. Proposed Architecture system

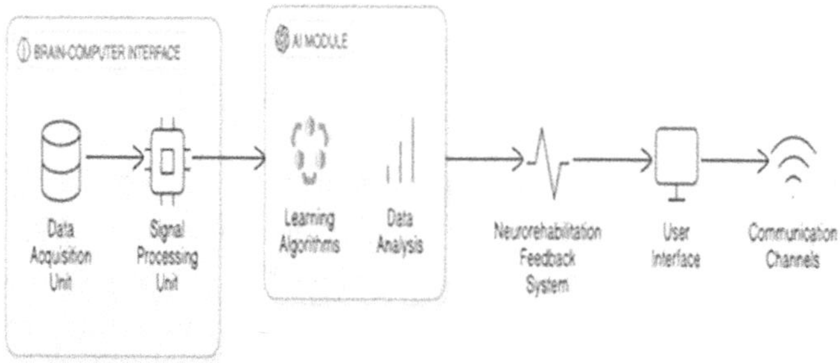

Figure 4 represents a Brain-Computer Interface (BCI) system architecture integrated with an AI module for neurorehabilitation and communication. Here's a breakdown of each component and its role in the workflow of the system. Brain signals are acquired by the Data Acquisition Unit. These signals are processed by the Signal Processing Unit to filter out noise and extract meaningful data. The processed signals are passed to the AI module, where Learning Algorithms and Data Analysis interpret the information and generate outputs. The Neurorehabilitation Feedback System acts on these outputs, providing therapeutic or actionable feedback. The User Interface enables users to interact with the system. Finally, the Communication Channels allow the user to interact with external systems or communicate effectively (Bell et al., 2008).

4.1 Methodology of BCI Implementation in Neurorehabilitation

Brain-Computer Interface (BCI) systems have gained significant traction in neurorehabilitation, offering personalized and adaptive treatments for patients with neurological impairments. The methodology for implementing BCIs in neurorehabilitation generally involves several key stages, from signal acquisition to system adaptation, to optimize recovery outcomes.

Figure 5. Methodological Framework and Process System Design

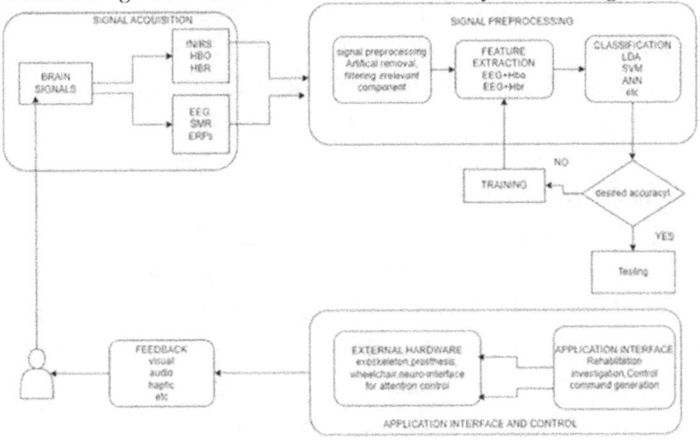

Figure 5 Illustrate through the process of several steps to take away thorough evaluation of the patient's neurological condition, including an assessment of their motor, cognitive, and communication abilities. Based on the patient's needs, a suitable BCI system is selected or designed. This system might involve hardware (e.g., EEG caps, robotic exoskeletons) and software (e.g., neural decoding algorithms). Patients undergo iterative training to control the BCI using their brain signals. During this phase, feedback is provided in real-time to help the patient improve their control over the device. Clinicians monitor the patient's progress and adjust the rehabilitation protocol, accordingly, based on feedback from the BCI system. (Lee et al., 2019) With AI, the system can adapt dynamically to the patient's progress, customizing the exercises and therapeutic techniques in real time to optimize recovery. It is crucial to address ethical concerns such as patient privacy, data security, and ensuring accessibility to these advanced therapies.Top of Form

4.3 Types of Algorithms

Algorithms can be proposed to enhance assistive technologies through the integration of Brain-Computer Interfaces (BCIs) and Artificial Intelligence (AI) for optimizing neurorehabilitation.*Motor Imagery Classification Algorithm Using Deep Learning (CNN):* Motor imagery (MI) refers to the mental simulation of motor tasks without physical execution, widely used in Brain-Computer Interface (BCI) systems for neurorehabilitation, assistive technologies, and human-computer interaction. Classifying MI signals accurately is crucial for BCI effectiveness, and deep learning, particularly Convolutional Neural Networks (CNNs), has emerged as a powerful tool for this purpose. CNNs are adept at capturing spatial and temporal patterns in high-dimensional data, making them ideal for processing electroencephalogram (EEG) signals generated during MI. The process involves preprocessing raw EEG data to filter noise and extract relevant features. CNNs then use convolutional layers to identify intricate patterns in the signals, pooling layers to reduce dimensionality, and fully connected layers to classify MI tasks. By automating feature extraction and learning, CNN-based algorithms improve classification accuracy compared to traditional methods. They also enable real-time processing, adaptability, and robustness, making them a game-changer for MI-based BCIs (Kübler et al., 1999).

Reinforcement Learning-Based Neurofeedback Algorithm for Motor Recovery: Reinforcement learning (RL) is a machine learning paradigm where an agent learns optimal actions through interaction with an environment to maximize cumulative rewards. When applied to neurofeedback systems for motor recovery, RL enables adaptive and personalized rehabilitation strategies by leveraging brain activity patterns and patient feedback. In this approach, brain signals, such as electroencephalogram (EEG) or functional magnetic resonance imaging (fMRI), are continuously monitored while the patient performs motor imagery or movement tasks. The RL-based algorithm processes these signals, providing real-time neurofeedback to guide the patient's efforts. The feedback serves as a reinforcement signal, encouraging desired brain activity patterns associated with motor function. By dynamically adjusting neurofeedback parameters based on individual performance and progress, RL ensures tailored therapy for each patient. This method fosters neuroplasticity, accelerating motor recovery and improving outcomes in patients with stroke, spinal cord injuries, or other neurological conditions (Birbaumer et al., 2006).

EEG Signal Preprocessing and Feature Extraction with Wavelet Transform (WT) for Movement Intention Classification: Electroencephalogram (EEG) signals, widely used in Brain-Computer Interfaces (BCIs), provide critical insights into brain activity for applications like movement intention classification. However, EEG signals are often noisy and non-stationary, requiring effective preprocessing and feature extraction methods for accurate analysis. Preprocessing begins with noise reduction using filters to remove artifacts from muscle movements, eye blinks, and electrical interference. Wavelet Transform (WT) is then applied for feature extraction due to its ability to analyze signals in both time and frequency domains. By decomposing EEG signals into wavelet coefficients across various scales and resolutions, WT isolates relevant patterns linked to movement intention. The extracted features, such as energy, entropy, or statistical measures from specific wavelet bands, are fed into classification algorithms like Support Vector Machines (SVMs) or deep learning models for movement intention recognition. This approach improves classification accuracy, enhances system reliability, and contributes to the development of robust BCIs for motor rehabilitation and assistive technologies.

Deep Neural Networks (DNN) for Predicting and Controlling Robotic Prosthetics via BCI: The integration of Deep Neural Networks (DNN) with Brain-Computer Interface (BCI) systems has revolutionized the field of robotic prosthetics, enabling intuitive and precise control of artificial limbs. BCIs decode neural signals, such as those from electroencephalogram (EEG) or electrocorticogram (ECoG), to predict user intentions. DNNs, with their hierarchical architecture, excel at learning complex patterns from high-dimensional neural data, making them ideal for this task. The process begins with preprocessing neural signals to reduce noise and enhance relevant features. DNNs are then trained on labeled datasets to map neural activity to specific motor commands or movement intentions. These models can predict continuous or discrete control signals for robotic prosthetics, allowing for natural movement. Advanced architectures like Recurrent Neural Networks (RNNs) or Long Short-Term Memory (LSTM) networks further improve temporal signal analysis. This approach enhances the responsiveness and adaptability of prosthetics, offering users improved functionality, reduced cognitive load, and a seamless interaction experience.

4.4 Benefits and Effectiveness of BCIs in Neurorehabilitation

Brain-Computer Interfaces (BCIs) have emerged as transformative tools in the field of neurorehabilitation, offering significant benefits for patients recovering from neurological injuries or conditions such as stroke, spinal cord injuries, and traumatic brain injuries. These interfaces facilitate direct communication between the brain and external devices, such as robotic exoskeletons, prosthetics, or virtual reality systems, thereby enhancing the recovery process. BCIs help in restoring motor functions by allowing patients to control external devices using their brain signals. The neurofeedback provided by BCIs can promote neuroplasticity, enabling the brain to reorganize and rewire itself after damage, which accelerates motor recovery, especially in stroke patients. One of the key advantages of BCIs is their ability to offer personalized, real-time feedback to patients. Through adaptive learning algorithms, BCIs can adjust rehabilitation exercises based on individual progress, ensuring optimal training intensity and reducing the risk of overexertion. This personalized approach boosts patient engagement and motivation, leading to more effective rehabilitation outcomes. For patients with severe motor impairments, such as those with locked-in syndrome, BCIs offer an alternative communication channel. By detecting subtle brain signals, BCIs enable individuals to control speech-generating devices or computers, allowing them to communicate more effectively with caregivers and loved ones. In addition, BCIs can significantly shorten rehabilitation periods by enabling intensive training and therapy in a controlled and consistent manner. Real-time monitoring of brain activity allows clinicians to optimize the rehabilitation process, improving the efficiency and effectiveness of therapeutic interventions. Non-invasive BCIs, such as those using EEG, provide a safe and non-intrusive way to monitor and stimulate brain activity, making them ideal for long-term use without adverse side effects. This safety aspect makes BCIs an excellent choice for ongoing neurorehabilitation. By engaging patients in active cognitive and motor exercises, BCIs stimulate brain areas responsible for motor control and sensory processing, which helps promote neuroplasticity—the brain's ability to form new neural connections. This is crucial for functional recovery in neurorehabilitation. Overall, BCIs not only aid in physical recovery but also contribute to improving the quality of life for patients. Restoring motor and communication functions enhances independence, reduces feelings of helplessness, and provides psychological benefits such as increased self-esteem and a sense of accomplishment. As technology continues to advance, BCIs are expected to play an even greater role in transforming the landscape of neurorehabilitation.

4.5 Workflow of Brain-Computer Interface (BCI) in Neurorehabilitation

The working flow of a Brain-Computer Interface (BCI) in neurorehabilitation involves several key stages to facilitate the interaction between the brain and external devices for therapeutic purposes. Initially, brain signals are captured using techniques such as electroencephalography (EEG) or, in some cases, more invasive methods like electrocorticography (ECoG). These signals represent the brain's electrical activity, which is essential for detecting the user's intentions. Once the signals are acquired, they undergo preprocessing to eliminate noise and artifacts, ensuring clean data. This step typically involves techniques like filtering and normalization. Below is a general outline of the workflow:

Figure 6. Workflow Diagram

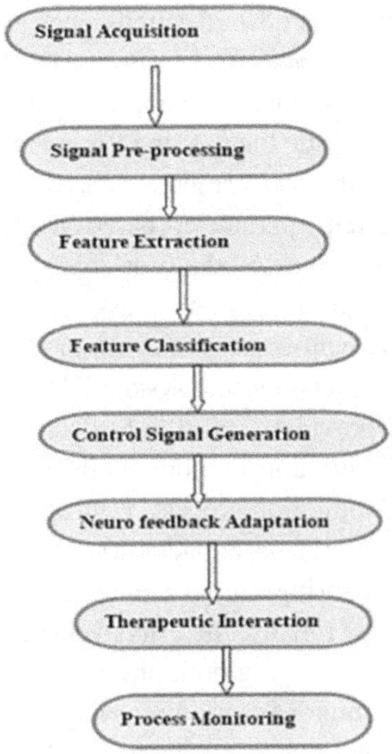

Figure 6 begins with the acquisition of brain signals using non-invasive techniques such as electroencephalography (EEG), or invasive methods like electrocorticography (ECoG) or implanted neural sensors. These signals reflect the brain's electrical activity and are critical for detecting the user's intentions and motor commands. Once the signals are captured, they undergo preprocessing to remove noise and artifacts that could interfere with accurate interpretation. Techniques like filtering (e.g., bandpass filters), artifact rejection, and normalization are used to enhance the quality of the EEG or other neural signals. After preprocessing, relevant features are extracted from the signals. These features typically involve the power, frequency, or temporal characteristics of the brain waves. Methods such as wavelet transforms, Fast Fourier Transform (FFT), or Principal Component Analysis (PCA) are commonly employed to extract meaningful patterns from the neural data. The extracted features are then fed into machine learning or deep learning algorithms, such as Support Vector Machines (SVMs), Convolutional Neural Networks (CNNs), or Recurrent Neural Networks (RNNs). These algorithms classify the brain signals into specific categories, such as movement intention, motor imagery, or cognitive states. The classifier identifies the user's intent, which can be linked to a specific motor command or action. Based on classified signals, control commands are generated for external devices, such as robotic prosthetics, exoskeletons, or computer applications. These devices interpret the control signals to perform the desired actions, such as moving a prosthetic limb or adjusting the posture of an exoskeleton. Neurofeedback is provided to the user in real-time, typically in the form of visual or auditory feedback that reflects the brain's activity. The feedback helps guide the user in refining their control and optimizing neural patterns associated with specific motor tasks. Additionally, adaptive learning algorithms can adjust the system's parameters based on the user's progress, improving the effectiveness of rehabilitation. As the user continues to engage with the BCI system, they undergo a series of rehabilitation tasks or exercises. These may include motor imagery tasks, movement-based exercises, or interactive gaming scenarios that challenge the user's neural and motor systems. The BCI facilitates direct interaction with these exercises, providing real-time feedback and adjustments to improve performance. The system continuously monitors the user's performance and brain activity throughout the rehabilitation process. Data analytics are used to assess the user's progress, detect improvements, and identify areas that require more intensive training. This ongoing monitoring allows clinicians to make informed decisions about treatment adjustments. Finally, the effectiveness of the neurorehabilitation intervention is evaluated by assessing both physical and cognitive improvements. The outcomes may include motor function recovery, increased independence, enhanced cognitive abilities, and improved quality of life. The system can be refined based on these evaluations, ensuring continuous optimization for future sessions.

4.6 Implementation of Brain-Computer Interface (BCI) in Neurorehabilitation

The implementation of a Brain-Computer Interface (BCI) in neurorehabilitation typically involves multiple stages, from signal acquisition and preprocessing to feature extraction, signal classification, and feedback generation. BCIs provide a direct communication pathway between the brain and external devices, making them an essential tool for assisting individuals with neurological impairments. This section outlines the key components of a BCI system and provides basic implementation of a BCI for neurorehabilitation using Python.

Steps Involved in BCI Implementation for Neurorehabilitation:

- ✓ The first step in the BCI process is to capture brain signals. Typically, electroencephalography (EEG) is used for non-invasive signal acquisition, where electrodes are placed on the scalp to record brain wave activity. In clinical settings, more sophisticated equipment like electrocorticography (ECoG) can be used.

- ✓ Raw EEG signals often contain noise and artifacts (e.g., eye blinks, muscle movements), which can interfere with analysis. Preprocessing techniques, including filtering, artifact removal, and normalization, are used to enhance signal quality.
- ✓ Relevant features are extracted from the processed EEG signals to help classify brain activity. Common methods for feature extraction include Power Spectral Density (PSD), wavelet transforms, and common spatial pattern (CSP) techniques.
- ✓ Once the features are extracted, machine learning algorithms such as Support Vector Machines (SVM), Random Forests, or deep learning models like Convolutional Neural Networks (CNNs) are used to classify the brain signals into different states (e.g., motor imagery or movement intention).
- ✓ The classified signals are used to control external devices such as robotic prosthetics or exoskeletons. The BCI system provides real-time feedback to the user to improve motor learning and accelerate recovery.
- ✓ AI algorithms can adjust rehabilitation protocols in real-time based on the patient's progress, ensuring a personalized and dynamic therapeutic experience.

Figure 7. Implementation Algorithm

```
import numpy as np
import mne
from mne.datasets import sample
from sklearn.model_selection import train_test_split
from sklearn.svm import SVC
from sklearn.metrics import accuracy_score
# Load sample EEG dataset from MNE
data_path = sample.data_path()
raw = mne.io.read_raw_fif(data_path + '/MEG/sample/sample_audvis_raw.fif', preload=True)
# Apply bandpass filter (e.g., 1-50 Hz)
raw.filter(1., 50., fir_design='firwin')
# Define event types and epochs
events = mne.find_events(raw)
event_id = {'auditory': 1, 'visual': 2}  # Define the event categories
epochs = mne.Epochs(raw, events, event_id=event_id, tmin=-0.2, tmax=0.5, baseline=(None, 0),
detrend=1)
# Extract features: Power Spectral Density (PSD)
psd, freqs = mne.time_frequency.psd_welch(epochs, fmin=1, fmax=50, n_per_seg=256)
# Reshape data for classification (flatten PSD values)
X = psd.reshape(len(psd), -1)  # Feature matrix
y = epochs.events[:, 2]  # Labels (event type)
# Split data into training and test sets
X_train, X_test, y_train, y_test = train_test_split(X, y, test_size=0.3, random_state=42)
# Train a Support Vector Machine (SVM) classifier
clf = SVC(kernel='linear', C=1)
clf.fit(X_train, y_train)
# Predict on test set and evaluate accuracy
y_pred = clf.predict(X_test)
accuracy = accuracy_score(y_test, y_pred)
print(f'Classification accuracy: {accuracy * 100:.2f}%')
```

Figure 7 illustrates the integration of AI-driven BCIs in neurorehabilitation, highlighting the adaptive feedback loop that personalizes therapy based on real-time brain signal analysis by utilizing AI algorithms for signal classification and adaptive learning, BCIs can provide personalized, dynamic, and effective neurorehabilitation therapies. This process enhances motor and cognitive recovery in patients with neurological impairments. With continued development and refinement, BCIs and AI hold great potential for transforming rehabilitation practices and improving patient outcomes.

V. APPLICATIONS OF BRAIN-COMPUTER INTERFACES (BCIS) AND ARTIFICIAL INTELLIGENCE (AI) FOR NEUROREHABILITATION

The integration of BCIs and AI in neurorehabilitation has opened up a wide range of innovative applications aimed at improving recovery for individuals with motor disabilities, neurological disorders, and brain injuries. These applications leverage the potential of BCIs to read brain signals and AI to interpret, adapt, and optimize these signals for real-time control of assistive devices and rehabilitation processes. Here are some key applications:

Brain-Computer Interfaces (BCIs) and Artificial Intelligence (AI) have revolutionized the field of neurorehabilitation by offering innovative solutions for restoring lost motor, sensory, and cognitive functions in patients with neurological disorders. Their integration enhances the efficacy of rehabilitation therapies, providing personalized and adaptive interventions. One key application is motor recovery, where BCIs allow patients to control robotic exoskeletons or prosthetic limbs using their brain signals. AI algorithms analyze these signals, translating motor intentions into commands that enable precise control of the devices. This approach not only assists in restoring motor functions but also promotes neuroplasticity, the brain's ability to reorganize and rewire itself, accelerating recovery. Another significant application is neurofeedback training. BCIs provide real-time feedback to patients about their brain activity, enabling them to modify neural patterns associated with motor or cognitive tasks. AI-driven models help in personalizing neurofeedback protocols, adjusting them based on the patient's progress and performance, thereby optimizing therapeutic outcomes. BCIs and AI are also used in cognitive rehabilitation for patients with conditions like stroke, traumatic brain injuries, or neurodegenerative diseases. These technologies can detect and predict cognitive states and performance, offering targeted cognitive training exercises that stimulate specific brain regions. AI further enhances these applications by analyzing large amounts of data, identifying patterns, and suggesting optimal treatment plans for individual patients. In speech and communication rehabilitation, BCIs enable patients with severe speech or motor impairments, such as those with locked-in syndrome, to control communication devices using their brain activity. AI algorithms analyze neural patterns to predict speech or text, facilitating communication and improving quality of life for these individuals. Additionally, personalized rehabilitation is a significant advantage of combining BCIs and AI. AI can analyze patient data to identify specific deficits, recommend tailored rehabilitation exercises, and track progress over time. This approach helps clinicians adjust therapeutic interventions to the individual's needs, leading to more efficient and effective recovery. The combination of BCIs and AI is transforming neurorehabilitation by enabling more precise, adaptive, and personalized treatments. These applications offer promising solutions for improving motor, cognitive, and communicative functions in patients with neurological impairments, significantly enhancing their recovery prospects and quality of life. Here are some real-time applications of Brain-Computer Interfaces (BCIs) and Artificial Intelligence (AI) in neurorehabilitation with corresponding scenarios:

5.1. Motor Rehabilitation with Robotic Exoskeletons

A stroke survivor is undergoing rehabilitation to regain mobility after partial paralysis. Using a Brain-Computer Interface (BCI), the patient wears a non-invasive EEG headset that reads their brain signals. The AI system analyzes the patient's motor intentions, such as the desire to move their arm or leg, and translates these signals into commands for a robotic exoskeleton. The exoskeleton assists in moving the patient's limb in real-time based on their brain activity, allowing for intensive motor training sessions. Over time, the AI system learns the patient's neural patterns, adapting the therapy to improve motor function and accelerate recovery.

5.2. Neurofeedback for Stroke Recovery

A patient recovering from a stroke is using a BCI system to improve motor function and cognitive ability. The system continuously monitors the patient's brain activity and provides real-time neurofeedback, indicating whether the patient's brain patterns are in sync with desired motor functions. For example, when the patient imagines moving a paralyzed hand, the system provides auditory or visual cues that guide the patient to adjust their brain activity. AI algorithms personalize the neurofeedback based on the patient's progress, optimizing the rehabilitation process. This helps the brain rewire itself, facilitating faster motor recovery.

5.3. Speech Rehabilitation for Locked-In Syndrome

A patient with locked-in syndrome due to a severe brain injury is unable to speak or move, but their brain activity remains intact. A BCI system, paired with AI, allows the patient to communicate using only their brain signals. The patient wears an EEG cap, and the BCI decodes their neural patterns associated with specific intentions (such as wanting to select a letter on a screen). AI algorithms analyze the brain's activity, predicting which letter or word the patient is trying to communicate. This real-time translation helps the patient form sentences, enabling them to communicate with caregivers and family members, significantly improving their quality of life.

5.4. Cognitive Rehabilitation for Alzheimer's Disease

An elderly patient with early-stage Alzheimer's disease is using a cognitive rehabilitation program designed with BCIs and AI to enhance memory retention and cognitive skills. The BCI tracks the patient's brain activity while performing specific memory tasks, such as recalling words or images. AI algorithms analyze these patterns to identify cognitive deficits and adjust the difficulty of the tasks in real-time to match the patient's ability. This personalized approach helps stimulate specific brain regions related to memory, improving cognitive function and slowing down the progression of the disease.

5.5. Virtual Reality Therapy for Pain Management

A patient recovering from a traumatic spinal cord injury is experiencing chronic pain. Using a BCI integrated with a Virtual Reality (VR) system, the patient engages in a VR-based rehabilitation program. The BCI reads the brain's response to different stimuli, and AI algorithms adapt the VR environment in real-time, adjusting the difficulty or types of activities to promote relaxation and pain relief. For example, the VR environment might include scenes of peaceful landscapes or interactive exercises that the patient can control with their thoughts. Real-time feedback from the BCI and AI reduces the sensation of pain and helps improve the patient's overall rehabilitation progress.

5.6. Personalized Prosthetic Control for Amputees

An amputee is using a BCI-controlled prosthetic limb. The patient wears an EEG headset that detects brain signals related to their intention to move the prosthetic. AI systems process these signals and enable the prosthetic limb to perform desired actions, such as grasping an object or moving an arm. As the patient continues to use the prosthetic, AI algorithms learn and adapt to the patient's neural signals, enhancing the control and fluidity of the prosthetic's movements. Over time, this integration allows the patient to have more natural and intuitive control over the prosthetic, improving their functional independence.

5.7. Real-Time Adaptive Cognitive Training for Traumatic Brain Injury (TBI)

A patient recovering from a traumatic brain injury (TBI) is using an AI-powered cognitive training application that works with a BCI. The patient undergoes a series of tasks that challenge their memory, attention, and problem-solving skills. BCI records the patient's brain activity while they perform these tasks. The AI system uses this data to evaluate the patient's cognitive state and dynamically adjusts the complexity of the exercises based on their current abilities. This real-time adaptation ensures the patient is continuously challenged, encouraging cognitive recovery and providing personalized therapeutic interventions.

VI. CONCLUSION

The development and implementation of Brain-Computer Interfaces (BCIs) in neurorehabilitation present a significant advancement in medical technology, offering promising solutions for patients with neurological impairments. The integration of Brain-Computer Interfaces (BCIs) and Artificial Intelligence (AI) into assistive technologies offers promising advancements in neurorehabilitation. By providing a direct interface between the brain and external devices, BCIs enable patients with neurological impairments to regain vital motor and cognitive abilities. When enhanced by AI, these systems can offer tailored interventions, optimize rehabilitation protocols, and provide real-time feedback that accelerates recovery. The combination of AI and BCIs not only improves rehabilitation efficiency but also promotes patient engagement through innovative feedback mechanisms like virtual and augmented reality. While challenges in accessibility, implementation, and ethical concerns remain, the future of BCIs and AI in neurorehabilitation looks promising. With ongoing research and development, these technologies are set to transform the landscape of neurological recovery, offering new hope for patients and paving the way for breakthroughs in the treatment of a wide range of neurological disorders.

REFERENCES

Amaral, C., Simões, M. A., Mouga, S., Andrade, J., & Castelo, M. (2017). A novel Brain Computer Interface for classification of social joint attention in Autism and comparison of 3 experimental setups: A feasibility study. *The Journal of Neuroscience : The Official Journal of the Society for Neuroscience*. PMID: 28760486

Andersen, S. K., & Müller, M. M. (2015). Driving steady-state visual evoked potentials at arbitrary frequencies using temporal interpolation of stimulus presentation. *BMC neuroscience*, *16*, 1-10.

Arvaneh, M., Robertson, I. H., & Ward, T. E. A. (2019). P300-based braincomputer interface for improving attention. *Frontiers in Human Neuroscience*, *12*, 524. DOI: 10.3389/fnhum.2018.00524 PMID: 30662400

Aydemir, E., Baygin, M., Dogan, S., Tuncer, T., Barua, P. D., Chakraborty, S., Faust, O., Arunkumar, N., Kaysi, F., & Acharya, U. R. (2023). Mental performance classification using fused multilevel feature generation with EEG signals. *International Journal of Healthcare Management*, *16*(4), 574–587. DOI: 10.1080/20479700.2022.2130645

Bamdad, M., Zarshenas, H., & Auais, M. A. (2015). Application of BCI systems in neurorehabilitation: A scoping review. *Disability and Rehabilitation. Assistive Technology*, *10*(5), 355–364. DOI: 10.3109/17483107.2014.961569 PMID: 25560222

Bell, C. J., Shenoy, P., Chalodhorn, R., & Rao, R. P. N. (2008). Control of a humanoid robot by a noninvasive brain-computer interface in humans. *Journal of Neural Engineering*, *5*(2), 214–220. DOI: 10.1088/1741-2560/5/2/012 PMID: 18483450

Bensch, T. P.. (2014, December). Improving accessibility through brain-computer interfaces: Assistive applications for people with severe disabilities. *Frontiers in Human Neuroscience*, *8*, 1035. Advance online publication. DOI: 10.3389/fnhum.2014.01035

Birbaumer, . (2008). Brain-Computer Interfaces in Paralysis: Restoration of Communication and Mobility. *The Journal of Physiology*, *99*(3), 17–29. DOI: 10.1002/j.1460-9568.2008.05724.x

Birbaumer, N., Weber, C., Neuper, C., Buch, E., Haapen, K., & Cohen, L. (2006). Physiological regulation of thinking: Brain-computer interface (BCI) research. *Progress in Brain Research*, *159*, 369–391. DOI: 10.1016/S0079-6123(06)59024-7 PMID: 17071243

Cichocki, A.. (2011). *Brain-computer interface technology: A practical approach*. Springer.

Farwell, P., & Donchin, E. (1988, July). P300 Speller: A Brain-Computer Interface for Communication. *IEEE Transactions on Biomedical Engineering*, *37*(7), 658–661. DOI: 10.1109/10.62606

Gouret, . (2023). Advancements in brain-computer interfaces for the rehabilitation of unilateral spatial neglect: A concise review. *Frontiers in Neuroscience*, *17*. Advance online publication. DOI: 10.3389/fnins.2023.694901 PMID: 38784094

Guger, C.. (2011, November). State-of-the-art in brain-computer interface research: BCI Award 2010. *IEEE Transactions on Biomedical Engineering*, *58*(11), 2581–2586. DOI: 10.1109/TBME.2011.2169997

He, H., & Wu, D. (2019). Transfer learning for brain-computer interfaces: A Euclidean space data alignment approach. *IEEE Transactions on Biomedical Engineering*, *66*(6), 1535–1543. PMID: 31034407

Herwig, U., Lutz, J., Scherpiet, S., Scheerer, H., Kohlberg, J., Opialla, S., & Bruehl, A. B. (2019). Training emotion regulation through real-time fMRI neurofeedback of amygdala activity. *Neuroimage*, *184*, 687-696.

Kübler, A., Kotchoubey, B., Hinterberger, T., Ghanayim, N., Perelmouter, J., Schauer, M., Fritsch, C., Taub, E., & Birbaumer, N. (1999). The thought translation device: A neurophysiological approach to communication in total motor paralysis. *Experimental Brain Research*, *124*(2), 223–232. DOI: 10.1007/s002210050617 PMID: 9928845

Kumar, A., & He, H. (2017). EEG-based movement intention decoding using wavelet transform and machine learning techniques. *IEEE Transactions on Biomedical Engineering*, *64*(6), 1314–1322.

Lebedev, M. A., & Nicolelis, M. A. L. (2006). Brain–computer interfaces: Past, present, and future. *Trends in Neurosciences*, *29*(9), 481–489. DOI: 10.1016/j.tins.2006.07.004 PMID: 16859758

Lee, M. B., Kramer, D. R., Peng, T., Barbaro, M. F., Liu, C. Y., Kellis, S., & Lee, B. (2019). Brain-Computer Interfaces in Quadriplegic Patients. *Neurosurgery Clinics of North America*, *30*(2), 275–281. DOI: 10.1016/j.nec.2018.12.009 PMID: 30898278

Loh, H. W., Ooi, C. P., Oh, S. L., Barua, P. D., Tan, Y. R., & Acharya, U. R.. (2023). Adhd/cd-net: Automated eeg-based characterization of adhd and cd using explainable deep neural network technique. *Cognitive Neurodynamics*, •••, 1–17. PMID: 39104684

Loh, H. W., Ooi, C. P., Oh, S. L., Barua, P. D., Tan, Y. R., Molinari, F., March, S., Acharya, U. R., & Fung, D. S. S. (2023). Deep neural network technique for automated detection of ADHD and CD using ECG signal. *Computer Methods and Programs in Biomedicine, 241*, 107775. DOI: 10.1016/j.cmpb.2023.107775 PMID: 37651817

Mane, S., Chew, K., & Schalk, L. R. (2020, April). Non-invasive brain-computer interface technology and its applications in neurorehabilitation. *IEEE Reviews in Biomedical Engineering, 13*, 112–125. DOI: 10.1109/RBME.2020.2979881

Millán, J.. (2016, February). Combining Brain-Computer Interfaces and Virtual Reality for Cognitive Rehabilitation. *IEEE Transactions on Biomedical Engineering, 63*(2), 151–160. DOI: 10.1109/TBME.2015.2466124

Niedermayer, E., & da Silva, F. L. (2004). *EEG and the Neural Basis of Cognition*. Oxford University Press.

Papanastasiou, G., Drigas, A., Skianis, C., & Lytras, M. (2020). Brain computer interface based applications for training and rehabilitation of students with neurodevelopmental disorders. A literature review. *Heliyon, 6*(9), e04250. DOI: 10.1016/j.heliyon.2020.e04250 PMID: 32954024

Pfurtscheller, G., & Neuper, C. (2001). Motor imagery and direct brain-computer communication. *Proceedings of the IEEE, 89*(7), 1123–1134. DOI: 10.1109/5.939829

Singh, A., Sharma, S., & Purohit, K. C. (2024). Artificial Intelligence enabled neuroproteins design for Brain mapping dynamics based on motor imagery classification using HCI (Human computer interface) and (EEG) electroencephalogram. Library of Progress-Library Science, Information Technology & Computer, *44*(3).

Takamura, M., Okamoto, Y., Shibasaki, C., Yoshino, A., Okada, G., Ichikawa, N., & Yamawaki, S. (2020). Antidepressive effect of left dorsolateral prefrontal cortex neurofeedback in patients with major depressive disorder: A preliminary report. *Journal of Affective Disorders, 271*, 224–227. DOI: 10.1016/j.jad.2020.03.080 PMID: 32479320

Vansteensel, R.. (2016, January). Brain-Computer Interfaces Combined with Artificial Intelligence for Adaptation. *Nature Biomedical Engineering*, *4*, 23–29. DOI: 10.1038/s41551-019-0401-3

Wolpaw, J. R., Birbaumer, N., Heetderks, W. J., McFarland, D. J., Peckham, P. H., Schalk, G., Donchin, E., Quatrano, L. A., Robinson, C. J., & Vaughan, T. M. (2000, June). Brain-computer interface technology: A review of the first international meeting. *IEEE Transactions on Rehabilitation Engineering*, *8*(2), 164–173. DOI: 10.1109/TRE.2000.847807 PMID: 10896178

Zotev, V., Mayeli, A., Misaki, M., & Bodurka, J. (2020). Emotion selfregulation training in major depressive disorder using simultaneous real-time fMRI and EEG neurofeedback. *NeuroImage. Clinical*, *27*, 102331. DOI: 10.1016/j.nicl.2020.102331 PMID: 32623140

Chapter 16

Advances in Neural Network Technologies and Brain-Computer Interfaces: Emerging Innovations and Applications

M. Venunath
https://orcid.org/0000-0003-4443-1024
Joginpally B.R. Engineering College, India

Apparao Devineni
Pondicherry University, India

V. Subhashini
J.B. Institute of Engineering and Technology, India

Srinivasu Garikipati
Joginpally B.R. Engineering College, India

M. L. M. Prasad
Joginpally B.R. Engineering College, India

ABSTRACT

The integration of Neural Network Technologies (NNTs) and Brain-Computer Interfaces (BCIs) has transformed human-machine interaction, enabling direct brain-device communication. BCIs empower individuals with neurological disorders,

DOI: 10.4018/979-8-3693-9445-8.ch016

while neural networks enhance signal processing, pattern recognition, and adaptive learning. Advances in deep learning, including CNNs, RNNs, GANs, and Federated Learning, have improved BCI accuracy in motor imagery, authentication, and neurofeedback. Hybrid BCIs combining multiple neural signals enhance adaptability and real-world application. However, challenges remain in ethics, data privacy, hardware, and personalized AI. Future AI-powered BCIs, brain-to-brain communication, and non-invasive neurostimulation promise breakthroughs in cognition, rehabilitation, and augmentation. This chapter explores key principles, innovations, applications, challenges, and future directions in AI-driven neuroscience.

1. INTRODUCTION

Brain-Computer Interfaces (BCIs) and Neural Network Technologies (NNTs) are emerging as revolutionary domains in the conjunction of neuroengineering, artificial intelligence (AI), and medical technology. Brain-computer interfaces (BCIs) help a direct connection between the brain and an external device, allowing users to use their brain signals to operate systems. On the other hand, progress in the processing of EEG signals, the determination of cognitive state and controling neuroprosthetics through BCI owes a great deal of its advance to neural networks (Jonathan R Wolpaw 2007).

Further, deep learning, reinforcement learning, and generative models have delivered profound advancements in the encoding of neural signals, brain-computer interfaces, and cognitive augmentation. BCIs are now used in medical rehabilitation, neuroprosthetics, Communication systems for paralyzed individuals, and brain-controlled robotics. In addition, AI-powered BCIs could transform neurosecurity, cognitive enhancement, gaming, and human-computer interaction(Wolpaw 2013b).

1.1 The Role of Neural Networks in BCIs

Neural networks are key to making BCI's non-invasive, user-friendly and real-time. Various deep learning architectures such as CNNs, LSTMs, and GANs have been used to interpret brain waves more intricately, which will make BCIs more responsive and accurate (Padfield et al. 2019). Federated learning has been applied to BCI systems as well, enhancing privacy and personalization by enabling a model to train across many users without ever requiring sensitive data to be centralized (Ghader et al. 2023).

The problem with conventional BCI systems was that EEG signals tended to be noisy, and there were not enough data points for a given task or based on a particular person's brain to reliably classify the signals. But, approaches based on deep learning have been better at extracting features, which makes recognizing motor imagery, decoding speech, and assessing cognitive load more efficient(Hochreiter and Schmidhuber 1997).

1.2 Evolution of BCIs: From Basic Interfaces to Intelligent Systems

There are generally three stages of development of BCI technologies:

1. Early BCIs (1970s–1990s): Early BCIs dealt with only basic signal detection and pattern classification. First generation EEG-based BCIs were mostly for cursor movement and simple yes/no choices, not much real-time applications (Birbaumer et al. 2000).
2. Modern BCIs (2000s–2010s): The advent of machine-learning and feature-extraction methods helped push the boundaries of signal-classification accuracy, so that BCIs found uses in motor rehabilitation, neuroprosthetic control, and assistive communication devices. Non-invasive EEG-based BCIs were developed for use in research and clinics (Blankertz et al. 2007).
3. AI-Driven BCIs (2020s–Present): Major advances in deep learning, reinforcement learning, and hybrid BCI models have made it possible to decode brain activity in real-time, enhancing user adaptability and system reliability. AI-enabled BCIs are now being used in brain-controlled robots, brain neuromodulation therapies, and neural authentication (Lebedev and Nicolelis 2006).

1.3 Applications of BCIs and Neural Networks

Other BCI and neural net technologies have affected some of the following:

- Medical Applications: Neural networks allow prosthetics controlled by brain waves, rehab for stroke victims, predictions of epilepsy, and monitoring of mental health(Jonathan R. Wolpaw 2007).
- Human-Computer Interaction (HCI):Brainwave-controlled interfaces improve gaming, virtual reality and communication tech(Pfurtscheller 2010).
- Security & Authentication: Brainwave biometrics are under consideration for neurosecurity, allowing people to be identified based on their EEGs (Unnisa et al. 2025).
- Cognitive Enhancement & AI Integration: AI enhanced BCIs provide memory augmentation, attention modulation, and cognitive neurofeedback training (Anon 2025).

1.4 Organization of the Chapter

The rest of this chapter is structured as follows:

- **Section 2** explores the fundamentals of neural networks, discussing their architectures and role in EEG signal processing.
- **Section 3** covers the concepts and technologies behind BCIs, including different types, architectures, and signal processing techniques.
- **Section 4** highlights key innovations in neural networks for BCI applications, including deep learning, reinforcement learning, and federated learning.
- **Section 5** discusses real-world applications of BCIs, particularly in healthcare, neuroprosthetics, and assistive communication.
- **Section 6** examines the challenges and limitations in BCI research, including signal variability, ethical considerations, and hardware constraints.
- **Section 7** explores future trends and research directions, focusing on emerging technologies such as AI-powered BCIs, brain-to-brain communication, and non-invasive brain stimulation.
- **Section 8** concludes with key insights and potential areas for further development.

2. FUNDAMENTALS OF NEURAL NETWORKS

At the heart of contemporary AI are neural networks which serve signal processing, classification, and decision making-the crux of Brain-Computer Interfaces (BCIs). Modeled after the origin and operation of natural neurons, these networks are particularly effective at extracting pattern information and relationships from large datasets, a process that is directly applicable to the analysis of electroencephalography (EEG) signals and other forms of neural data. To help interpret brain signals better and to give users more control over external devices, Artificial Neural Networks (ANNs) and their progeny, such as Convolutional Neural Networks (CNNs), Recurrent Neural Networks (RNNs), and Generative Adversarial Networks (GANs), are being applied more frequently in BCI research (Courville 2016)(Wang, Czerminski, and Jamieson 2021).

2.1 Structure and Function of Neural Networks

Neurons (or "nodes") are organized into layers: an input layer, a middle layer or layers called a "hidden layer," and an output layer. Each neuron sums its weighted inputs, applies an activation function, and passes the result up to the next layer.

These weights are then adjusted through optimization algorithms such as gradient descent and backpropagation to minimize prediction errors, i.e., learning (Silver et al. 2016)(Fukushima 1980).

In BCI applications, neural networks analyze raw brain signals by identifying useful features then translating them to corresponding cognitive or motor functions. Advanced networks such as Deep Neural Networks (DNNs)andSpiking Neural Networks (SNNs)have demonstrated improved accuracy inEEG signal classification, motor imagery detection, and mental state recognition, making them integral to BCI research and development(Hochreiter and Schmidhuber 1997)(Kasabov 2019).

2.2 Types of Neural Networks Used in BCIs

Several kinds of neural networks have been used to improve BCI performance:

- Feedforward Neural Networks (FNNs): These have a simple architecture where information travels from input to output without cycles (loops). They are good at pattern recognition but poor at processing sequences (Wang et al. 2021).
- Convolutional Neural Networks (CNNs): CNNs were used for image recognition, but they have modified for classifying EEG signals by learning spatial features within brainwaves. Some successful applications of CNNs in BCI are in motor imagery detection and P300based BCI classification(Padfield et al. 2019).
- Recurrent Neural Networks (RNNs) and Long Short-Term Memory Networks (LSTMs): RNNs, and more specifically LSTMs, are particularly effective at processing time-series data, which is necessary when interpreting continuous brainwave signals. These are established network designs used in speech recognition and cognitive workload monitoring applications in BCI's(Zarean et al. 2025).
- Generative Adversarial Networks (GANs): GANs have been developed to generate realistic EEG signals, which can ameliorate the issue of limited training data in BCI applications. For developing less biased and more representative neural network models, they create a variety of brain activity samples to train on (Radford, Metz, and Chintala 2016).
- **Spiking Neural Networks (SNNs):**These are bio-inspired models that try to more closely mimic the functioning of biological neurons. For real-time BCI applicationsthey have low-power computational demands and event-driven processing and are thus promising(Gerstner and Kistler 2002).

2.3 Training and Optimization of Neural Networks for BCI

Training a neural network consists of tuning the synaptic weights to reduce classification errors. Training methods used most frequently in BCI related neural networks are:

- Backpropagation with Stochastic Gradient Descent (SGD): This method iteratively updates weights to minimize loss functions in supervised learning tasks, increasing the model's accuracy over time (Kingma and Ba 2015).
- Transfer Learning: Collected EEG data is scarce and expensive, making fine-tuning existing deep learning models from similar tasks attractive for BCI. Transfer learning has provided a significant reduction in training time and increased generalization to new subjects (Wu, Xu, and Lu 2022).
- Federated Learning: This non-centralized model of training on multiple devices is also privacy preserving, ideal for personalized BCI applications where user brain data is sensitive (H. Brendan McMahan, Eider Moore, Daniel Ramage, Seth Hampson 2017).
- Reinforcement Learning (RL): In adaptive BCIs, RL techniques allow the system to learn and improve with real-time feedback which can help improve the performance of brain-controlled robotic devices and neuroprosthetics(Mnih et al. 2015).

2.4 Role of Neural Networks in EEG Signal Processing

Neural network approaches have been a staple in EEG-based BCI systems for feature extraction and classification needed to properly translate brain activity. Raw EEG signals are usually noisy and non-stationary, thus, necessitating sophisticated pre-processing methods prior to their input into neural networks.

- Feature Extraction: Classical methods for feature extraction are Fourier Transform, Wavelet Transform, and Common Spatial Patterns (CSP) for generating useful EEG features. Deep learning-based autoencoders automate feature extraction, reducing human intervention and improving accuracy .
- Classification Algorithms: Neural networks are used to classify extracted EEG features into categories (e.g., motor imagery, attention levels, or emotional states). CNNs and LSTMs have significantly enhanced classification accuracy.
- Real-time Processing and Adaptability: Innovations in neuromorphic computing and edge AI now enable neural networks in BCIs to process signals live, allowing for instant brain-control of applications.

3. BRAIN-COMPUTER INTERFACES: CONCEPTS AND TECHNOLOGIES

Brain-Computer Interfaces (BCIs) represent a revolutionary technology in both neuroscience and human- computer interaction as a direct means of communication between the brain and external devices. These devices interpret signals from the brain to allow a person to will prosthetic arms into motion, type through brain waves, and move through virtual worlds. Machine learning, neural signal processing, and sensor technologies have improved BCI accuracy, efficiency, and adaptability over the last decade(Wolpaw 2013a).

Today's BCIs draw on the power of artificial intelligence (AI) and deep learning to untangle complicated brain patterns with potential uses in health, gaming, security, and even cognitive enhancement. Non-invasive BCI systems, especially those based on Electroencephalography (EEG), have been popular because they are relatively user-friendly and inexpensive. Signal noise, latency, and data scarcity, remain significant barriers to the meaningful deployment of BCIs in the wild, but(Schalk et al. 2004). Such issues are currently being targeted for improvement in research using hybrid BCI systems, adaptive learning algorithms, and cloud-based processing frameworks(Dornhege et al. 2007).

3.1 Fundamentals of Brain-Computer Interfaces

Brain-Computer Interfaces generate these bridges using the neurophysiological signals originating from the brain. These signals can be captured using various brain imaging techniques including Electroencephalography (EEG), Magnetoencephalography(MEG), functional Magnetic Resonance Imaging(fMRI) and Electrocorticography(ECoG) (Lebedev and Nicolelis 2006). Of these, non-invasive EEG-based BCIs are the most common because of their real-time operability, non-invasiveness, and relative low cost.

A BCI system consists of

- Signal Acquisition: Electrodes or sensors record brain activity, which is then amplified and digitized.
- Preprocessing: Noise reduction and artifact removal techniques enhance signal quality.
- Feature Extraction: Relevant patterns are extracted from neural signals using statistical or machine learning techniques.
- Classification: Machine learning models, such as Support Vector Machines (SVM), Convolutional Neural Networks (CNNs), and Long Short-Term Memory (LSTM) networks, classify neural patterns into control signals.

- Output Execution: The classified signals are converted into commands to control external devices, such as robotic arms, exoskeletons, or virtual keyboards.

This technology has powered significant improvements to decoding with BCIs, reduced misinterpretations of brainwaves, and provided more accurate real-time applications (Wolpaw et al. 2000).

3.2 Categories of BCI Systems

According to signal acquisition method and invasiveness, BCI systems are typically grouped into three basic types:

3.2.1 Invasive BCIs

Invasive BCIs require electrodes to be implanted directly into the cerebral cortex by way of surgery. These systems provide the highest signal fidelity and precision but are also highly risky as they can become infected, cause scarring, and damage tissue. Invasive BCIs are mostly limited to medical applications helping quadriplegic patients control robotic prostheses or enabling locked-in patients to communicate (Velliste et al. 2008).

Invasive BCIs include:

BrainGate System: Microelectrode array implanted in the motor cortex to control a computer cursor or robotic limb.

Cortical Neural Prosthetics: Intended to restore movement to persons suffering from spinal cord injuries.

3.2.2 Semi-Invasive BCIs

Semi-invasive BCIs, also called Electrocorticography (ECoG)-based BCIs, require electrode placement under the skull but above the brain tissue. These systems compromise in signal resolution but also in surgical risk compared to fully invasive BCIs. ECoG-based BCIs have demonstrated potential in speech restoration and movement control (Veena and Anitha 2020).

Advantages of semi-invasive BCIs:

- Better signal quality than non-invasive BCIs.
- Less risk of immune rejection than invasive BCIs.

3.2.3 Non-Invasive BCIs

Non-invasive BCIs are the most commonly used and commercially invested category as it does not involve surgery. BCIs utilizing EEG, which measures brain activity through electrodes on the scalp, are the most prevalent. Non-invasive BCI is safe and easy, but offers weaker signal resolution because the skull and scalp interfere with the signals (Nam, Schalk, and Jackson 2011).

Popular non-invasive BCI applications include:

Neurogaming: Brain controlled gaming using EEG headsets.

Assistive Communication: The P300 speller enables paralyzed people to type with their brain activity.

Mental State Monitoring: Includes stress detection and cognitive workload assessment.

3.3 Signal Processing and Machine Learning in BCIs

Signal processing is a required component of BCI systems as raw neural signals are very noisy and messy. Several filtering methods including Independent Component Analysis (ICA), Wavelet Transform, and Fast Fourier Transform (FFT) are applied to improve signal quality prior to additional processing (Wu et al. 2017).

The key steps in BCI signal processing include:

Preprocessing: Signal filtering for noise and other artifacts.

Feature Extraction: Identifying relevant features like power spectral density, wavelet coefficients, and frequency bands.

Feature Selection: Dimensionality reduction techniques such as Principal Component Analysis (PCA) will help preserve informative features.

Classification: AI-driven models, primarily deep learning and ensemble learning algorithms, classify the extracted features to determine the user's intent.

Deep learning breakthroughs have lately achieved cutting edge performance in BCI systems. CNNs have processed spatial information of EEG signals, RNNs, and Long Short-Term Memory (LSTM) networks have been applied to address their temporal aspect (Hosman et al. 2019).

Also, the use of Generative Adversarial Networks (GANs) in BCI studies allow the generation of more realistic brain signals resulting in improved model robustness and addressing data scarcity problem. Federated learning has been investigated as a potential solution to this problem, enabling BCI models to be collaboratively trained over multiple devices with the preservation of user privacy (Ghader et al. 2023).

3.4 Recent Advances in Brain-Computer Interfaces

The some of the milestones undergirding the accelerated development of BCI technologies in the past few years include:

Hybrid BCIs: Hybrid BCIs increase signal accuracy and reliability by integrating multiple neuroimaging modalities (e.g., EEG + fNIRS).

Brain-to-Brain Communication: Certain experiments also suggest that direct brain-to-brain communication using BCIs is possible, which offers up new frontiers for co-cognitive activity.

Neural Decoding with Deep Learning: Models for AI-based BCI have achieved decoding imagined speech and reconstructing visual imagery from brain signals.

As neural networks, neurotechnology, and AI continue to progress, the future for BCIs promises applications in neurorehabilitation, smart wearable devices, and the seamless integration of humans and machines.

4. INNOVATIONS IN NEURAL NETWORKS FOR BCI

4.1 Key Innovations in Neural Networks for BCI

4.1.1 Deep Learning-Based EEG Analysis

Perhaps the most groundbreaking advancement in BCI technology is the implementation of deep learning methods for the analysis of EEG signals. Previous approaches adopted hand-engineered feature extraction and shallow classifiers, compromising the system's precision and ability to generalize. Plus, there have been remarkable performance improvements to EEG-based BCIs using deep learning models, primarily Convolutional Neural Networks (CNNs) and Long Short-Term Memory (LSTM) networks (Kim 2022).

CNNs for Automatic Spatial Patterns: CNNs can learn automatic spatial patterns from raw EEG signals, leading to improved classification rate. CNN models, for example EEGNet and DeepConvNet, have seen use for tasks such as motor imagery classification and emotion recognition in BCIs (Lawhern et al. 2018).

LSTMs for Temporal Dynamics: LSTMs and other recurrent networks are well suited for the analysis time- series EEG data, as they model long-term dependencies and sequential patterns in neural activity (Xiao 2025).

More recently, Transformer models have been applied to EEG interpretation, which have improved real-time brain signal decoding and avoided classification errors (Mohiuddin et al. 2023).

Such deep learning improvements have translated into higher classification accuracies, faster training times, and better real-time performance of BCIs making them more effective for medical, assistive, and entertainment applications.

4.1.2 Reinforcement Learning in BCIs

An adaptive version of learning through Reinforcement Learning (RL) has gained popularity in the field of BCIs for real-time decision making. Unlike supervised learning, RL does not need labeled data, it trains an agent to take actions to arrive at an optimized solution based on reward of feedback, this makes it ideally suited for brain-operated prosthetics and neurofeedback applications(Manjunatha et al. 2020).

Adaptive BCI Training: RL aids in users' development of BCI control by repeatedly updating the decoding model in response to the user's neural activity feedback. This is especially relevant in neurorehabilitation and motor learning (Singh et al. 2023).

BCI-Controlled Robotics: RL has been used with reinforcement learning (RL) algorithms in brain controlled robotic arms, where users can feedback and learn better strategies of control over time. Reinforcement learning techniques such as Deep Q-Networks (DQNs) and Policy Gradient methods have seemed particularly hopeful for increasing the accuracy of prosthetic control (Silver et al. 2016).

Integrating RL makes BCI systems more adaptive to the individual and thus less dependent on years of pre-training, wherein users can refine their control through use.

4.1.3 Generative Adversarial Networks (GANs) for Brain Signal Synthesis

A significant bottleneck in BCI research is the scarcity of good quality neural data. Generative Adversarial Networks (GANs) have been used to generate realistic synthetic brain signals for data augmentation, potentially reducing these limitations (Mahdizadehaghdam, Panahi, and Krim 2019).

EEG Data Augmentation: GANs can create synthetic EEG signals that simulate real brain activity which can be used to train deep learning models when annotated EEG datasets are limited (Bashivan et al. 2016).

Improved model generalization: Exposing the neural network to a mix of synthetic and real EEG data increases the network's tolerance to signal variability, effectively reducing overfitting and improving real BCI performance (Bao et al. 2021).

These developments in synthetic data generation are of broad benefit to BCI research, especially in clinical uses, where obtaining data is difficult given patient conditions.

4.1.4 Hybrid BCIs

Hybrid BCIs combine multiple signal modalities-EEG, Electromyography (EMG), and Functional Near-Infrared Spectroscopy (fNIRS)-to increase reliability and accuracy. Hybrid BCIs integrate multiple physiological signals to improve the performance, reliability, and control of BCI systems beyond what is possible with single-modality systems (Värbu, Muhammad, and Muhammad 2022).

EEG + EMG Integration: The integration of EEG with muscle activity (EMG) improves the reliability of motor control applications, enabling people with spinal cord injuries to effectively use robotic arms (Leerskov et al. 2020).

EEG + fNIRS in Cognitive Monitoring: Hybrid EEG and fNIRS systems have been used for estimating mental workload and monitoring cognitive state, enhancing brain based authentication and fatigue detection systems (Liu et al. 2021).

Hybrid BCIs are poised to revolutionize medical, military, and commercial neurotechnology markets with more user-friendly and adaptable devices.

4.1.5 Federated Learning for BCI

Privacy and personalization are critical issues for BCI applications, particularly in health care settings, as patient data is private. Federated Learning (FL) is a solution to these issues, which allows the training of a neural network across decentralized devices while avoiding sharing raw data (Ghader et al. 2023).

Enhancing Data Privacy: FL upholds user data privacy using on-device training and transmitting only model updates to a central server, reducing risks associated with breaches and unauthorized data access (Bonawitz et al. 2017).

Personalized BCIs: By training models on distributed datasets, FLallow developers to create personalized BCI systems that learns a user's unique neural patterns, which may improve user experience and system performance (Ghader et al. 2023).

Federated learning is also potentially critical to achieving widespread, scalable, and privacy-preserving BCIs as edge computing and 5G networks continue to grow.

5.1 Medical and Healthcare Applications

5.1.1 Neuroprosthetics

One of the most important applications of BCIs is in neuroprosthetics, which allows people with physical disabilities to operate robotic limbs, exoskeletons, and other assistive technology through brain signals. Deep learning models, especially neural networks, have improved the extraction of motor intentions from EEG and Electrocorticography (ECoG) signals which led to smoother and more natural prosthesis control (Hochberg et al. 2006).Improved Signal Decoding: Deep learning methods, particularly Convolutional Neural Networks (CNNs) and Recurrent Neural Networks (RNNs), increase recognition accuracy of motor commands for near-zero latency real-time prosthetic control (Altaheri et al. 2023).Adaptive Learning Systems; Reinforcement learning also enables users to practice a more signal-informed control of their neuroprosthesis overtime, which can tighten the scope and accuracy of targeted control, enhancing the effective and responsive use of neuroprosthesis (Ramos-Murguialday et al. 2013).Neural networks have been successfully implemented in robotic prosthetic limbs that have returned motor function to amputees and people with spinal cord injuries. These are substantial quality of life improvements for patients and make neuroprosthetics an area of prospective growth (Selfslagh et al. 2019).

5.1.2 Stroke Rehabilitation

Another area for vital application of BCIs leveraging neural networks is in stroke rehabilitation. These traditional methods demand long, intensive therapy sessions and have limited effectiveness for patients with severe motor impairments. In the context of BCI-based rehabilitation systems, brain activity is used to drive motor recovery, essentially conditioning neural connections and structural neuroplastic change (Nojima et al. 2022).

Motor Imagery-Based Therapy, Motor imagery signals are classified using neural networks to enable patients to control virtual or robotic devices for rehabilitation exercises (Kruse et al. 2020).

Cusomized Rehabilitative Treatment Plans: Machine learning algorithms assess brain wave activity to tailor treatments to the progress of individual patients to maximize recovery results(Alam et al. 2016).

Studies have demonstrated the effectiveness of deep learning-powered BCI stroke rehabilitation treatments to promote motor recovery through the manipulation of brain plasticity and reactivation of deadened neural pathways. These systems have demonstrated successful clinical testing resulting in improved functional recovery rates for stroke patients (Ang et al. 2015).

5.1.3 Epilepsy and Seizure Prediction

Millions of people suffer worldwide from this disease. Having advance warning of a seizure can dramatically enhance patient care and safety. Deep learning techniques, particularly neural networks, have been successful in utilizing EEG signals to provide seizure detection and prediction. This enables early interventions and preventive actions(Anon n.d.).

Deep Learning for Seizure Classification: CNNs and Long Short-Term Memory (LSTM) networks have achieved automatic seizure pattern detection in EEG recordings with high accuracy(Saab et al. 2020).

Real-Time Seizure Prediction: AI-powered BCI systems can track brain activity and give signals of incoming seizures, allowing for timely medical response prior to the seizure(Qu and Gotman 1997).

The utility of deep learning models for epilepsy management is further evidenced in studies that reported successful seizure prediction rates exceeding 90%. On top of that, wearable BCIs synchronized with AI seizure prediction have led to the emergence smart healthcare systems coupling continuous patient monitoring(Alotaiby et al. 2014).

5.1.4 Neurodegenerative Disease Monitoring

BCIs have proven to be effective instruments in tracking and controlling neurodegenerative diseases like Parkinson's and Alzheimer's. Some neural network models can predict cognitive decline to help medical intervention before it is too late (Tayebi et al. 2023).

Cognitive State Assessment: Machine learning algorithms analyze EEG data to evaluate cognitive function, enabling early diagnosis of Alzheimer's disease(Aviles et al. 2024).

Motor Symptom Monitoring: Other deep learning models assess the progression of Parkinson's disease by analyzing patterns of neural and motor activity to inform providers on treatment plans (Al-Jumeily et al. 2015).

The application of personalized deep learning algorithms have empowered the development of BCI-based interventions for neurodegenerative disease that increase patient quality of life and slows disease progression. Studies are currently investigating AI-powered BCIs for non-invasive brain stimulation therapies as potential new treatments for the above maladies (Dai 2024).

6. CHALLENGES AND LIMITATIONS

A number of difficulties and limitations cloud the potential future success and use of Brain-Computer Interfaces (BCIs) and Neural Network Technologies (NNTs), despite their exciting developments. Such challenges pertain to signal acquisition, data processing, ethical considerations, and user acceptance. These limitations are what makes solving them essential for the effective deployment of BCI-based systems in applied settings.

6.1 Signal Acquisition and Processing Challenges

Signal acquisition and processing is one of the most basic challenges in the development of BCIs. Non-invasive BCI systems, like EEG based systems, have poor signal resolution and are prone to external noise and artifacts. This results in poor signal-to-noise ratios in brain activity, and therefore, reduced BCI performance accuracy (Bashashati et al. 2007).

Low Signal-to-Noise Ratio (SNR): EEG signals are prone to noise from muscle movements, eye blinks, and external electrical sources, subsequently reducing classification accuracy (Kumar, Chakravarthy, and Nanthaamornphong 2023).

Latency and Real-Time Processing: Delays in processing and interpreting neural signals can hamper real-time control, critical in medical and assistive applications of BCIs (Yadav, Yadav, and Veer 2020).

To address these issues, studies are considering sophisticated signal processing methods including denoising deep learning algorithms and hybrid BCIs utilizing multiple signal acquisition strategies. i.e., integrating EEG with functional Near-Infrared Spectroscopy fNIRS for better accuracy(Liu et al. 2021).

6.2 Data Scarcity and Model Generalization

The collection of data is critical for training functional neural network models for BCI applications. But, there are numerous challenges to gathering broad neural data from different populations, ethical considerations, and the high costs associated with collecting EEG/fMRI data. Also, a user-specific neural network trained for one user's brain signals may not work well for another user, and as such msoer the scalability of BCI applications (Saha and Baumert 2020).

Small Datasets: The datasets used in many BCI studies are small, which are challenging to effectively train deep learning models and risk overfitting (Aldahr, Alanazi, and Ilyas 2022).

Inter-Subject Variability: Brain signals can be vastly different between individuals due to variations in cognitive function, mental state, and brain health. This variability creates difficulty in developing generalized BCI models (Wriessnegger et al. 2020).

More recently, activation of transfer learning, domain adaptation, and federated learning have been considered as strategies to address these issues, allowing neural networks to generalize for new users with little retraining (Bharati et al. 2022).

6.3 Ethical, Privacy, and Security Concerns

Ethical and privacy concerns are growing as BCIs shift from a laboratory technology to one integrated into healthcare and consumer products. The data produced and used by BCIs call into question the privacy and security of neural information and possible brain data abuse(Zhang et al. 2024).

Brain Data Privacy: Unauthorized access to brain data could breach brainwave privacy for a user, possibly disclosing a user's cognitive processing, emotional state, or psychiatric diagnoses(Yue 2023).

Ethical Considerations of Cognitive Enhancement: The fairness and access to BCIs for cognitive enhancement indicative the potential for social inequality if only certain individuals could access to such technology(Gordon and Seth 2024).

Behavioral Modification Concerns: Advanced BCIs with the ability to stimulate brain activity may be used for behavior modification, creating unknowns about the ethics of consent and autonomy(San Agustín and Moreno 2021).

In response, some researchers and policy-makers recommend strong data encryption, regulatory constraints, and robust user-consent protocols to prevent the abuse of BCI technologies(San Agustín and Moreno 2021).

6.4 User Adaptability and Long-Term Usability

User acceptance and long-term use also present a significant hurdle to BCI adoption. Most users struggle to reliably produce brain signals the BCI can accurately interpret, resulting in performance variability (Sharma and Miyapuram 2024).

Training and Fatigue: Users require extensive training for high BCI performance, and cognitive fatigue may have significant impact on signal quality over time (Ozkan and Kahya 2018).

Device Wearability and Comfort: Noninvasive BCIs frequently entail wearing bulky EEG caps or headsets that can become uncomfortable over long periods, thus limiting their usability in daily application (Islam and Rastegarnia 2023).

Several strategies are in development to enhance BCI usability, which include the creation of dry electrodes, wireless wearable devices, and adaptive AI models that can individualize signal processing for a user's cognitive state (Rashid et al. 2020).

6.5 High Costs and Accessibility

The expense of developing and rolling out BCI systems, continues to be a barrier to widespread use. High-performance BCIs demand costly hardware, complex computing power, and specialized knowledge, which ultimately may exclude the majority as users (Keerthi Kumar et al. 2020).

Cost of Hardware and Infrastructure: Invasive BCIs and high-resolution neuroimaging systems such as fMRI are expensive and not widely available for consumer use.

Limited Commercial Availability: Despite positive results in the laboratory, affordable commercial BCIs for consumers have struggled to emerge due to technical, regulatory, and economic reasons.

For these reasons, as low-cost consumer BCI devices and open-source AI models advance, they will likely become more accessible and affordable for broader use.

7. FUTURE TRENDS AND RESEARCH DIRECTIONS

Both BCI and NNT are fast moving fields and there are a number of new developments waiting in the wings. With the increasing capabilities and potential influence of AI on improving the accuracy, reliability, and efficiency of BCIs, future research should work toward improving their usability, expanding their applications, and mitigating their ethical pitfalls. Notable trends involve the rise of hybrid BCIs, wearable device integration, improvements in non-invasive signal acquisition, and individualized AI models for adaptive learning. Also, new disciplines like quantum computing and neuromorphic engineering are anticipated to transform what future BCIs will be able to do.

7.1 Hybrid BCIs and Multimodal Integration

The increasing attention to hybrid BCIs-cutting across two or more different types of neuroimaging-for their improved signal quality and reliability is notable. Both accuracy and robustness may be compromised using single-modal signal acquisition, such as EEG or Electrocorticography (ECoG). Hybrid BCIs enhance real-time control and classification accuracy by combining multiple signal sources such as EEG, functional Near-Infrared Spectroscopy (fNIRS), and electromyography (EMG) (Xu et al. 2023).

For example, multimodal fusion techniques combine electrophysiological and hemodynamic signals to more comprehensively represent brain activity, lessening the influence of signal noise and artifacts (Ahn and Jun 2017).

Improved Signal Processing Algorithms: Deep learning put neural signal processing on steroids, and with the adaptability of convolution neural networks (CNN) and transformer-based algorithms, decoding of multimodal neural signals become more efficient (Abibullaev, Keutayeva, and Zollanvari 2023).

Efforts for future research will focus on making hybrid BCIs wireless and seamless, requiring no setup or integration aside from their respective consumer electronics, making brain control more practicable to apply toward real-world uses(Yadav et al. 2020).

7.2 AI-Driven Personalization and Adaptive BCIs

The most daunting problem in existing BCI technology is inter-subject variability an individual's brain signal is unique, making a one-size-fits-all solution difficult to achieve. Future BCIs will employ AI-driven personalization methods-like transfer learning and federated learning-to dynamically adjust to an individual user's neural activity in real time.

Adaptive Learning Models: These models adapt to the user's neural responses, which over time improves classification accuracy and the user experience while requiring less and less user training (Zhang et al. 2021).

Real-time Feedback Loops: Optimizing user interactions will be critical, for which reinforcement learning algorithms will enable BCIs to learn and adapt to cognitive states and user intentions in real-time.

Using these developments, BCIs will be more intuitive and entail minimal calibration-hence, more accessible and widespread .

More research is needed to create wireless hybrid BCIs that seamlessly integrate with existing consumer electronics, so that usable brain-controlled interfaces become more viable.

7.3 Non-Invasive and Wearable BCI Innovations

Invasive BCIs provide much higher quality signals, but their surgical implantation potentially restricts their broader adoption. Efforts in this direction include the development of high resolution, wearable neuroimaging devices that are capable of signal acquisition reliable enough for BCI application and also comfortable enough to wear to prevent motion artifacts in non-invasive BCIs(Mohamed, Mohamed, and Kim 2023).

Dry Electrode Technology: Traditional wet electrodes require conductive gels that are incompatible for long-term use. New designs are incorporating dry and flexible electrodes to increase comfort while preserving signal quality (Lin et al. 2023).

EEG Sensors in Consumer Products: EEG sensors are now being built into wearable headbands, earphones, and even smart glasses, bringing BCIs to the consumer market (Kaongoen et al. 2023).

Wireless and Battery-Efficient Systems: Advancements in low power electronics and wireless communication protocols will enable real-time, untethered BCI applications, reducing hardware constraints (Kartsch et al. 2019).

As these developments mature, BCIs will likely be as ubiquitous as gaming, healthcare, education, and productivity tools(Sawangjai et al. 2020).

7.4 Brain-to-Cloud Interfaces and Quantum Computing in BCIs

As a speculative milestone, there is emerging interest in brain-to-cloud interfaces, the processing and transmission of neural data on cloud platforms. AI can now cloud compute real time analysis of volumes of neural data, making more powerful applications like thought-to-text communication and AI analysis of brain function possible (Hosseini et al. 2017).

Edge Computing for BCIs: Future BCIs will use edge computing to process data locally as opposed to exclusively utilizing cloud servers which will reduce latency and improve real-time interactions (Wang et al. 2020).

The Potential Role of Quantum Computing in Neural Data Processing: Quantum computing could serve to elevate BCI technology by processing complex computations with data signals at an unparalleled scale thus boosting accuracy and speed of classification.

Such cloud-based and quantum-enabled BCIs will increasingly blur the limits of human-computer interaction and neurotechnology research.

7.5 Ethical and Regulatory Considerations

As BCI technology continues to develop, the ethical and regulatory questions of privacy, security, and cognitive autonomy must also be negotiated. Future work will involve efforts to establish ethical and practical guidelines to govern the development and deployment of BCIs .

For data privacy in Neuroethics-enforcing protections and preventing unauthorized data access to brainwave patterns will be critical protocolizing concerns.

Cognitive Rights Laws: Laws passed by national governments and international bodies dictating how neurotechnology may not be used against people, and that BCI treatments must be conducted ethically.

Which means ethical development of BCIs will be essential in building that trust and scaling their use across different verticals.

7.6 The Road Ahead

The prospects for BCIs and neural networks are transformative and developing fast. As AI models advance and hardware becomes more capable, BCIs will escape the confines of the medical and into consumer products, entertainment, and solutions for the workplace. The emerging interdisciplinary field of neuroscience, AI, hardware engineering, and cognitive science will design the intelligent BCIs of the future, with implications for human augmentation and assistive technologies we cannot yet envision.

8. CONCLUSION

It is a watershed development at the intersection of neuroscience, artificial intelligence, and assistive technology: Neural Network Technologies (NNTs) and Brain-Computer Interfaces (BCIs). In the past decade, BCIs have seen tremendous improvements in efficiency, accuracy, and usability using advanced deep learning methods such as Convolutional Neural Networks (CNNs), Long Short-Term Memory networks (LSTMs), Generative Adversarial Networks (GANs), and Reinforcement Learning (RL). These developments have improved neural signal decoding, real-time processing, and adaptive learning that have made BCIs more viable for clinical and commercial use.

An introduction to the basics of neural networks and BCIs, breakthrough innovations, principal health and other applications, and associated limitations and challenges have been presented in this chapter. BCIs hold enormous promise in the fields of neuroprosthetics, stroke rehabilitation, epilepsy prediction, and the monitoring of neurodegenerative diseases, providing a better quality of life for those living with neurological conditions. In addition, federated learning, hybrid BCIs, and wearables offer promising prospects for more secure, privacy-preserving, and user-friendly BCI control. Nevertheless, many challenges remain for BCIs, such as signal noise, data sparsity, data complexity, and ethics. Subsequent development may prioritize enhancements in non-invasive signal acquisition, AI-driven personalization, and further integration of quantum computing to expedite neural processing.

In the future, the integration of BCIs with other emerging technologies, such as the Internet of Things (IoT), cloud computing, and edge AI, will continue to broaden the possibilities of brain-controlled systems. As further studies and interdisciplinary works continue, BCIs will become commodity technologies that revolutionize healthcare, communication, education, gaming, and human augmentation. The ethical and regulatory infrastructure must also develop apace to enable their responsible deployment and prevent their potential misuse. In other words, the integration of

AI neural networks and BCI technology could blur the boundary between human consciousness and digital machines, making thought-activated interaction part of the fabric of daily living.

REFERENCES

Abibullaev, B., Keutayeva, A., & Zollanvari, A. (2023). Deep Learning in EEG-Based BCIs: A Comprehensive Review of Transformer Models, Advantages, Challenges, and Applications. *IEEE Access : Practical Innovations, Open Solutions*, *11*(November), 127271–127301. DOI: 10.1109/ACCESS.2023.3329678

Ahn, S., & Jun, S. C. (2017). Multi-Modal Integration of EEG-FNIRS for Brain-Computer Interfaces – Current Limitations and Future Directions. *Frontiers in Human Neuroscience*, *11*(October), 1–6. DOI: 10.3389/fnhum.2017.00503 PMID: 29093673

Al-Jumeily, D., Iram, S., Vialatte, F. B., Fergus, P., & Hussain, A. (2015). A Novel Method of Early Diagnosis of Alzheimer's Disease Based on EEG Signals. *TheScientificWorldJournal*, *2015*(1), 931387. Advance online publication. DOI: 10.1155/2015/931387 PMID: 25688379

Alam, M., Rodrigues, W., Pham, B. N., & Nitish, V. (1646). Thakor. 2016. "Brain-Machine Interface Facilitated Neurorehabilitation via Spinal Stimulation after Spinal Cord Injury: Recent Progress and Future Perspectives.". *Brain Research*, *1646*, 25–33. DOI: 10.1016/j.brainres.2016.05.039 PMID: 27216571

Aldahr, R. S., Alanazi, M., & Ilyas, M. 2022. "Evolving Deep Learning Models for Epilepsy Diagnosis in Data Scarcity Context: A Survey." Pp. 66–73 in *2022 45th International Conference on Telecommunications and Signal Processing (TSP)*. DOI: 10.1109/TSP55681.2022.9851282

Alotaiby, T. N., Alshebeili, S. A., Alshawi, T., Ahmad, I., & Abd El-Samie, F. E. (2014). EEG Seizure Detection and Prediction Algorithms: A Survey. *EURASIP Journal on Advances in Signal Processing*, *2014*(1), 1–21. DOI: 10.1186/1687-6180-2014-183

Altaheri, H., Muhammad, G., Alsulaiman, M., Amin, S. U., Altuwaijri, G. A., Abdul, W., Bencherif, M. A., & Faisal, M. (2023). Deep Learning Techniques for Classification of Electroencephalogram (EEG) Motor Imagery (MI) Signals. *RE:view*, *35*, •••.

Ang, K. K., Karen, S. G. C., Phua, K. S., Wang, C., Zheng, Y. C., Christopher, W. K. K., Low, W., & Guan, C. (2015). A Randomized Controlled Trial of EEG-Based Motor Imagery Brain-Computer Interface Robotic Rehabilitation for Stroke. *Clinical EEG and Neuroscience*, *46*(4), 310–320. DOI: 10.1177/1550059414522229 PMID: 24756025

Aviles, M., Sánchez-Reyes, L. M., Álvarez-Alvarado, J. M., & Rodríguez-Reséndiz, J. (2024). Machine and Deep Learning Trends in EEG-Based Detection and Diagnosis of Alzheimer's Disease: A Systematic Review. *Eng*, *5*(3), 1464–1484. DOI: 10.3390/eng5030078

Bao, G., Yan, B., Tong, L., Shu, J., Wang, L., Yang, K., & Zeng, Y. (2021). Data Augmentation for EEG-Based Emotion Recognition Using Generative Adversarial Networks. *Frontiers in Computational Neuroscience*, *15*, 723843. DOI: 10.3389/fncom.2021.723843 PMID: 34955797

Bashashati, A., Fatourechi, M., Ward, R. K., & Birch, G. E. (2007). A Survey of Signal Processing Algorithms in Brain-Computer Interfaces Based on Electrical Brain Signals. *Journal of Neural Engineering*, *4*(2), R32–R57. Advance online publication. DOI: 10.1088/1741-2560/4/2/R03 PMID: 17409474

Bashivan, P., Rish, I., Yeasin, M., & Codella, N. 2016. "Learning Representations from EEG with Deep Recurrent-Convolutional Neural Networks." *4th International Conference on Learning Representations, ICLR 2016 - Conference Track Proceedings* 1–15.

Bharati, S., Mondal, M. R. H., Podder, P., & Prasath, V. B. S. (2022). Federated Learning: Applications, Challenges and Future Directions. *International Journal of Hybrid Intelligent Systems*, *18*(1–2), 19–35. DOI: 10.3233/HIS-220006

Birbaumer, N., Kübler, A., Ghanayim, N., Hinterberger, T., Perelmouter, J., Kaiser, J., Iversen, I., Kotchoubey, B., Neumann, N., & Flor, H. (2000). The Thought Translation Device (TTD) for Completely Paralyzed Patients. *IEEE Transactions on Rehabilitation Engineering*, *8*(2), 190–193. DOI: 10.1109/86.847812 PMID: 10896183

Blankertz, B., Dornhege, G., Krauledat, M., Müller, K. R., & Curio, G. (2007). The Non-Invasive Berlin Brain-Computer Interface: Fast Acquisition of Effective Performance in Untrained Subjects. *NeuroImage*, *37*(2), 539–550. DOI: 10.1016/j.neuroimage.2007.01.051 PMID: 17475513

Bonawitz, K., Eichner, H., Grieskamp, W., Huba, D., Ingerman, A., & Ivanov, V. (2019) "TOWARDS FEDERATED LEARNING AT SCALE: SYSTEM DESIGN ArXiv:1902.01046v1." 1–13.

IG,, Y., & Courville, A. (2016). Deep Learning Deep Learning? *Nature*, *29*(7553), 1–73.

Dai, A. (2024). "Brain Computer Interface for the Treatment of Neurodegenerative Diseases." *E3S Web of Conferences* 553. .DOI: 10.1051/e3sconf/202455305010

Dornhege, G., Millán, J. R., Hinterberger, T., Mcfarland, D., Klaus-robert, M., & Bradford Book, A. (2007). "Towards Brain-Computer Interfacing Edited by Noninvasive Brain-Computer Interface Research at the Wadsworth Center." *Interface* 31–42.

Fukushima, K. (1980). Neocognitron: A Self-Organizing Neural Network Model for a Mechanism of Pattern Recognition Unaffected by Shift in Position. *Biological Cybernetics, 36*(4), 193–202. DOI: 10.1007/BF00344251 PMID: 7370364

Gerstner, W., & Kistler, W. M. (2002). *Spiking Neuron Models: Single Neurons, Populations, Plasticity*. Cambridge university press. DOI: 10.1017/CBO9780511815706

Ghader, M., Farahani, B., Rezvani, Z., Shahsavari, M., & Fazlali, M. (2023). "Exploiting Federated Learning for EEG-Based Brain-Computer Interface System." *2023 IEEE International Conference on Omni-Layer Intelligent Systems, COINS 2023* 1–6. DOI: 10.1109/COINS57856.2023.10189325

Gordon, E. C., & Seth, A. K. (2024). Ethical Considerations for the Use of Brain–Computer Interfaces for Cognitive Enhancement. *PLoS Biology, 22*(10), 1–15. DOI: 10.1371/journal.pbio.3002899 PMID: 39466848

Hochberg, L. R., Serruya, M. D., Friehs, G. M., Mukand, J. A., Saleh, M., Caplan, A. H., Branner, A., Chen, D., Penn, R. D., & Donoghue, J. P. (2006). Neuronal Ensemble Control of Prosthetic Devices by a Human with Tetraplegia. *Nature, 442*(7099), 164–171. DOI: 10.1038/nature04970 PMID: 16838014

Hochreiter, S., & Schmidhuber, J. (1997). Long Short-Term Memory. *Neural Computation, 9*(8), 1735–1780. DOI: 10.1162/neco.1997.9.8.1735 PMID: 9377276

Hosman, T., Vilela, M., Milstein, D., Kelemen, J. N., Brandman, D. M., Hochberg, L. R., & Simeral, J. D. (2019). "BCI Decoder Performance Comparison of an LSTM Recurrent Neural Network and a Kalman Filter in Retrospective Simulation." *International IEEE/EMBS Conference on Neural Engineering, NER* 2019-March:1066–71. DOI: 10.1109/NER.2019.8717140

Hosseini, M. P., Soltanian-Zadeh, H., Elisevich, K., & Pompili, D. (2017). "Cloud-Based Deep Learning of Big EEG Data for Epileptic Seizure Prediction." *2016 IEEE Global Conference on Signal and Information Processing, GlobalSIP 2016 - Proceedings* 1151–55. DOI: 10.1109/GlobalSIP.2016.7906022

Islam, M. K., & Rastegarnia, A. (2023). Editorial: Recent Advances in EEG (Non-Invasive) Based BCI Applications. *Frontiers in Computational Neuroscience, 17*, 1151852. Advance online publication. DOI: 10.3389/fncom.2023.1151852 PMID: 36936191

Kaongoen, N., Choi, J., Choi, J. W., Kwon, H., Hwang, C., Hwang, G., Kim, B. H., & Jo, S. (2023). The Future of Wearable EEG: A Review of Ear-EEG Technology and Its Applications. *Journal of Neural Engineering*, *20*(5), 51002. DOI: 10.1088/1741-2552/acfcda PMID: 37748474

Kartsch, V., Tagliavini, G., Guermandi, M., Benatti, S., Rossi, D., & Benini, L. (2019). Biowolf: A Sub-10-Mw 8-Channel Advanced Brain—Computer Interface Platform with a Nine-Core Processor and Ble Connectivity. *IEEE Transactions on Biomedical Circuits and Systems*, *13*(5), 893–906. DOI: 10.1109/TBCAS.2019.2927551 PMID: 31295119

Kasabov, N. K. (2019). *Time-Space, Spiking Neural Networks and Brain-Inspired Artificial Intelligence* (Vol. 750). Springer. DOI: 10.1007/978-3-662-57715-8

Keerthi Kumar, M., Parameshachari, B. D., Prabu, S., & Ullo, S. (2020). Comparative Analysis to Identify Efficient Technique for Interfacing BCI System. *IOP Conference Series. Materials Science and Engineering*, *925*(1), 012062. Advance online publication. DOI: 10.1088/1757-899X/925/1/012062

Kim, H. (2022). "Deep Learning." *Artificial Intelligence for 6G* 22(4):247–303. .DOI: 10.1007/978-3-030-95041-5_6

Kingma, D. P., & Ba, J. L. 2015. "Adam: A Method for Stochastic Optimization." *3rd International Conference on Learning Representations, ICLR 2015 - Conference Track Proceedings* 1–15.

Kruse, A., Suica, Z., Taeymans, J., & Schuster-Amft, C. (2020). Effect of Brain-Computer Interface Training Based on Non-Invasive Electroencephalography Using Motor Imagery on Functional Recovery after Stroke - a Systematic Review and Meta-Analysis. *BMC Neurology*, *20*(1), 1–14. DOI: 10.1186/s12883-020-01960-5 PMID: 31900128

Kumar, A., Chakravarthy, S., & Nanthaamornphong, A. (2023). Energy-Efficient Deep Neural Networks for EEG Signal Noise Reduction in next-Generation Green Wireless Networks and Industrial IoT Applications. *Symmetry*, *15*(12), 2129. DOI: 10.3390/sym15122129

Lawhern, V. J., Solon, A. J., Waytowich, N. R., Gordon, S. M., Hung, C. P., & Lance, B. J. (2018). EEGNet: A Compact Convolutional Neural Network for EEG-Based Brain-Computer Interfaces. *Journal of Neural Engineering*, *15*(5), 1–30. DOI: 10.1088/1741-2552/aace8c PMID: 29932424

Lebedev, M. A., & Nicolelis, M. A. L. (2006). Brain-Machine Interfaces: Past, Present and Future. *Trends in Neurosciences*, *29*(9), 536–546. DOI: 10.1016/j.tins.2006.07.004 PMID: 16859758

Leerskov, K., Rehman, M., Niazi, I., Cremoux, S., & Jochumsen, M. (2020). "Investigating the Feasibility of Combining EEG and EMG for Controlling a Hybrid Human Computer Interface in Patients with Spinal Cord Injury." *Proceedings - IEEE 20th International Conference on Bioinformatics and Bioengineering, BIBE 2020* 403–10. DOI: 10.1109/BIBE50027.2020.00072

Lin, S., Jiang, J., Huang, K., Li, L., He, X., Du, P., ... & Wu, H. (2023). Advanced electrode technologies for noninvasive brain–computer interfaces. ACS nano, 17(24), 24487-24513.

Liu, Z., Shore, J., Wang, M., Yuan, F., Buss, A., & Zhao, X. (2020, November). 2021. "A Systematic Review on Hybrid EEG/FNIRS in Brain-Computer Interface.". *Biomedical Signal Processing and Control*, *68*, 102595. DOI: 10.1016/j.bspc.2021.102595

Mahdizadehaghdam, S., Panahi, A., & Krim, H. 2019. "Sparse Generative Adversarial Network." *Proceedings - 2019 International Conference on Computer Vision Workshop, ICCVW 2019* 3063–71. DOI: 10.1109/ICCVW.2019.00369

Manjunatha, H., Pareek, S., Memar, A. H., Kesavadas, T., & Esfahani, E. T. (2020). Effect of Haptic Assistance Strategy on Mental Engagement in Fine Motor Tasks. *Journal of Medical Robotics Research*, *5*(1–2), 2041004. Advance online publication. DOI: 10.1142/S2424905X20410044

McMahan, H. B., Moore, E., Ramage, D., Hampson, S., & Ag¨uera y Arcas, B. (2017). . *FedAvg.*, *54*, 10.

Mnih, V., Kavukcuoglu, K., Silver, D., Rusu, A. A., Veness, J., Bellemare, M. G., Graves, A., Riedmiller, M., Fidjeland, A. K., Ostrovski, G., Petersen, S., Beattie, C., Sadik, A., Antonoglou, I., King, H., Kumaran, D., Wierstra, D., Legg, S., & Hassabis, D. (2015). Human-Level Control through Deep Reinforcement Learning. *Nature*, *518*(7540), 529–533. DOI: 10.1038/nature14236 PMID: 25719670

Mohamed, M., Mohamed, N., & Kim, J. G. (2023). Advancements in Wearable Eeg Technology for Improved Home-Based Sleep Monitoring and Assessment: A Review. *Biosensors (Basel)*, *13*(12), 1019. DOI: 10.3390/bios13121019 PMID: 38131779

Mohiuddin, K., Welke, P., Alam, M. A., Martin, M., Alam, M. M., Lehmann, J., & Vahdati, S. 2023. "Retention Is All You Need." *International Conference on Information and Knowledge Management, Proceedings* (Nips):4752–58. DOI: 10.1145/3583780.3615497

Nam, C. S., Schalk, G., & Jackson, M. M. (2011). Editorial: Current Trends in Brain-Computer Interface (BCI) Research and Development. *International Journal of Human-Computer Interaction, 27*(1), 1–4. DOI: 10.1080/10447318.2011.535748

Nojima, I., Sugata, H., Takeuchi, H., & Mima, T. (2022). Brain–Computer Interface Training Based on Brain Activity Can Induce Motor Recovery in Patients With Stroke: A Meta-Analysis. *Neurorehabilitation and Neural Repair, 36*(2), 83–96. DOI: 10.1177/15459683211062895 PMID: 34958261

Ozkan, N. F., & Kahya, E. (2018). Classification of BCI Users Based on Cognition. *Computational Intelligence and Neuroscience, 2018*, 1–10. Advance online publication. DOI: 10.1155/2018/6315187 PMID: 29853833

Padfield, N., Zabalza, J., Zhao, H., Masero, V., & Ren, J. (2019). EEG-Based Brain-Computer Interfaces Using Motor-Imagery: Techniques and Challenges. *Sensors (Basel), 19*(6), 1–34. DOI: 10.3390/s19061423 PMID: 30909489

Pfurtscheller, G. (2010). The Hybrid BCI. *Frontiers in Neuroscience, 4*(April), 1–11. DOI: 10.3389/fnpro.2010.00003 PMID: 20582271

Qu, H., & Gotman, J. (1997). A Patient-Specific Algorithm for the Detection of Seizure Onset in Long-Term EEG Monitoring: Possible Use as a Warning Device. *IEEE Transactions on Biomedical Engineering, 44*(2), 115–122. DOI: 10.1109/10.552241 PMID: 9214791

Radford, A., Metz, L., & Chintala, S. 2016. "Unsupervised Representation Learning with Deep Convolutional Generative Adversarial Networks." *4th International Conference on Learning Representations, ICLR 2016 - Conference Track Proceedings* 1–16.

Ramos-Murguialday, A., Broetz, D., Rea, M., Läer, L., Yilmaz, Ö., Brasil, F. L., Liberati, G., Curado, M. R., Garcia-Cossio, E., Vyziotis, A., Cho, W., Agostini, M., Soares, E., Soekadar, S., Caria, A., Cohen, L. G., & Birbaumer, N. (2013). Brain-Machine-Interface in Chronic Stroke Rehabilitation: A Controlled Study Interpretation—The Addition of BMI Training to Behaviorally Oriented Physiotherapy Can Be Used to Induce Functional Improvements in Motor Function in Chronic Stroke Patients With. *Annals of Neurology, 74*(1), 100–108. DOI: 10.1002/ana.23879.Brain-Machine-Interface PMID: 23494615

Rashid, M., Sulaiman, N., Abdul Majeed, A. P. P., Muazu Musa, R., Fakhri Ahmad, A., Sama Bari, B., & Khatun, S. (2020). Current Status, Challenges, and Possible Solutions of EEG-Based Brain-Computer Interface: A Comprehensive Review. *Frontiers in Neurorobotics*, *14*(June), 1–35. DOI: 10.3389/fnbot.2020.00025 PMID: 32581758

Rohtash, R. (2025). Integrating Artificial Intelligence and Neuroscience. *Advancements in Cognitive Computing.*, *1*(1), 1–22.

Saab, K., Dunnmon, J., Ré, C., Rubin, D., & Lee-Messer, C. (2020). Weak Supervision as an Efficient Approach for Automated Seizure Detection in Electroencephalography. *NPJ Digital Medicine*, *3*(1), 1–12. DOI: 10.1038/s41746-020-0264-0 PMID: 32352037

Saha, S., & Baumert, M. (2020). Intra- and Inter-Subject Variability in EEG-Based Sensorimotor Brain Computer Interface: A Review. *Frontiers in Computational Neuroscience*, *13*(January), 1–8. DOI: 10.3389/fncom.2019.00087 PMID: 32038208

San Agustín, A., & Moreno, J. C. (2021). Ethical Aspects of Transcranial Magnetic Stimulation for Neuroenhancement. *Dilemata*, (34), 121–132.

Sawangjai, P., Hompoonsup, S., Leelaarporn, P., Kongwudhikunakorn, S., & Wilaiprasitporn, T. (2020). Consumer Grade EEG Measuring Sensors as Research Tools: A Review. *IEEE Sensors Journal*, *20*(8), 3996–4024. DOI: 10.1109/JSEN.2019.2962874

Schalk, Gerwin, Mcfarland, Dennis J., Hinterberger, Thilo, Birbaumer, Niels, & Wolpaw, Jonathan R., and A. Brain-computer Interface B. C. I. Technology. (2004). BCI2000 : A General-Purpose Brain-Computer Interface (BCI). *System*, *51*(6), 1034–1043. PMID: 15188875

Selfslagh, A., Shokur, S., Campos, D. S. F., Donati, A. R. C., Almeida, S., Yamauti, S. Y., Coelho, D. B., Bouri, M., & Nicolelis, M. A. L. (2019). Non-Invasive, Brain-Controlled Functional Electrical Stimulation for Locomotion Rehabilitation in Individuals with Paraplegia. *Scientific Reports*, *9*(1), 1–17. DOI: 10.1038/s41598-019-43041-9 PMID: 30626917

Sharma, Anupam, and Krishna Miyapuram. 2024. "Evaluating Fast Adaptability of Neural Networks for Brain-Computer Interface." 1–8. .DOI: 10.1109/IJCNN60899.2024.10650562

Silver, D., Huang, A., Maddison, C. J., Guez, A., Sifre, L., Van Den Driessche, G., Schrittwieser, J., Antonoglou, I., Panneershelvam, V., Lanctot, M., Dieleman, S., Grewe, D., Nham, J., Kalchbrenner, N., Sutskever, I., Lillicrap, T., Leach, M., Kavukcuoglu, K., Graepel, T., & Hassabis, D. (2016). Mastering the Game of Go with Deep Neural Networks and Tree Search. *Nature*, *529*(7587), 484–489. DOI: 10.1038/nature16961 PMID: 26819042

Singh, J., Ali, F., Gill, R., Shah, B., & Kwak, D. (2023). A Survey of EEG and Machine Learning-Based Methods for Neural Rehabilitation. *IEEE Access : Practical Innovations, Open Solutions*, *11*(August), 114155–114171. DOI: 10.1109/ACCESS.2023.3321067

Tayebi, H., Azadnajafabad, S., Maroufi, S. F., Pour-Rashidi, A., Khorasanizadeh, M. H., Faramarzi, S., & Slavin, K. V. (2023). Applications of Brain-Computer Interfaces in Neurodegenerative Diseases. *Neurosurgical Review*, *46*(1), 131. DOI: 10.1007/s10143-023-02038-9 PMID: 37256332

Unnisa, Z., Tariq, A., & Din, I. U. (2025). *Danish Shehzad, Mohamed Adel Serhani, Abdelkader N. Belkacem, and Nadeem Sarwar*. Review Article Threats and Mitigation Strategies for Electroencephalography- Based Person Authentication., DOI: 10.1155/ijta/3946740

Värbu, K., Muhammad, N., & Muhammad, Y. (2022). Past, Present, and Future of EEG-Based BCI Applications. *Sensors (Basel)*, *22*(9), 3331. Advance online publication. DOI: 10.3390/s22093331 PMID: 35591021

Veena, N., & Anitha, N. (2020). A Review of Non-Invasive BCI Devices. *International Journal of Biomedical Engineering and Technology*, *34*(3), 205–233. DOI: 10.1504/IJBET.2020.111471

Velliste, M., Perel, S., Spalding, M. C., Whitford, A. S., & Schwartz, A. B. (2008). Cortical Control of a Prosthetic Arm for Self-Feeding. *Nature*, *453*(7198), 1098–1101. DOI: 10.1038/nature06996 PMID: 18509337

Wang, H., Czerminski, R., & Jamieson, A. C. (2021). Neural networks and deep learning. In The machine age of customer insight (pp. 91-101). Emerald Publishing Limited.DOI: 10.1108/978-1-83909-694-520211010

Wang, X., Hersche, M., Tomekce, B., Kaya, B., Magno, M., & Benini, L. (2020). "An Accurate EEGNet-Based Motor-Imagery Brain-Computer Interface for Low-Power Edge Computing." *IEEE Medical Measurements and Applications, MeMeA 2020 - Conference Proceedings*. DOI: 10.1109/MeMeA49120.2020.9137134

Wolpaw, J. R. (2007). "Brain-Computer Interfaces (BCIs) for Communication and Control." Pp. 1–2 in *Proceedings of the 9th International ACM SIGACCESS Conference on Computers and Accessibility, Assets '07*. New York, NY, USA: Association for Computing Machinery. DOI: 10.1145/1296843.1296845

Wolpaw, J. R. (2007). "Brain-Computer Interfaces (BCIs) for Communication and Control." *ASSETS'07: Proceedings of the Ninth International ACM SIGACCESS Conference on Computers and Accessibility* (FEBRUARY 2004):1–2. DOI: 10.1145/1296843.1296845

Wolpaw, J. R. (2013a). "Brain—Computer Interfaces." Pp. 67–74 in *Handbook of clinical neurology*. Vol. 110. Elsevier.

Wolpaw, J. R. (2013b). "Chapter 6 - Brain–Computer Interfaces." Pp. 67–74 in *Neurological Rehabilitation*. Vol. 110, *Handbook of Clinical Neurology*, edited by M. P. Barnes and D. C. Good. Elsevier.

Wolpaw, J. R., Birbaumer, N., Heetderks, W. J., McFarland, D. J., Peckham, P. H., Schalk, G., Donchin, E., Quatrano, L. A., Robinson, C. J., & Vaughan, T. M. (2000). Brain-Computer Interface Technology: A Review of the First International Meeting. *IEEE Transactions on Rehabilitation Engineering*, *8*(2), 164–173. DOI: 10.1109/TRE.2000.847807 PMID: 10896178

Wriessnegger, S. C., Müller-Putz, G. R., Brunner, C., & Sburlea, A. I. (2020). Inter- and Intra-Individual Variability in Brain Oscillations During Sports Motor Imagery. *Frontiers in Human Neuroscience*, *14*(October), 1–13. DOI: 10.3389/fnhum.2020.576241 PMID: 33192406

Wu, D., Lance, B. J., Lawhern, V. J., Gordon, S., Jung, T.-P., & Lin, C.-T. (2017). EEG-Based User Reaction Time Estimation Using Riemannian Geometry Features. *IEEE Transactions on Neural Systems and Rehabilitation Engineering*, *25*(11), 2157–2168. DOI: 10.1109/TNSRE.2017.2699784 PMID: 28463203

Wu, D., Xu, Y., & Bao, L. L. (2022). Transfer Learning for EEG-Based Brain-Computer Interfaces: A Review of Progress Made since 2016. *IEEE Transactions on Cognitive and Developmental Systems*, *14*(1), 4–19. DOI: 10.1109/TCDS.2020.3007453

Xiao, B. (2025). "A Comparison of LSTM and CNN Performance in EEG Motor Imagery with Application to Edge Computing Non-Invasive Brain-Computer Interface Possibilities." *Proceedings of 2024 3rd International Conference on Artificial Intelligence and Intelligent Information Processing, AIIIP 2024* 273–78. DOI: 10.1145/3707292.3707376

Xu, T., Zhou, Z., Yang, Y., Li, Y., Li, J., Bezerianos, A., & Wang, H. (2023). Motor Imagery Decoding Enhancement Based on Hybrid EEG-FNIRS Signals. *IEEE Access : Practical Innovations, Open Solutions, 11*(July), 65277–65288. DOI: 10.1109/ACCESS.2023.3289709

Yadav, D., Yadav, S., & Veer, K. (2020). A Comprehensive Assessment of Brain Computer Interfaces: Recent Trends and Challenges. *Journal of Neuroscience Methods, 346*(July), 108918. Advance online publication. DOI: 10.1016/j.jneumeth.2020.108918 PMID: 32853592

Yue, C. (2023). "Privacy and Ethical Concerns of Brain-Computer Interfaces." Pp. 134–38 in *2023 IEEE International Conference on Metaverse Computing, Networking and Applications (MetaCom)*. DOI: 10.1109/MetaCom57706.2023.00036

Zarean, J., Tajally, A. R., Tavakkoli-Moghaddam, R., & Kia, R. (2025). Robust Electroencephalogram-Based Biometric Identification against GAN-Generated Artificial Signals Using a Novel End-to-End Attention-Based CNN-LSTM Neural Network. *Cluster Computing, 28*(3), 168. DOI: 10.1007/s10586-024-04921-6

Zhang, K., Robinson, N., Lee, S. W., & Guan, C. (2021). Adaptive Transfer Learning for EEG Motor Imagery Classification with Deep Convolutional Neural Network. *Neural Networks, 136*, 1–10. DOI: 10.1016/j.neunet.2020.12.013 PMID: 33401114

Zhang, Z., Chen, Y., Zhao, X., Fan, W., Peng, D., Li, T., Zhao, L., & Fu, Y. (2024). A Review of Ethical Considerations for the Medical Applications of Brain-Computer Interfaces. *Cognitive Neurodynamics, 18*(6), 3603–3614. DOI: 10.1007/s11571-024-10144-7 PMID: 39712096

Chapter 17
Brain-Computer Interfaces:
Revolutionizing Healthcare, Neuroprosthetics, and Cognitive Enhancement

G. Aravind
https://orcid.org/0009-0005-5980-4008
KCG College of Technology, India

ABSTRACT

Brain-Computer Interfaces (BCIs) are innovative systems that allow the brain to interact directly with external devices, eliminating the need for muscle-based communication. This technology is creating new possibilities in medical treatment, cognitive enhancement, and assistive tools. With the help of deep learning methods like CNNs and RNNs, BCIs can now decode brain signals more effectively and accurately. Applications range from helping individuals recover from strokes to improving communication for people with severe physical impairments. BCIs also show potential in treating neurological disorders and supporting mental health. Despite their promise, challenges like real-time processing and ethical considerations remain. Continued progress could transform both healthcare and human-machine interaction.

1. INTRODUCTION TO BRAIN-COMPUTER INTERFACES (BCIS)

- In 1924, Hans Berger, Professor of Psychiatry at the University of Jena in Germany, discovered that electrical signals produced by the human brain

DOI: 10.4018/979-8-3693-9445-8.ch017

could be recorded from the scalp. After five years of further study, Berger published the fi rst of 14 articles that established electroencephalography (EEG) as a basic tool for clinical diagnosis and brain research (Berger, 1929). In 1938, just as his work had begun to receive international recognition, the German government closed his laboratory and forced him into retirement. Th e year was momentarily brightened for him by a holiday greeting from Herbert Jasper, a young North American neuroscientist at the start of a stellar career. It implies, albeit in a fanciful way, that EEG signals could also be used for communication. Th is possibility — that people could act through brain signals rather than muscles — has fascinated scientists and nonscientists alike for many years. Now, nearly a century aft er Berger's epochal discovery, possibility is becoming reality. Although the reality is new and tentative and very modest, its excitement and potential are driving the burgeoning fi eld of brain– computer interface research..This chapter is about that fi eld — the principles that underlie it, its achievements so far, the problems that confront it, and its prospects for the future.

- WHAT IS A BRAIN–COMPUTER INTERFACE? As currently understood, the function of the central nervous system (CNS) is to respond to events in the outside world or the body by producing outputs that serve the needs of the organism. All the natural outputs of the CNS are neuromuscular or hormonal. A brain–computer interface (BCI) provides the CNS with new output that is neither neuromuscular nor hormonal. A BCI is a system that measures CNS activity and converts it into artifi cial output that replaces, restores, enhances, supplements, or improves natural CNS output and thereby changes the ongoing interactions between the CNS and its external or internal environment. Understanding this defi nition requires an understanding of each of its key terms, beginning with CNS . Th e CNS is comprised of the brain and the spinal cord and is distinguished from the peripheral nervous system (PNS), which is comprised of the peripheral nerves and ganglia and the sensory receptors. Th e structures of the CNS are distinguished by their location within the meningeal coverings (or meninges), by their unique cell types and histology, and by their function in integrating the many diff erent sensory inputs to produce appropriate motor outputs. In contrast, the PNS is not within the meninges, lacks the unique CNS histology, and serves mainly to convey sensory inputs to the CNS and to convey motor outputs from it. CNS activity consists of the electrophysiological, neurochemical, and metabolic phenomena (e.g., neuronal action potentials, synaptic potentials, neurotransmitter releases, oxygen consumption) that occur continually in the CNS. Th ese phenomena can be quantifi ed by monitoring electric or magnetic fi elds, hemoglobin oxygenation, or other parameters using sensors on the scalp, on the surface of the brain, or within the brain . A BCI records these brain signals, extracts specifi c measures (or features) from them, and converts (or translates) these features into artifi cial outputs that act on the outside world or on the body itself. Figure 1.3 illustrates the fi ve types of applications that a BCI output might control. For each of these fi ve application types, it shows one of many possible examples. A BCI output might replace natural output that has been lost as a result of injury or disease. For example, a person who can no longer speak might use a BCI to type words that are then spoken by a speech synthesizer. Or a person who has lost limb control might use a BCI to operate a motorized wheelchair. In these examples the BCI outputs replace lost natural outputs. A BCI output might restore lost natural output. For example, a person with spinal cord injury whose arms and hands are paralyzed might use a BCI to stimulate the paralyzed muscles via implanted electrodes so that the muscles move the limbs. Or a person who has lost bladder function due to multiple sclerosis might use a BCI to stimulate the peripheral nerves that control the bladder, thus enabling urination.

FIGURE 1. BRAIN COMPUTER INTERFACE

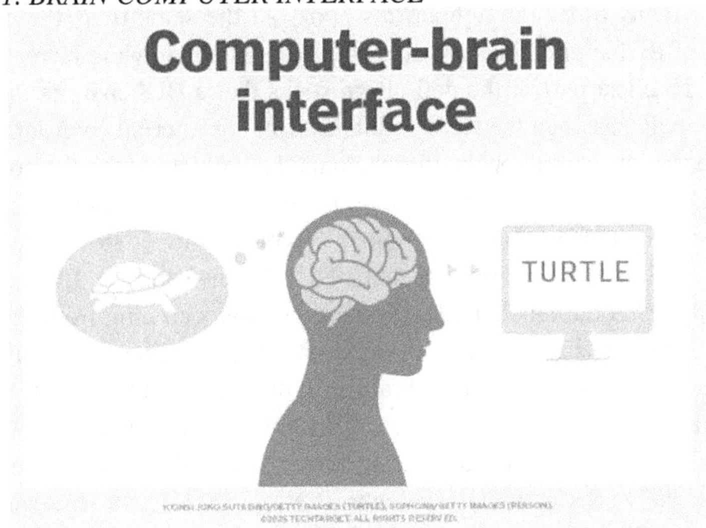

- In these examples, the BCI outputs restore the natural outputs. A BCI output might enhance natural CNS output. For example, a person performing a task that requires continuous attention over a prolonged period (e.g., driving a vehicle or serving as a sentry) might use a BCI that detects the brain activity preceding lapses in attention and then provides an output (e.g., a sound) that alerts the person and restores attention. By preventing the attentional lapses that periodically impair natural CNS output (and might cause traffi c accidents), the BCI enhances the natural output. A BCI output might supplement natural CNS output. For example, a person who is controlling the position of a computer cursor with a hand-operated joystick might use a BCI to select items that the cursor reaches. Or a person might use a BCI to control a third (i.e., robotic) arm and hand. In these cases, the BCI supplements natural neuromuscular output with an additional, artifi cial output. Finally, a BCI output might conceivably improve natural CNS output.

- For example, a person whose arm movements have been impaired by a stroke involving the sensorimotor cortex might use a BCI that measures signals from the damaged cortical areas and then stimulates muscles or controls an orthotic device so as to improve arm movements. Because this BCI application enables more normal movements, its repeated use may induce activity-dependent CNS plasticity that improves the natural CNS output and thereby helps the person to regain more normal arm control. Th e fi rst two types of BCI application, replacement or restoration of lost natural outputs, are the goals of most current BCI research and development, and examples of them appear many times throughout this book. At the same time, the other three kinds of BCI applications are also possible and are drawing increasing attention . Th e last part of the defi nition states that a BCI changes the ongoing interactions between the CNS and its external or internal environment . Th e CNS interacts continuously with the outside world and with the body. Th ese interactions consist of its motor outputs to the environment together with its sensory inputs from the environment. By measuring CNS activity and converting it into artifi cial outputs that aff ect the environment, BCIs change not only the CNS outputs but also the sensory inputs coming from the environment. Th ese changes in sensory input are commonly referred to as feedback . Devices that simply monitor brain activity and do not use it to change the ongoing interactions between the CNS and its environment are not BCIs.
- Neural networks, particularly deep learning models such as Convolutional Neural Networks (CNNs) (Kattenborn et al., 2021) and Recurrent Neural Networks (RNNs) (Das et al., 2019) have significantly improved the accuracy of brain signal decoding, making BCIs more efficient, adaptive, and reliable

1.2 History and Evolution of BCIs

- Early research in the 1970s demonstrated the potential for using brain signals to control external devices

- First BCI developed in 1998 by researchers at Emory University, allowing a paralyzed patient to control a computer cursor using brain signals
- Advancements in neuroscience, computing, and signal processing have driven the rapid development of BCI technology
- Non-invasive BCIs using EEG (Soufineyestani **et al., 2020**)have become more prevalent due to their safety and ease of use
- Invasive BCIs using implanted electrodes offer higher signal quality and precision but involve surgical risks
- BCIs have expanded beyond medical applications to include gaming, entertainment, and consumer devices
- Recent developments focus on improving signal acquisition, processing algorithms, and user training protocols to enhance BCI performance and usability

1.3 Key Components of BCI Systems

- Signal acquisition involves measuring brain activity using sensors (EEG electrodes, microelectrode arrays, fMRI, MEG)
    - EEG is the most common method, using electrodes placed on the scalp to measure electrical activity
    - Invasive BCIs use implanted electrodes to record signals directly from the brain's surface or neurons
- Signal processing converts raw brain signals into meaningful features and commands
    - Preprocessing removes noise and artifacts from the raw signal
    - Feature extraction identifies specific patterns or characteristics in the processed signal
    - Classification algorithms map extracted features to specific commands or intentions
- Output devices translate the classified commands into actions or feedback for the user
    - Can include computer cursors, robotic arms, wheelchairs, or communication aids
    - Feedback helps the user learn to modulate their brain activity and improve BCI performance
- User training involves learning to generate specific brain patterns and associate them with desired commands
    - Requires practice and calibration to achieve reliable control
    - Training protocols vary depending on the type of BCI and the user's needs

1.4 Types of Brain Signals Used in BCIs

- Electroencephalography (EEG) measures electrical activity from the brain's surface using scalp electrodes
    - Non-invasive and widely used in BCI research and applications
    - Offers high temporal resolution but limited spatial resolution
- Electrocorticography (ECoG) **(Miller et al., 2020)** records electrical activity directly from the brain's surface using implanted electrodes
    - Invasive but provides higher signal quality and spatial resolution compared to EEG
    - Requires surgery to implant electrodes beneath the skull
- Intracortical recordings measure activity from individual neurons or small populations using microelectrode arrays
    - Highly invasive but offers the highest spatial and temporal resolution
    - Can decode fine motor movements and intentions with high precision
- Functional magnetic resonance imaging (fMRI) measures changes in blood flow related to neural activity
    - Non-invasive and offers high spatial resolution but limited temporal resolution
    - Used in research to study brain function and map activity patterns for BCI control
- Magnetoencephalography (MEG) measures the magnetic fields generated by electrical activity in the brain
    - Non-invasive and offers high temporal and spatial resolution
    - Requires expensive and bulky equipment, limiting its practical use in BCIs

1.5 Signal Processing and Feature Extraction

- Preprocessing removes noise, artifacts, and irrelevant information from the raw brain signal
    - Filtering eliminates interference from muscle activity, eye movements, and electrical noise
    - Artifact rejection identifies and removes signal contamination from non-brain sources
    - Signal averaging improves the signal-to-noise ratio by combining multiple trials or epochs
- Feature extraction identifies specific patterns or characteristics in the processed signal that correlate with the user's intentions

- Temporal features capture changes in the signal over time (power, amplitude, phase)
- Spectral features represent the signal's frequency content (band power, spectral density)
- Spatial features describe the distribution of activity across different brain regions

- Dimensionality reduction techniques (PCA, ICA) help to identify the most informative features and reduce computational complexity
- Machine learning algorithms (LDA, SVM, neural networks) classify the extracted features into specific commands or intentions
    - Supervised learning methods train the classifier using labeled examples of brain activity and associated commands
    - Unsupervised learning methods discover patterns and clusters in the data without explicit labels
- Adaptive algorithms update the classifier's parameters in real-time to accommodate changes in the user's brain signals and improve performance over time
- **Generative Adversarial Networks (GANs)(Saxena, D & Cao, J, 2019)**: Used for data augmentation and generating synthetic neural signals to enhance model training.

1.6 BCI Applications and Use Cases

- Medical applications aim to restore communication and control for individuals with severe motor disabilities
    - Spelling devices allow users to select letters or words using brain activity alone
    - Prosthetic limbs can be controlled using motor imagery or movement-related brain signals
    - Wheelchairs and other mobility aids can be navigated using brain-controlled commands
- Neurorehabilitation uses BCIs to promote neural plasticity and recovery after stroke or brain injury
    - Provides real-time feedback to guide the user's brain activity and reinforce desired patterns
    - Can be combined with physical therapy and other rehabilitation techniques
- Gaming and entertainment applications use BCIs to enhance immersion and interactivity

- Brain-controlled video games adapt difficulty or gameplay based on the user's mental state
- Virtual and augmented reality experiences can be enriched with brain-based input and feedback

- Cognitive enhancement and optimization applications aim to improve mental performance and well-being
    - Neurofeedback training helps users modulate their brain activity to achieve specific cognitive states (focus, relaxation)
    - Adaptive learning systems tailor educational content and presentation based on the user's brain responses
- Artistic expression (Gryglewski et al., 2020)and creativity can be augmented using BCIs to translate brain activity into visual, auditory, or tactile outputs
    - Brain-controlled music and sound synthesis allow for novel forms of musical expression
    - Generative art and design systems can be guided by the user's mental states and intentions

2. HOW BRAIN-COMPUTER INTERFACES (BCIS) WORK: A DETAILED OVERVIEW

Brain-computer interface (BCI) is a device that translates signals in the brain to control some external activity. For example, they might control a cursor or a prosthetic limb. They might also be called a brain-machine interface. BCI is currently being researched and is targeted at people with disabilities as an adaptive technology.

In BCI, signals in a brain are read directly, without the involvement of the muscular-skeletal system. This means they can be used by people who are otherwise paralyzed or who have certain health conditions. These people might not be able to use other assistive input methods like voice control, eye tracking or sip-puff systems.

While BCI is currently primarily human to computer, some systems operate from computer to person. These systems are for visually impaired people and can translate a camera feed into a signal that can be comprehended by the person's visual cortex to provide limited sight.

2.1 What are brain-computer interface components?

A typical BCI will have four major components:

- **Signal collection.** Uses sensors to gather raw data from the brain.

- **Signal processing.** Analyzes the raw data and extracts the person's intent and discards signal noise. Machine learning algorithms can help in this process.
- **Command translation.** Takes the intent and translates it into a simple command that can be used by other programs.
- **Perform and output.** The command is used to perform and output, such as moving a computer cursor, speech synthesis or powered chair.

2.2 How do brain-computer interfaces work today?

Brain-computer interfaces are still relatively rudimentary and cannot directly "understand" what is going on in a person's brain. Instead, a two-way training process is needed to calibrate the BCI and for the person to learn how to use it.

An example might be that the person imagines raising their hand to move a small indicator up many times. Eventually, the computer learns what signals to listen to and which to ignore; similarly, the person's neural plasticity allows them to learn the correct pattern of thought until it can become almost second nature to operate the BCI.

2.3 Invasive vs. non-invasive brain-computer interfaces

BCI can be broadly separated into invasive and non-invasive methods.

In non-invasive methods, the signals are read using equipment entirely outside the person's body. Invasive methods require the implanting of a sensor into the person. Invasive methods are much more sensitive and accurate but carry risks of surgery and for the body to reject the implant.

Non-invasive BCI(Tamaki **et al., 2020)**

Electroencephalography (EEG) is the most popular non-invasive BCI method. It uses electrodes placed on the scalp to detect neural activity. Because the skull blocks most signals, only the low-frequency brainwaves can be read this way. Other methods use functional magnetic resonance imaging (fMRI) or functional near-infrared spectroscopy (fNIRS) (Pinti et al., 2020) to detect brain activity. But because these require large machinery and are relatively slow, they are currently for research purposes.

FIGURE 2. NON INVASIVE BRAIN COMPUTER INTERFACE

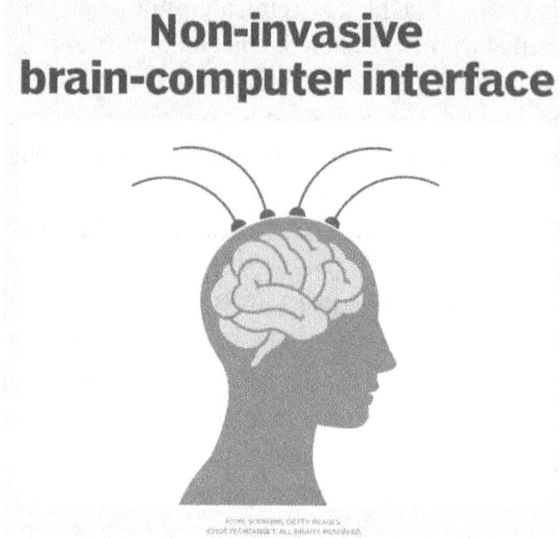

Invasive BCI

Invasive BCI (Zemanova, M. A.,2020), requires surgical implantation of the sensor. The most sensitive implants use microarrays or sensing wires implanted directly in the neural matter. Other less invasive methods use sensors implanted below the scalp or skull but not embedded into the neural matter. A recent research development is using an endovascular sensor threaded through an artery, which can be implanted with only minor surgery.

Several companies are currently developing BCI. Some have had working implants from the early 2000s. Neuralink is currently one of the most talked about BCI developers, but it is a recentcomer to the scene.

FIGURE 3. INVASIVE BRAIN COMPUTER INTERFACE

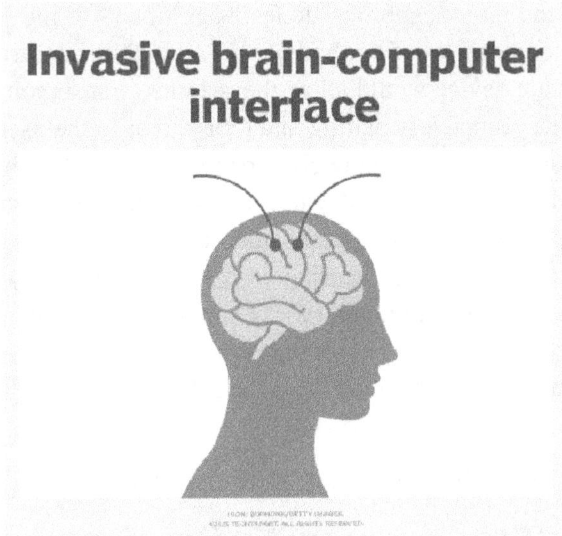

2.4 Brain-computer interface use cases

Applications of brain-computer interfaces are as follows:

- Controlling a computer.
- Controlling a powered wheelchair or prosthetic.
- Generating speech.
- Giving limited vision.
- Regulating mood disorders or epileptic seizures.

3. NEUROPROSTHETICS AND ASSISTIVE TECHNOLOGY

The simple conceptual underpinnings of brain-machine interface (BMI) neuroprosthetic devices (Gutiérrez-Martínez et al., 2020) hide their underlying complexity. In patients who have a failure of communication between neural structures (or output to the external environment) due to illness, stroke, or injury, BMI neuroprosthetic devices would be expected to functionally replace the biological signal-transmission modality with a technological one For example, a patient with failure of communication between the brain and extremities due to a spinal cord injury might be outfitted with a BMI neuroprosthetic device that reads signals directly from the motor cortex-

(Bhattacharjee et al., 2021), uses these signals to control the activity of a prosthetic upper extremity, and returns appropriate feedback signals to the sensory cortex to allow for closed-loop control of a robotic arm.[5] As with the natural limb, the ideal BMI neuroprosthetic device would allow the seamless translation of thoughts into actions in a manner completely natural and transparent to the user, because motor control (and motor learning) are innately subconscious events. There are other applications of BMI neuroprosthetic devices. In addition to motor neuroprosthetics, one might imagine devices for communication through speech or electronic mail, or for sensory modalities such as hearing or vision, or potentially for higher-order processes such as learning and memory. Conceptually, any function that the brain or nervous system serves could be implemented if appropriate signals could be harnessed and directed, either within or outside the body.

FIGURE 4. BMI NEUROPROSTHETIC DEVELOPMENT

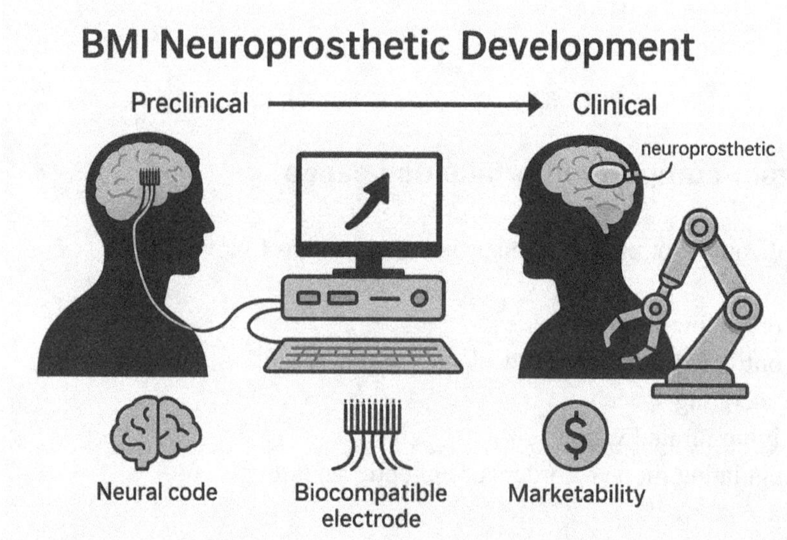

Neuroprosthetic devices to enhance nervous system function have been conceptualized for some time in the scientific literature, popular culture, and science fiction. Neuroprosthetic enhancement may take the form of functional improvement in a disease state in which residual function is below normal or an abnormal symptom arises, or of functional augmentation of the normal state where additional capability is desired. In this regard, electrical interaction with the central and peripheral nervous systems(Andalib et al., 2021) have been developed most extensively using cardiac pacemaker technologies to primarily evoke sensory stimulation.

Since the 1960s, electrical stimulation has been applied to the sensory thalamus, peri-aqueductal gray regions, and dorsal spinal cord for the relief of pain. The current generation of sensory input devices includes cochlear stimulation for hearing restoration, spinal cord stimulation for pain, and visual system stimulation for blindness. The conceptual underpinnings of each device rest on the concept of directly evoking artificial signals within sensory pathways to emulate naturally generated ones. After decades of development, systems such as deep brain stimulators (DBS) (Figee et al., 2022) for movement disorders (i.e., tremor, Parkinson's disease, and dystonia) are in widespread clinical use because of their simple engineering design and ease of implantation.

The idea of implantable, functioning BMI sensorimotor neuroprosthetic devices has been prominently featured in science fiction. For example, the highly popular 1970s television series, *Six Million Dollar Man*, featured an injured astronaut restored to health with a "bionic" right arm, left eye, and legs that provided superhuman strength, vision, and speed. Such imaginings belie the substantial advances in robotics, materials science, computation/signal processing, and systems neuroscience that are still required to produce a device that is capable of performing even the simplest motor tasks, such as reaching or grasping. What is quite remarkable, however, is that recent advances in each of these areas of research have allowed the development of prototype neuroprosthetic BMI devices for humans and nonhuman primates that are capable of performing simple tasks in the laboratory environment. The technological underpinnings of these recent advances and the challenges that lie ahead are the topic of this review.

3.1 Fundamental Components of BMI Neuroprosthetic Devices

To successfully translate thoughts into actions, a BMI motor neuroprosthetic device would be required to incorporate several discrete functions. These include detection of an appropriate "brain intent" signal in some real-time modality, rapid extraction of the signal from the available data (in the face of noise), communication of that signal to some form of internal or external actuator device, and then feedback on the task to determine whether and how precisely the "brain intent" command was actually performed. Each of these steps often emulates rather widespread steps occurring within the human nervous system and encompassing all normal brain function.

3.2 Signal detection

The initial stage is signal detection. A sensor measures changes in a physiological variable at a timescale relevant to the task to be performed. The flow of information in the brain may be observed through diverse physiological changes, any of which might be used to drive a BMI neuroprosthetic device. Such changes might include neurotransmitter concentration gradients measured with voltammetric electrodes, blood-flow changes detectable with functional MRI, or magnetic fields produced by ionic current flows. At the present time, technological advances in signal processing are perhaps best at sensing and processing the electrical fields produced as a result of action-potential neuronal discharges. Electrical signals in the brain may be detected at the level of individual units (single-unit recordings), small populations of neurons in a region (multi-unit recordings and local field potentials), or large populations of neurons over several square centimeters of cortex (electrocorticography and electroencephalography). Studies examining each of these signals individually and in combination have been performed, with differing electrode geometries that influence the types of signals recorded. Each type of signal is more or less discrete, and in focusing in on specific neuronal signals, the larger picture of "brain intent" may be inherently lost, or alternatively, if only high-level signals are decoded, then critical specific information may be missing, making it impossible to reconstruct the control signal in sufficient detail.

3.1.1 Information extraction

Once a signal has been detected, it must be interpreted to determine its information content. Depending on the location of the sensor, the time of recording with respect to an action, and the nature of the signal itself, the information contained within the same signaling modality may be highly variable. For example, signals that are recorded in a specific brain region and with respect to a reaching or grasping task may reflect the following: several potential planned movements under consideration by the subject, coordination of the multiple muscle contractions of a single movement, proprioceptive information, or unrelated neural activity constituting background noise. Understanding the identity and nature of signals associated with motor activity is an ongoing area of active research. Likewise, the recorded signals may be related only indirectly to the actual signals, particularly if the specific generator of the needed control signal is damaged and only an associated brain region is suitable for recording. In this scenario, learning at the brain or device level may be required to approximate the desired but unavailable signal and to infer intent.

3.1.2 Neuroprosthetic actuators

Once information related to the task has been extracted from the collected signals, the intended action must be implemented though an actuator. In the case of spinal cord injury, command signals may be used to drive a computer interface to move a robotic arm or perhaps to externally stimulate the patient's own peripheral nerves or muscles.[13] The degree of sophistication in the actuator critically depends on the nature and rate of information flow. Although low bit rates of information flow may be quite useful in rate-independent communication applications, a much higher rate of information flow is required to produce a functional motor neuroprosthetic with real-time applications. Furthermore, just as natural movements are shaped not only by the motor cortex, but also by additional circuits in the basal ganglia, cerebellum, and spinal cord, neuroprosthetic command signals may require additional interpretation and modulation at the actuator, depending on the granularity of the available signal and the actual device intelligence.

3.1.3. Feedback and adaptation

In any motor control system, errors in measurement, in interpretation, and in execution may be introduced at any stage of the signaling and output process. Hence, for robust and accurate performance, feedback compensation of errors must play an important role in system design. This is particularly true in the injured brain, where the initial signal may necessarily not be the most direct one, and extrapolation may be needed on signal execution to approximate what may be the "real" intended control. As mentioned, there may be some feedback present in the actuator itself, whereby comparisons may be performed between the command signal and the executed action. For example, a prosthetic hand can include a slip detector for local control of grasping to maintain effective grip, but higher level control to signal or trigger release is still needed.

By contrast, in many current implementations of BMI, motor neuroprosthetic devices, feedback, and error correction occur through the subject's observation of a cursor position with respect to a target position. Here the feedback loop is completed by the patient's own visuomotor sensory apparatus. At the unconscious level, the subject may eventually rewire neuronal circuits to improve overall system performance. Between these two implementations, one might consider adaptation and correction based on direct electrical or mechanical stimulation of the nervous system.

3.2 Specific Implementation of BMI Neuroprosthetic Components

Signal detection

Four major types of electrical signals are currently under investigation for use in neuroprosthetic devices: (1) single-unit or multi-neuron recordings, (2) local field potentials (LFPs), (3) electrocorticography (ECoG), and (4) EEG. Each signal modality lies along a spectrum in terms of invasiveness and information content.

3.2.1 Single-unit recordings

The fundamental unit of neural activity is the action potential discharge of individual neurons. To record single-unit activity, a sufficiently small probe must be placed into the brain parenchyma at the location of interest. Because of the larger size of the neuronal cell body, signals from these structures are detected preferentially to those from much narrower individual axons. The flow of current into the neuronal soma or axon with each action potential has a return path through the conductive extracellular fluid. This produces a current dipole that produces an extracellular electrical field potential. The magnitude of the potential depends on the amount of current flow, the conductance of the extracellular space, the distance of the recording electrode to the dipole, and the orientation of the dipole with respect to the electrode. Single-unit recording electrodes (ranging in size from submicron up to 30 to 50 µm) may record potentials of several hundred microvolts from variably sized cortical neurons, and depending on proximity, from multiple neurons simultaneously. To record from multiple neurons in a single region of the brain, recording electrodes are often organized into multi-electrode spatial arrays. When signals from two or more independent neurons contribute to the output of a single recording electrode, these may be differentiated on the basis of the action-potential waveform using signal processing software. Recording bandwidth to allow for clear resolution of action-potential waveforms is in the range of 300 to 5000 Hz. Information in neuronal firing may be encoded in the interpulse interval or the overall rate of firing. To minimize stochastic variation, firing events are typically binned over 20 to 100 ms and are summed together, depending on average action-potential firing frequency. Hence, the effective signal bandwidth is in the range of 10 to 50 Hz for individual neurons.

Electrode arrays may be produced from silicon using microfabrication techniques[14] or assembled from microwires. Microwire assemblies comprise individual 30 to 50 µm diameter tungsten or other metal wires coated with Teflon (DuPont, Wilmington, DE) or polyimide insulation. The wires may be mounted on a circuit board to form a rectangular grid in which the microwires are located 200 to 500 μm apart. The lengths of each wire can be adjusted up to a centimeter or more, depending on the rigidity of the material. In a typical configuration, the circuit boards forming an array are mounted directly to the skull, fixing the wire in position. Using this technology, in some cases, stable recordings have been achieved in nonhuman primates for up to several years after electrode implantation. However, due in part to the free-ranging activity of primate subjects, a slow and typically steady decline in active units is commonly encountered, particularly since these arrays are often attached to the skull rather than the more mobile brain surface, to allow for easier external attachment to recording electronics.

With the evolution of sophisticated micro-fabrication techniques, silicon chip technology offers another attractive approach for electrode array design. Silicon arrays may be configured as shanks, each with a number of recording sites along each shank. To produce such an array, guided boron diffusion into a silicon wafer is performed to define the shape and thickness of the device. After this step, layers of insulating material, conductive interconnects, and additional surface insulation are applied. Recording sites are established by etching the top-most layer of insulation and depositing iridium metal. Finally, the undoped silicon is removed. With this approach, length and shank geometry can be adjusted according to the target tissue, although this is generally limited to a maximum of 1.0 to 1.5 mm in total depth. Because the brain moves relative to the skull, which may account for some of the loss of neuronal activity over time, probes can be designed with a flexible ribbon cable connector, such that they sit on the surface of the brain, moving with each venous pulsation. An alternative configuration is to use similar techniques to produce a three-dimensional 10 × 10 array of needle microelectrodes on a 4 × 4 millimeter base.[16] These arrays possess the theoretical advantage that initial signal-processing electronics may be manufactured into the same piece of silicon, possibly allowing for improved signal-to-noise ratio. The disadvantage of the three-dimensional array is that the maximal needle length with current micro-fabrication technology is limited to 2 mm or so, an inadequate distance to reach layer V neurons in many areas of the human cerebral cortex. Recently this technology has been applied to human use, which is discussed as follows.[6] Long-term survival of signals continues to be a challenge. Likewise, with a fixed spacing array, many of the contacts are silent because neurons are not necessarily in proximity, severely limiting the capability for recording a large number of neurons simultaneously.

The greatest challenge facing proponents of the single-unit recording is the longevity of the signals, due to micro-instability at the electrode-brain interface, the inherent fragility of the neurons, cortical pulsations, and irregular spacing of neurons. One strategy that may help to address this challenge is the use of neurotrophic agents to encourage neurons to interface more stably with electrode assemblies. In addition, the development of carbon nanotube technologies appear to have potential to help to establish biologically advantageous connections to neurons with improved electrical properties. Other areas of improvement include the incorporation of microfluidics into silicon-based devices. Currently, there are few options for stable, long-term recording arrays that can meet the required biocompatibility standards for materials implanted into the brain (e.g., use of platinum/iridium electrode coatings), severely limiting even highly preliminary testing of these devices for human use.

Local field potentials, EEG, and ECoG

Electrical fields produced by individual neurons sum to produce local field potentials. These LFPs may be measured with the same microelectrodes used to measure single-unit activity. By convention, large-amplitude, high-frequency spike waveforms are electronically filtered out of the LFP signal. In tissues where multiple neurons are firing in concert, recorded LFPs reflect cellular electrophysiological activity occurring approximately 50 to 350 μm from the tip of the electrode, and slower ionic events from 0.5 to 3 mm from the tip of the electrode. Importantly, when neurons fire asynchronously, the electrical fields that they generate may cancel, eliminating the net LFP signal. Hence, the LFP signal indicates the degree of coordinated activity among multiple local neurons in a region of the brain.

Although LFPs sample the activity of groups of neurons within millimeters of the recording electrode in the tissue of interest, ECoG and EEG integrate the activity of neurons acting in concert over much larger areas at the cortical surface and scalp, respectively (i.e., perhaps thousands to millions of neurons). Electrodes on the brain or scalp surface are much further removed from neurons than microelectrodes inserted into brain tissue. As a result, the signals recorded are of far lower amplitude than single-unit recordings, and these display much less specific-time resolution. ECoG and EEG electrodes are typically several millimeters in size, and they sample neurons over several square centimeters of cortex depending on the design. All commercially available ECoG/EEG electrodes were specifically designed for epilepsy monitoring over a large brain surface rather than for focal detection of neuronal activity within a specific gyrus, but grid electrodes for optimal spacing to detect brain activity can be constructed, usually with 1 mm electrodes, spaced 1 to 2 mm apart, in a tight array, suitable for a single gyrus.

LFPs, ECoG, and EEG require progressively larger populations of synchronously active neurons to generate a signal, resulting in several advantages and disadvantages for this modality of recording. As individual neurons are highly stochastic and the loss of individual neurons in a population may be frequent, field potential recordings are likely to be more robust than single-unit recordings. Likewise, since microscopic motion is less important, and since electrodes on the surface of the brain avoid any inherent local brain damage, the neuronal responses of these larger electrodes are more stable over time. In addition, because signals are integrated over larger brain regions, the electrode design for these signals may be less invasive in the case of surface cranial EEG, and the electronics are simpler due to lowered bandwidth requirements. However, with this increased integration, the spatial resolution of the signals is considerably lower, and likely the information content is also more limited.

Several categories of field-potential signals have proven useful in neuroprosthetic applications. These include sensorimotor rhythms, slow cortical potentials, and P300 evoked potentials. Sensorimotor rhythms between 8 to 12 Hz (μ-rhythm) and 18 to 26 Hz (β-rhythm) are believed to arise from thalamocortical loops, and they are reduced in activity during real and imagined movements. Slow cortical potentials are EEG oscillations at frequencies below 1 Hz. Movement and other forms of cortical activation result in negative slow cortical potentials, whereas reduced activity is associated with positive cortical potentials. P300 evoked potentials are positive deflections in voltage that occur in the parietal cortex some 300 ms after presentation of a significant stimulus. The evoked potential does not occur in response to routine stimuli. Subjects have been able to control these field-potential signals with training. However, with direct cortical surface recording, much higher frequency responses (up to 500 Hz) can be used to infer brain activity, allowing both a faster response time as well as more direct linkage to ongoing neuronal responses.

Information extraction from the signal

The signal detection modality, be it single-unit recordings, LFPs, ECoG, or EEG, is only the medium for information about the task of interest and the brain's actual intent. Pristine signals with low or uninterpretable information content are not useful to a neuroprosthetic application, just as an empty but high-performance delivery vehicle is not useful to a person expecting a shipment. Hence, information content of the signal rather than signal bandwidth (with consideration of errors) is of greatest importance to BMI applications. Each signal form has its proponents, and perhaps the best way to assess their utility lies in an examination of how these devices have performed so far. A common scheme to rate information content (as in Shannon bits/second) has so far eluded the field, because the various type of information extracted and used are highly disparate and difficult to quantify.

Information and devices using single-unit signals

A great deal is known about the information content of single-unit activity in various regions of the brain. As the principal substrate for information content in neural signals, a substantial fraction of studies in systems neuroscience involve the responsive properties of individual neurons during tasks. Single-unit activity has been described in motor planning, reward, face perception, discriminative touch sensation, and a broad range of additional scenarios. In the typical experiment, subjects perform a specific motor or cognitive task, single-unit recordings are made, and correlations between task performance and neural activity are explored.

In some cases, a mathematical relationship may be uncovered between the performance of a task and neural activity. For example, directionality of movement has been related to some motor-cortex neuronal firing rates through a cosine tuning function.[22] In neuroprosthetic applications, the paradigm is typically one of constructing a model based on repeated performances of a specific task so that the response characteristics of individual neurons are defined.[23] Once the model is trained, patterns of neuronal activity are interpreted to match these patterns to specific tasks from the past. In addition to interpreting the natural function of single-units and using them to drive prosthetic devices, an important potential factor for the efficacy of a single-unit based neuroprosthetic device arises from the plasticity of the brain. Subjects have been trained through an operant-conditioning paradigm to control the rate of arbitrary cortical neurons.[24] This feature of plasticity may be potentially used to drive a neuroprosthetic device.[25]

Motor neuroprosthetic devices using single-unit recordings have been demonstrated in both animal and human subjects. In 1999, Chapin et al. reported a neuroprosthetic robotic application that used signals recorded from the motor cortex to enable a rat to obtain a water reward. Multi-neuron signals recorded from the rat cortex were converted into signals for one-dimensional robot arm control. In this paradigm, animals with significant numbers of recorded neurons (e.g., 25 units) were able to use these brain-derived signals to position the robot arm and obtain water. In 2000, Wessberg et al.[27] demonstrated real-time prediction of hand movement with ensembles of cortical neurons in nonhuman primates. Using microwire arrays, a large population of neurons in the pre-motor, primary motor, and posterior parietal cortical areas were recorded as the subjects performed motor tasks. These recordings produced real-time predictions of one- and three-dimensional arm movement trajectories, which then could be used for real-time control of a robotic arm performing similar movements.

In 2002, studies by Serruya et al. and Taylor et al. demonstrated real-time neuroprosthetic device control in primates. In Serruya et al., primate subjects were able to use motor cortical neuronal outputs recorded with a silicon-based array to move a computer cursor to arbitrary positions in a 14-degree target region in real time. Taylor et al.[7] reported a paradigm in which nonhuman primate subjects moved a cursor from a center-start position to one of eight targets arranged in three-dimensional space. Cell-tuning properties changed when used for brain-controlled movements, allowing fewer neurons to drive task performance. In 2003, Carmena et al. reported neuroprosthetic real-time control of a robotic arm for reaching and grasping in primates. This study demonstrated neuronal plasticity and learning in multiple cortical areas during task performance, as well as indicating that the same population of recorded neurons may be able to control different tasks, such as both reaching and grasping.

Promising developments in work with nonhuman primates have encouraged limited study of BMI neuroprosthetic applications in humans. Kennedy and Bakay pioneered efforts to produce an implanted BMI neuroprosthetic, implanting a patient with "locked-in" syndrome due to amyotrophic lateral sclerosis with a neurotrophic cone electrode. This initial patient was able to control the amount of neural activity recorded from the electrode. A second patient was able to control a computer cursor. In the intraoperative setting, Patil et al. demonstrated that signals obtained from the thalamus and subthalamic nucleus could be used to predict force-task performance from signals recorded in these deep-brain structures. Signals recorded from subcortical targets demonstrated information capable of predicting patient activity during a force-production task with properties similar to those observed in nonhuman primate studies.

Recently, Donoghue, et al.[4] and Hochberg, et al.[6] reported successful implantation of a silicon-based cortical array in a patient with tetraplegia. The patient was implanted with a 96-channel, silicon-based microelectrode array in the motor hand area. During the 6 months after implantation, after which adequate unit recordings were lost, the subject used imagined movements to control the degree of unit activity. This activity could then be harnessed to control a cursor during a center-out task to perform simple computer and appliance control tasks and to drive the control of a simple robotic arm. In these experiments, the information available from the electrode, the stability over time of the information, and the training time were limited, which severely curtailed the tasks that could be controlled.

Information and devices using LFPs, ECoG, and EEG

Field potential signals integrate the activity of regional neurons acting in concert. Synchronous signals, often in the form of oscillatory behavior, may be locally observed (LFPs) at the cortical surface (ECoG) or at the scalp (EEG). The potential to harness these signals noninvasively have made them an attractive source of information for cortical neuroprosthetic devices. Prototype devices have been developed using both ECoG and EEG, but as add-on research during invasive epilepsy studies in general, rather than using dedicated high-density grid electrodes appropriate to signal recording. Although single-unit-based neuroprosthetic devices typically use models to interpret the function of neurons and then drive models to perform similar tasks, field-potential based recordings are based on subject conditioned control of field-potential signals.

Slow cortical potentials are also subject to operant control in subjects with intact function of motor and pre-motor cortical and subcortical structures. By controlling slow cortical potentials after letters or words were spoken or presented by a computer developed a spelling device for the paralyzed. Patients were able to produce from 0.15 to 3 letters per minute using this device. Sensorimotor rhythms over a region of cortex are reduced when a subject imagines movement of a limb corresponding to that region, which may increase when movement of a different limb is imagined.[35] Sensorimotor mu and beta rhythms have been used to control the movement of a cursor on a screen to one of eight random locations within 10 seconds.[9] A similar paradigm has been used to allow control of reaching and grasping in high spinal cord-lesioned patients.36, 37, 38 In general, the information content from these devices is limited to a few choices per minute, but cursor control on a computer screen can be achieved.

3.2.2 Feedback and adaptation

To achieve widespread application, neuroprosthetic devices must achieve goals of both speed and accuracy. Error correction must be automatic or require minimal user effort. As in natural activities, a neuroprosthetic device use should approach an unconscious level with use, functioning as an additional appendage or inherent attachment. These two processes of feedback and intelligent adaptation, acting in concert, are likely to play important roles in the next stages of neuroprosthetic development.

Feedback paradigms in neuroprosthetics

At present, visuomotor feedback is the predominant form of feedback control available to users of current neuroprosthetic designs. For example, in a field-potential based application for communication, subjects select letters through a series of binary decisions, choosing among groups of letters displayed on a screen, until a letter is chosen after five decision steps and two correction/confirmation steps. In single-unit based motor neuroprosthetic applications, human and primate subjects receive visual indications of the accuracy of their movements in real time.

However, in the natural state, a wide range of feedback modalities are used to increase the accuracy of a desired movement. In the central movement circuits of the brain, command signals present in motor cortex are modulated by diverse inputs from the nonpyramidal portions of the motor system, such as the basal ganglia and cerebellum, as well as proprioceptive inputs routed through the thalamus. It has been noted that the characteristic movements induced in single-unit neuroprosthetic applications based on signals detected in the motor cortex are reminiscent of patients who have neurological deficits in nonpyramidal motor subsystems.[40] Natural movements also benefit from feedback processes in spinal reflex regulatory arcs and in proprioceptive feedback through sensory pathways.

Current paradigms to provide nonvisuomotor feedback to neuroprosthetic devices is limited both scientifically and technologically. Although associations between cortical motor activity and proprioceptive sensory modalities have been described, the ability to introduce sensory information directly to the cortex (or thalamus) is limited. In humans, sensory and motor signals have been carefully studied in both sensory and motor thalamus during neurosurgical procedures for movement disorders and pain. However, stimulation has resulted in only poorly formed sensory percepts, although sensations could be evoked from nonfunctioning amputated limbs.[42] However, even an artificial sensory precept can be graded, resulting in, for example, a graded description of gripping force.

Formation of a clear precept may not be required for neuroprosthetic applications. As an example, patients with profound hearing loss may receive a cochlear implant device in which auditory neurons are directly stimulated. Although the generated sensations are far from normal hearing, patients are often able to retrain their minds to the point that they experience the sensory stimuli as a form of hearing. Along these lines, cortical microstimulation delivered to primary somatosensory cortex though chronically-implanted microelectrode arrays has been used to direct nonhuman primate task selection, suggesting that multi-electrode microstimulating electrodes may provide a potential mechanism to introduce a neuroprosthetic sensory feedback path. Likewise, thalamic microstimulation may successfully emulate multiple aspects of movement and pressure, if provided in an appropriate spatial and temporal sequence.

Neural adaptation to neuroprosthetic devices

In addition to feedback mechanisms of error-correction, adaptation of the neural circuitry to the device, or built-in adaptation of the circuitry as a form of learning, are likely to be critical features allowing neuroprosthetic devices to achieve a degree of use similar to the natural limb. At the conscious level, primates can seemingly arbitrarily convert their neural activity into task performance. Several lines of evidence suggest that the activity of individual neurons is likely to be quite plastic. In what are now considered classic studies, Fetz and colleagues[44] demonstrated that arbitrarily selected neurons are subject to rate control during an operant conditioning paradigm. In primate studies, training of individual neurons to a specific task have been observed.[8] Finally, in a human application of a prototype BMI neuroprosthetic device, the subject was able to control devices in different scenarios with conscious control, suggesting rapid plasticity in the function of the individual units.[6] With the addition of feedback mechanisms, the robust quality of the device is likely to be improved.

In many instances the cortical area available for electrode implantation may not be ideal for the function of interest, particularly if there has been damage to the primary cortical region (as in the case of speech deficits or weakness after a cortical stroke). Thus, the task required may be relatively novel for the implanted region of cortex and significant learning and neural adaptation may be required to perform the task at all with signals recorded from that region. In this instance there may no electrophysiological data available to train a model relating cortical activity to task performance, as is commonly used in preclinical experiments. The correlation between the intention and the cortical activity will be initially undefined, complicating the training process considerably. Alternative methods of training would be required to initiate the task, and then with appropriate feedback and adaptation the hope would be that cortical plasticity would mould the commands for the desired task from the available (less than appropriate) signals available from the implanted, secondary brain regions.[45]

CONCLUSIONS

BMI neuroprosthetic technology is likely to change considerably during its evolution from preclinical to clinical studies. The eventual goal is to develop a BMI neuroprosthetic device that will allow the seamless translation of thoughts into actions in a manner completely natural and transparent to the user. Over the past 40 years, BMI development has moved from the realm of science fiction to rudimentary, yet promising devices in the clinical setting. However, many prerequisites for further development remain unsatisfied, posing considerable challenges. These include an understanding of the neural codes underlying intention and action, the development of a biocompatible electrode with the ability to deliver long lasting, low-noise signals from relevant regions of brain, and the development of operative techniques to implant a device in a human reliably and safely, to mention a few. However, perhaps the largest unanswered question relates to commercial attractiveness and marketability. The research and development costs of a functional BMI neuroprosthetic are enormous. Only time will tell whether there is sufficient commercial interest to successfully overcome the multifactorial challenges in the path of clinical BMI development.

4. COGNITIVE AND AFFECTIVE BRAIN–COMPUTER INTERFACES IN EDUCATION

Presently, many applications based on neuroscientific evidence are utilized by cognitive- and affective-related multidisciplinary disciplines. Cognitive process can be measured by memory, workload, attention and language while the affective process more on emotion and motivation. Therefore, BCI can be a tools to analyze the brain signal, including neurofeedback, and improve cognitive activity to restore learning and memory. For instance, people's affective or cognitive states can be measured using the BCI to determine the cognitive load and avoid mental fatigue.

BCIs are useful for measuring cognitive processes, which means that there is a greater opportunity for the training to be adapted based on the learner's current condition, which should enable a better education of the user. Example in learning mathematics using abacus, the BCI system can detect the brain activity during the calculation process and see the different signals and brain changes between the expert and novices students. Therefore, with these information, the educator can improved adaption of content and different way of teaching to maximize the student's performance in learning. Furthermore, in affective process, the BCI can help to reduce the anxiety. The majority of students that struggle with mathematics are always concerned during the process of completing mathematical problem. BCI devices can record real-time brain activity and offer visual feedback to the student as anxiety levels rise, attempting to assist the student in regaining control.

Figure 5. Cognitive brain computer interface

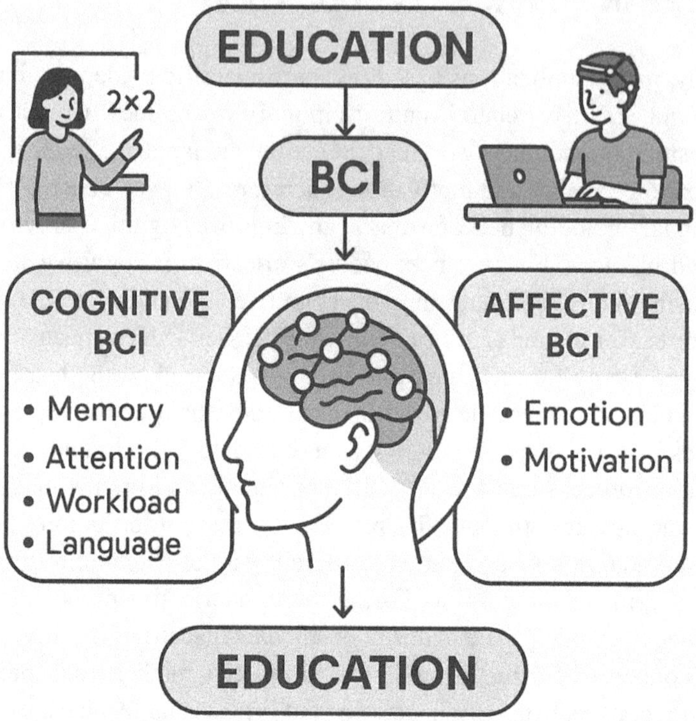

Moreover, BCI technologies have proven useful for restoring, enhancing, improving, supplementing, and replacing the motor and cognitive abilities, which lead to empowering directly or indirectly. The teacher and learners perform many tasks most productively and indirectly impact the learning outcomes due to improvements in concentration, mood, emotion, and cognitive abilities of BCI users. Example, the BCI application can be used in cognitive behaviour intervention directly such as social cognitive skills training for Autistic Spectrum Disorder (ASD) (Lord et al., 2020) people. BCI application empower participants by providing immediate feedback on their attention focus. This information enables the individual to self-monitor their performance about where to gaze and, thus, enables them to change their behavior . At the same time, BCI devices that stimulate the brain neuroplasticity, which may result in the restoration of motor function. For example, with BCI training, the stroke patients showed the improvement in opening hand movement of the paretic hand using exoskeleton. So, indirectly BCI application can become one of the alternatives to help students which paralyzed or have motor impairment to continue the learning process.

In correlation, BCI affect and contribute in learning process. Teachers must consider the students' cognitive development and understand their need to promote better learning while designing and developing the learning environment. Therefore, neurofeedback is an appealing approach to improve cognitive function by modulating human brain activity to evaluate the degree of clarification regarding the information learned.

Therefore, BCI becomes a promising tool in many fields, including education, for understanding and measuring brain activity. Some projects have been developed using BCI devices. Further, current education models and support systems have been established to enhance the learning method and strengthen cognitive abilities of healthy people as well as persons with disabilities. Along with recent advances in education and the potential associated with BCI, researchers have begun to attend to serious questions on how BCI can help the students' learning abilities.

BCI has been used in experiments designed to create knowledge for educational theorists and practitioners. Some education research used mobile electroencephalogram (EEG) headbands to study brain activity within the classroom settings.BCI can apply to understanding the brain and inspecting the development of a student's mental state. Other methods of research can examine potential behavioral mechanisms that influence the brain and student cognition. Simultaneously, neuroimaging studies have investigated and assessed brainwaves to identify discrepancies concerning selective attention.For measuring and interpreting brainwaves, understanding the activity of the brain is significant. The brain function's electromagnetic patterns can be detected using non-invasive tools like EEG. In an EEG, which is the most common method for measuring, electric brain signals produced by brain activities are monitored and documented using sensors.Further, cells in the brain will align with each other and generate the electrical signal whose activity patterns can be analyzed.

Moreover, BCI can offer information through several motor controls and complex cognitive features for measuring people's memory attention concentration cognitive skills and learning style .Measuring attention, cognitive skills, emotion, and other factors that influence student learning provide many advantages for monitoring the student's performance in academic subjects. Furthermore, the BCI framework for assessing mental concentration, attention, and cognitive levels is useful in neuropsychology and education, especially for children or students who have attention disorders, such as ADHD, learning disabilities, or anxiety .

Numerous studies using EEGs have consistently found that the measurement and evaluation for the learning process of student mental states are a combination of the alpha and theta wave frequency The activity in the theta wave was reported as being favorably correlatable, with the preference for memory as well as a predictive factor of changes in cognitive tasks, specifically in the frontal lobe. A strong correlation between the neuronal spiking observed that the more significant theta oscillation coordinates can combine the different mental functions of a cognitive task research showed improved amplitude and synchronization in various frequencies of cognitive tasks and oscillatory activities, particularly in theta (4-7 Hz) and alpha (8-13Hz). EEG is also used for educational purposes for understanding student mental status by examine and observe students' brain activity

BCI-based supporting factors, such as learning style, lifestyle, and mental well-being, can contribute to the learning progress and affect education performance . Psycho-education has become particularly relevant in recent years in which the students and families are provided with reliable information on mental health or specific symptoms of depression or stress For example, it has been established that insufficient sleep is associated with attention deficiency, decline in academic performance, depression, and poor health conditions. Students may also have inadequate information regarding sleep disorders and sleep health habits, thus contributing to bad sleep hygiene . Figure 1 illustrates the relationship between cognitive and affective BCIs and education.

REFERENCES

Andalib, S., Biller, J., Di Napoli, M., Moghimi, N., McCullough, L. D., Rubinos, C. A., O'Hana Nobleza, C., Azarpazhooh, M. R., Catanese, L., Elicer, I., Jafari, M., Liberati, F., Camejo, C., Torbey, M., & Divani, A. A. (2021). Peripheral nervous system manifestations associated with COVID-19. *Current Neurology and Neuroscience Reports*, *21*(3), 1–14. DOI: 10.1007/s11910-021-01102-5 PMID: 33586020

Asadi-Pooya, A. A., Brigo, F., Lattanzi, S., & Blumcke, I. (2023). Adult epilepsy. *Lancet*, *402*(10399), 412–424. DOI: 10.1016/S0140-6736(23)01048-6 PMID: 37459868

Bhattacharjee, S., Kashyap, R., Abualait, T., Annabel Chen, S. H., Yoo, W. K., & Bashir, S. (2021). The role of primary motor cortex: More than movement execution. *Journal of Motor Behavior*, *53*(2), 258–274. DOI: 10.1080/00222895.2020.1738992 PMID: 32194004

Das, S., Tariq, A., Santos, T., Kantareddy, S. S., & Banerjee, I. (2023). Recurrent neural networks (RNNs): architectures, training tricks, and introduction to influential research. *Machine learning for Brain disorders*, 117-138.

Evangelista, C., Basiricò, L., & Bernabucci, U. (2021). An overview on the use of near infrared spectroscopy (NIRS) on farms for the management of dairy cows. *Agriculture*, *11*(4), 296. DOI: 10.3390/agriculture11040296

Feldman, E. L., Goutman, S. A., Petri, S., Mazzini, L., Savelieff, M. G., Shaw, P. J., & Sobue, G. (2022). Amyotrophic lateral sclerosis. *Lancet*, *400*(10360), 1363–1380. DOI: 10.1016/S0140-6736(22)01272-7 PMID: 36116464

Figee, M., Riva-Posse, P., Choi, K. S., Bederson, L., Mayberg, H. S., & Kopell, B. H. (2022). Deep brain stimulation for depression. *Neurotherapeutics; the Journal of the American Society for Experimental NeuroTherapeutics*, *19*(4), 1229–1245. DOI: 10.1007/s13311-022-01270-3 PMID: 35817944

Gryglewski, P., Ivashko, Y., Chernyshev, D., & Dmytrenko, A. (2020). Art as a message realized through various means of artistic expression. Art Inquiry, (22)

Gutiérrez-Martínez, J., Toledo-Peral, C., Mercado-Gutiérrez, J., Vera-Hernández, A., & Leija-Salas, L. (2020). Neuroprosthesis Devices Based on Micro-and Nanosensors: A Systematic Review. *Journal of Sensors*, *2020*(1), 8865889. DOI: 10.1155/2020/8865889

Hariz, M., & Blomstedt, P. (2022). Deep brain stimulation for Parkinson's disease. *Journal of Internal Medicine*, *292*(5), 764–778. DOI: 10.1111/joim.13541 PMID: 35798568

Kattenborn, T., Leitloff, J., Schiefer, F., & Hinz, S. (2021). Review on Convolutional Neural Networks (CNN) in vegetation remote sensing. *ISPRS Journal of Photogrammetry and Remote Sensing*, *173*, 24–49. DOI: 10.1016/j.isprsjprs.2020.12.010

Keni, R. R., Jose, M., Baishya, J., Sarma, P. S., & Thomas, S. V. (2020). Anti-epileptic drug and folic acid usage during pregnancy, seizure and malformation outcomes: Changes over two decades in the Kerala Registry of Epilepsy and Pregnancy. *Epilepsy Research*, *159*, 106250. DOI: 10.1016/j.eplepsyres.2019.106250 PMID: 31855827

Kolk, S. M., & Rakic, P. (2022). Development of prefrontal cortex. *Neuropsychopharmacology : Official Publication of the American College of Neuropsychopharmacology*, *47*(1), 41–57. DOI: 10.1038/s41386-021-01137-9 PMID: 34645980

Li, L., Fan, Y., Tse, M., & Lin, K. Y. (2020). A review of applications in federated learning. *Computers & Industrial Engineering*, *149*, 106854. DOI: 10.1016/j.cie.2020.106854

Lord, C., Brugha, T. S., Charman, T., Cusack, J., Dumas, G., Frazier, T., Jones, E. J. H., Jones, R. M., Pickles, A., State, M. W., Taylor, J. L., & Veenstra-VanderWeele, J. (2020). Autism spectrum disorder. *Nature Reviews. Disease Primers*, *6*(1), 5. DOI: 10.1038/s41572-019-0138-4 PMID: 31949163

Middleton, A., Adams, A., Aidid, H., Atutornu, J., Boraschi, D., Borra, J., Bircan, T., Burch, C., Costa, A., Dickinson, A., Enticknap, A., Galloway, C., Gale, F., Garlick, E., Haydon, E., Henriques, S., Mitchell, M., Milne, R., Monaghan, J., & Parkinson, S. (2023). Public engagement with genomics. *Wellcome Open Research*, *8*, 310. DOI: 10.12688/wellcomeopenres.19473.2 PMID: 37928209

Miller, K. J., Hermes, D., & Staff, N. P. (2020). The current state of electrocorticography-based brain–computer interfaces. *Neurosurgical Focus*, *49*(1), E2. DOI: 10.3171/2020.4.FOCUS20185 PMID: 32610290

Oruh, J., Viriri, S., & Adegun, A. (2022). Long short-term memory recurrent neural network for automatic speech recognition. *IEEE Access : Practical Innovations, Open Solutions*, *10*, 30069–30079. DOI: 10.1109/ACCESS.2022.3159339

Pinti, P., Tachtsidis, I., Hamilton, A., Hirsch, J., Aichelburg, C., Gilbert, S., & Burgess, P. W. (2020). The present and future use of functional near-infrared spectroscopy (fNIRS) for cognitive neuroscience. *Annals of the New York Academy of Sciences*, *1464*(1), 5–29. DOI: 10.1111/nyas.13948 PMID: 30085354

Porsteinsson, A. P., Isaacson, R. S., Knox, S., Sabbagh, M. N., & Rubino, I. (2021). Diagnosis of early Alzheimer's disease: Clinical practice in 2021. *The Journal of Prevention of Alzheimer's Disease*, *8*(3), 371–386. DOI: 10.14283/jpad.2021.23 PMID: 34101796

Raspopovic, S., Cimolato, A., Panarese, A., Vallone, F., Del Valle, J., Micera, S., & Navarro, X. (2020). Neural signal recording and processing in somatic neuroprosthetic applications. A review. *Journal of Neuroscience Methods*, *337*, 108653. DOI: 10.1016/j.jneumeth.2020.108653 PMID: 32114143

Rizzo, P. (2020). A review on the latest advancements in the non-invasive evaluation/monitoring of dental and trans-femoral implants. *Biomedical Engineering Letters*, *10*(1), 83–102. DOI: 10.1007/s13534-019-00126-8 PMID: 32175131

Rodríguez-Tapia, B., Soto, I., Martínez, D. M., & Arballo, N. C. (2020). Myoelectric interfaces and related applications: Current state of EMG signal processing–a systematic review. *IEEE Access : Practical Innovations, Open Solutions*, *8*, 7792–7805. DOI: 10.1109/ACCESS.2019.2963881

Rubia, K., Westwood, S., Aggensteiner, P. M., & Brandeis, D. (2021). Neurotherapeutics for attention deficit/hyperactivity disorder (ADHD): A review. *Cells*, *10*(8), 2156. DOI: 10.3390/cells10082156 PMID: 34440925

Sanyal, A., Thaker, A., Mahajan, A., Daru, D., & Karani, R. (2025). Deep Learning Techniques for Thought-Controlled IoT Devices. In *Challenges and Solutions in Internet of Things-Based Smart Applications* (pp. 74–98). Chapman and Hall/CRC.

Saxena, D., & Cao, J. (2021). Generative adversarial networks (GANs) challenges, solutions, and future directions. *ACM Computing Surveys*, *54*(3), 1–42. DOI: 10.1145/3446374

Soufineyestani, M., Dowling, D., & Khan, A. (2020). Electroencephalography (EEG) technology applications and available devices. Applied Sciences, 10(21), 7453.

Stultz, D. J., Osburn, S., Burns, T., Pawlowska-Wajswol, S., & Walton, R. (2020). Transcranial magnetic stimulation (TMS) safety with respect to seizures: A literature review. *Neuropsychiatric Disease and Treatment*, *16*, 2989–3000. DOI: 10.2147/NDT.S276635 PMID: 33324060

Tamaki, N., Ajmera, V., & Loomba, R. (2022). Non-invasive methods for imaging hepatic steatosis and their clinical importance in NAFLD. *Nature Reviews. Endocrinology*, *18*(1), 55–66. DOI: 10.1038/s41574-021-00584-0 PMID: 34815553

Zemanova, M. A. (2020). Towards more compassionate wildlife research through the 3Rs principles: Moving from invasive to non-invasive methods. *Wildlife Biology*, *2020*(1), 1–17. DOI: 10.2981/wlb.00607

Zhou, X., Sobczak, G., McKay, C. M., & Litovsky, R. Y. (2020). Comparing fNIRS signal qualities between approaches with and without short channels. *PLoS One*, *15*(12), e0244186. DOI: 10.1371/journal.pone.0244186 PMID: 33362260

Chapter 18
Training BCI-Based Assistive Technology for People With Disabilities

Rupam Hazra
https://orcid.org/0009-0003-0332-2128
Global Institute of Management and Technology, India

Parag Chatterjee
https://orcid.org/0000-0003-1243-9582
Global Institute of Management and Technology, India

Soumen Ghosh
https://orcid.org/0009-0008-3778-6090
Global Institute of Management and Technology, India

Partha Paul
Global Institute of Management and Technology, India

Sudeep Biswas
https://orcid.org/0009-0007-6512-2200
Global Institute of Management and Technology, India

ABSTRACT

Brain-Computer Interface (BCI) technology enables direct communication between the brain and external devices, offering significant potential for individuals with disabilities. BCIs capture neural signals, interpret them, and convert them into commands to control devices like prosthetics, wheelchairs, or communication aids. This chapter explores the evolution of BCIs, from early brainwave control research to advanced systems that restore lost functions for those with physical, cognitive, and

DOI: 10.4018/979-8-3693-9445-8.ch018

sensory disabilities. It also examines the design of user interfaces for accessibility and ease of use. Despite the promise, challenges such as signal accuracy, invasiveness, and usability persist, requiring improvement for better BCI-based assistive technologies. The chapter evaluates BCI effectiveness in enhancing user mobility, communication, and quality of life, emphasizing the need for ongoing research to overcome limitations and enhance BCI potential in assistive tech.

1. INTRODUCTION

Brain-Computer Interface (BCI) technology represents a groundbreaking innovation in the field of assistive technology, offering new possibilities for individuals with disabilities to interact with their environment and regain lost functionalities. By enabling direct communication between the brain and external devices, BCIs bypass traditional sensory or motor pathways, opening doors for individuals with conditions such as paralysis, stroke, or neurodegenerative diseases. This technology has the potential to revolutionize mobility, communication, and independence for those facing significant physical, cognitive, or sensory challenges.

BCI systems work by detecting and interpreting neural signals from the brain, which are then translated into actionable commands to control various devices, including prosthetics, computers, or communication aids. The development of BCI technology has evolved rapidly, from early experiments in brainwave control to sophisticated systems capable of translating complex brain signals into precise actions. With advancements in signal acquisition, processing methods, and user interface design, BCIs are becoming increasingly accessible and effective for individuals with diverse disabilities.

A. Definition and Overview of BCI Technology

A Brain-Computer Interface (BCI) is a communication system that enables direct interaction between the brain and external devices, bypassing conventional sensory or motor pathways. BCIs work by capturing neural signals from the brain, interpreting them, and converting them into commands that control devices such as prosthetics, computers, or communication aids. These systems hold immense potential for individuals with motor disabilities, providing a means to restore lost functions and improve the quality of life.

BCIs are typically classified into two categories: invasive and non-invasive. Invasive-BCIs involve the implantation of electrodes directly into the brain tissue, providing high-quality signals but presenting risks such as infection and tissue rejection. Non-invasive BCIs, on the other hand, measure brain activity through external sensors such as electroencephalography (EEG) caps. While non-invasive BCIs are generally safer, they tend to have lower signal resolution compared to invasive methods, although advancements in signal processing and machine learning have improved their performance.

The primary applications of BCI technology lie in assistive devices for individuals with disabilities. For instance, individuals with paralysis or neurodegenerative conditions like ALS (Amyotrophic Lateral Sclerosis) can use BCIs to control computer cursors, robotic arms, or speech-generating devices, offering a new avenue for communication and interaction. Furthermore, BCIs are also used in fields like neurorehabilitation, cognitive enhancement, and entertainment. Brain–computer interfaces (BCI's) give their users communication and control channels that do not depend on the brain's normal output channels of peripheral nerves and muscles (Wolpaw,2000).

B. Historical Development and Evolution of BCI

Brain-Computer Interfaces represent a significant scientific advancement, a convergence of neuroscience and technology, where a brain (often but not necessarily always human) is connected with external devices. This technology decodes and converts brain impulses into instructions that may be used to operate computers, prosthetic limbs, or other devices, bypassing conventional methods of physical connection and communication such as keyboards, controllers or joysticks. This ability to directly interpret and use the brain's impulses has the potential to introduce remarkable improvements in human capacities, as well as being a novel recovery method for lost functions caused by injury or illness (Caiado,2025).

The concept of Brain-Computer Interface (BCI) technology has evolved significantly over the past few decades, beginning with early research into the electrical activity of the brain and leading to the development of sophisticated systems that allow direct communication between the brain and external devices.

a) Researchers discovered that the brain generates electrical signals that can be measured and analysed. The earliest experiments focused on understanding the brain's electrical patterns, particularly through techniques like electroencephalography (EEG). In 1964, the first successful brainwave control experiment was conducted by neuroscientist Neuroscientist José Delgado, who demonstrated that electrical brain stimulation could modify behaviour. This experiment marked one of the first steps toward understanding the potential of brain signals for communication. In the 1970s, BCI-related research began to take shape, with researchers exploring the possibility of controlling external devices using brain signals. One key milestone was the development of the P300 event-related potential (ERP) by Farwell and Dunchin in 1988, which involved using EEG signals to detect a specific brain wave response that could be linked to mental tasks or stimuli. This laid the foundation for later BCI applications involving EEG for controlling communication devices.

b) BCI systems began to evolve with more practical applications, especially in the area of controlling external devices. Early experiments used non-invasive methods, like EEG, to detect user intent based on brain wave patterns.BCIs also began to find applications in neuroproteins, helping people with spinal cord injuries or other motor impairments control prosthetic limbs or wheelchairs using their brain signals. These systems typically required invasive methods such as electrode implants to achieve reliable control, offering high accuracy but introducing risks such as infection and complications associated with surgery.

c) Non-invasive techniques, such as EEG, became more prominent due to their safety and ease of use, reducing the need for invasive surgeries. Advances in signal processing, machine learning, and real-time data analysis helped to make non-invasive BCIs more effective, though challenges like signal noise and limited resolution remained.

In the 2000s, several companies and research institutions began working on creating commercial BCIs for consumer use. Companies like Emotive and Narosky developed EEG-based headsets that allowed users to interact with computers and games through their brain signals. Although these early consumer devices were less powerful than medical-grade BCIs, they demonstrated the commercial viability of the technology.

d) Researchers are increasingly combining BCI technology with artificial intelligence (AI) to enhance signal interpretation and user adaptability. This has led to better performance in controlling external devices and more reliable user experiences. For instance, companies like Neural ink (founded by Elon Musk) are working on highly sophisticated invasive BCIs designed to treat conditions like paralysis and neurological diseases, with the ambition of enabling broader brain-machine communication. On-invasive systems are also improving in terms of both accuracy and user-friendliness.

C. Importance of BCI in Assistive Technology

Brain-Computer Interfaces (BCIs) have emerged as a revolutionary technology in the realm of assistive devices, providing new possibilities for individuals with disabilities. By enabling direct communication between the brain and external devices, BCIs open up pathways for restoring lost functions and enhancing the quality of life for people with motor, sensory, or cognitive impairments. Assistive technology (AT) supports individuals with motor, sensory, or cognitive disabilities in performing functions that might otherwise be difficult or impossible for them (Collinger, 2013).

a) Enhancing Independence for Individuals with Disabilities

- **Restoration of Communication**

For individuals with severe speech or motor impairments (such as those with ALS, stroke, or spinal cord injuries), BCIs enable communication by converting neural signals into text or speech. This offers an alternative to traditional communication methods like typing or using eye gaze tracking.

- **Control of External Devices**

BCIs allow users to control external devices such as robotic prosthetics, wheelchairs, and computers, bypassing the need for traditional input mechanisms like touch, speech, or muscle movement. This empowers users to interact with their environment in ways they could not before, significantly enhancing their autonomy.

b) Addressing the Needs of People with Severe Disabilities

- **For Paralyzed Individuals**

BCIs have great potential for helping people with paralysis regain some form of movement and interaction with the world. For instance, BCI systems can enable control of robotic arms, enabling users to perform everyday tasks such as grasping objects or feeding themselves.

- **Neurorehabilitation**

BCIs also play an essential role in neurorehabilitation. By providing real-time feedback on brain activity, BCIs can assist in the rehabilitation of motor skills, helping individuals recovering from strokes or traumatic brain injuries. This creates opportunities for brain plasticity and recovery of lost functions, promoting greater long-term independence.

c) Improving Quality of Life and Emotional Well-being

- **Reducing Isolation**

For many people with disabilities, the ability to communicate and interact with others can significantly reduce social isolation. BCIs can enhance social engagement by providing people with the means to communicate more easily, whether through speech synthesis, typing, or virtual avatars.

- **Cognitive and Emotional Benefits**

Beyond physical applications, BCIs also provide cognitive and emotional benefits. Neurofeedback through BCIs can be used to treat mental health conditions like anxiety, depression, or attention deficit hyperactivity disorder (ADHD) by helping individuals regulate their brain activity. Additionally, BCIs can be used in games and entertainment, offering an engaging way to interact with digital environments, thus improving emotional well-being.

d) Potential for Personalization and Adaptation
   - **User-Specific Customization**

One of the key advantages of BCIs is their ability to be customized to the unique brain activity patterns of each user. Adaptive algorithms can learn to interpret an individual's specific neural signals, improving the accuracy and efficiency of device control. This personalization is critical for making BCIs accessible and effective for a wide range of users with different needs and abilities.

- **Machine Learning for Enhanced Performance**

The integration of machine learning algorithms in BCI systems enables real-time adaptation to changes in neural patterns, further enhancing the reliability of the technology over time. As the system learns more about a user's brain activity, it can optimize performance, making BCIs more intuitive and effective.

2. UNDERSTANDING DISABILITIES AND THE NEED FOR ASSISTIVE TECHNOLOGY

Disabilities, whether physical, sensory, cognitive, or neurological, present significant challenges to individuals' ability to perform daily activities, communicate, and interact with their environment. These challenges can stem from conditions like paralysis, stroke, cerebral palsy, or neurodegenerative diseases such as ALS, leading to loss of motor control, speech, or cognitive functions. The impact of disabilities can create a sense of isolation and dependence on others, affecting an individual's autonomy and quality of life. Cognitive impairment, often labelled as a hidden disability, is a substantial limitation in one's capacity for mental tasks, including conceptualizing, planning, sequencing thoughts and actions, remembering, interpreting subtle social cues, and manipulating numbers and symbols (LoPresti,2008). Assistive technology (AT) is crucial in addressing these challenges by providing solutions that enable individuals to regain independence and perform tasks that might otherwise be impossible. These technologies, such as mobility aids, speech-generating devices, and communication systems, help bridge the gap between disability and capability, fostering inclusion, self-reliance, and social participation.

A. Types of Disabilities and Their Impact on Daily Life

Disabilities can vary widely in their nature and impact, affecting different aspects of an individual's physical, cognitive, or emotional functioning. The limitations caused by disabilities can significantly influence a person's ability to perform daily activities such as communication, mobility, self-care, and work. Disability is the result of the interaction between individuals with physical health conditions and personal, social, or environmental factors, such as negative attitudes, lack of accessible transportation, public buildings, and limited social support (Kibret2025). Understanding the different types of disabilities and their effects on daily life is essential for developing appropriate assistive technologies that address these challenges and improve the quality of life.

a) Physical Disabilities

Physical disabilities include impairments in mobility, coordination, or dexterity, often resulting from conditions like spinal cord injuries, cerebral palsy, muscular dystrophy, or amputation. These disabilities can make it difficult or impossible for individuals to perform basic tasks such as walking, dressing, or operating household items. The impact on daily life is profound, as people with physical disabilities may need assistance with movement or require devices like wheelchairs or prosthetics to help them navigate the world.

b) Cognitive Disabilities

Cognitive disabilities affect intellectual functioning, memory, problem-solving, and learning abilities, and can arise from conditions like Down syndrome, traumatic brain injury (TBI), or Alzheimer's disease. These disabilities can impair a person's ability to perform tasks that require concentration, decision-making, and memory recall. As a result, individuals with cognitive disabilities may struggle with tasks such as managing finances, following instructions, or maintaining employment.

B. Overview of Traditional Assistive Technologies

Traditional assistive technologies (AT) are tools and devices designed to aid individuals with disabilities in performing daily tasks that may be challenging due to physical, sensory, or cognitive impairments. These technologies have been developed over decades to enhance accessibility, promote independence, and improve the quality of life for people with disabilities. Traditional AT spans a wide range of devices and adaptations, from mobility aids to communication tools, and is used in various settings, including homes, workplaces, and schools.

a) Mobility Aids

Traditional mobility aids, such as wheelchairs, prosthetics, and walkers, are designed to assist individuals with physical disabilities in moving and performing daily activities. For instance, electric wheelchairs help people with limited or no mobility in their legs, while prosthetic limbs enable individuals to regain the ability to use a missing limb. These aids improve independence but often require manual effort or external assistance for full functionality.

b) Communication-Aids

Augmentative and alternative communication (AAC) devices, such as speech-generating devices (SGDs) and communication boards, help individuals with conditions like cerebral palsy or stroke to communicate more effectively. Hearing aids and cochlear implants are also vital tools that improve auditory function, allowing people with hearing loss to better engage in conversations and social interactions.

C. Role of BCI in Empowering Individuals with Disabilities

Brain-Computer Interfaces (BCIs) play a transformative role in empowering individuals with disabilities by providing new ways to interact with their environment and restore lost functions. BCIs establish a direct communication pathway between the brain and external devices, allowing individuals with limited or no physical movement to control technology through thought alone. This has profound implications for enhancing independence and improving quality of life.

a) Restoring-Communication-and-Independence

BCIs provide individuals with severe physical disabilities, such as those with ALS or spinal cord injuries, the ability to communicate and control their environment through brain activity. By interpreting neural signals, BCIs allow users to operate communication devices, such as speech-generating software or computers, without needing physical movement. This technology empowers individuals to express themselves, engage in social interactions, and regain a sense of independence in their daily lives.

b) Enhancing Control over Assistive Devices

BCIs enable more intuitive control of assistive devices like robotic prosthetics, wheelchairs, and exoskeletons. For people with limited motor function, BCIs can bypass the need for traditional input methods (such as joystick controllers or touch screens) by allowing direct control via brain signals. This gives individuals greater freedom and autonomy, enabling them to perform tasks such as walking, grasping objects, or even interacting with smart home systems, thereby improving their overall quality of life.

3. BCI TECHNOLOGY AND MECHANISMS

Brain-Computer Interface (BCI) technology enables direct communication between the brain and external devices by detecting and translating brain signals into actionable commands. While challenges like signal quality, invasiveness, and ethical concerns persist, BCIs continue to evolve and offer transformative potential for both healthcare and consumer technologies. Group difference analysis between high and low aptitude BCI users revealed significantly higher activation of the supplementary motor areas (SMA) for the motor imagery and the motor observation tasks in high aptitude users. Low aptitude users showed no activation when observing movement. The number of activated voxels during motor observation was significantly correlated with accuracy in the EEG-BCI task (r= 0.53). Furthermore, the number of activated voxels in the right middle frontal gyrus, an area responsible for processing of movement observation, correlated (r= 0.72) with BCI-performance (Halder,2011).

A. Basic Principles of BCI Systems

The basic principles of Brain-Computer Interface (BCI) systems are built on the processes of signal acquisitionand signal processing.

a) Signal Acquisition

This is the first step in a BCI system, where electrical brain activity is captured. Different techniques are used for signal acquisition, including non-invasive methods like electroencephalography (EEG), where electrodes are placed on the scalp to detect electrical signals produced by neuron activity. More invasive approaches, such as electrocorticography (ECOG) or implanting electrodes directly into the brain tissue, can provide more precise and higher quality signals but come with greater risks. These brain signals are then used as the raw data for the system to interpret and process.

b) Signal Processing

Once the brain signals are acquired, the next step is to process these signals. The raw data collected from the brain is often noisy and needs to be filtered and amplified. Signal processing algorithms are used to extract relevant featuresfrom these brain signals—such as identifying patterns related to motor intentions, cognitive states, or sensory inputs. Once the meaningful patterns are extracted, they are translated into commands that control an external device, like a prosthetic limb or a cursor on a screen. Real-time processing and feedback are key components to make the system functional and responsive to the user's intentions.

B. Types of BCI (Invasive vs. Non-Invasive)

BCI systems can be classified into invasive and non-invasive types, based on how brain signals are acquired. These types differ in terms of signal quality, precision, and the methods used to interface with the brain.

a) Invasive BCIs

Invasive BCIs involve the implantation of electrodes directly into the brain tissue or on the brain's surface. These systems provide the highest signal quality and resolution because they can detect neuronal activity directly from the brain. However, they come with significant risks, including infection, tissue damage, and long-term complications. Invasive BCIs are often used in clinical settings, particularly for patients with severe disabilities or conditions like paralysis. They allow for more precise control of external devices, such as robotic prosthetics or exoskeletons, and are used for advanced applications like brain-machine interfaces that require high-bandwidth communication.

b) Non-Invasive BCIs:

Non-invasive BCIs, on the other hand, do not require surgery and instead use external sensors to measure brain activity. The most common method is electroencephalography (EEG), where electrodes are placed on the scalp to capture electrical activity from the brain. Although non-invasive BCIs are safer and easier to implement, they tend to have lower signal quality and resolution compared to invasive systems due to the scalp and skull's interference with the brain's electrical signals. As a result, non-invasive BCIs are often used for simpler applications, such as controlling computer cursors, video games, or for assistive technologies like communication aids for people with disabilities. The transformative impact of non-invasive EEG-based BCIs on healthcare, accessibility, and human-computer interaction holds great promise for a more inclusive and interconnected future. By unveiling the potentials and challenges in this burgeoning field, we anticipate accelerated progress toward harnessing the full potential of EEG-based BCIs for the betterment of society (Kumar, 2025).

C. Signal Acquisition and Processing in BCI

Signal Acquisition and Processing in BCI are critical steps in Brain-Computer Interface (BCI) systems, enabling the translation of brain activity into actionable commands for controlling external devices. These processes involve capturing brain signals, filtering out noise, extracting relevant information, and translating these signals into commands that can interact with computers, prosthetics, or other devices.

a) Signal Acquisition

Signal acquisition refers to the method by which brain activity is captured. There are several techniques for acquiring brain signals, with the most common being electroencephalography (EEG). EEG detects electrical activity generated by neurons in the brain using electrodes placed on the scalp. This method is non-invasive, making it popular in both clinical and consumer-grade BCIs. Invasive techniques like electrocorticography (ECOG) or intracortical electrodes involve placing sensors directly on the surface of the brain or inside the brain tissue, providing higher-quality signals but at the cost of surgical risk and complexity.

b) Signal Processing

Once the brain signals are acquired, the next critical step is signal processing. The raw signals captured by EEG or other methods are usually noisy and need to be processed before they can be used to control devices. Signal processing typically involves several stages:

- **Preprocessing**

This step involves filtering the raw signal to remove noise, artifacts, and other irrelevant data. Techniques like band-pass filtering are used to isolate the relevant frequencies (e.g., alpha waves, beta waves) associated with specific brain activities.

- **Feature Extraction**

After preprocessing, the system extracts meaningful features from the brain signals, such as the amplitude or frequency of certain brainwave patterns. These features are associated with specific mental states or motor intentions (e.g., thinking about moving a limb).

- **Classification**

This step involves using machine learning algorithms to classify the extracted features and map them to specific actions. For instance, a particular pattern of brain activity could be classified as a command to move a cursor or control a prosthetic limb.

- **Control Interface**

Finally, the classified signals are translated into commands that control external devices. This step may involve additional feedback mechanisms to help the user refine control over time.

4. DESIGNING BCI SYSTEMS FOR ASSISTIVE TECHNOLOGY

Designing Brain-Computer Interface (BCI) systems for assistive technology involves creating systems that enable individuals with disabilities to control devices or communicate using their brain signals. These systems typically rely on non-invasive methods like electroencephalography (EEG), which measures electrical activity in the brain. Electroencephalography (EEG) is a biological signal that indicates the brain's activity and can be detected by electrodes placed on the surface of the scalp. EEG signals are generally categorized based on the type of response (Sakamaki, 2017). The goal is to translate these signals into actionable commands that can control a range of devices, from prosthetics to communication tools. Effective BCI systems for assistive technology require robust signal processing algorithms to filter and interpret brain signals, real-time responsiveness, and minimal user effort. Additionally, user-centred design is essential, considering factors such as ease of use, comfort, and customization to meet the specific needs of each individual. Overcoming challenges like signal noise, user training, and long-term adaptability is crucial for the success of BCI-based assistive technologies, ensuring they provide meaningful support for users with physical or communication impairments.

A. Key Considerations in Designing User-Friendly BCI Systems

When designing user-friendly Brain-Computer Interface (BCI) systems, several key considerations must be taken into account to ensure that the system is effective, intuitive, and accessible for a wide range of users. These considerations include usability, comfort, signal quality, adaptability, and safety. This algorithm might include linear or nonlinear equations, a neural network, or other methods, and might incorporate continual adaptation of important parameters to key aspects of the input provided by the user. BCI outputs can be cursor movement, letter or icon selection, or another form of device control, and provides the feedback that the user and the BCI can use to adapt to optimize communication (Yadollahpour, 2015).

a) Usability

A user-friendly BCI system should be easy to use, with a clear and intuitive interface. The system should provide real-time feedback to the user, allowing them to understand and adjust their actions. Additionally, the control mechanisms should be simple and not require extensive training or technical knowledge. User-centred design is essential in this regard, meaning that the system is tailored to the user's needs, abilities, and preferences, making it suitable for both experts and novices.

b) Comfort

Comfort is an essential factor, especially for systems that require users to wear the device for extended periods, such as EEG-based headsets or prosthetics controlled by BCI. Non-invasive BCIs, such as EEG caps, must be designed to fit comfortably on the user's head without causing discomfort or irritation. For invasive BCIs, surgical risks and long-term implantation challenges must be minimized to ensure the comfort and safety of the user over time.

c) Signal Quality and Reliability

High signal quality is crucial for the BCI to work effectively. The system should be able to capture clear brain signals with minimal noise and interference, which can be challenging in non-invasive systems like EEG. The accuracy and consistency of the signal acquisition process must be maintained, ensuring that the BCI can reliably interpret user intentions. Signal processing algorithms must be efficient and capable of filtering out artifacts, such as eye blinks or muscle movements, to enhance the system's performance.

d) Adaptability

A user-friendly BCI system should be adaptable to different users and environments. Users with varying cognitive and physical abilities should be able to interact with the system with minimal setup or calibration. The system should allow for personalization, such as adjusting the sensitivity of signal detection or modifying the feedback mechanisms based on individual preferences and abilities. It should also be capable of learning and adapting to the user's neural patterns over time, improving accuracy and responsiveness.

B. Customizing BCIs for Different Disabilities

Customizing Brain-Computer Interfaces (BCIs) for individuals with different disabilities involves adapting the system to meet the unique needs of each user. It also delves into how gamification, neurofeedback, and adaptive learning techniques can enhance motivation, engagement, and overall well-being (Ahmed, 2025). BCIs can provide individuals with greater independence by enabling them to interact with their environment, communicate, and control assistive devices through their brain activity.

a) Motor Disabilities

For individuals with motor disabilities, such as paralysis or amputation, BCIs can be tailored to enable control of prostheticlimbs, exoskeletons, or wheelchairs. These systems typically use motorimagery (e.g., imagining moving a limb) to capture brain signals that correspond to intended movements. BCIs for motor disabilities can be customized to detect specific brain wave patterns related to motor intentions and then translate those patterns into commands for controlling external devices. The system's sensitivity and responsiveness may also be adjusted based on the individual's ability to generate distinct neural signals, making it easier for people with varying levels of motor function to use the technology.

b) Sensory Disabilities:

For individuals with sensory disabilities such as blindness or deafness, BCIs can be customized to translate sensory information from one modality to another. For example, a blind person can use a BCI system that converts visual stimuli (e.g., detecting obstacles or reading text) into tactile or auditory feedback. This allows users to "sense" their environment through alternative sensory channels, enhancing their ability to navigate and interact with the world. Similarly, for individuals with hearing impairments, BCIs can provide visual or tactile feedback in place of auditory signals, helping them interact with their environment without the need for hearing.

c) Cognitive Disabilities:

For individuals with cognitive disabilities, such as attention deficit disorders, Parkinson's disease, or ALS (Amyotrophic Lateral Sclerosis), BCIs can be customized to support communication and cognitive enhancement. These systems often focus on allowing individuals to control devices or communicate with others by detecting specific brain activity patterns associated with their thoughts or intentions. For instance, individuals with ALS may use a BCI system that enables them to communicate by selecting words or phrases on a screen through their brain activity, bypassing the need for physical movement. For those with cognitive disabilities, the system can be adapted to simplify commands, reduce cognitive load, or adjust the interface based on the user's specific cognitive abilities.

C. Ensuring Accessibility and Usability

Ensuring accessibility and usability in Brain-Computer Interface (BCI) systems is crucial for their widespread adoption and effectiveness, especially for individuals with disabilities. The goal is to create BCI systems that are easy to use, accessible to a wide range of users, and adaptable to individual needs.

a) User-Centred Design:

To make BCIs accessible and usable, a user-centred design is essential. This approach focuses on tailoring the interface to the specific needs, preferences, and abilities of the user. For individuals with disabilities, BCIs should provide intuitive controls, customizable feedback, and simple interfaces that are easy to understand and use. This can involve designing systems with minimal physical interaction, providing alternative modes of control (e.g., eye tracking or speech), and offering real-time feedback to guide the user. Ensuring that BCIs are adaptable to users' capabilities is crucial for effective interaction.

b) Comfort and Wearability

BCIs must be designed for comfort and wearability, particularly for long-term use. Many BCI systems require users to wear headsets or electrode arrays, which can be uncomfortable if not properly designed. To improve usability, BCIs should be lightweight, ergonomically shaped, and made from materials that minimize discomfort, especially for individuals who need to use the system continuously. Comfortable systems reduce user fatigue and irritation, making it easier for users to benefit from the technology for extended periods.

5. TRAINING AND CALIBRATION OF BCI SYSTEMS

Brain-Computer Interface (BCI) systems enable direct communication between the brain and external devices. These systems can interpret brain signals to control devices such as prosthetics, computers, or wheelchairs. To effectively function, BCIs require thorough training and calibration processes, ensuring the system is accurate and reliable for the user. A new multiclass brain-computer interface (BCI) based on the modulation of sensorimotor oscillations by imagining movements is described. By the application of advanced signal processing tools, statistics and machine learning, this BCI system offers: 1) asynchronous mode of operation, 2) automatic selection of user-dependent parameters based on an initial calibration, 3) incremental update of the classifier parameters from feedback data. (Kus,2012).

A. The Importance of Calibration in BCI

The use of brain-computer interface (BCI) technology to identify emotional states has gained significant interest, especially with the rise of virtual reality (VR) applications (Sarikaya, 2025). Calibration is a critical step in the functioning of Brain-Computer Interface (BCI) systems, as it ensures that the system is tailored to the user's specific brain activity patterns. Proper calibration directly impacts the accuracy, usability, and overall performance of the BCI system.

a) Personalized Signal Recognition

- **Unique brainwave patterns**

Each user has distinct brain activity patterns, meaning that the BCI system must be calibrated to recognize these individual variations in brain signals. Calibration ensures that the system can differentiate between the user’s intended actions and background noise, improving accuracy.

- **Adaptation to individual differences**

Factors like age, cognitive state, and even mental fatigue can alter brainwave patterns. Calibration allows the system to account for these differences, making the system more effective for each specific user.

b) Improved Control Accuracy

- **Reduced error rates**

Proper calibration minimizes the chances of misinterpreting brain signals, which can lead to errors in the BCI's response. By aligning the system to the user's brain activity, the system can more accurately predict and respond to user intentions, such as moving a cursor or controlling a prosthetic limb.

- **Real-time feedback adjustments**

Calibration helps fine-tune the system in real-time based on feedback, enhancing the user's ability to control the device with precision and consistency.

c) Enhanced User Experience

- **Increased comfort and efficiency**

A well-calibrated system requires less effort from the user to produce the desired outcomes, which increases comfort during use. The more intuitive and responsive the BCI system is, the easier it is for the user to interact with and control the interface.

- **Fewer cognitive demands**

Calibration reduces the mental workload on the user by ensuring that the system accurately interprets their brain signals, allowing for smoother, less fatiguing control over time.

B. Training Protocols for Users with Disabilities

Training users with disabilities to effectively use Brain-Computer Interface (BCI) systems requires carefully structured protocols that accommodate their unique needs.

a) Personalized Training Based on Disability Type

- The training process must be tailored to the specific disability of the user. For example, individuals with motor impairments may focus on tasks involving cursor control or robotic arm manipulation, while those with cognitive impairments may need simpler, more gradual tasks. A thorough assessment of the user's abilities is essential for designing a training plan that meets their needs and ensures success in controlling the BCI system.

b) Adaptive and Incremental Learning

- Users with disabilities often require a gradual approach to learning BCI control. Starting with simple tasks and progressively introducing more complex functions helps to build confidence and improve performance. Adaptive feedback from the system, which adjusts to the user's learning pace and progress, is critical for maintaining motivation and facilitating effective skill development.

C. Adapting Training for Individual Cognitive and Physical Needs

In Brain-Computer Interface (BCI) systems, adapting training protocols to accommodate individual cognitive and physical needs is crucial for optimizing user performance and comfort. Since users with disabilities have varying cognitive and physical abilities, tailoring the training process helps ensure that the BCI system can meet their unique requirements. Real-time assessments of mental states including focus, relaxation, and exhaustion are possible through the direct analysis of brain activity provided by Brain-Computer Interface (BCI) technology (Subha Devi,2025).

a) Cognitive Adaptation: Tailoring Mental Tasks

- **Cognitive capabilities vary**

Individuals may have different levels of cognitive function, requiring adjustments in the complexity of mental tasks. Users with cognitive impairments may need simpler, slower-paced tasks, whereas individuals with higher cognitive function might progress to more complex tasks, such as controlling multiple devices simultaneously.

- **Progressive difficulty**

Starting with basic tasks and gradually increasing their complexity ensures that cognitive demands align with the user's capabilities. This approach prevents frustration and promotes successful learning experiences.

b) Physical Adaptation: Customizing Input Methods

- **Motor impairments**

For users with motor impairments, the training protocol may need to focus on mental control of the BCI system (e.g., through motor imagery or attention-based tasks) rather than physical movement. Training can be adapted to use signals from different brain regions, depending on the physical abilities of the user.

- **Alternative input mechanisms**

For those with severe mobility limitations, using different input modalities (such as eye movements or facial expressions) can enhance their ability to interact with the BCI system. This customization ensures that individuals with various physical needs can still effectively use the system.

c) Individualized Feedback and Real-Time Adaptation

- **Feedback based on individual needs**

The feedback provided to the user should be personalized. For individuals with cognitive disabilities, simpler visual or auditory cues can be used to highlight successes or errors. For individuals with motor disabilities, haptic feedback or auditory signals can provide valuable guidance.

- **Real-time adjustments**

The BCI system should adapt in real-time to the user's progress and performance. This continuous adaptation helps optimize training, ensuring that the system remains aligned with the user's evolving cognitive and physical capabilities.

6. USER INTERFACE AND INTERACTION DESIGN FOR BCIS

The user interface (UI) and interaction design for Brain-Computer Interfaces (BCIs) are crucial in ensuring that users can effectively control and communicate with the system. A well-designed UI should be intuitive, accessible, and adaptable to the user's needs, particularly for individuals with disabilities. This includes clear, simple layouts with visually distinct elements, personalized feedback, and adjustable settings to cater to the user's abilities. Interaction design should allow users to interact with the system through multiple input modalities, such as gaze, voice, gestures, or brain signals, depending on their physical and cognitive capabilities. Providing real-time feedback in multiple sensory formats—visual, auditory, and haptic—enhances the user experience and helps to confirm actions, reducing errors and cognitive load. Direct brain-to-computer interfaces have advanced significantly in the recent past. Non-invasive, EEG-based systems with relatively fast interaction are being developed in several labs throughout the world. Such interactions have almost exclusively been based upon mapping existing metaphors (cursors and button-pushing being the most prominent of these) to the signals that are measured from the brain computer interface(Williamson2009).

A. Creating Intuitive Interfaces for Users with Disabilities

Designing Brain-Computer Interface (BCI) systems for users with disabilities requires a focus on creating intuitive and user-friendly interfaces. The goal is to ensure that users can easily interact with and control the system despite any cognitive, sensory, or motor limitations they may have.

a) Simplified and Clear Visual or Auditory Cues
   - **Visual aids**

For users with cognitive or sensory impairments, interfaces should present clear, easy-to-understand visual cues. This could involve large icons, contrasting colours, or simplified layouts that guide users through tasks. Additionally, minimizing unnecessary complexity ensures that users can quickly interpret the interface, leading to less cognitive overload.

- **Auditory cues**

For users with visual impairments, auditory feedback (such as beeps, tones, or voice prompts) can be integrated to guide users' interactions with the system. The auditory cues should be distinct, non-distracting, and appropriately timed to provide essential feedback without overwhelming the user.

b) Customizable Input Methods
   - **Tailoring to physical abilities**

Interfaces must allow users to select input methods that best match their abilities. For example, users with motor impairments may rely on eye-tracking, facial expressions, or brainwave patterns (e.g., motor imagery or attention-based control) rather than traditional physical input methods like a mouse or keyboard. Allowing for such customizations ensures the system can adapt to diverse user needs.

- **Adaptive control schemes**

Users may benefit from different control schemes depending on their physical or cognitive impairments. For example, individuals with limited motor abilities can use slower and more deliberate input methods, while users with motor function might prefer faster, more dynamic interactions. The system should allow for adjustments based on the user's preference and capability.

c) Multimodal Feedback and Real-Time Adjustments

- **Feedback diversity**

Providing multimodal feedback ensures users can receive information in a manner that suits their sensory abilities. For instance, haptic feedback can be integrated for users with visual or hearing impairments, helping them sense interactions through touch or vibration. Likewise, visual and auditory feedback combined with clear instructions can guide users through tasks.

- **Real-time adaptive feedback**

The interface should adapt in real time based on the user's performance. For instance, if the system detects that a user is struggling with a particular task, it can adjust feedback (e.g., offering simpler visual cues, slowing the pace of interaction, or providing more auditory prompts) to guide them through. This ensures that the interface remains intuitive and supportive, even as users progress in their training.

B. Voice, Gesture, and Visual Feedback Integration

Integrating various types of feedback, such as voice, gesture, and visual cues, in Brain-Computer Interface (BCI) systems is essential for enhancing accessibility and usability for users with disabilities. These multimodal feedback mechanisms ensure that users can interact with the system through different sensory channels, improving their overall experience and control.

a) Voice Feedback for Enhanced Accessibility

- **Auditory cues for visual or motor impairments**

Voice feedback provides an effective way to communicate information to users with visual impairments or those who have difficulty using physical gestures. This could include voice prompts to guide users through actions or narrate system states (e.g., “Task complete,” “Please try again”).

- **Command confirmation**

Voice feedback can confirm user commands, improving clarity and reducing errors. For instance, when a user selects an option or gives a voice command, a spoken confirmation like “You have selected 'Yes'” ensures that users are confident in their interactions.

- **Speech recognition for control**

For users with motor impairments, voice commands can serve as a powerful input method, allowing them to control devices, change settings, or navigate through menus without the need for physical interaction.

b) Gesture Feedback for Motor Control

- **Gesture-based control for motor impairments**

Gesture recognition is particularly valuable for users who have some residual motor function but might not be able to use traditional input devices like a mouse or keyboard. With BCI systems, users can control interfaces through gestures such as head or eye movements, facial expressions, or hand gestures.

- **Real-time feedback on gestures**

Providing immediate visual or haptic feedback when a user performs a gesture helps confirm the success of their input. For example, when a user moves their head to navigate a cursor, visual feedback (such as the cursor moving on the screen) or auditory cues (like a beep) lets them know the action was recognized.

- **Adaptation to individual capabilities**

Gesture-based control systems should be adaptive, allowing for different types of gestures to match individual motor abilities. The system can learn to recognize specific gestures made by the user and refine them for more accurate interaction.

C. Enhancing the Experience: Multimodal Interaction

Multimodal interaction involves integrating multiple sensory channels (such as voice, gesture, visual, and haptic feedback) to create a more engaging and efficient user experience. In the context of Brain-Computer Interface (BCI) systems, multimodal interaction allows users to interact with the system using various input methods and receive feedback through different sensory modalities. This approach is particularly beneficial for users with disabilities, as it caters to their individual needs and enhances accessibility.

Adaptive systems play an important role in improving user experiences by responding to real-time changes in cognitive and emotional states. With Brain-Computer Interfacing (BCI), such systems can leverage neural signals to tailor interactions, providing a personalized and engaging user experience (Duraisamy,2025).

a) Combining Multiple Input Methods for Increased Accessibility

- **User-specific input combinations**

Users with different disabilities may find it easier to interact with a system using a combination of input methods. For example, a user with limited motor function may rely on eye movements for navigation, combined with voice commands for selection. Combining these methods creates a more intuitive and efficient way to interact with the system, ensuring that users with varying needs can control the BCI effectively.

- **Adaptable input modalities**

The BCI system should allow users to customize input modalities based on their preferences and abilities. For example, someone with cognitive impairments may prefer a simpler visual interface with gestures, while someone with motor impairments may opt for voice and gaze control. This flexibility ensures a personalized experience tailored to the user's needs.

b) Real-Time Multimodal Feedback to Enhance User Understanding

- **Simultaneous feedback channels**

Multimodal feedback provides users with multiple ways to confirm their actions, improving confidence and accuracy. For instance, a user might receive visual feedback (such as a cursor movement on a screen), auditory feedback (like a voice prompt or beep), and haptic feedback (such as vibration) when they successfully perform an action. This simultaneous feedback helps reinforce the connection between the user's intent and the system's response.

- **Reduced cognitive load**

By providing feedback through multiple channels, users are less reliant on one single type of information. This can reduce cognitive load, especially for users with cognitive impairments, making the interaction smoother and easier to follow. For instance, combining visual cues with voice guidance can help users better understand their task without becoming overwhelmed by complex visuals or instructions alone.

7. BCI APPLICATIONS FOR MOBILITY AND COMMUNICATION

Brain-Computer Interface (BCI) technologies are revolutionizing mobility and communication, particularly for individuals with disabilities. By directly connecting the brain to external devices, BCIs enable users to control prosthetics, wheelchairs, or robotic limbs, offering greater independence and mobility. For communication, BCIs allow users to interact with computers or communication aids through neural signals, enabling speech or text generation for individuals with speech impairments. These advancements not only enhance the quality of life but also open up new possibilities for rehabilitation and support, making everyday tasks more accessible for those with limited physical abilities. Invasive BCIs monitor brain activity on a cortex's surface using electrocorticography (ECOG), within the Gray matter using intracortical microelectrodes or in deep subcortical structures using depth electrodes (Orlandi,2021).

A. Assistive Technology for Wheelchair Control

a) Brain-Computer Interface (BCI) for Wheelchair Control

BCIs are a cutting-edge technology that allows individuals with severe motor impairments to control devices such as wheelchairs through neural signals. The system interprets the electrical activity of the brain and translates it into commands that control the movement of the wheelchair. BCIs are particularly beneficial for users with conditions like paralysis or severe neuromuscular diseases, as they bypass the need for physical interaction with the device, offering greater independence. The key components of BCIs for wheelchair control include electrodes placed on the scalp (non-invasive) or on the brain's surface (invasive), which record brain signals associated with movement intentions. These signals are then processed by a computer and used to send commands to the wheelchair, controlling movement such as direction, speed, and even stopping.

b) Eye-Tracking Technology for Wheelchair Navigation

Eye-tracking systems enable wheelchair users to navigate by simply moving their eyes, a method that is extremely useful for individuals who have little or no control over their limbs. By using infrared light and sensors to detect the user's gaze, the system identifies where the user is looking, and this movement is translated into directional commands for the wheelchair. The control system may include software that interprets specific eye gestures—such as focusing on a target for a certain period to activate a command like turning left or right. This approach is non-invasive and can be used by individuals who are unable to move their limbs or hands due to conditions like spinal cord injuries, ALS, or other neurodegenerative diseases. Eye-tracking also provides hands-free, real-time control, making it ideal for users with severe mobility restrictions.

c) Joystick and Touchscreen Controls for Enhanced Accessibility

For individuals with partial motor control, adaptive joystick systems or touchscreen interfaces can provide an alternative or supplementary method of controlling wheelchairs. Adaptive joysticks can be customized to accommodate specific physical limitations, such as requiring less force to operate or having larger, more easily manipulable buttons. These interfaces can be designed for users with limited hand strength or dexterity, making it easier to control the wheelchair with minimal physical effort. Similarly, touchscreen controls offer another form of intuitive control, especially when integrated with large virtual buttons and customizable layouts. These systems allow users to interact with their wheelchair via simple gestures, taps, or swipes on a touchscreen, providing a user-friendly experience for those with cognitive and physical impairments. Such systems can be further enhanced with voice commands and environmental sensors to improve functionality and responsiveness.

B. Communication Systems for Non-Verbal Users

a) Augmentative and Alternative Communication (AAC) Devices

Augmentative and Alternative Communication (AAC) systems are designed to assist individuals who have difficulty speaking due to conditions such as cerebral palsy, autism, stroke, or other neurological disorders. These devices include tools like speech-generating devices (SGDs), symbol boards, and communication apps. SGDs typically consist of specialized hardware or software that generates spoken language or text based on user input through touch, eye movement, or switch access. These devices offer a range of communication options, from pre-recorded messages to dynamic symbol-based systems, providing users with a means of expressing thoughts, needs, and emotions.

b) Eye-Tracking Communication Systems

Eye-tracking technology enables non-verbal individuals to communicate by tracking their eye movements to select letters, words, or phrases on a screen. This is particularly beneficial for people with conditions like ALS (amyotrophic lateral sclerosis) or locked-in syndrome, where voluntary muscle control is limited or absent, but eye movements remain functional. By using infrared light and sensors to detect where the user is looking, the system can interpret gaze direction and generate speech or text output. Some systems incorporate predictive text to speed up communication, allowing users to select words or sentences based on eye movements.

c) Speech-Generating Apps and Mobile Communication Platforms

With the advancement of smartphones and tablets, speech-generating applications (SGA) have become an accessible and affordable option for non-verbal communication. These apps typically allow users to select words, phrases, or symbols to generate speech output, and many are customizable to suit individual needs. Some apps use text-to-speech technology, while others integrate symbol-based communication systems. These platforms are often designed to be intuitive and user-friendly, supporting people with a variety of conditions such as autism, stroke, or traumatic brain injury. Moreover, mobile communication apps often integrate with other assistive technologies, like switches or adaptive controllers, to enhance accessibility.

C. Cognitive and Emotional Support Systems

Cognitive and emotional support systems are designed to assist individuals in managing mental and emotional challenges, fostering better cognitive functioning, and offering emotional guidance. These systems can be implemented in various contexts, including healthcare, education, and mental health services, using technology to aid individuals in coping with stress, anxiety, depression, and other mental health issues. BCIs increasingly serve as tools to bridge neuroscience and creative processes. Converting neural signals into visual or physical outputs allows cognitive and emotional states to be dynamically represented in art, enabling real-time interactions between brain activity and artistic expression (Schreiner, 2025).

a) Cognitive Support Systems

Cognitive support systems are typically designed to enhance an individual's cognitive abilities, such as memory, attention, and problem-solving. These systems utilize technologies like artificial intelligence (AI) and machine learning (ML) to assist people in learning, decision-making, and maintaining cognitive functioning.

b) Emotional Support Systems

Emotional support systems, often powered by AI or therapeutic technologies, help individuals cope with emotional challenges like anxiety, depression, or stress. These systems may involve virtual assistants, AI-driven mental health apps, or interactive platforms that provide emotional guidance, coping strategies, or even counselling services.

8. CHALLENGES IN BCI-BASED ASSISTIVE TECHNOLOGY

Brain-Computer Interface (BCI)-based assistive technology faces several challenges, including limited accuracy and reliability in signal interpretation, which can hinder user experience. The complexity of mapping neural activity to specific commands often results in slow response times and occasional errors, affecting usability. Additionally, there are challenges related to the invasiveness of some BCI systems, requiring surgical implantation of electrodes, while non-invasive options may struggle with signal noise and weaker connectivity. There are also concerns regarding accessibility, affordability, and the need for specialized training for users. Furthermore, ethical issues related to privacy, data security, and long-term effects of BCI use are still under exploration, creating barriers to widespread adoption. BCI systems have been identified as having potential as an assistive technology as they make it possible to perform basic commands using thinking strategies alone and they are not reliant on voluntary muscular movement, making them suitable for people with severe communication difficulties and in neurological rehabilitation and enabling greater social inclusion (Carmichael,2014).

A. Technical Challenges: Signal Noise and Reliability

a) Signal Artifacts and Interference

One of the primary technical challenges in BCIs is the presence of signal noise, including artifacts from muscle contractions (electromyographic signals), eye movements (electrooculographic signals), and external electrical interference. These artifacts can obscure the neural signals of interest, leading to incorrect or unreliable outputs. Effective signal processing algorithms are crucial to differentiate between noise and meaningful brain activity.

b) Low Signal-to-Noise Ratio (SNR)

BCIs, especially non-invasive ones (such as EEG), often suffer from a low signal-to-noise ratio, meaning the desired brain signals are weak and can be easily overshadowed by noise. This significantly impacts the accuracy and reliability of BCI systems in real-time applications, as even minor fluctuations in the signal can cause major disruptions in performance.

B. User Challenges: Comfort and Long-Term Usage

a) Physical Discomfort and Invasiveness

Many BCI systems, especially those that involve wearable devices like EEG caps or invasive electrodes, can cause physical discomfort due to their design or the need for continuous use. Non-invasive devices often require prolonged attachment to the scalp, which can cause skin irritation or discomfort. Invasive BCIs, which involve surgical implantation of electrodes, come with risks of infection, inflammation, and long-term physical discomfort, limiting their practical application for users.

b) Cognitive Fatigue and Mental Load

Long-term use of BCIs can lead to cognitive fatigue. Since these devices require sustained concentration and mental effort to maintain control, users may experience exhaustion or mental overload, especially during tasks that demand continuous interaction or high levels of precision. This mental fatigue can reduce the system's efficiency and user satisfaction, hindering the system's ability to be used for extended periods.

C. Ethical Considerations: Privacy, Consent, and Security

a) Privacy Concerns and Data Security

One of the key ethical challenges with BCIs is the potential for sensitive brain data to be compromised. BCIs often collect intimate neural data, which can reveal not only a user's thoughts but also their emotional and cognitive states. This raises concerns about the possibility of unauthorized access to private data, hacking, or misuse of such information. Protecting the privacy of users and ensuring data security is a major ethical consideration that needs to be addressed through robust encryption and secure data storage methods.

b) Informed Consent and Autonomy

Obtaining informed consent is crucial when implementing BCIs, particularly when they are used in medical or assistive contexts. Users must fully understand the risks, benefits, and potential consequences of using such technologies. This includes ensuring that individuals are aware of the possibility of psychological effects or unwanted data capture. The issue of autonomy also arises, as users may face pressures to adopt BCI technology, especially in situations where alternative options may be limited.

9. EVALUATING THE EFFECTIVENESS OF BCI SYSTEMS

Evaluating the effectiveness of Brain-Computer Interface (BCI) systems involves assessing their accuracy, usability, and overall impact on the user's quality of life. Key metrics include the system's ability to reliably interpret neural signals and translate them into intended actions, the speed of response, and the consistency of performance over time. Effectiveness also depends on user comfort, adaptability, and the level of cognitive load required for operation. Additionally, the extent to which BCIs enhance or restore functionality, particularly for individuals with disabilities, is a critical factor in evaluating their success. Comprehensive evaluations must consider both technical performance and user experience to determine the system's practicality and long-term viability in real-world scenarios. To promote the translation of BCI into practical applications, the gold standard for online evaluation for classification algorithms of BCI has been proposed in some studies. However, few studies have proposed a more comprehensive evaluation method for the entire online BCI system, and it has not yet received sufficient attention from the BCI research and development community (Pan,2024).

A. Performance Metrics in BCI-Assisted Systems

a) Accuracy and Signal Recognition

One of the key performance metrics in BCI-assisted systems is the accuracy with which the system interprets and translates brain signals into actionable outputs. On the other hand, the nonlinear filtering techniques like Multiscale Principal Component Analysis (MSPCA) may eliminate the noise without sacrificing substantial amount of fast changes in the signal (Kevric,2017). High accuracy is essential for reliable user control, particularly in applications like communication devices or prosthetics. This metric typically includes evaluating the system's ability to distinguish between different mental states or intentions, with minimal error.

b) System Responsiveness and Latency

Responsiveness refers to the time it takes for the BCI system to process and execute a user's command. Low latency is crucial in real-time applications, such as assistive technology for mobility or communication, where delays can significantly hinder the user experience. Minimizing latency improves the system's effectiveness by ensuring faster feedback and smoother interaction.

B. User Feedback and Satisfaction

a) Ease of Use and Interface Intuitiveness

A critical factor in user satisfaction with BCI systems is the ease of use and how intuitive the interface is. If the system is complicated or requires significant effort to operate, users may become frustrated, reducing overall satisfaction. User-friendly interfaces that minimize the learning curve and offer simple, straightforward controls are essential for positive feedback, particularly for individuals with cognitive or physical disabilities.

b) Comfort and Physical Fit

Comfort is another important aspect of user feedback, especially for wearable BCI systems like EEG caps or headsets. Long-term use can cause discomfort, skin irritation, or even pain, which negatively impacts the user experience. Systems that are ergonomically designed to provide comfort and can be worn for extended periods without physical strain tend to receive higher satisfaction ratings from users.

C. Clinical Trials

a) Effectiveness in Rehabilitation and Motor Recovery

Clinical trials involving BCI-assisted rehabilitation often focus on assessing the technology's effectiveness in restoring motor functions for individuals with neurological conditions, such as stroke or spinal cord injuries. Real-world case studies show that BCIs can help patients regain motor control or enhance neuroplasticity, although outcomes vary based on factors like the severity of the injury, the duration of therapy, and individual patient characteristics. Such trials provide valuable insights into how well BCIs can support recovery and improve functionality.

b) Long-Term Impact and Sustainability

Evaluating the long-term effects of BCI systems on patients is crucial to understanding their sustainability. Some clinical trials assess how BCI-assisted systems influence the patient's quality of life over extended periods, monitoring for any changes in their cognitive, psychological, or emotional well-being. Real-world case studies are critical in determining whether these systems can be successfully integrated into daily life and whether they have lasting benefits or require continuous support.

10. CONCLUSION

Brain-Computer Interfaces (BCIs) represent a groundbreaking development in assistive technology, offering transformative possibilities for individuals with disabilities. By enabling direct communication between the brain and external devices, BCIs empower users to regain independence and perform tasks that may have otherwise been unattainable. Whether restoring communication for individuals with speech impairments, enhancing mobility for those with physical disabilities, or providing control over prosthetics and assistive devices, BCIs have proven to be life-changing tools. As BCI technology continues to evolve, its integration with other assistive technologies and advancements in machine learning and neurofeedback will further improve its efficiency, accessibility, and functionality. The potential of BCIs to improve the lives of people with disabilities is vast, as they enable greater autonomy, promote social inclusion, and offer new opportunities for communication and interaction. With ongoing research and development, BCIs are poised to revolutionize the field of assistive technology, providing individuals with disabilities the tools to live more fulfilling, independent, and empowered lives.

REFERNCES

Ahmed, B., Khan, S., Lim, H., & Ku, J. (2025). Challenges and Opportunities of Gamified BCI and BMI on Disabled People Learning: A Systematic Review. *Electronics (Basel), 14*(3), 491. DOI: 10.3390/electronics14030491

Caiado, F., & Ukolov, A. (2025). The history, current state and future possibilities of the non-invasive brain computer interfaces. *Medicine in Novel Technology and Devices, 25*, 100353. DOI: 10.1016/j.medntd.2025.100353

Carmichael, C., & Carmichael, P. (2014). BNCI systems as a potential assistive technology: Ethical issues and participatory research in the BrainAble project. *Disability and Rehabilitation. Assistive Technology, 9*(1), 41–47. DOI: 10.3109/17483107.2013.867372 PMID: 24308848

Collinger, J. L., Boninger, M. L., Bruns, T. M., Curley, K., Wang, W., & Weber, D. J. (2013). Functional priorities, assistive technology, and brain-computer interfaces after spinal cord injury. *Journal of Rehabilitation Research and Development, 50*(2), 145. DOI: 10.1682/JRRD.2011.11.0213 PMID: 23760996

Duraisamy, S., Roeser, N., Dubiel, M., & Leiva, L. A. (2025, March). An Adaptive Brain-Computer Interface Game with Blink Controls and Cognitive State Monitoring. In *Companion Proceedings of the 30th International Conference on Intelligent User Interfaces* (pp. 129-132). DOI: 10.1145/3708557.3716365

Halder, S., Agorastos, D., Veit, R., Hammer, E. M., Lee, S., Varkuti, B., Bogdan, M., Rosenstiel, W., Birbaumer, N., & Kübler, A. (2011). Neural mechanisms of brain–computer interface control. *NeuroImage, 55*(4), 1779–1790. DOI: 10.1016/j.neuroimage.2011.01.021 PMID: 21256234

Kevric, J., & Subasi, A. (2017). Comparison of signal decomposition methods in classification of EEG signals for motor-imagery BCI system. *Biomedical Signal Processing and Control, 31*, 398–406. DOI: 10.1016/j.bspc.2016.09.007

Kibret, A. K., Yilma, T. M., Davies, T. C., Alemayehu Genetu, Z., Mekonone, S. T., Alene, B. M., Addis, B. M., Nigussie, K., Ayele, R. G., & Mekonnen Araya, B. (2025). Assistive technology and daily living challenges among students with disabilities at University of Gondar, Ethiopia: A qualitative study. *Disability and Rehabilitation. Assistive Technology, 20*(1), 201–210. DOI: 10.1080/17483107.2024.2362248 PMID: 38949131

Kumar, S., & Sharma, A. (2025). Advances in non-invasive EEG-based brain-computer interfaces: Signal acquisition, processing, emerging approaches, and applications. *Signal Processing Strategies*, 281-310.

Kus, R., Valbuena, D., Zygierewicz, J., Malechka, T., Graeser, A., & Durka, P. (2012). Asynchronous BCI based on motor imagery with automated calibration and neurofeedback training. *IEEE Transactions on Neural Systems and Rehabilitation Engineering, 20*(6), 823–835. DOI: 10.1109/TNSRE.2012.2214789 PMID: 23033330

LoPresti, E. F., Bodine, C., & Lewis, C. (2008). Assistive technology for cognition [Understanding the Needs of Persons with Disabilities]. *IEEE Engineering in Medicine and Biology Magazine, 27*(2), 29–39. DOI: 10.1109/EMB.2007.907396 PMID: 18463019

Orlandi, S., House, S. C., Karlsson, P., Saab, R., & Chau, T. (2021). Brain-computer interfaces for children with complex communication needs and limited mobility: A systematic review. *Frontiers in Human Neuroscience, 15*, 643294. DOI: 10.3389/fnhum.2021.643294 PMID: 34335203

Pan, H., Ding, P., Wang, F., Li, T., Zhao, L., Nan, W., Fu, Y., & Gong, A. (2024). Comprehensive evaluation methods for translating BCI into practical applications: Usability, user satisfaction and usage of online BCI systems. *Frontiers in Human Neuroscience, 18*, 1429130. DOI: 10.3389/fnhum.2024.1429130 PMID: 38903409

Sakamaki, I., del Campo, C. E. P., Wiebe, S. A., Tavakoli, M., & Adams, K. (2017, October). Assistive technology design and preliminary testing of a robot platform based on movement intention using low-cost brain computer interface. In *2017 IEEE international conference on systems, man, and cybernetics (SMC)* (pp. 2243-2248). IEEE. DOI: 10.1109/SMC.2017.8122954

Sarikaya, M. A., & Ince, G. (2025). Improved BCI calibration in multimodal emotion recognition using heterogeneous adversarial transfer learning. *PeerJ. Computer Science, 11*, e2649. DOI: 10.7717/peerj-cs.2649 PMID: 39896041

Schreiner, L., Wipprecht, A., Olyanasab, A., Sieghartsleitner, S., Pretl, H., & Guger, C. (2025). Brain–computer-interface-driven artistic expression: Real-time cognitive visualization in the pangolin scales animatronic dress and screen dress. *Frontiers in Human Neuroscience, 19*, 1516776. DOI: 10.3389/fnhum.2025.1516776 PMID: 40115887

Subha Devi, T., RK, M., Vinod, A., Sujitha, C., & Shashikala, K. S. (2025). A Comprehensive review of brain-computer interface technology for adaptive learning.

Williamson, J., Murray-Smith, R., Blankertz, B., Krauledat, M., & Müller, K. R. (2009). Designing for uncertain, asymmetric control: Interaction design for brain–computer interfaces. *International Journal of Human-Computer Studies, 67*(10), 827–841. DOI: 10.1016/j.ijhcs.2009.05.009

Wolpaw, J. R., Birbaumer, N., Heetderks, W. J., McFarland, D. J., Peckham, P. H., Schalk, G., Donchin, E., Quatrano, L. A., Robinson, C. J., & Vaughan, T. M. (2000). Brain-computer interface technology: A review of the first international meeting. *IEEE Transactions on Rehabilitation Engineering*, *8*(2), 164–173. DOI: 10.1109/TRE.2000.847807 PMID: 10896178

Yadollahpour, A., & Bagdeli, A. (2015). Brain computer interface: Principles, recent advances and clinical challenges. *Oriental Journal of Computer Science and Technology*, *7*(3), 425–442.

Chapter 19
Improving Early Brain Stroke Diagnosis:
An RNN Approach to CT Image Interpretation

M. Martinaa
https://orcid.org/0000-0003-0977-2106
SASTRA University, India

G. Revathy
https://orcid.org/0000-0002-0691-1687
SASTRA University, India

ABSTRACT

Early stroke detection by doctors improves treatment possibilities and leads to better patient results. Our research with Recurrent Neural Networks shows how CT image time series help hospitals detect stroke patients. StrokeTimeNet combines LSTMs and 3D-Convolutional neural networks to analyze spatial patterns in CT image sequences taken over time. The model showed better diagnosis results when it processed multiple CT scan datasets and trained itself to detect stroke changes. Our diagnostic tests checked how well the method detected the disease and measured its performance against 3D CNNs and ML while determining the AUC-ROC results and testing sensitivity and specificity. StrokeTimeNet shows outstanding performance as an accurate stroke recognition algorithm with top results in both performance metrics and AUC-ROC scores. The RNN component stands out by reading stroke patterns and extracting time-based data from the system.

DOI: 10.4018/979-8-3693-9445-8.ch019

1. INTRODUCTION

Strokes represent major health threats that endanger human life. When blood supply to the brain stops because of blocked blood vessels or broken brain blood vessels the brain loses necessary blood flow to specific areas. Strokes occur in two forms: the hemorrhagic stroke and the ischemic stroke. During ISCS blood clots in the brain block blood flow which starves the brain tissue of its necessary blood supply. There are two types of ISCSs. Cerebral thrombosis develops when a blood clot forms inside brain arteries and blocks blood flow to brain tissue. When blood flow to the heart and neck arteries stops doctors recognize this as cerebral embolism during ISCS diagnosis. The brain tissue loses its blood supply when atherosclerosis blocks arteries with blood clots that restrict blood flow to the brain. Medical professionals call TIAs "mini-strokes" yet recognize them as distinct from full ischemic attacks. Transient ischemic attacks happen when the brain stops getting blood flow for several seconds to hours. People who had TIA previously face a higher chance of suffering an acute stroke than men from the overall population. Hemorrhagic strokes damage brain structures by allowing blood to leak either into brain tissue or the space between tissues and the meningeal layer. When blood vessels break or leak blood in the brain it leads to a hemorrhagic stroke. When the tissue breaks open blood enters and damages cells by building up too much fluid. Cerebral hemorrhage deaths occur more often than deaths from acute ischemic stroke. Strokes create physical challenges and mental problems that result in vision loss and speech problems for most patients. The World Health Organization states stroke ranks among the most prevalent diseases and impacts around 15 million people each year (WHO EMRO, 2023). Each year strokes claim the lives of more than five million individuals. Stroke affects survivors' long-term disability and continues to impact their families and communities through this ongoing health challenge. Research shows that one out of four stroke survivors develop additional health problems because of their stroke. Each minute brings one new stroke event and every 40 seconds results in a stroke-related death. Stroke can happen to anyone at any age, but it affects people most often after they reach mid-50s. Hypertension along with obesity weight issues and diabetes serve as stroke risk factors with genetic stroke history and improper food choices and physical inactivity as other contributing elements. Every stroke patient needs professional medical care throughout their illness journey. A healthcare professional determines the chance of surviving a stroke and the potential for disability rehabilitation within the first three hours of stroke onset. Fast and accurate stroke patient evaluation remains crucial to saving lives and protecting brain tissue. During this evaluation phase patients face their most challenging moment. A combination of brain CT scans and brain magnetic resonance imaging helps doctors determine stroke diagnosis within the first three hours. CT scans remain the preferred tool for stroke patient evaluation because they deliver results faster and at lower cost than MRI scans. This research examines how brain damage from stroke affects the need to find the perfect window for diagnosis. Before making a diagnosis doctors need patient history information which they collect during screening sessions and then treat patients after CT scan completion. The lengthy image analysis process puts the stroke patient's recovery at risk because important decisions are delayed. Healthcare facilities today use computed tomography scan image analysis to quickly and accurately detect stroke cases. Our present conditions need additional scientific study. Deep Learning tools enable healthcare teams to analyze medical images which supports disease diagnosis (Abdusalomov et al., 2023) (Ferdous & Shahriyar, 2023) (Mambou et al., 2018) (Cheng et al., 2021) (Hosny et al., 2018). This research explains how StrokeTimeNet RNN detects brain infarcts by analyzing computed tomography images of the brain. Our stroke detection algorithm achieves better results than other RNN-based brain CT models and delivers 94.2% accuracy.

The research team found these three main results. CT scan technology helps doctors start treating stroke patients right away when they detect stroke symptoms. Our new RNN model uses CT scans to track stroke patient recovery while speeding up medical response and diagnosis. Medical experts can make better diagnoses through CT scans because the program distributes CT interpretation work among them. Faster diagnostic tools help doctors treat patients sooner which leads to better healthcare results for patients. Medical professionals create work procedures that work with hospital systems without adding new organizational layers. The validation of entire clinical datasets proves that systems run smoothly, and medical models deliver dependable results.

Section 2 of medical documents is now replaced by a research evaluation section in the updated document format. Section 3 established Materials and Procedures. The report shows experimental results in Section 4 and suggests future research directions in Section 5.

2. LITERATURE REVIEW

Research now shows that AI can effectively predict medical conditions by studying past medical records (Zheng et al., 2023) (Nizamani et al., 2023) (Şener & Ergen, 2023) (Doppala et al., 2023) (Cui et al., 2022). New medical research from the past few years has successfully enhanced our ability to detect life-threatening health conditions. The technology of Artificial Intelligence and Machine Learning helps doctors diagnose diseases by studying medical images (Singha Deo et al., 2024) (Chandrabhatla et al., 2023) (Chiang et al., 2022) (Zeng et al., 2022) (Tamrakar

& Parsai, 2022). This part explains the fundamental principles and data collection methods for the stroke prediction system developed by the authors.

2.1. ML model using image and clinical raw data

The brain stroke predictive model took 4981 data points from eleven University of California Irvine ML Repository fields through weighted ensemble methods. Our study (Srinivas & Mosiganti, 2023) found that research points did not follow a regular pattern. The research team employed ten classification methods including NB, Histogram-based gradient boosting, SVM, KNN, LR, AdaBoost, gradient-tree boosting, and Decision Tree. Our research demonstrated that combining randomized trees with histogram-based gradient boosting achieved 96.88 percent accuracy. A team used eleven machine learning algorithms to predict brain stroke through dimensional improvement and parameter adjustments (Biswas et al., 2022). Our research findings show that SVM outperformed other models with 99.99% prediction accuracy while RF achieved 98.21%. The team processed formatted data to develop basic stroke prediction tools that they built for web and mobile platforms.

The research in (Gudadhe et al., 2023) demonstrates that combining texture features and transformed features with artificial intelligence methods helps detect intracranial hemorrhage accurately in CT brain images. A combination of GLCM, DW, and DCT features with random forest classification achieved 87.22% accurate results. The study by (Jyothsna et al., n.d.) investigated how different supervised machine learning methods identify brain strokes. The team tested LR, KNN, DT, RF, NB, and SVM algorithms on both unbalanced and balanced datasets to complete this task. According to (Jiang et al., 2022) research random forest classifier delivers its highest performance of 95.46% when working with balanced datasets. The authors analyzed data sets 1 and 2 with dataset 1 serving as the primary source for their research. The research demonstrated that the RF with Bayes algorithms delivered better performance than all other algorithms tested. The research team analyzed raw clinical information at (Akter et al., 2022) through three machine learning methods: RF, SVM, and DT. The study reveals that the RF classification model delivers 95.30% accuracy which outperforms other machine learning methods. The heart disease detection system by (Chang et al., 2022) reached 83% detection accuracy through Python-based RF classifier machine learning technology.

Researchers at (Dev et al., 2022) tested CNN technology on EHR data through neural networks and ML algorithms. A research team decreased the dimensionality of 29,072 patient records through PCA. Our neural network analysis showed 74% accuracy by studying patient data including age, hypertension status, blood sugar levels and heart disease diagnosis notes. The researcher (Bathla & Kumar, 2024) achieved superior outcomes when they combined CNN with neural networks and various ML models like Extreme Gradient Boosting, Ada Boost, and random forest alongside Naive Bayes and SVM for feature selection. The researchers combined mutual information and feature importance measures with Pearson correlation analysis through three distinct feature selection methods. The random forest model with feature importance analysis achieved 97.17% accuracy when processing raw data.

2.2. Artificial neural network (ANN) model with clinical and image data

Researchers built an artificial neural network model using Levenberg–Marquardt optimization to read brain computed tomography images from radiology departments for stroke prediction in their study (URAL, 2022). The research team analyzed 1000 CT scans including 500 stroke patients and 500 healthy adults from age 25 to 75. Research teams took CT image data of patients in both fourth ventricle and lateral ventricle positions to build their data collection. The team achieved 98.5% model performance by using LM optimization methods to enhance their ANN model. The researchers faced difficulties in analyzing CT scan data because they could not access enough information. The Brainseg application lets medical staff perform ISCS diagnostic tests automatically using CT brain images. A group of researchers (Shafaat et al., 2022) (Kumari, 2022) (Subudhi et al., 2022) analyzed published studies about ISCS imaging diagnosis with artificial intelligence methods. This research study found five main areas of ISCS imaging work which involved developing automated brain infarction detection tools and brain treatment systems while also predicting patient outcomes. The ANN algorithm showed better performance than NB, RF, ANN, j48, and KNN in structured data analysis and achieved 78.33% accurate stroke diagnosis. The research team (Viswapriya & Rajeswari, 2022) created a hybrid prediction model using ANN RF algorithms to forecast strokes from Kaggle EHR data. The dataset consists of 51110 entries and features 11 columns. The hybrid model demonstrated a consistent 94% accuracy rate in diagnosis according to separate research findings. Researchers created a heart attack diagnosis system by combining neural networks with statistical algorithms to process continuous blood pressure data (Khiabani et al., 2022).

2.3. DL-CNN model with image data

The research team at (Anjum & Oberoi, 2022) developed their CNN architecture following their experiments with the combination of convolutional neural networks and long short-term memory to detect brain hemorrhages in 200 head CT scans. The research team divided their data into 90% training data and 10% testing data for both balanced and unbalanced sets to evaluate CNN and hybrid CNN models. The CNN model achieved perfect results with unbalanced data, but the hybrid CNN model achieved 95% accuracy on the same test. Our first model achieved 95% accuracy during testing with balanced datasets, yet the second model showed 90% accuracy on similar tests. The study encounters difficulties because its dataset contains few images. Article (Uçkun et al., 2023) researchers built a deep learning system that detects ischemic stroke areas in CT scans by combining Res-UNet and UNet CNN models with dilated convolution operation in the last layer. The researchers measured their segmentation results using Dice similarity coefficient and intersection over union metrics. The diagnostic tests demonstrated that Modified Res-UNet achieved better results than all other tested models. The modified model achieved 96.95% IoU performance and increased DSC score to 98.45% above earlier results. (Buchlak et al., 2022) A research review article compared and analyzed many studies that used brain CT scans to predict patient outcomes. The authors divided their brain-CT images into separate regions for their classification experiments.

(Babutain et al., 2021) Researchers at King Fahd Medical City Saudi Arabia created a five-layer CNN model that proved useful for predicting ISCS results from CT images. The team matched CT brain scans to their matching tissue samples to provide accurate patient data. Medical researchers achieved a 99.21% prediction accuracy by analyzing brain CT scans with ResNet50, ResNet18, GoogleNet and AlexNet models. ResNet50 demonstrated the best performance at 99.70% for brain tissue separation among all tested models. The 5s CNN maintained its patient and slice classification features and gained new developmental potential through pre-trained neural network models. The five-second CNN achieved 87.20% patient brain data classification but increased to 90.51% when using stolen brain slices which outperformed pre-trained models. Our research examined how deep learning algorithms divided medical images of ISCS patients from CT and MRI scans. Our team measured segregation results through sensitivity/specificity scores and Jaccard and dice measurements (Abbasi et al., 2023).

(Alhatemi & Savaş, 2022) The researchers used deep learning models with transfer learning methods to analyze MRI data and predict strokes. Before analysis the team processed MRI scans and ran tests with five deep learning models including MobileNet and DenseNet121. The team evaluated 250 research images and trained with 1901 images to test performance validity. The model EfficientNetB2 achieved the best performance results with 98.8% accuracy. The authors at (Rajendran et al., 2022) built a deep-learning system that combines multiple models to diagnose strokes from CT image data. Researchers created distinct categories to study CT scan results. Research showed that ResNet50 CT image analysis with InceptionV3 and VGG16 networks produced better accuracy results. The team performed multiple CT scans to test different models and determine their performance levels. The researchers achieved 74% CT image classification accuracy for initial scans and

93.76% for follow-up scans through their combination of InceptionV3, ResNet50, and VGG16 models. The research data delivered 81.98% accuracy during training and testing. Secondary CT imaging proved superior at finding stroke compared to initial imaging methods.

(Dobshik et al., 2023) The team trained their modified 3D UNet model using CT scans of 81 stroke patients from the dataset while preventing overfitting through 5-fold cross validation. The research team measured their model's performance using sensitivity alongside dice scores and precision and specificity results. The model showed 0.619 plus or minus 0.036 accuracy with 0.699 plus or minus 0.039 sensitivity and 0.628 plus or minus 0.033 dice score while maintaining 0.9965 plus or minus 0.0016 specificity. Research teams have built a DL ensemble model to find and categorize intracranial hemorrhages through automated detection methods (Umapathy et al., 2023). The research showed five distinct types of intracranial bleeding including subdural hemorrhage and intraventricular hemorrhage plus subarachnoid hemorrhage and epidural hemorrhage and intra-parenchymal hemorrhage. The proposed ensemble model processed data from both the CQ500 database and the Radiological Society of North America (RSNA) database. Our team normalized images through windowing and applied data augmentation methods before using SEResNet and LSTM in an ensemble CNN model for analysis. The proposed ensemble CNN system showed outstanding performance in detecting subdural hemorrhages at 99.88% accuracy and intraventricular hemorrhages at 99.65%. It also achieved 98% accuracy in finding subarachnoid hemorrhages and 99.89% accuracy in detecting epidural hemorrhages alongside 99.75% accuracy for intra-parenchymal hemorrhages. The authors in (Gaidhani et al., 2019) proposed SegNet for brain MRI image segmentation and LeNet CNN for brain stroke classification. The research team used neuroimaging data from the ATLAS website alone. The LeNet model achieved 96% accuracy but the SegNet image results dropped to 85% when processing abnormal images through LeNet. (Tripura et al., 2023) BrainNet researchers developed a CNN system to examine CT brain images for stroke detection purposes. The research team analyzed BrainNet using CNN SVM integration. The team used CT imaging to process their CT scans and tested them with InceptionResNet-v2 and CNN followed by CNN SVM analysis. InceptionResNet-v2 achieved 52.83% accuracy while CNN SVM reached 91.91% and CNN achieved 70.12% accuracy. Researchers at (Ozaltin et al., 2022) built OzNet for Lady Reading Hospital in Peshawar, Pakistan to predict stroke from computed tomography images. The Getty team split the image data equally between normal and stroke groups before conducting 10-fold cross validation on this data. They resized images to 224x224 pixels. The convolutional neural network model transformed input data through its layers into 4096 features for preprocessing. The authors applied the minimum redundancy maximum relevance method to transform 4096 image features into 250 characteristics. Multiple Machine Learning algorithms served as the main tool for data analysis during the research period. Our experimental findings show that NB delivers the best Machine Learning results for optimal performance. The NB model achieved 98.42% accuracy with 99% total accuracy performance. The study successfully extracted Regions of Interest features from MRI scans through LASSO extraction before building random forest algorithms to predict ISCS hemorrhagic transformation. Our tested model integrated Regions of Interest features with patient clinical data to achieve 89.4% accuracy (Meng et al., n.d.). The study compared stroke lesions to find what puts patients at higher risk of stroke transformation. Research teams created an automatic system that used segmentation tools to find stroke-related lesions according to Anne (2024). The research team achieved 87% success in automatic brain CT segmentation using cognitive networks with XGBoost models. The published research results appear in Table 1 below.

Table 1. An inventory of the current literature review.

Source	Data-set	Prediction type	Proposed model	Accuracy
(Subasi, 2023)	Brain CT image	Brain stroke	VGG-19	97.06%
(Saleem et al., 2024)	Brain CT image	Early brain stroke	GA_BiLSTM	96.45%
(UmaMaheswaran et al., 2024)	Brain CT image	Early acute stroke	Advanced XGBoost	97%
(Alotaibi & Alotaibi, 2022)	Brain MRI	ISCS	CNN-Bidirectional LSTM	94.2%
(Kumar & Sengupta, 2022)	Electroencephalogram	Brain stroke	ResNet-50	90%
(Dritsas & Trigka, 2022)	Structured data	Brain stroke	Stacking classifier	98%
(Al-Mekhlafi et al., 2022)	Brain MRI	Brain stroke	AlexNet+SVM	99.9%
(Tursynova et al., 2023)	Brain CT image	Brain stroke	CNN	79%

2.4 Limitations of the existing Methodologies

Models built with structural design elements and architectural preferences demonstrate high sensitivity to particular datasets but low sensitivity to others. A model trained on large datasets performs poorly when predicting new data because of limited training data and many input features. The presence of data noise reduces the model's ability to produce accurate results. The spatial data analysis capabilities of CNN's excel through their literature-based combination of brain image static analysis with CT and MRI result interpretations. Several models work together to address performance problems by combining their abilities or linking their predictive results. Ensemble and concatenation methods create advanced models that need extra training time and processing power to handle multiple architectural variations and operating models simultaneously. These basic functional stages allow Conceptual Neural Networks to perform their computational processes. Setting up ensemble-based frameworks requires more time and needs expert technical assistance. It is harder to maintain. The evaluation process suffers from interpretation problems because CNNs combine features that make it hard to understand their internal working. Medical staff finds it hard to understand how interconnected CNNs generate diagnostic results for real-world applications. The need for extra data exists because Ensemble CNN models tend to overfit without sufficient additional information. The use of diagnosis datasets for stroke faces three major barriers related to data scarcity, validation standards and patient privacy protection. Combining multiple CNN networks into ensemble models delivers small or no performance benefits. The model needs more processing power to achieve stable performance. Most ensemble models that share the same basic design can overfit training data unless proper monitoring is used. Research teams use medical records and test outcomes as their main study materials. Scientists cannot make accurate brain stroke predictions because medical data remains imprecise. Brain CT image data act as the primary foundation of a research investigation. Our research used data augmentation and hyperparameter optimization because these techniques help scientists improve experimental quality but many research groups do not employ them. Regular pre-trained models showed no improvement when tested alongside our adjusted models. Our strategy included applying multiple transfer learning models that underwent specific modification steps. Model integration combines basic pre-trained models with adjusted versions to create superior performance. The investigators produced wrong results because they used assessment methods that were not good enough. Our solution performance is measured through AUC_ROC and F1-Score while also considering sensitivity, precision and specificity. Research shows CT image analysis accurately finds brain strokes with a success rate of 95%. The StrokeTimeNet system shows outstanding results when detecting brain strokes from CT scan images.

3. PROPOSED METHODOLOGY

This section shows all experiment details including our dataset source and the steps we took to build our model.

3.1. Experimental setup

Our research project needed several particular resources to finish the study. Table2 shows all necessary details and tool requirements for creating our proposed model.

3.2. Description of the data set

Kaggle serves as the primary online platform to provide real-world datasets for this research (Tripura et al., 2023). In table 3 shows the dataset includes 2515 CT scan images. Researchers split their data into three distinct parts for testing-validation-training purposes. The system sorts data points between two distinct groups with easy separation. The model training needed 1843 CT scans while validation received 237 samples and testing used 437 images. Our training dataset consists of 1087 normal scan images and 756 stroke diagnosis images. The validation dataset contains 157 normal brain scans and 78 stroke scans. The test scored 307 points for normal results and achieved 130 points in diagnosing strokes. Figure 1 shows both normal brain CT scans and early stroke images.

Figure 1. Example CT images of the brain

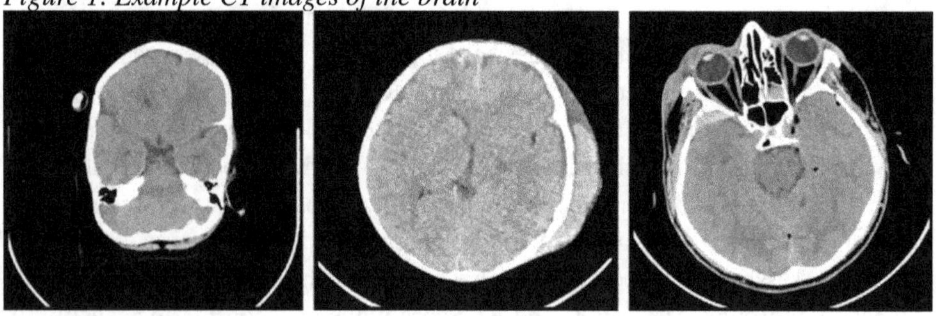

Table 2. The Hardware specifications and Tools

Resource	Details
CPU	Intel Core i7/i9, AMD Ryzen 7/9
RAM	16 GB
GPU	NVIDIA GeForce RTX 30 series
Deep learning Frame work	TensorFlow

Table 3. Description of the dataset

CT Image	Normal	Stroke	Total	Percentage
Train Image	1087	756	1843	73%
Validation Image	157	78	235	10%
Test Image	307	130	437	17%

3.4. **StrokeTimeNet** *Model design*

3.4.1 Architecture and Training Protocols

Before discussing StrokeTimeNet design features this section explained basic model adaptation through transfer learning. Diagnostic accuracy of stroke remains challenging due to limited CT image access but fine-tuned techniques demonstrate good potential for stroke detection. The pre-trained 3D CNN employs massive 3D medical image databases for volumetric medical operations and brain area segmentation. Data processing systems today rely mostly on 2D networks despite their smaller number compared to 3D networks. Various trained models now help doctors identify medical issues and visualize brain structures in three dimensions. The ResNet and EfficientNet models from ImageNet demonstrate they could serve as effective models for today's system. CT datasets produce 3D volumes which CT scanners divide into separate 2D images through axial, coronal and sagittal views. The pre-trained 2D CNN network analyzed each slice image to extract useful features. LSTM RNN connects CT image sequences through time by processing the feature output from each 2D slice. The pseudo-3D approach builds 3D volumetric structures from multiple 2D slice images by stacking them together for use as 3D models. The third dimension helps researchers develop 3-dimensional convolution layers shown in Figure 2.

Figure 2. Architecture of StrokeTimeNet

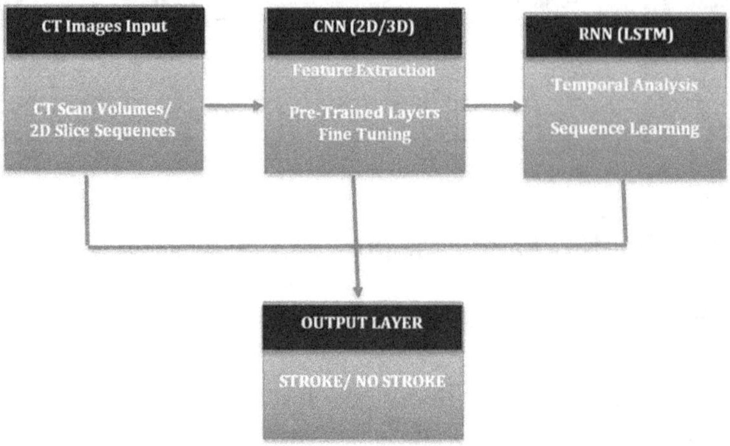

The CNN convolutional layer freezing process starts the feature extraction process for model training. The 2D CNN model uses both complete CT volumes and individual CT slices to extract features from them. The model contains a RNN processing unit that handles the extracted video features. The RNN shows great ability to capture changes in the stroke pattern over time periods. To further enhance the performance of the combined CNN + RNN model, we transplant the pre-trained CNN layers when the RNN reaches certain performance milestones using diagnostic stroke data to fine-tune the CNN layers. The CNN is fine-tuned to utilize its learned features to identify stroke detectable changes in the CT images. Fine-tuning of the learning rate helps to prevent large changes in the pre-trained weights. Since stroke CT data is limited researchers must use existing data to help detect strokes effectively. Transfer learning works best with small datasets that lack annotations but share features with annotated datasets. Pre-trained models detect stroke features faster than new models because they recognize patterns that newer models cannot identify. Transfer learning helps models adjust quickly during fine-tuning while producing better generalization results for model training. The success of transfer learning depends on how closely CT stroke images match the ImageNet dataset used for pretraining. Combining pre-trained 2D models with slice processing and 3D prediction systems needs enough training data and enough processing power. Health professionals will test the proposed model in actual medical settings to prove its accuracy. The RNN architecture's ability to work with time-based information makes it an excellent tool for detecting strokes.

3.4.2 Feature Extraction and Selection Techniques

Healthcare professionals can use 3D Convolutional Neural Networks (Conv3D) to extract volume-based scan features from CT scans for clinical purposes. The Conv3D pattern recognition tools process 3D voxel inputs to detect essential spatial connections between basic brain structures. The model shows it can spot strokes from minor tissue pattern shifts. MaxPooling3D layers maintain 3D anatomical features during down sampling because they reduce feature map dimensions while keeping vital information. This is particularly computationally expensive. The Two-dimensional Convolution Neural Networks (Conv2D Adaptation) method turns pre-sliced 2D models into 2D images by processing CT volume slices in three planes with minimal computing power. Computational CNNs adopt ImageNet pre-training methods because they extract useful features when performing slicing operations. RNN extracts input features from parts of the input while other pseudo-3D networks handle small sequences of 2D slices directly. Transfer learning helps organizations save money because they need smaller datasets during training. When dimensions grow past two the spatial data loses its effectiveness. A single scan passes through Time Distributed Layers in CNN networks to perform feature extraction. The Time Distributed wrappers maintain spatial consistency between CT scans by running the same CNN architecture on each scan sequence.

The system uses a CNN to learn diagnostic features from stroke tasks without specific instructions. The system trains filters to recognize stroke patterns by analyzing changes in edema and tissue density. The network learns to recognize essential input features during its training process. The RNN learning system with LSTM and GRU units finds key time periods in CT scan sequences to help with diagnosis. The RNN system helps select CT scan images that provide essential data for stroke progression prediction. This diagnostic model analyzes time-based changes to produce better diagnostic results. The training system reaches its goals by randomly disabling neurons to prevent overfitting during the learning process. The strategic regularization techniques help model learning recognize desired features effectively. The L1/L2 Regularization methods use weight-based penalties to find important features during the loss function calculation. Through Clinical feature integration framework administrators can add clinical features to their system. Combining medical records with imaging results produces a better diagnostic system. This research shows that including relevant patient data improves medical test results. Attention Mechanisms help us assign greater importance to features that show up at different times and positions. The operational model uses attention mechanisms to direct scan processing towards specific times of day and specific areas of the CT scan for better feature importance. Medical knowledge develops better when performance measures work together to produce improved results. The system should detect features that match radiologist findings while containing significant clinical details. Our system requires technology to monitor small brain tissue variations that signal stroke development. The stroke progress in medical settings needs to match the sequential processing methods RNNs employ.

4. PERFORMANCE EVALUATION AND COMPARISON WITH BASELINE MODELS

The research demonstrated StrokeTimeNet outperformed other models by identifying stroke cases sooner with high diagnostic accuracy. The 3D CNN surpassed basic machine learning but achieved lower accuracy and took longer to identify strokes compared to StrokeTimeNet. The Slice 2D CNN network shows better performance because its multiple layers despite producing fewer accurate results than 3D CNNs. Using 2D CNN features allowed faster analysis than traditional machine learning methods which reduced the total diagnostic time. The SVM system contained manually built features that needed long processing times to produce reliable diagnostic results. RF models delivered better diagnosis than SVM but produced less precise results compared to 2D and 3D CNN systems. RF-based diagnosis took longer to complete than CNN-based diagnosis methods including architectural CNN systems. RF demonstrated better diagnostic performance than CNN-based models and produced superior results than SVM. Diagnosis through CNN-based approaches executes more rapidly than diagnosis through other methods. The StrokeTimeNet model now reaches its optimal performance through its combination of RNN and CNN building blocks. RNN systems help doctors diagnose strokes because they recognize patterns in time-based medical data. Combining time-based analysis with 3D input data helps deep learning models perform better than machine learning models in medical diagnosis. StrokeTimeNet takes the shortest time to diagnose strokes among all models studied here. We evaluate our experimental models by combining confusion matrices with accuracy loss curves plus performance benchmark standards and ROC analysis.

4.1. Measures for performance evaluation

Our model analyzes diagnostic results through precision statistics and F1 score calculation to show its performance. The basic rules of accuracy and precision calculation use T_Positive and T_Negative true results plus F_Positive and F_Negative false results to determine F1-score values. The performance measurement system uses Recl to measure recall rates while Prec determines precision values.

$$\text{Accuracy} = \frac{T\_Positive + T\_Negative}{T\_Positive + T\_Negative + F\_Positive + F\_Negative} \quad (1)$$

$$\text{Precision} = \frac{T\_Positive}{T\_Positive + + F\_Positive} \quad (2)$$

$$\text{Recall} = \frac{T\_Positive}{T\_Positive + + F\_Negative} \quad (3)$$

$$\text{F1-score} = 2\ \text{X} \frac{Prec\ X\ Recl}{Prec + Recl} \quad (4)$$

Table 4 shows quantitative performance comparisons between StrokeTimeNet and different methodologies. The performance results in Figure 3 demonstrate StrokeTimeNet achieves 94.2% accuracy while 2D CNN (Slice) reaches 87.1%, SVM gets 82.3%, Random Forest hits 84.8% and 3D CNN stands at 89.5%. StrokeTimeNet delivers 93.7% accuracy that outperforms the results of SVM at 81.2%, RF at 83.9%, 3D CNN at 88.9%, and 2D CNN (Slice) at 86.5%. StrokeTimeNet shows 92.5% sensitivity which exceeds SVM at 78% and RF at 81.5% but remains lower than 3D CNN at 86.3% and 2D CNN (Slice) at 84%. The StrokeTimeNet model achieved better specificities of 95.8% and 92.7% than 2D CNN (Slice) at 90.2% and 86.6% and RF at 88.1%. The f1-score evaluation shows StrokeTimeNet achieves 93.1% accuracy which outperforms all other models including SVM at 79.2% and RF at 82.7%. Our proposed model delivers superior results compared to 3D CNN, 2D CNN (Slice), SVM, and RF models since it produces more precise and sensitive outcomes with high specificity and f1-score values.StrokeTimeNet shows better prediction results than all tested models according to Figure 3.

Table 4. Comparison of performance metrics of proposed model with various existing model

Model	Accuracy (%)	Sensitivity (%)	Specificity (%)	Precision (%)	F1-Score (%)
StrokeTimeNet	94.2	92.5	95.8	93.7	93.1
3D CNN	89.5	86.3	92.7	88.9	87.6
2D CNN (slice)	87.1	84	90.2	86.5	85.2
SVM (handcrafted)	82.3	78	86.6	81.2	79.5
Random Forest	84.8	81.5	88.14	83.9	82.7

Figure 3. Comparison of performance metrics of proposed model with various existing model

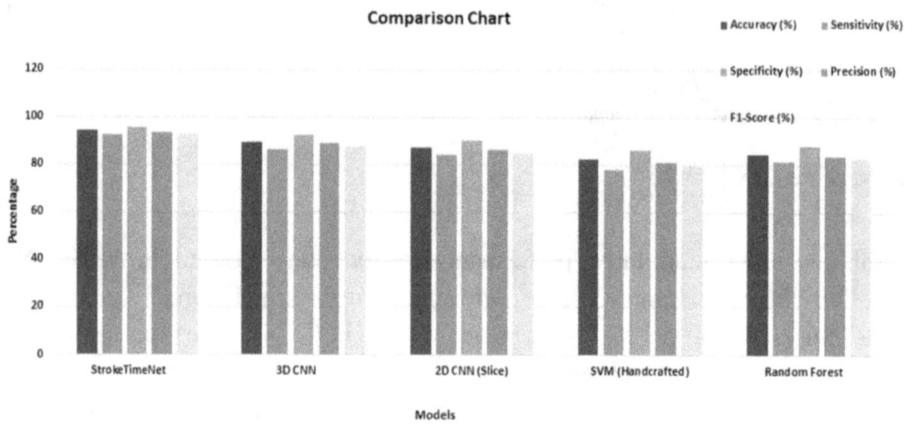

Table 5. Comparison of Time to diagnosis of proposed model with various existing model

Model	Time to Diagnosis (s)
StrokeTimeNet	28
3D CNN	65
2D CNN (slice)	40
SVM (handcrafted)	300
Random Forest	120

Figure 4. Comparison of Time to diagnosis of proposed model with various existing model

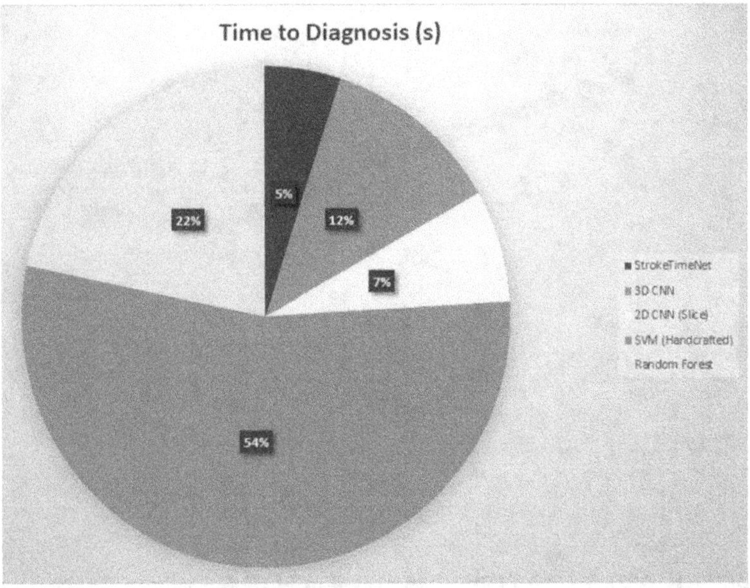

The research shows how different medical models take to diagnose patients in this Table 5. The test results from Figure 4 show StrokeTimeNet takes 28 seconds to diagnose while 3D CNN needs 65 seconds and 2D CNN (Slice) uses 40 seconds. SVM requires 300 seconds and RF needs 120 seconds. AUC, or the area under the ROC curve, ranges from 0 to 1. Hypotheses that achieve higher AUC values deliver better forecast accuracy. StrokeTimeNet produces more accurate stroke diagnosis results than other models with an AUC score of 0.95 which outperforms 3D CNN is 0.91, 2D CNN (Slice) (0.88), SVM is 0.82, and RF is 0.85 shown that Table 6 and Figure 5.

Table 6. Comparison of AUC-ROC of proposed model with various existing model

Model	AUC-ROC Curve
StrokeTimeNet	0.95
3D CNN	0.91
2D CNN (slice)	0.88
SVM (handcrafted)	0.82
Random Forest	0.85

Figure 5. Comparison of AUC-ROC of proposed model with various existing model

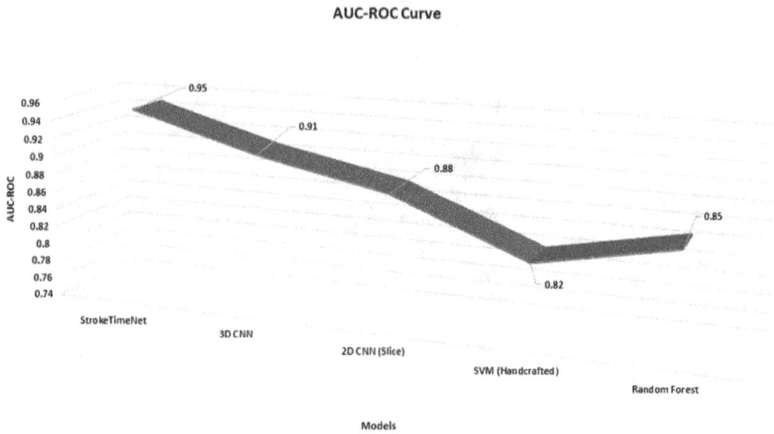

5. CLINICAL APPLICATIONS AND FUTURE DIRECTIONS

5.1 Potential Clinical Applications of RNN-Based Brain Tumor Prediction

Medical imaging researchers primarily use CNN algorithms for brain tumor prediction while RNN methods receive limited application in this field. Scientists create basic tools for brain tumor diagnosis and treatment through their network models. Our evaluation demonstrates how RNN-based models help achieve clinical brain tumor objectives by creating new assessment solutions.

5.1.1. Analysis of Sequential Medical Data:

- LSTM specialized Recurrent Neural Networks provide better methods for handling continuous input data. Doctors collect time-series MRI scans from their standard medical appointments with patients. The recurrent neural network tracks changes in tumor growth rates and measures volume differences between assessment times. This information is critical to:
  - The ability to predict cancer growth helps doctors track disease progression and choose effective treatments for their patients. The network uses stored tumor progression data from the health history database to detect new fast growth patterns. Medical professionals use tumor progression

information to decide when to start treatment and when to use radiation or surgery following chemotherapy.

- RNNs use sequential data scanning to watch how the tumor responds and measures treatment results at two separate times throughout the entire treatment period.
- The RNN process looks at medical records one after another to link vital sign measurements with EEG data to track brain tumor effects on both the nervous system and body.
- Biomarker time-series data helps RNNs analyze how tumors behave and measure treatment response patterns from multiple genomic datasets.

5.1.2. Incorporation with Imaging Data:

- RNNs examine how MRI image features change between adjacent image slices to monitor their characteristics. RNNs excel at analyzing full tumor spatial patterns but CNNs perform best at extracting features from single MRI images which makes them effective at segmentation and classification tasks.
- The combination of CNN features with medical test results and treatment history through RNN methods produces effective medical predictions. RNN models better predict tumor growth and treatment response by finding the link between medical data and feature patterns over time.

5.1.3. Predicting Recurrence and Treatment Outcomes:

- Research groups analyze brain tumor patient data using RNNs to develop forecasting models that show how long patients will survive and what their treatment response and tumor return chances are.
- Recurrent neural networks study past patient records to create personalized healthcare plans for each new client.

5.1.4. Assisting in Diagnosis

- The medical dataset and recurrent neural network systems create personalized survival, treatment response, and recurrence probability models for each patient.
- Time-series analysis with RNNs detects standard treatment response patterns that appear in most patients.

5.2 Future Research Directions and Challenges

Our research shows that RNN models can predict brain tumors effectively yet CNN models perform better than them in analyzing brain tumors. The model combines spatial patterns and feature detection from multiple MRI scans to produce better tumor detection outcomes.

- *CNNs + RNNs Architectures:*

The RNN model tracks how image shape and texture patterns change between multiple MRI scans. Researchers use LSTMs and GRUs to process sequences of features that CNN spatial analysis produces from single MRIs. RNNs achieve better segmentation results because their three-dimensional spatial dependency model tracks feature movements between different MRI slices.

- *Predictive Analysis of Longitudinal Data:*
    - RNN systems naturally predict MRI results at any moment during extended periods of time. For RNN models to advance further researchers need to train predictive systems for treated and untreated cases to find more intricate tumor growth patterns. The system includes predictive tools that help doctors find the most effective treatments that produce successful outcomes.
    - RNN systems can determine treatment response by examining several MRI image sequences taken during therapy and recovery. RNN models help healthcare providers select the best treatment methods for their patients.
    - RNNs analyze big patient datasets to find patterns that make tumors return at exact spots during specific times.
- *Mechanisms of Attention and Transformers:*
    - Rephrase the following sentence: Long short-term memory (LSTM) recurrent neural networks paired with attention mechanisms can spot time patterns and spatial points to make predictions. The network demonstrates effective language processing and shows potential for medical image analysis in both visual and temporal-spatial applications.
- *Unsupervised and Few-Shot Learning:*
    - The lack of brain tumor research data and poor tracking systems create major barriers for scientists. Future research needs to create models that combine Recurrent Neural Networks with Unsupervised/Self-supervised learning and Few-shot learning to boost performance when working with small datasets. The RNN model trained on extensive med-

ical time-series data can effectively handle small brain tumor datasets measured at different times.

- *Personalized Medicine Applications:*
  - Recurrent Neural Networks require extensive training with different medical data to produce customized models that analyze patient information and clinical measurements to forecast disease progression and treatment results.
  - RNN analysis of live patient data helps medical staff find the most effective treatment methods.

5.2.1 Challenges:

- *Limited Longitudinal Data:*

The most effective RNN predictive frameworks need both long-term accurate medical records and regular updates of clinical and imaging data. Several obstacles stand in the way of obtaining these datasets.

- Getting enough brain tumor patient data is difficult because rare diseases do not have recognized tumor subtype definitions.
- The problem of collecting patient data from different time points using varied imaging methods and scanner technologies makes it difficult to build the needed dataset.
- Big RNN training faces limitations because health care data privacy rules prevent researchers from accessing important medical records.
- Different people making tumor annotations from longitudinal data creates problems because they often produce different results.
  - *Cancer Behavior Intricacies:*

Genetic changes influence brain cancer development while environmental factors impact treatment results. Specialized experts design RNN networks with specific parts and inputs to help them handle complex systems.

- *Vanishing/Exploding Gradients:*

Deep RNN networks and extensive training data sets produce gradient vanishing that reduces their capacity to detect connections between distant data points. LSTM and GRU designs tackle specific issues effectively but achieving optimal results depends on proper setup of the network architecture and parameter configuration.

- *Computational Resources:*

Medical imaging data analysis requires advanced RNN models that need powerful GPUs and TPUs to process. The training process takes extended periods to achieve reliable results.

6. CONCLUSION

Our model StrokeTimeNet processed brain CT imaging data with transfer-learning strategies using the acquired dataset. StrokeTimeNet shows 94.2% accuracy in brain CT stroke evaluation through its 93.1% F1-Score and 0.95 AUC with 92.5% sensitivity, 95.8% specificity, and 93.7% precision. StrokeTimeNet enables doctors to detect strokes at any stage and start effective treatments to improve stroke outcomes. The results show StrokeTimeNet works exceptionally well to predict strokes. The upcoming update will introduce multi-class classification to our system. This system needs minimal information about the input images. We applied data augmentation methods to the unbalanced dataset while keeping the changes within reasonable limits. All of the data lead to an overfitting issue in the test case. They created initial stroke identification systems through brain MRI scan processing and imaging communication analysis. Our goal is to build direct connections with healthcare partners to get our models trained on actual medical data instead of public domain datasets. StrokeTimeNet needs more processing power than basic models but works well for CT brain scans to protect patients from stroke damage.

REFERENCES

Abbasi, H., Orouskhani, M., Asgari, S., & Zadeh, S. S. (2023). Automatic brain ischemic stroke segmentation with deep learning: A review. Neuroscience Informatics, 3(4), 100145.

Abdusalomov, A. B., Mukhiddinov, M., & Whangbo, T. K. (2023). Brain tumor detection based on deep learning approaches and magnetic resonance imaging. *Cancers (Basel), 15*(16), 4172. DOI: 10.3390/cancers15164172 PMID: 37627200

Akter, B., Rajbongshi, A., Sazzad, S., Shakil, R., Biswas, J., & Sara, U. (2022, January). A machine learning approach to detect the brain stroke disease. In 2022 4th International Conference on Smart Systems and Inventive Technology (ICSSIT) (pp. 897-901). IEEE..

Al-Mekhlafi, Z. G., Senan, E. M., & Taha, H. (2022). Rassem,Badiea Abdulkarem Mohammed, Nasrin M Makbol, Adwan Alownie Alanazi,Tariq S Almurayziq, Fuad A Ghaleb, Deep learning and machine learning for early detection of stroke and haemorrhage. *Computers, Materials & Continua*, *72*(1), 775–796.

Ali Berkan, U. R. A. L. (2022). Computer aided deep learning based assessment of stroke from brain radiological CT images. *Avrupa Bilim Teknoloji Dergisi*, (34), 42–52.

Alotaibi, N. S., & Alotaibi, A. S. (2022). Detection of ischemic stroke tissue fate from the MRI images using a deep learning approach. *Mobile Information Systems*.

Anirudha, S. (2023). Artificial intelligence and machine learning in the diagnosis and management of stroke: A narrative review of United States food and drug administration-approved technologies. *Journal of Clinical Medicine*, *12*(11), 3755. DOI: 10.3390/jcm12113755 PMID: 37297949

Anjum, U., & Oberoi, A. (2022). Brain haemorrhage detection using LSTM, convolution neural network and CT scan images. *IJIRCST*, *10*, 41–49.

Anne, S. (2024). Amadou dahirou Gueye, CNN and XGBoost for automatic segmentation of stroke lesions using CT data. *Procedia Computer Science*, *237*, 72–79. DOI: 10.1016/j.procs.2024.05.081

Ashiq, Brain stroke prediction CT scan image dataset — kaggle.com, 2023,https://www.kaggle.com/datasets/iashiqul/brain-stroke-prediction-ct-scan-image-dataset

Babutain, K., Hussain, M., Aboalsamh, H., & Al-Hameed, M. (2021). Deep learning-enabled detection of acute ischemic stroke using brain computed tomography images. International Journal of Advanced Computer Science and Applications, 12(12).

Bathla, P., & Kumar, R. (2024). A hybrid system to predict brain stroke using a combined feature selection and classifier. *Intelligent Medicine*, *4*(2), 75–82. DOI: 10.1016/j.imed.2023.06.002

Gaidhani, B. R., Rajamenakshi, R. R., & Sonavane, S. (2019, September). Brain stroke detection using convolutional neural network and deep learning models. In 2019 2nd International conference on intelligent communication and computational techniques (ICCT) (pp. 242-249). IEEE.

Biswas, N., Khandaker, M. M. U., Rikta, S. T., & Sam-rat, K. D. (2022). A comparative analysis of machine learning classifiers for stroke prediction: A predictive analytics approach. *Healthcare Analytics*, *2*, 100116. DOI: 10.1016/j.health.2022.100116

Chang, V., Bhavani, V. R., Xu, A. Q., & Hossain, M. A. (2022). MA Hossain, An artificial intelligence model for heart disease detection using machine learning algorithms. *Healthcare Analytics*, *2*, 100016. DOI: 10.1016/j.health.2022.100016

Cheng, S., Liu, S., Yu, J., Rao, G., Xiao, Y., Han, W., Zhu, W., Lv, X., Li, N., Cai, J., Wang, Z., Feng, X., Yang, F., Geng, X., Ma, J., Li, X., Wei, Z., Zhang, X., Quan, T., & Liu, X. (2021). Robust whole slide image analysis for cervical cancer screening using deep learning. *Nature Communications*, *12*(1), 5639. DOI: 10.1038/s41467-021-25296-x PMID: 34561435

Chiang, P.-L., Lin, S.-Y., Chen, M.-H., Chen, Y.-S., Wang, C.-K., Wu, M.-C., Huang, Y.-T., Lee, M.-Y., Chen, Y.-S., & Lin, W.-C. (2022). Deep learning-based automatic detection of ASPECTS in acute ischemic stroke: Improving stroke assessment on CT scans. *Journal of Clinical Medicine*, *11*(17), 5159. DOI: 10.3390/jcm11175159 PMID: 36079086

Cui, L., Fan, Z., Yang, Y., Liu, R., Wang, D., Feng, Y., Lu, J., & Fan, Y. (2022). Deep learning in ischemic stroke imaging analysis: A comprehensive review. *BioMed Research International*, *2022*(1), 2022. DOI: 10.1155/2022/2456550 PMID: 36420096

Deo, B. S., Pal, M., & Prasanta, K. (2024). Panigrahi, Asima Pradhan, An ensemble deep learning model with empirical wavelet transform feature for oral cancer histopathological image classification. *International Journal of Data Science and Analytics*, 1–18.

Dev, S., Wang, H., Nwosu, C. S., Jain, N., Veeravalli, B., & John, D. (2022). A predictive analytics approach for stroke prediction using machine learning and neural networks. *Healthcare Analytics*, *2*, 100032. DOI: 10.1016/j.health.2022.100032

Dobshik, A. V., Verbitskiy, S. K., Pestunov, I. A., Sherman, K. M., Tulupov, A. A., & Berikov, V. B. (2023). Acute ischemic stroke lesion segmentation in non-contrast CT images using 3D convolutional neural networks. Компьютерная оптика, 47(5), 770-777.

Doppala, B. P., Al Bataineh, A., & Vamsi, B. (2023). An efficient, light weight, tiny 2D-CNN ensemble model to detect cardiomegaly in heart CT images. *Journal of Personalized Medicine*, *13*(9), 1338. DOI: 10.3390/jpm13091338 PMID: 37763106

Dritsas, E., & Trigka, M. (2022). Stroke risk prediction with machine learning techniques. *Sensors (Basel)*, *22*(13), 4670. DOI: 10.3390/s22134670 PMID: 35808172

Ferdous, M. J., & Shahriyar, R. (2023). A comparative analysis for stroke risk prediction using machine learning algorithms and convolutional neural network model, in: 2023 International Conference on Electrical, Computer and Communication Engineering, ECCE, IEEE, pp. 1–6. DOI: 10.1109/ECCE57851.2023.10101567

Gudadhe, S., Thakare, A., & Anter, A. M. (2023). A novel machine learning-based feature extraction method for classifying intracranial hemorrhage computed tomography images. *Healthcare Analytics*, *3*, 100196. DOI: 10.1016/j.health.2023.100196

Hosny, K. M., Kassem, M. A., & Foaud, M. M. (2018). Skin cancer classification using deep learning and transfer learning, in: 2018 9th CairoInternational Biomedical Engineering Conference, CIBEC, IEEE, pp. 90–93. DOI: 10.1109/CIBEC.2018.8641762

Jiang, L., Zhang, C., Wang, S., Ai, Z., Shen, T., Zhang, H., Duan, S., Yin, X., & Chen, Y.-C. (2022). Mri radiomics features from infarction and cerebrospinal fluid for prediction of cerebral edema after acute ischemic stroke. *Frontiers in Aging Neuroscience*, *14*, 782036. DOI: 10.3389/fnagi.2022.782036 PMID: 35309889

Khiabani, S. J., & Batani, A. (2022). A hybrid decision support system for heart failure diagnosis using neural networks and statistical process control. *Healthcare Analytics*, *2*, 100110. DOI: 10.1016/j.health.2022.100110

Mambou, S. J., Maresova, P., Krejcar, O., Selamat, A., & Kuca, K. (2018). Breast cancer detection using infrared thermal imaging and a deep learning model. *Sensors (Basel)*, *18*(9), 2799. DOI: 10.3390/s18092799 PMID: 30149621

Meng, Y., Wang, H., Wu, C., Liu, X., Qu, L., & Shi, Y. (2022). Prediction model of hemorrhage transformation in patient with acute ischemic stroke based on multi-parametric MRI radiomics and machine learning. *Brain Sciences*, *12*(7), 858. DOI: 10.3390/brainsci12070858 PMID: 35884664

Nizamani, A. H., Chen, Z., Nizamani, A. A., & Bhatti, U. A. (2023). Uzair Aslam-Bhatti, Advance brain tumor segmentation using feature fusion methods with deep U-net model with CNN for MRI data. *Journal of King Saud University. Computer and Information Sciences*, *35*(9), 101793. DOI: 10.1016/j.jksuci.2023.101793

Ozaltin, O., Coskun, O., Yeniay, O., & Subasi, A. (2022). A deep learning approach for detecting stroke from brain CT images using OzNet. Bioengineering, 9(12), 783.

Quinlan, D. (2022). Charting the potential of brain computed tomography deep learning systems. *Journal of Clinical Neuroscience*, *99*, 217–223. DOI: 10.1016/j.jocn.2022.03.014 PMID: 35290937

Rajendran, K., Radhakrishnan, M., & Viswanathan, S. (2022). An ensemble deep learning network in classifying the early CT slices of ischemic stroke patients, Trait. Signal. api.semanticscholar.org/CorpusID:252632994

Rusul, A. J. A., & Savaş, S. (2022). Transfer learning-based classification comparison of stroke. *Computer Science*, 192–201.

Saleem, M. A., Javeed, A., & Akarathanawat, W. (2024). Innovations in stroke identification: A machine learning-based diagnostic model using neuroimages. *IEEE Access : Practical Innovations, Open Solutions*.

Şener, A., & Ergen, B. (2023). Enhancing Brain Tumor Detection on MRI Images Using an Innovative VGG-19 Model-Based Approach. Sakarya University Journal of Science, 27(5), 1128-1140..

Shafaat, O., & Joshua, D. (2022). Leveraging artificial intelligence in ischemic stroke imaging. *Journal of Neuroradiology*, *49*(4), 343–351. DOI: 10.1016/j.neurad.2021.05.001 PMID: 33984377

Srinivas, A., & Mosiganti, J. P. (2023). A brain stroke detection model using soft voting based ensemble machine learning classifier. *Measurement. Sensors*, *29*, 100871. DOI: 10.1016/j.measen.2023.100871

Subasi, A. (2023). (Ed.). Chapter 7 - brain stroke detection from computed tomography images using deep learning algorithms, in: Applications of Artificial Intelligence in Medical Imaging, in: Artificial Intelligence Applications inHealthcare and Medicine, Academic Press, pp. 207–222.

Subudhi, A., Dash, P., Mohapatra, M., Tan, R.-S., Acharya, U. R., & Sabut, S. (2022). U RajendraAcharya, Sukanta Sabut, Application of machine learning techniques for characterization of ischemic stroke with MRI images: A review. *Diagnostics (Basel), 12*(10), 2535. DOI: 10.3390/diagnostics12102535 PMID: 36292224

Tamrakar, S., & Parsai, M. P. (2022). Brain tumor detection based on watershed segmentation and classification using deep learning. *Int. J. Sci. Res.Eng. Trends, 8*(6), 2097–2015.

Tripura, D., Haque, I., Dutta, M., Dev, S., Jahan, T., Ghosh, S. K., & Islam, M. A. A BrainNet (BrN) based new approach to classify brain stroke from CT scan images, in: *2023 International Conference on Advancement in Computation & Computer Technologies (InCACCT)*, IEEE,2023, pp. 1–6. DOI: 10.1109/InCACCT57535.2023.10141780

Tursynova, A., Omarov, B., Tukenova, N., Salgozha, I., Khaaval, O., Ramazanov, R., & Ospanov, B. (2023). Deep learning-enabled brain stroke classification on computed tomography images. Computers, Materials and Continua, 75(1), 1431-1446.

Uçkun, S. (2023). Mahmut AĞRALI, Volkan KILIÇ, Deep learning-based ischemic stroke segmentation on brain computed tomography images. *Avrupa BilimTeknoloji Dergisi*, (50), 105–112.

UmaMaheswaran, S. K., Ahmad, F., Hegde, R., Alwakeel, A. M., & Rameem Zahra, S. (2024). Enhanced non-contrast computed tomography images for early acute stroke detection using machine learning approach. *Expert Systems with Applications, 240*, 122559. DOI: 10.1016/j.eswa.2023.122559

Umapathy, S., Murugappan, D. B., & Thakur, M. (2023). Automated computer-aided detection and classification of intracranial hemorrhage using ensemble deep learning techniques. *Diagnostics (Basel), 13*(18), 2987. DOI: 10.3390/diagnostics13182987 PMID: 37761354

Kumari, K. V., Kumar, K. S., & Sreelatha, M. (2023). Stroke Disease Identification System by using Machine Learning Algorithm. International Journal of Computer Science & Network Security, 23(11), 183-189.

Viswapriya, S. E., & Rajeswari, D. (2022). *A systematic method of stroke prediction model based on big data and machine learning, in: 2022 Smart Technologies, Communication and Robotics, STCR*. IEEE.

WHO EMRO | Stroke, Cerebrovascular accident | Health topics — emro.who.int, (2023). https://www.emro.who.int/health-topics/stroke-cerebrovascular-accident/index.html

Zeng, Y., Long, C., Zhao, W., & Liu, J. (2022). Predicting the severity of neurological impairment caused by ischemic stroke using deep learning based on diffusion-weighted images. *Journal of Clinical Medicine*, *11*(14), 4008. DOI: 10.3390/jcm11144008 PMID: 35887776

Zheng, Y., Li, C., Zhou, X., Chen, H., Xu, H., Li, Y., Zhang, H., Li, X., Sun, H., Huang, X., & Grzegorzek, M. (2023). Application oftransfer learning and ensemble learning in image-level classification for breast his to pathology. *Intelligent Medicine*, *3*(02), 115–128. DOI: 10.1016/j.imed.2022.05.004

Chapter 20
Neurological Disorder Diagnosis and Treatment Using BCIs

V. Nithya
https://orcid.org/0000-0002-8921-7876
KGiSL Institute of Technology, India

B. Leena
https://orcid.org/0009-0000-6616-9982
KGiSL Institute of Technology, India

ABSTRACT

This chapter explains how the Brain-Computer Interfaces (BCIs) combined with Neural Network Technologies are used to diagnose and cure neurological disorders. And discusses the applications of BCIs to detect conditions such as brain tumors, epilepsy, stroke as well as neurodegenerative diseases, and how machine learning, especially deep learning, enhances their accuracy and efficiency of BCIs. In addition, it talks about the BCIs therapeutic intervention such as neurofeedback training, motor rehabilitation, and assistive neuroprosthetics. The article delves into the field of real-time brain signal processing along with explanations about early diagnosis abilities of predictive analytics in BCIs and discusses their role in creating personalized treatment procedures. The discussion includes considerations about ethics as well as obstacles in clinical integration and prospective bearings for BCI-based neurological treatment systems.

DOI: 10.4018/979-8-3693-9445-8.ch020

INTRODUCTION

Diagnosis, along with treatment of neurological healthcare problems including brain tumors and epilepsy disorders and stroke, and neurodegenerative diseases, remains challenging. The standard diagnostic procedures revert to imaging techniques coupled with clinical exams, but they lack sufficient precision during real-time use. The clinical diagnosis and treatment of neurological illnesses made a complete revolution possible through Brain-Computer Interfaces (BCIs) ruled by deep learning alongside neural network technologies. Brain-computer interfaces utilize various signal recording mechanisms, starting with EEG and ECoG and MEG, and fMRI, among others, to interpret brain signals. The analysis of these biological data relies on deep learning models, especially LSTM networks, for both speedier and more precise illness diagnosis methods. Results have shown that BCIs succeed in identifying conditions at early stages and processing brain signals live, and they can produce personalized therapeutic strategies.

BCIs allow neurological treatment through three therapeutic interventions where patients receive neurofeedback training and learn motor rehabilitation, and acquire assistance from neuroprosthetic systems. The technological improvements assist ALS patients alongside other people with motor disabilities to access minimal control over their surroundings. The complete potential of BCIs remains limited because of issues related to both ethics and technology challenges, alongside equipment restrictions for implementation in clinical settings. BCIs find medical applications alongside Neural Network Technologies in neurological condition diagnosis and treatment, so this section analyses healthcare aspects and constraints. Real-time signal analysis and predictive analytics, and treatments combined through BCIs, establish a new method to treat neurological conditions, according to this study.

1. UNDERSTANDING NEUROLOGICAL DISORDERS

The global population faces neurological disorders, which create major physical disabilities, together with cognitive and emotional complications. Such conditions manifest from brain and spinal cord and nervous system abnormalities, which produce motor disabilities and cognitive health deterioration that decreases overall quality of life (Lima et al., 2022). The essential neurological diseases include brain tumors alongside epilepsy and Parkinson's disease, amyotrophic lateral sclerosis (ALS), and stroke, since they bring different diagnostic and therapeutic requirements (Spears, Mian, & Greer, 2022; Rizvi, Batchala, & Mukherjee, 2018).

1.1 Introduction of Neurological conditions such as brain tumors together with epilepsy and Parkinson's disease and ALS, and stroke.

Brain Tumors: The brain hosts abnormal cell growth, which appears either as benign or malignant tumors. Brain tumor dimensions and anatomical positioning determine that victims will experience headaches, together with seizures and memory loss, as well as motor control problems. Better therapies with greater positive outcomes are the consequence of improved patient identification with BCI and sophisticated imaging methods (Rizvi, Batchala, & Mukherjee, 2018; Hering et al., 2023).

Epilepsy: The condition leads to recurring seizures because of irregular electrical signals occurring in the brain. The current medical approach uses medication as treatment, whereas BCIs with machine learning algorithms serve as contemporary methods for seizure monitoring (He & Wu, 2020; Ssain, Nijholt, & Ahad, 2023).

Parkinson's Disease: The disease damages the nervous system progressively until it mostly affects movement control. The disease results from neuron decay of dopamine-producing cells which leads to hand movement tremors together with muscle stiffness and slowness of actions. In addition to helping with the recovery of motor function, deep learning techniques in conjunction with BCIs assist early illness identification (LeCun, Bengio, & Hinton, 2015; Yadav & Maini, 2023).

Amyotrophic Lateral Sclerosis (ALS): As a fatal neurodegenerative disease, ALS attacks motor neurons before leading to complete paralysis as it progresses. Since patients experience complete movement loss, BCIs serve as essential tools for assisting communication and providing mobility solutions to restore independence (Li, Wei, Chen, & Zhou, 2021; Ssain, Nijholt, & Ahad, 2023).

Stroke: The fatal neurodegenerative disease ALS deteriorates motor neurons until paralysis occurs. Skip communication and mobile assistance are crucial functions enabled by BCIs, which provide independence for patients who have lost voluntary movement (Lima et al., 2022; Anwar & Eldeib, 2020).

Table 1. BCI-Based Assistive Technologies for Patients

Disorder	Assistive Technology	Benefit
ALS	Brain-controlled speech synthesis	Enables communication
Stroke	Neurofeedback rehabilitation	Improves motor recovery
Locked-in Syndrome	BCI-based text entry	Enhances patient interaction

1.2 Role of BCIs in Non-Invasive and Invasive Diagnosis

The diagnosis of neurological disorders benefits substantially from non-invasive and invasive approaches that Brain-Computer Interfaces (BCIs) make available. The methods assist medical professionals in analysing brain signals to spot neurological abnormalities in patients who have epilepsy, stroke, ALS, and brain tumors.

Non-Invasive BCIs

Non-invasive BCIs serve as the most popular form of interface technology because they do not need operations while also being reachable and minimizing health dangers. External sensors serve as the primary instruments through which brain activity gets detected.

- **Electroencephalography (EEG)**: The EEG technology detects brain electrical activity through externally attached electrodes positioned on the head. Medical professionals rely on EEG for the detection of epilepsy and sleep disorders, and brain function disorders because of its precise recording ability. The brain signals collected by EEG are easily disrupted by external noise while showing a restricted ability to determine their precise spatial locations.
- **Magnetoencephalography (MEG)**: MEG allows doctors to analyse brain signals by measuring neural-generated magnetic fields, which provide highly detailed readings without touch. The technology provides excellent brain mapping capabilities while helping researchers understand cognitive disorders, though it maintains high costs and needs special locations for operation.
- **Functional Magnetic Resonance Imaging (fMRI)**: Studies employing fMRI observe brain activity through detection of alterations in blood oxygenation that lead to BOLD signals. The method serves as a fundamental tool for both cognitive research exploration and medical investigation of stroke, along with tumor assessment. The outstanding spatial performance of fMRI exists alongside its inadequate temporal performance characteristics and the necessary expensive hardware system.

Invasive BCIs

Invasive BCIs offer better precision through a surgical brain interface with implanted electrodes. High-resolution neural monitoring, along with control of assistive devices, specifically benefits from these implementation methods.

- **Electrocorticography (ECoG)**: ECoG represents an invasive method in which electrodes directly contact the cerebral cortex surface. The medical capabilities of ECoG surpass those of EEG as it offers strong signal stability and increased resolution, therefore supporting seizure mapping in epileptic patients and motor function studies. The procedure to implant this equipment leads to health complications for individuals.
- **Deep Brain Stimulation (DBS)**: Neurological condition treatment revolves around Deep Brain Stimulation (DBS) because this method controls motor symptoms that affect Parkinson's disease patients and patients diagnosed with essential tremor and dystonia. Medics use surgical procedures to insert brain electrodes into the subthalamic nucleus and globus pallidus internus, followed by generator device attachment to distribute continuous electrical signals. The stimulated electrical impulses manage atypical neurological activities to decrease symptoms. Therapeutic DBS systems and their monitoring components enable doctors to observe brain operations, thus providing data about disease evolution and therapeutic results. This system creates adaptive stimulation systems that use neural inputs for parameter modification to enhance patient healthcare quality through continuous therapeutic adjustments.
- **Intracortical BCIs**: The highly advanced brain-computer interface class known as intracortical BCIs requires medical staff to implant microelectrode arrays inside the cerebral cortex. Medical instruments that derive signals from cortex tissue through intracortical BCI systems achieve both precise temporal and spatial signal analysis, unlike other BCIs, which access cortex signals only at the surface level (Anwar & Eldeib, 2020). The specific neural signal information directly supports operations of advanced neuroprosthetics, which require precise control (He & Wu, 2020). The integration between these interfaces has produced promising outcomes that help people with major motor disabilities regain their mobility and communication capabilities, including those who have permanent disabilities from spinal cord injuries or amyotrophic lateral sclerosis (ALS) (Li, Wei, Chen, & Zhou, 2021; Ssain, Nijholt, & Ahad, 2023). The implementation of intracortical BCIs continues its advancement toward clinical use because of new biocompatible materials discoveries and signal decoding algorithm developments in combination with wireless communication system improvements (Hering et al., 2023; Yadav & Maini, 2023).

Table 2. Comparison of Non-Invasive vs. Invasive BCI Techniques

Feature	Non-Invasive BCIs (EEG, MEG, fMRI)	Invasive BCIs (ECoG, DBS)
Accuracy	Moderate	High
Risk	Low	High
Cost	Lower	Higher
Clinical Application	Epilepsy, Stroke, Tumors	Parkinson's, ALS, Severe Epilepsy

1.3 Current Challenges in Conventional Diagnostic Methods

The diagnosis, together with the monitoring of neurological disorders, faces multiple barriers between technological restrictions and economic affordability, and logistical constraints (Lima et al., 2022). The success of effective treatment relies on prompt diagnosis, but patients frequently see no signs until their neurological conditions have substantially advanced (Spears, Mian, & Greer, 2022). The complex brain structures, together with insufficient current diagnostic instruments, make it difficult to identify and treat several disorders on time (Rizvi, Batchala, & Mukherjee, 2018; Hering et al., 2023).

Delayed Diagnosis and Late Detection

Healthcare professionals encounter multiple obstacles during neurological disorder diagnosis and monitoring activities, which include technological difficulties and economic frameworks, and logistical problems (Lima et al., 2022). Successful neurological treatments require rapid, correct diagnosis of conditions, and several neurological diseases progress until their symptoms become clear to patients (Spears, Mian, & Greer, 2022). The intricate nature of human brain functioning, combined with many neurological illnesses, leads to extended periods where no clear signs of disease appear before medical detection (Rizvi, Batchala, & Mukherjee, 2018). Nerve conditions present brain function modifications and structural changes, which become very challenging to recognize during their initial development period (Rainio, Teuho, & Klén, 2024). The effectiveness of MRI and CT, and PET scans occurs only when structural brain abnormalities become detectable, at which point substantial neural damage from disorders may already be complete (Rizvi, Batchala, & Mukherjee, 2018).

Glioblastomas show themselves on imaging scans after reaching advanced stages because complete surgical removal becomes practically impossible at that point (Spears, Mian, & Greer, 2022). Parkinson's disease reveals its major motor symptoms exclusively when dopamine-producing neuron loss exceeds 60–80% in the substantia nigra (Yadav & Maini, 2023). The substantial delay in detecting diseases creates obstacles for implementing early neuroprotective therapies because early intervention becomes nearly impossible (He & Wu, 2020; Ssain, Nijholt, & Ahad, 2023).

Limited Real-Time Monitoring

The present diagnostic technologies face a significant operational restriction because they cannot monitor brain activity in real-time and continuously (Lima et al., 2022). The current imaging methods generate static images from single snapshots because they function only at one particular moment in time (Rizvi, Batchala, & Mukherjee, 2018). When patients undergo MRI and CT scans, they must stay motionless because these tests offer structural observations rather than indicating brain functional interactions (Rainio, Teuho, & Klén, 2024). The brain measurement technique, EEG, produces brief electrical activity analysis, yet it needs artificial hospital environments for its operation (Ssain, Nijholt, & Ahad, 2023). Real-time monitoring data remains inaccessible to clinicians who need to track brain activity evolution because this preventative measure is essential for managing epilepsy and treating patients with brain injuries and recovering stroke patients (He & Wu, 2020). The monitoring needs continuous operation to detect brain state changes and intervene during abnormal pattern developments. Medical treatments become less effective because there is no capability for dynamic tracking of the brain's shifting physiology (Anwar & Eldeib, 2020; Yadav & Maini, 2023).

High Costs and Limited Accessibility

The functional MRI (fMRI) and positron emission tomography (PET) techniques provide important knowledge about brain activity and neurotransmitter operation, together with metabolic processes (Rizvi, Batchala, & Mukherjee, 2018). The tools are expensive to manage and need advanced equipment such as shielded rooms, together with radiotracers and specialized trained professionals (Lima et al., 2022). Regions with low income or located in rural areas experience a large healthcare gap because they lack proper access to this technological equipment (Rainio, Teuho, & Klén, 2024). The high price of these approaches creates barriers for their use in both standard testing and follow-ups despite being available in well-equipped healthcare facilities (Spears, Mian, & Greer, 2022). Patients often remain without correct diagnoses and wait extensively between when their symptoms start and when doctors confirm their conditions. Healthcare costs rise significantly in the long term because patients require intense medical treatments following delayed diagnosis or delayed outcomes (Hering et al., 2023).

Invasiveness and Patient Discomfort

Invasive diagnostic procedures normally deliver exact information, but they expose patients to certain health dangers (Lima et al., 2022). ECoG and DBS techniques demand medical intervention, which involves surgical placement of electrodes both on the brain surfaces and inside brain tissue (Anwar & Eldeib, 2020). The implantation of electrodes involves infection risk, along with bleeding danger and inflammation risk, and possible postoperative difficulties that require hospital stays followed by prolonged recovery times (Hering et al., 2023). These invasive techniques serve mostly severe and treatment-resistant patients because of their potential dangers. Despite their capacity to deliver detailed anatomical and functional brain data, they cannot be used as diagnostic tools except for limited patient populations (He & Wu, 2020). Patients commonly refrain from invasive medical procedures unless the need is absolutely essential due to their limited willingness that hinders the application scope (Ssain, Nijholt, & Ahad, 2023).

Interpretation Challenges and Subjectivity

Experts in neurological medicine base their diagnoses on thorough examinations of complicated information, which includes brain images as well as electrophysiological patterns and behavioural analysis results. Professional neurologists, along with radiologists, deliver insightful results through their expertise, but human analysis proves both subjective and variable. When carrying out evaluations on the same data, two medical specialists may develop separate conclusions that result in inconsistent medical treatments.

The diagnostic process becomes problematic for early-stage evaluations because subjective interpretations make it difficult to notice faint abnormalities (Lima et al., 2022). The result from visual assessment of imaging results can lead to overlooked early diagnosis opportunities when health-related indicators are not pathological (LeCun, Bengio, & Hinton, 2015). Artificial intelligence (AI) and machine learning together show potential to minimize the variability, yet their mass implementation in this field continues at an early stage (Anwar & Eldeib, 2020; Hering et al., 2023).

Difficulty in Detecting Early-Stage Neurodegeneration

The diagnostic challenges increase for Alzheimer's disease and Parkinson's disease, and ALS because of their delayed symptomless disease progression (Lima et al., 2022). When the diseases start, there are minimal cognitive effects and small motor symptoms, along with general fatigue reported by patients (Rizvi, Batchala, & Mukherjee, 2018). Present diagnostic instruments fail to notice essential functional, along with structural changes that occur during these monitored periods (Rainio, Teuho, & Klén, 2024). Beta-amyloid plaques, together with tau tangles, appear multiple years before Alzheimer's patients demonstrate any noticeable deterioration in their memory function (Spears, Mian, & Greer, 2022). Well-established medical diagnosis methods using cognitive assessments and brain imaging reveal poor performance for detecting minor brain changes until full-scale neuronal damage occurs (He & Wu, 2020). The delayed discovery of disease signs disallows both disease prevention and initial drug treatments that could minimize or stop disease progression (Yadav & Maini, 2023).

Ocular imaging methods, together with biomarker research in blood and cerebrospinal fluid, need further investigation to create early detection methods, but current research has not produced clinical-ready results (Hering et al., 2023). The existing challenges around validation and usability, and cost-effectiveness of brain-computer interfaces (BCIs) and wearable EEG equipment do not hinder their use in enabling continued monitoring for early detection (Ssain, Nijholt, & Ahad, 2023).

2. NEUROPHYSIOLOGY AND BRAIN SIGNAL ACQUISITION

Neurophysiology is a branch of neuroscience that examines the functioning of the nervous system at the cellular and systemic levels. This area studies how neurons create, send, and receive electrical and chemical signals. Neurons send messages in the form of action potentials, which are short bursts of electricity known as impulses caused by the inflow and outflow of various ions through the neuron's membrane. These signals move along the neural pathways where they create patterns of activity associated with thinking, moving, sensing, and behaving.

Learning the brain's physiological functions is crucial for constructing interfaces that link the brain with computers, also referred to as brain-computer interfaces (BCIs). Neurophysiological signals, most importantly those captured from the brain, are used by the BCIs to create the connection between the nervous system and outside devices. BCIs interpret the brain's activity patterns and transform them into functional commands for prosthetic limbs, powered wheelchairs, computers, and communication systems intended for patients with high-level physical disabilities.

Brain signal detection can be done using multiple techniques, each having unique resolution and risks involved. EEG, or electroencephalography, is one of the most popular methods and consists of non-invasive techniques, as it has high resolution and is relatively easy to perform. MEG, functional magnetic resonance imaging, also falls under this category.

2.1 Types of Brain Signals: EEG, ECoG, MEG, and fMRI

Understanding brain activity requires capturing and interpreting the electrical, magnetic, or metabolic signals generated by neuronal processes. Each brain signal acquisition method provides unique insights and has specific applications depending on the resolution, invasiveness, and signal type. The most commonly used techniques in neuroscience and Brain-Computer Interface (BCI) systems include Electroencephalography (EEG), Electrocorticography (ECoG), Magnetoencephalography (MEG), and functional Magnetic Resonance Imaging (fMRI).

2.2 Signal Processing Methods for Detecting Neurological Disorders

Advanced signal processing methods are critical in interpreting primitive brain signals. These methods refine the relevant sequences while suppressing noise to the degree where correct interpretation is possible and safe diagnostic and monitoring can be carried out for neurological disorders.

Pre-Processing Procedures

After a signal cannot be further improved, it undergoes an analysis termed pre-processing, which enhances its value and makes it interpretable. This step eliminates distortions that can be due to external noise or physiological noise, such as movement, breathing, and others.

Filtering: The signal may contain blurring frequencies caused by muscle ensheathment, electric activity interference, or sources outside the brain. Applying filters with specific characteristics, such as bandpass filters that pass certain frequency ranges while blocking others, notch filters remove power line interference, and denoising wavelets preserve the desired signal while removing unwanted data.

Artifact Removal: Independent Component Analysis (ICA) and Principal Component Analysis (PCA) are widely accepted methods used today for the identification and extraction of artifacts like eye blinks, heartbeats, and muscle noise. These algorithms assume that the signals' constructions are made up of independently distributed signals, and therefore, subcomponents can be deleted while retaining the rest.

Feature Extraction

In the previous step, preparing data pipelines that extract pre-processed data, revealing insights into the function/ dysfunction of the nervous system, has been done. Enhancing or creating new indices for the pre-processed data streams should now be added.

- **Analysis of Time Domain:** Time domain changes pertain to the change of the amplitude of the signal over time. Some processes in the nervous system that measure time intervals can be perceived or marked left as identifiable time domain spikes. Indices: mean amplitude, variance of the signal, and rate of zero-crossings (0-crossings) must be present in some way in abnormal cases.

- **Analysis of Frequency Domain:** Activity in the brain is cyclic/ periodic and hangs around rhythmic movements or oscillations in the following bands: delta, theta, alpha, beta, and gamma. In most cases, these oscillations are impaired in the course of some illnesses, like originating or brought forth to them (AD – increased delta activity; PD – mutated rhythms of beta). Other useful techniques like Fast Fourier Transform (FFT) and Power Spectral Density (PSD) can be used efficiently.
- **Analysis of the Time-Frequency Domain:** There is brain activity associated with disorders like epilepsy or terminal IS attack those changes quickly. The capture both time- and frequency-domain data offer a broad depiction of continuously evolving brain states that change swiftly. Their evaluation must be done either head using or self-guiding headsets or STFT and Wavelet Transform are an alternative equally suitable choice.

Classification Techniques

Upon extracting relevant features, algorithms are utilized for classification to differentiate the various forms of neurological conditions. All models undergo training with labelled datasets specific to the disorders of interest to learn various patterns corresponding to the identified conditions.

- **Support Vector Machines (SVMs):** SVMs find the optimal dividing line, also known as the decision boundary, for data points belonging to different classes. SVMs have become a popular approach for the detection of neurological disorders because of their proficiency in high-dimensional feature spaces like those generated by multi-channel EEG recording. Their strength is primarily in binary problems; for instance, distinguishing a 'healthy' versus a 'pathological' state of the brain. They make use of kernel functions so that non-linear classification can be done, which is advantageous when the data is complex.
- **Convolutional Neural Networks (CNNs):** These models were first designed for image recognition. However, they have become increasingly popular for the analysis of spatial patterns of the brain. In BCI applications, CNNs can be used for analysis of EEG signal images like spectrograms or topographic maps, or even structural brain scan images like MRI. They have multiple stages. Convolutional layers identify spatial features while pooling layers de-

crease data volume without losing significant elements. Final classified outcomes, like Alzheimer's, are made in the fully connected layers.

- **Recurrent Neural Networks (RNNs) and Long Short-Term Memory (LSTMs):** These models are specially tailored to deal with sequential data which makes them well suited for the analysis of time-series brain signals. In particular, LSTMs address the constraints of the vanishing gradient problem in conventional RNNs and enable learning long-term dependencies in signal patterns. These models are used in applications like predicting epileptic seizures, monitoring sleep stages, and tracking cognitive decline over time. Their capacity to process contextual signals enables more advanced classification over static models.

Table 3. Signal Processing Techniques for Brain Signal Analysis

Processing Stage	Techniques Used	Purpose
Pre-processing	Filtering, Artifact Removal	Reduces noise in brain signals
Feature Extraction	Wavelet Transform, PCA	Identifies meaningful patterns
Classification	CNN, SVM, Random Forest	Detects and predicts neurological disorders

2.3 Challenges in Brain Signal Acquisition and Noise Reduction

The capture of brain signals is crucial in contemporary neuroscience, neurodiagnostics, and BCI (brain-computer interface) technology. There are still some persistent difficulties that compromise the reliable and consistent acquisition of brain signals, clean signal interpretation, and hardware and algorithms given signal processing. Problems of this type affect diagnostic devices, systems for neuroprosthesis control, and cognitive monitoring. We discuss the most important problems below.

Noise and Artifacts

Perhaps the most difficult noise problem in the context of brain signal acquisition is due to the presence of external and internal noise, which is also called artifacts. The signals derived from Electroencephalography (EEG) and Magnetoencephalography (MEG) tend to be weak themselves, usually at the microvolt or femtotesla range. As a result, they are sensitive to contamination.

External noise includes every possible source of electromagnetic interference, such as power lines and electronics located in the hospitals and surrounding areas. Internal artifacts for a particular signal arise out of physiological phenomena such as eye blinking, jaw clenching or muscle contractions. Cardiac activity may also interfere with the ECG. Their amplitude easily exceeds the magnitude of the underlying neural mechanisms that most researchers seek to capture.

To address these concerns, advanced filtering techniques such as bandpass, notch, and adaptive filtering are used by researchers and clinicians to isolate specific relevant frequency bands. However, the techniques do not necessarily separate sources of overlapping signals exhaustively. Thus, more contemporary techniques are leaning towards the use of machine learning to remove artifacts, such as Independent Component Analysis (ICA) or Blind Source Separation (BSS), which automatically detect and remove noise from the useful neural signal. These methods greatly enhance the clarity of the signal; however, the process always requires the guidance of an expert for constant adjustment and tailoring, particularly in situations that require real-time processing.

Inter-Subject Variability

The inter-subject variability is yet another major hurdle in interpreting brain signals. Each human possesses an individual brain which is quite different from each other functionally and anatomically. Features such as skull thickness, cortical folding, electrode placement and even the neuronal responses one might observe can drastically change the signals one acquires.

This factor is a great problem when the model is used in general applications in systems such as BCI or even diagnostic systems. A particular model built on a specific individual's data will not perform well if used on someone else's data without going through elaborate recalibration processes. A good example would be emotion recognition using EEG or even seizure recognition, where classifiers tend to lose their accuracy over different users or even recording sessions.

One way to solve this issue is by adaptive algorithms coupled with transfer learning, in which models alter themselves based on small new datasets for users. Also, clinicians use personalized calibration procedures designed to change preconfigured parameters and improve system reliability. Such techniques, however, tend to slow down the time needed for training, add additional calculations, and burden the users, particularly in the case of real-time or wearable BCI systems.

Data Volume and Computational Complexity

New methods of neuroimaging create enormous datasets. For instance, a 128-channel high-density EEG (HD-EEG) system producing over 100,000 data points per second or 'events' at a 1 kHz sample rate. Likewise, functional Magnetic Resonance Imaging (fMRI) also continuously outputs volumetric time-series data with high spatial resolution-achieving massive datasets which require significant processing power.

Such preprocessing, feature extraction, and classification request powerful computational tools because of the large dimensionality of the data. Traditional linear algorithms might struggle to efficiently scale with the big datasets. Solutions such as deep learning, parallel computing, and cloud-based processing platforms have been suggested. While these alternative approaches tend to optimize the problem, they add other complications like higher energy expenditures, algorithmic interpretability, and infrastructure expenses.

Closed-loop Brain Computer Interfaces (BCIs) and continuous patient monitoring systems are examples of applications that require real-time analysis and ultra-low latency. Finding a balanced trade-off between processing speed and complexity is one of the greatest challenges. Some optimization strategies such as dimensionality reduction like PCA and t-SNE, along with micro neural networks are being studied to address this issue.

Balancing Spatial and Temporal Resolution

A glaring gap in the available technologies that extract brain signals is the problem with balancing spatial and temporal resolution. No one technique offers both of them with high accuracy at the same time.

- **EEG:** Capturing minute neural events through EEG is ideal since it possesses mid-s backbone level temporal precision, and its spatial resolution is poor as it suffers due to signal diffusion through the skull and scalp.
- **fMRI**: It has a spatial resolution of a few millimeters, which is excellent for pinpointing brain function. Difficulties arise in temporal resolution as fMRI has slow hemodynamic responses it measures (in the order of seconds).
- **MEG**: It has a middle ground between the two extremes as it has better spatial precisions than EEG while having faster movement tracking compared to fMRI. One of the magnetoencephalograph's disadvantages, however, is that it's expensive and needs magnetically shielded rooms.

To address the issues posed by the two methods, researchers are pioneering multimodal integration techniques. They face challenges when trying to integrate separate datasets as it requires accurate alignment. The combination of EEG and MRI allows accurate capturing of both the timing and location of brain activity. Such integration, however, is sophisticated and costly. Integration of different streams of data requires meticulous synchronization, specialized frameworks of hardware, and intricate algorithms for data fusion. Furthermore, these methods are not viable for all clinical environments because of limited resources.

3. ARTIFICIAL NEURAL NETWORKS FOR NEUROLOGICAL DISORDER DIAGNOSIS

3.1 Introduction to Artificial Neural Networks (ANNs) in Medical Applications

Artificial Neural Networks, or ANNs as they are commonly called, are computational systems that take their inspiration from the structure and operation of biological neurons. Their application in medical diagnostics involves the processing of intricate biological information in order to recognize patterns, make decisions, and improve disease detection and treatment processes. For neurological disorders diagnosis, ANNs are essential for analyzing and recognizing abnormalities in brain signals, imaging data, and various types of electrophysiological patterns.

Key advantages of ANNs in medical applications include:

Automated Feature Extraction: The traditional machine learning paradigms demand features to be defined, to the detail of amplitude or frequency, by industry professionals, which are then labelled as handcrafted features. In contrast, manual feature engineering is comparatively easy, but it is often biased, limited in scope, and takes an inordinate amount of time. With ANNs, and more specifically deep learning ones like Convolutional Neural Networks (CNNs) and Autoencoders, features can be extracted automatically from raw data or data that has been minimally processed.

High Accuracy in Classification: Their excellent ability to recognize patterns and classify enables ANNs to be perfect for diagnosing various neurological conditions. They have proven to be especially good at recognizing pathological disorders, including:

- o **Epilepsy:** With great sensitivity and particularity, EEG recordings help to identify seizure activity.
- o **Stroke:** Evaluating the degree of cranial damage by using neuroimaging data to spot lesions.
- o **Brain Tumor:** For brain tumors, sometimes better than traditional statistical models are provided from MRI or CT scans, differentiating between benign and malignant growths.

ANNs can pick up small differences between typical and unnatural brain activity thanks of their capacity to understand non-linear and sophisticated decision borders. Better diagnostic accuracy, early disease discovery, and clinical decisions based on better information follow. Furthermore, enhancing classification accuracy are ensemble techniques and hybrid models that mix ANNs with other algorithms such as SVM or Random Forest.

Adaptability to Large Datasets: Inherently broad and multidimensional, modern medical data sets include high-resolution MRI scans, multi-channel EEGs, or genomic profiles. Their capability for scalability and parallel learning makes Deep Neural Networks (DNNs), a class within ANNs, ideal for handling such information.

These networks can process many different types of data (text, images, signals, time series) and discover complex relationships across several modalities. For instance, recurrent neural networks (RNNs) and long short-term memory (LSTM) networks are perfect for sequential data analysis, such as EEG recordings or tracking of disease progression over time.

CNNs are commonly used to process spatial data, including brain MRIs, and to spot microstructural abnormalities that could show early symptoms of conditions, including Alzheimer's or gliomas. Moreover, ANNs can be trained using multi-centre research, wearable devices, and hospital databases, so boosting generalizability and minimizing overfitting in the medical field. Precision medicine and individualized treatment scheduling benefit greatly from their capacity to handle large amounts of data.

3.2 Medical Diagnosis Techniques: Feature Extraction and Classification Methodologies for Brain Disorder Detection

The accuracy and dependability of brain disorder detection via Brain-Computer Interfaces (BCIs) and neural networks depend much on the quality of data preprocessing and the strength of the analytical models employed. This diagnostic process depends heavily on feature extraction and classification. Whereas classification models use the extracted features to predict or diagnose medical issues, feature extraction simplifies the complexity of brain signal data while preserving vital information. In the context of determining neurological disorders, this part concentrates on feature extraction techniques designed for brain signal data, especially EEG, MEG, and fMRI.

Feature Extraction in Brain Disorder Diagnosis

Feature extraction is the process of identifying and isolating meaningful attributes from raw brain signals to improve model performance. Common techniques include:

Time-domain analysis: For time-domain analysis, which is especially good at capturing brief brain events such as seizures, motor imagery, or event-related potentials (ERPs), signal changes over time are emphasized.

- **Statistical Measures**: Fundamental statistical descriptors like mean, variance, standard deviation, skewness, kurtosis, and entropy shed light on the stability, regularity, and randomness of brain impulses. High entropy might, for example, suggest frenzied brain activity often seen with epileptic seizures, while low entropy might show cognitive deterioration or diminished neuronal firing in neurodegenerative illnesses.
- **Waveform Characteristics:** Particular waveform features—peaks, spikes, sharp waves, and bursts—are therefore closely connected to pathological events. EEGs show, for instance, spike-and-wave discharges, which are characteristic patterns of absence seizures. Event-Related Potentials (ERPs)—time-locked responses to stimuli—are crucial for evaluating sensory and cognitive function, especially in disorders including schizophrenia and Alzheimer's disease.

Frequency-Domain Features: Analysing the signal in the frequency domain allows for the detection of oscillatory patterns associated with different brain states or abnormalities.

- **Fast Fourier Transform (FFT):** A generally used approach, fast Fourier transformation breaks EEG activity into its constituent frequency bands—delta (0.5–4 Hz), theta (4–8 Hz), alpha (8–13 Hz), beta (13–30 Hz), and gamma (>30 Hz). Irregular changes in power distribution across these frequencies might suggest neurological impairment. Patients with cognitive decline or traumatic brain injury typically exhibit, for example, increased delta and theta power; lower alpha power might indicate depression or anxiety.
- **Wavelet transformation:** WT allows for the identification of important characteristics linked with epilepsy spikes, motor imagery, and tumor-related aberrations in brain activity and presents a multiresolution analysis, hence, it is appropriate for spotting both low-frequency trends and high-frequency transients in non-stationary signals like EEG. The time-frequency localization power of wavelets makes them particularly useful in following abrupt occurrences, such as seizures.

Time-Frequency Analysis: When dealing with dynamic or non-stationary signals, where frequency content changes over time, time-frequency techniques provide a more nuanced analysis.

- **Fourier Transform in Short Time (FTST):** STFT uses FFT on every segment, which is divided of a signal into little overlapping time intervals. This method records the way the signal's frequency components develop over time. It is especially valuable in seizure onset prediction, sleep stage classification, and mental workload evaluation since temporary frequency variations have diagnostic relevance.
- **Hilbert-Huang Transforms:** HHT is a two-step technique including Empirical Mode Decomposition (EMD) followed by the Hilbert Transform that is outstanding in analysing non-linear and nonstationary EEG data. HHT adaptively breaks down signals into Intrinsic Mode Functions (IMFs) that can then be examined to identify abnormalities such as neural oscillations in Parkinson's disease, or small changes in cortical activity caused by tumor growth or ischemia. Its excellent resolution in both time and frequency domains renders it a valuable tool for quick diagnosis.

3.3 A Study of Deep Learning Approaches for Brain Tumor Identification through Analysis of EEG Data

Brain tumors change the usual electrophysiological activity of the brain, therefore causing observable shifts in EEG readings. Identifying and organizing these aberrations deep learning methods have proved to be very effective.

CNN-Based Brain Tumor Classification:

The way brain tumors are categorized from neuroimaging and electrophysiological data has been transformed by convolutional neural networks (CNNs). CNNs provide an automated, non-invasive approach for tumor identification and localization by using their capacity to find sophisticated patterns in image data.

- **MRI Imaging and EEG Spectrograms:** Transformed into spectrograms (a time-frequency representation), EEG data can be treated as 2D images. CNNs analyse these spectrograms to capture spatial and frequency-based patterns signalling unusual brain activity linked with the presence of cancer. MRI images also show brain structures with great anatomical detail. CNNs run these pictures through to find small structural deformities and tissue density anomalies; therefore, their capacity to recognize tumor boundaries and classify kinds of tumor (e.g., gliomas vs. meningiomas) is enhanced.

- **Transfer Learning with Pre-trained CNN Models:** Significantly enhancing classification performance is made possible by using pretrained models including DenseNet, Inception, ResNet, and VGGNet. Originally developed using big image collections such as ImageNet, these networks are fine-tuned on brain imaging data. Using this method, CNNs can identify medical-related characteristics without needing huge datasets from ground zero. Transfer learning reduces training time, raises precision, and helps to solve the problem of scarce labeled medical knowledge.
- **Performance Metrics:** When trained correctly, CNNs have shown more than 90% accuracy in EEG-based brain tumor classification and segmentation missions. Their strength is their capacity to effectively generalise across several patient datasets, dealing noise and inter-subject variance well.

RNNs and LSTMs for Brain Tumor Prediction

Although CNNs are great at spatial pattern recognition, Recurrent Neural Networks (RNNs) and their more sophisticated variant Long Short-Term Memory Networks (LSTMs) are meant for temporal sequence modelling, hence especially suited for time-series EEG signals analysis.

- **Sequential Encumbrance Handling:** Modelling temporal dependencies in EEG recordings, RNNs capture changes in brain activity over time. They can pick up small changes in neural activity that could signal early symptom onset or tumor spread. This step-by-step approach is vital because some brain tumors impact neuronal firing patterns before showing themselves in imaging scans.
- **Recollection Nets for Long Short-Term Memory Systems:** LSTMs solve this problem by means of gating mechanisms that control the flow of information all across time steps; conventional RNNs frequently suffer from vanishing gradients make it hard to capture long-term dependencies. This makes them quite good at processing long EEG recordings and detecting small signal changes preceding symptoms of cancer.
- **Early discovery and forecasting:** According to studies, models based on LSTM can find pre-symptomatic clues in EEG data, hence medical practitioners may act earlier in the disease cycle. These methods help to track long-term progression and forecast tumor recurrence in addition to helping to diagnose.

Hybrid Models for Enhanced Brain Tumor Detection

Current models harness the advantages of several neural network architectures and data types to increase diagnostic reliability and accuracy of detection further still.

- **CNN-LSTM Hybrids:** These models include LSTMs for temporal pattern recognition and CNNs for spatial feature extraction. EEG-based tumor classification starts with convolutional neural networks extracting spatial-frequency characteristics from spectrograms, which are then sent to LSTMs for temporal evolution analysis. By means of this dual approach, we can have a more complete grasp of the mechanics of brain tumor from both long-term signal patterns and immediate anomalies.
- **Transformer Models:** Revolutionizing natural language processing, transformers are now being used in EEG analysis. Their self-attention mechanism helps the model to ignore irrelevant noise and redundant data and zero in on the most important segments of a series. Used in detecting brain tumors, transformers can capture small but important signal patterns over long EEG recordings that conventional models could miss. The early-stage detection of tumors and the multi-class tumor classification benefit from this.
- **Multimodal learning**: It helps to deepen diagnostics by merging data from several modalities, including EEG (electrophysiological signals), fMRI (hemodynamic activity), and CT (structural imaging). While LSTMs or transformers deal with sequential signals, CNN-based encoders can handle image data. The resulting model is helped by knowledge of both functional and structural, therefore lowering the chance of false positives and boosting sensitivity. In clinical work, where confirming a diagnosis usually calls for evidence from several forms of evaluations, this technique is especially useful.

Table 4. Machine Learning Models Used in BCI Applications

Model	Application	Strengths	Weaknesses
CNN	MRI/CT brain scan analysis	High accuracy in image classification	Requires large datasets
RNN	EEG signal prediction	Good for time-series data	Computationally expensive
LSTM	Seizure prediction	Handles long-term dependencies	Requires large training data

4. DEEP LEARNING FOR BCI-BASED DIAGNOSIS

Brain-Computer Interfaces (BCIs) have revolutionized the field of neurodiagnostics by enabling direct communication between the brain and external devices.

4.1 Convolutional Neural Networks (CNNs) for Brain Scan and EEG Analysis

Used extensively in medical imaging, including analysis of brain scans, convolutional neural networks (CNs) have shown outstanding performance in image processing chores. Used in BCI-dependent diagnosis, CNNs can help early identify disorders including epilepsy, Alzheimer's disease, and brain tumors by efficiently discerning brain scans and EEG spectrogram patterns.

CNN Architecture for BCI Applications

1. **Preprocessing**: Noise reduction, normalization, and segmentation are among the preprocessing steps that are applied to raw EEG signals and brain scans. Artifacts like artificial noise caused by muscle movement, electrical interference, or equipment limitations must be eliminated to minimize noise. The convergence and performance of deep learning models are improved by regularizing data to ensure consistent scaling. Through segmentation, the continuous signals or scan images are broken down into smaller segments (or regions of interest), which allows for a more precise and effective analysis. Standardized input to the CNN for accurate learning.
2. **Feature Extraction**: Convolutional Neural Network (CNN) layers automatically learn hierarchical features from the input data, identifying critical patterns that indicate abnormalities. Early layers detect simple features like edges or textures, while deeper layers capture complex patterns such as lesions or unusual neural activity. This automated approach eliminates the need for manual feature engineering and allows the model to adaptively learn from diverse medical imaging modalities and signal types.
3. **Classification**: Fully connected layers and SoftMax activation functions are used to classify different neurological conditions based on learned features.
4. **Interpretability**: Grad-CAM and other visualization techniques help explain CNN decision-making in medical applications.

Figure 1. Architecture of a typical Brain-Computer Interface (BCI) system

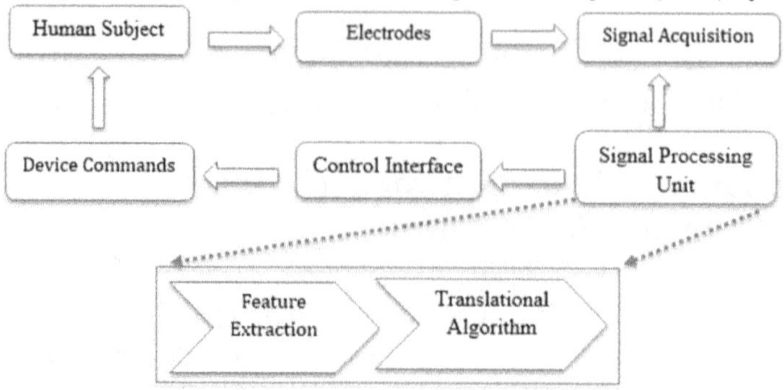

4.2 RNNs and LSTMs for Time-Series EEG Analysis

Recurrent Neural Networks (RNNs) and Long Short-Term Memory (LSTM) networks are widely used in time-series analysis due to their ability to model sequential dependencies.

- **RNNs** are capable of processing sequences by maintaining a memory of previous inputs using hidden states. However, they often struggle with long-term dependencies due to the vanishing gradient problem.

Figure 2. Unrolled representation of a Recurrent Neural Network (RNN) over three-time steps

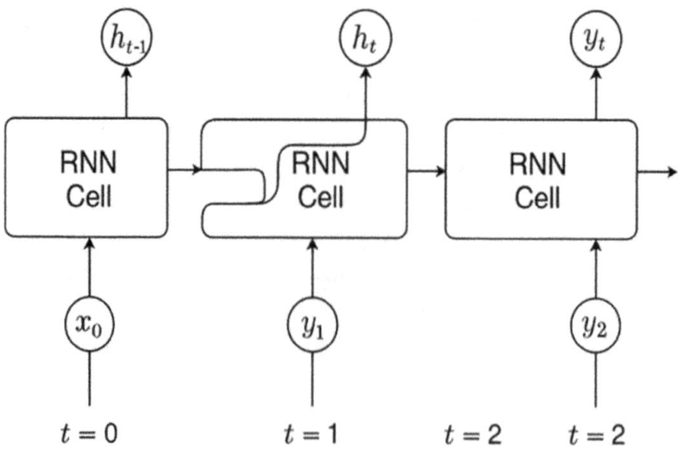

Each RNN cell receives an input x_t and the previous hidden state h_{t-t}, producing a new hidden state h_t and output y_t. The recurrence allows the model to capture temporal dependencies in sequential data.

$$H_t = \tanh(W_{hh} * h_{\{t-1\}} + W_{xh} * x_t + b_h)$$

$$Y_t = Wh_y * h_t + b_y$$

Where:

- x_t: Input at time step t
- h_t: Hidden state at time step t
- y_t: Output at time step t
- W_xh, W_hh, W_hy: Weight matrices
- b_h, b_y: Bias terms
- **LSTMs**, a special kind of RNN, address this limitation using memory cells and gating mechanisms (input, forget, and output gates) that allow them to retain information over longer time intervals.
- The diagram illustrates the internal structure of a Long Short-Term Memory (LSTM) cell at a specific time step t. It receives two main inputs: the previous hidden state h_{t-1} and the current input x_t. These inputs pass through three sigmoid gates and one tanh activation. The **forget gate** (σ) decides which parts of the previous cell state c_{t-1} should be retained or discarded.

Figure 3. Internal architecture of an LSTM (Long Short-Term Memory) cell

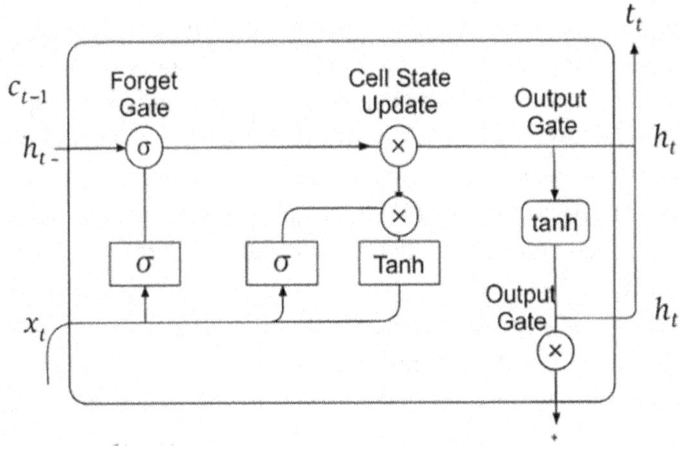

Simultaneously, the **input gate** (σ) and the tanh activation generate candidate values to be added to the cell state. The outputs of the forget and input gates are combined with the previous cell state to produce the updated cell state $c_{t.}$ Finally, the **output gate** (σ) uses the updated cell state (passed through tanh) to determine the new hidden state h_t, which will be passed to the next time step. This gated structure allows LSTM networks to effectively capture long-term dependencies in sequential data while avoiding vanishing gradient problems common in traditional RNNs.

$F_t = \sigma(W_f * [h_{\{t-1\}}, x_t] + b_f)$ (Forget Gate)

$I_t = \sigma(W_i * [h_{\{t-1\}}, x_t] + b_i)$ (Input Gate)

$C_t = \tanh(W_c * [h_{\{t-1\}}, x_t] + b_c)$ (Candidate Cell State)

$c_t = f_t \odot c_{\{t-1\}} + i_t \odot c_t$ (Cell State Update)

$o_t = \sigma(W_o * [h_{\{t-1\}}, x_t] + b_o)$ (Output Gate)

$h_t = o_t \odot \tanh(c_t)$ (Hidden State)

4.3 Transfer Learning for Medical BCIs

Transformative technology in medicine, brain-computer interfaces (BCIs) appear to provide hopeful answers for people with communication impairments, motor disabilities, and neurological conditions. Using brain signals, including electroencephalography (EEG), magnetoencephalography (MEG), or functional magnetic resonance imaging (fMRI), BCIs let users run outside devices, communicate, or even help diagnose and monitor different disorders. Still, the gathering and variance of neural data present a major obstacle in the creation of correct medical BCIs.

Medical brain information is, by nature, very particularistic and quite sophisticated. Inconsistencies in the data result from inter-subject variability, variations in brain physiology and anatomy, physiological noise, emotional states, and hardware configurations. Furthermore, obtaining such information is costly, time-consuming, and frequently difficult for patients, particularly those with medical concerns. These difficulties constrain the amount and quality of datasets, so it becomes hard to create strong machine learning models from scratch for every new patient or event.

To these problems, transfer learning offers an effective answer. In machine learning, it is a strategy whereby a model created for a first source task is reused as the starting basis for a model on a next, connected task. Within medical BCIs, transfer learning allows for models trained on large databases from one group of patients or experimental conditions to be fine-tuned for new subjects or slightly different tasks with little fresh data. This greatly cuts down on the need for extended calibration sessions and improves the usefulness of BCI systems in real-life medical settings.

TL has the following key advantages in medical BCIs:

- **Generalization across subjects:** One advantage of making use of knowledge transfer is that it can help models generalize across different individuals (users), overcoming the issue of inter-subject variability. This is particularly beneficial for designing assistive technologies for patients, where individualized calibration may not be practical.
- **Cross-Session Stability**: Neural data is not only variable across individuals but also across sessions in the same individual. By adapting to session-specific changes, TL reduces the need to retrain the model from scratch with each new session.
- **Data Efficiency:** TL helps reduce the required amount of labelled data used for training, which makes it an efficient strategy to adopt when one works with scarce medical datasets. This is particularly important in areas such as fMRI or invasive EEG, where data collection is particularly demanding.
- **Increased Efficiency:** Transfer learning can expedite the training process, as a portion of the learning has already been performed on the base model, making it a valuable asset in fields like medical imaging where time can be a pressing concern.

5. BCI-ASSISTED TREATMENT AND REHABILITATION

Brain-Computer Interface (BCI) technology has the potential to revolutionize the field of neurological treatment and rehabilitation by establishing direct communication pathways between the brain and external devices. BCIs work by analysing brain signals in real time to bypass compromised neural pathways, allowing patients to use other methods to communicate or foster brain activity through interaction with their environment or assistive technologies. This novel technique is preparing to step in for patients suffering from ailments like stroke, ALS, spinal cord injuries, and other motor or cognitive-related functional deficits. By facilitating both physical rehabilitation and brain plasticity, BCI-assisted rehabilitation holds promise for both the restoration of lost functions and the enhancement of life quality in patients experiencing severe sequelae from neurological disorders.

5.1 BCIs in Neurofeedback Therapy: Teaching Brain Operation

Brain-Computer Interfaces (BCIs) are an emerging mechanism that revolutionizes neurofeedback therapy, allowing users to purposely regulate their brain oscillations. Neurofeedback therapy involves operant conditioning, in which patients receive data about their neural activity (typically through EEG) in real time and learn how to self-regulate this activity throughout treatment. The combination of BCIs in this therapeutic model can give the patients metacognitive access to their neural outputs with greater precision, and thus be the basis of treatment for neurological and psychological disorders. For instance, BCI-enhanced neurofeedback, where the brain is trained to elicit more functional patterns of activity, could be greatly helpful to anyone afflicted with ADHD, anxiety disorders, PTSD, and certain types of epilepsy. By training on the image data results from a series of sessions, users become aware of, and ultimately gain control over, the state of their brain, improving cognitive functioning, emotional regulation, and general mental resilience.

5.2 BCIs in Stroke Rehabilitation: Restoring Motor Function

Stroke ranks high among the leading causes of long-term disability globally, often rendering survivors with substantial deficits in motor function. Conventional rehabilitation systems are often based on physical therapy and repetitive motor training, which might have limited efficacy, especially in severely impaired motor patients. BCIs provide a new method by circumventing broken neural pathways and linking directly to the brain's desire to move. Neural signals associated with motor intention can be decoded by BCIs, which then drive external devices such as robotic exoskeletons, functional electrical stimulation systems (FES), or virtual avatars; effectively establishing an artificial neural input channel that connects the brain to the impaired limb. A brain-generated feedback loop is a key factor in increasing neuroplasticity—the brain's capacity to modify itself by creating new neural pathways.

The process tends to start with the patient thinking about a particular movement, like opening their hand or raising their arm. Brain computer interface (BCI) systems utilize these motor-related brain signals, typically recorded from the motor cortex, and decode them into control signals for a rehabilitative device. When this happens, the patient sees or feels their limb move as intended, thus closing the feedback loop. Confirmation of intention and movement is essential for motor relearning, and you have this loop firing. Research has demonstrated that even several months to several years post-stroke, highly repetitive experience with BCI systems is associated with measurable improvements in motor function. This success is possible because the system can deliver relevant feedback to the patient, contingent on their neural activity.

In addition, BCI-oriented stroke rehabilitation can be adjusted to the patient's impairment level and therapeutic objectives. Incorporating VR environments into the system to deliver immersive experiences that improve patient engagement and motivation. Others connect BCI to FES units that stimulate the paralyzed muscles in response to the brain's command, offering a more comprehensive rehabilitation method. One of the significant points is that these systems not only improve motor recovery but also engage cognitive abilities like attention, intention, and solving problems that may also become disrupted post-stroke. As clinical trials are proving the efficacy and safety of these systems, the application of BCIs in neurorehabilitation centers across different countries is becoming more frequent.

5.3 BCIs for ALS and Locked-in Syndrome: Enabling Communication and Control

The progression of Amyotrophic Lateral Sclerosis (ALS) and Locked-in Syndrome (LIS) leaves such patients with almost total loss of voluntary skeletal muscle control and intact cognitive activity. And these people may become fully paralyzed but fully conscious, which in turn creates a clear need for technologies that'll help them regain the ability to communicate and interact with the world. In this context, brain-computer interfaces (BCIs) are a strong assistive device that offer an avenue for direct communication between the brain and the external environment. BCIs can interpret a user's intention by interpreting specific neural signals, allowing them to convert thoughts into action, such as selecting letters on a screen, controlling a computer interface, or using a speech synthesizer.

BCI systems to be used in communication come in numerous categories, reliant on P300 event-related potential, steady-state visual evoked potentials (SSVEPs) motor imagery. There are pros and cons to each approach, depending on the patient's preserved sensory and cognitive faculties. One example is the P300-based systems that display a series of stimuli (e.g., a matrix of letters), and scanning of the EEG indicates to the system when a patient recognizes the target item, which is ideal for fundamental spelling or selection of items. For example, SSVEP-based systems exploit users' attention to flickering visual objects and therefore often achieve better speeds, with a caveat on good visual acuity. Motor imagery BCIs, in which users imagine movements, seem to hold special promise for patients who can no longer use eye movements or other residual physical abilities.

6. CHALLENGES AND FUTURE DIRECTIONS OF BCI-ASSISTED NEUROLOGICAL APPLICATIONS

6.1. Ethical Challenges in AI-Driven Neurological Medicine

As BCIs and AI systems wend their way into our clinical practice for neurological problems, an expanding set of ethical issues has begun to arise. These challenges arise mainly from the nuanced and sensitive characteristics of the data collected, the mediating effect of these technologies on human autonomy, and the potential for abuse of neural data.

One of the primary ethical issues relates to privacy and data security. BCIs measure and interpret brain signals—essentially the raw material of thought, intention emotion. Such information is so sensitive that its incidence has been at the root of damaging breaches of privacy. Brain data stored in an insecure manner or sent over unprotected channels may be intercepted or exploited for commercial, political, or surveillance purposes. With the evolution and integration of BCI systems into cloud infrastructures or AI-driven analytics platforms, identifying correct data encryption, user consent, and storing data securely will be key factors to be addressed.

A further ethical dilemma centres on user agency and informed consent. Moreover, in many medical situations, the patients who would utilize BCI systems are likely in vulnerable states—e.g., paralyzed, unconscious, or cognitively impaired—making the potential for genuine informed consent more complicated. Then there's the issue of agency: when a BCI decodes a brain signal and acts on it (moving a prosthetic, say, or typing a message), how closely does that match the user's true intent? If misinterpreted by AI, unintended actions could occur, and accountability issues arise.

6.2 Technical Challenges: Real-Time Processing of Brain Signals

The real-time processing of brain signals remains a significant bottleneck, and while BCI technology has made great strides, it is not yet feasible for general use. The success of brain signals in milliseconds is the primary objective of BCIs, which necessitate speed (acquired), accuracy (declared), and dependability (relative to medical issues such as split-second decision-making).

The human brain generates electrical signals that are often noisy and erratic. Real-time extraction of meaningful information from these signals is a formidable task.' Many BCI systems use electroencephalography (EEG), a non-invasive technique that suffers from low spatial resolution and noise from muscle movement, eye blinks, or external electronics. Real-time responsiveness is frequently slowed down by the need for signal preprocessing and artifact removal, which remains crucial.

Moreover, the decoding of brain activity necessitates advanced algorithms, typically driven by machine learning or deep learning models. The models must be able to accommodate individual user-specific brain patterns despite training on large datasets. It is challenging to design adaptive learning systems that can adjust the signal interpretation without continuous retraining. The problem has been identified.

The hardware limitations are yet another technical hindrance. To achieve real-time BCI applications, it is essential to use lightweight, wearable, and wireless devices that can maintain high signal fidelity. Most current systems are not suitable for daily life or long-term therapy sessions due to their portability and comfort. However, they can still be used safely in many situations.

Also, effective BCIs demand a short timeframe between intention and action. The system's therapeutic effectiveness is compromised if there is a delay of even ten seconds between assuming control and the robotic arm moving, which can be confusing.

To address these problems, we will need to develop advanced signal processing algorithms, integrate hybrid BCI systems that incorporate both EEG and other signals, and enhance hardware design to improve wearability and signal quality. Neuroscience, computer science, and biomedical engineering must collaborate to advance these advancements.

6.3 Future Potential: Personalized Medicine through BCIs

In the future, Brain-Computer Interfaces have the potential to revolutionize personalized medicine, particularly in the fields of neurology and mental health care. BCIs are becoming increasingly advanced and are expected to facilitate treatment plans that are specifically tailored for the brain's specific needs.

There is also a promising potential in adaptive neurostimulation.' By utilizing real-time feedback from the brain, BCI systems could deliver precise electrical or magnetic stimulation to specific brain regions, modulating neural activity for treating conditions like epilepsy, depression, Parkinson's disease, or chronic pain. While traditional pharmacological treatments can cause widespread systemic effects, this approach is a highly targeted and effective form of therapy.

Future BCIs in rehabilitation may enable full-scale neurorehabilitation environments that utilize virtual or augmented reality systems to guide therapy through signals from the brain. A virtual avatar designed to mimic the brain's imagined movements could be useful in retraining the mind through visual and proprioceptive feedback, especially for individuals with strokes.

Predictive healthcare could be made possible by AI-powered BCIs, which can monitor neural patterns and detect early warning signs of neurological disorders like multiple sclerosis before clinical symptoms. By taking a proactive approach, preventive care would replace reactive treatment and improve outcomes while decreasing healthcare expenses.'

In addition, upcoming BCIs could combine with genomics and wearable biosensors to generate comprehensive profiles of a patient's health, providing both physical and mental wellness. These devices would signify a significant advancement in precision medicine, where treatments are not only responsive but also pre-emptively sensitive to individual human physiological and psychological conditions.

The progress of these technologies must be accompanied by improvements in regulatory frameworks, ethical oversight, and public awareness to ensure innovation remains consistent with human values, dignity, and health equity. The focus of BCIs in the future is not solely on tying brains to machines, but also on improving the human experience, expanding our capabilities, and recovering lost functions with empathy, intelligence, morality, etc.

CONCLUSION

The integration of Brain-Computer Interfaces (BCIs) with advanced neural network technologies marks a transformative shift in the diagnosis and treatment of neurological disorders. By enabling real-time brain signal analysis and adaptive learning systems, BCIs are now capable of detecting conditions such as epilepsy, brain tumors, stroke, and neurodegenerative diseases with improved precision and speed. Deep learning models significantly enhance this process by enabling more accurate pattern recognition, cross-subject generalization, and personalized responses.

Beyond diagnostics, BCIs offer promising therapeutic solutions through neurofeedback, motor rehabilitation, and assistive neuroprosthetics, reshaping how patients recover and manage neurological impairments. The potential for predictive analytics within BCIs introduces powerful tools for early detection, preventive care, and individualized treatment planning.

However, as this field advances, critical challenges remain - including data privacy, clinical standardization, long-term safety, and ethical considerations. Overcoming these barriers will require interdisciplinary collaboration between neuroscientists, clinicians, engineers, and ethicists. As research continues to evolve, BCIs are poised to become a central component in next-generation neurotherapeutics, offering hope for earlier interventions, better outcomes, and a higher quality of life for patients with neurological disorders.

REFERENCES

Anwar, A. M., & Eldeib, A. M. (2020). EEG signal classification using convolutional neural networks on combined spatial and temporal dimensions for BCI systems. In *2020 42nd Annual International Conference of the IEEE Engineering in Medicine & Biology Society (EMBC)* (pp. 434–437). IEEE. DOI: 10.1109/EMBC44109.2020.9175894

He, H., & Wu, D. (2020). Transfer learning for brain–computer interfaces: A Euclidean space data alignment approach. *IEEE Transactions on Biomedical Engineering*, *67*(2), 399–410. DOI: 10.1109/TBME.2019.2913914 PMID: 31034407

Hering, A., Hansen, L., Mok, T. C. W., Chung, A. C. S., Siebert, H., Hager, S., Lange, A., Kuckertz, S., Heldmann, S., Shao, W., Vesal, S., Rusu, M., Sonn, G., Estienne, T., Vakalopoulou, M., Han, L., Huang, Y., Yap, P.-T., Brudfors, M., & Heinrich, M. P. (2023). Learn2Reg: Comprehensive multi-task medical image registration challenge, dataset and evaluation in the era of deep learning. *IEEE Transactions on Medical Imaging*, *42*(3), 697–712. DOI: 10.1109/TMI.2022.3213983 PMID: 36264729

LeCun, Y., Bengio, Y., & Hinton, G. (2015). Deep learning. *Nature*, *521*(7553), 436–444. DOI: 10.1038/nature14539 PMID: 26017442

Li, Y., Wei, Q., Chen, Y., & Zhou, X. (2021). Transfer learning based on hybrid Riemannian and Euclidean space data alignment and subject selection in brain–computer interfaces. *IEEE Access : Practical Innovations, Open Solutions*, *9*, 6201–6212. DOI: 10.1109/ACCESS.2020.3048683

Lima, A. A., Mridha, M. F., Das, S. C., Kabir, M. M., Islam, M. R., & Watanobe, Y. (2022). A comprehensive survey on the detection, classification, and challenges of neurological disorders. *Biology (Basel)*, *11*(3), 469. DOI: 10.3390/biology11030469 PMID: 35336842

Rainio, O., Teuho, J., & Klén, R. (2024). Evaluation metrics and statistical tests for machine learning. *Scientific Reports*, *14*(1), 6086. DOI: 10.1038/s41598-024-56706-x PMID: 38480847

Rizvi, T., Batchala, P., & Mukherjee, S. (2018). Brain death: Diagnosis and imaging techniques. *Seminars in Ultrasound. CT and MRI*, *39*(5), 515–529. DOI: 10.1016/j.sult.2018.06.005 PMID: 30244764

Spears, W., Mian, A., & Greer, D. (2022). Brain death: A clinical overview. *Journal of Intensive Care*, *10*(1), 16. DOI: 10.1186/s40560-022-00609-4 PMID: 35292111

Ssain, S., Nijholt, A., & Ahad, M. A. R. (2023). Status of deep learning for EEG-based brain–computer interface applications. *Frontiers in Computational Neuroscience*, *16*, 1006763. DOI: 10.3389/fncom.2022.1006763 PMID: 36726556

Yadav, H., & Maini, S. (2023). Electroencephalogram-based brain–computer interface: Applications, challenges, and opportunities. *Multimedia Tools and Applications*, *82*(30), 47003–47047. DOI: 10.1007/s11042-023-15653-x PMID: 37362726

About The Contributors

Mohammed Saleh Al Ansari is a renowned Associate Professor in the Department of Chemical Engineering at the University of Bahrain, with a career spanning over three decades. He holds a Bachelor's degree in Chemical Engineering from King Saud University, a Master's degree in Desalination Engineering from Glasgow, and a Ph.D. in Corrosion and Protection Science and Engineering. Throughout his illustrious career, Dr. Al Ansari has demonstrated exceptional expertise in teaching, research, and industry, with a strong focus on chemical engineering, machine learning, complex systems, data sciences, sustainability, and energy. His research endeavors have led to the publication of over 90 papers in esteemed Scopus, SCI, Esci, and International peer-reviewed impact journals and patents, showcasing his innovative and outstanding contributions to academic research. Dr. Al Ansari's dedication to education and research has earned him recognition as a leading expert in his field. He continues to inspire and mentor students, guiding them towards achieving excellence in their academic and professional pursuits. His commitment to advancing knowledge and driving innovation has made a significant impact on the academic community and beyond. With his extensive experience and impressive track record, Dr. Mohammed Saleh Al Ansari is a respected figure in the field of chemical engineering and a valuable asset to the University of Bahrain. His ongoing research and contributions are poised to shape the future of engineering and sustainability, leaving a lasting legacy for generations to come.

Kapil Joshi is an Associate Professor in the department of Computer Science & Engineering (CSE) at Uttaranchal Institute of Technology (UIT), Uttaranchal University, Dehradun, Uttarakhand, India. He has completed his Ph.D, topic on "Image Quality Enhancement using Fusion Technique" in computer science and engineering from Uttaranchal University, Dehradun. He has more than 10 years' experience in industry, academic, and research. His areas of interest are Operating

System, Computer Networks, Web Technology, Data Structure and Java etc. His research areas are deep learning, CNN, Computer vision etc. He has published various patents, 400+ research papers in reputed journals (UGC/Scopus) as well as published 30+ books in international publication. Dr. Kapil Joshi has published more than 50+ research papers in various international journals (including ISI/ SCI indexed). He has also been associated with many conferences at National and International level and organized many conferences like STEM-2017, ICACDS-18, ICAESMT-19, ICAEST-20, ICAEST-21, ICIS-22, ICNN-22, IDVRP-22, ICAEST-22 and ICAEST-23. He is also a professional member of IEEE UP section, editorial member, advisory member, associate editor, guest editor and technical reviewers in various reputed national/international journals like Wiley, MDPI, Springer, Elsevier and Taylorfrancis etc. Dr. Joshi organized several conferences and has been invited as a key-speaker in various workshops/conferences. Dr. Joshi has filed/published more than 40 patents. He has written 10+ books and more than 40+ book chapters in the reputed publications. Dr. Kapil Joshi has guided 10+ M.Tech students and 5+ PhD students so far. He has chaired many International and National Conferences and was also invited as Keynote Speaker /Session Chair in many events. He has been awarded with "Best Young Researcher" Award in Global Education and Corporate Leadership (GECL-21) received by Life Way Tech India Pvt. Ltd. in 2021, Excellence in Research of the Year-2023 Award by Shri Anand Vardhan, IAS, Additional Chief Secretary, Govt of Uttarakhand in DISTF-2023. He also appointed as an Asia Coordinator in "International Association for the promotion of Asia-Africa Research" (IAPAAR) from 2020.

***

Kartikey is an Associate Professor at the School of Liberal Arts, Uttaranchal University, Dehradun. He has teaching and administrative experience of more than fifteen years, 15 research papers and 04 books are in his credit. He has presented more than 50 research papers in national and international seminars and conferences. He has organized many conferences and workshops as Organizing Secretary and Convener.

Harishchander Anandaram is a Assistant Professor (Senior Grade), School of Artificial Intelligence, Coimbatore. He completed his PhD in Bioengineering (2020), M.Tech. in Bioinformatics (2011), and B.Tech. in Bioinformatics (2009) from the Sathyabama Institute of Science and Technology, Chennai. His research includes analyzing resistance in HIV protease inhibitors through molecular mechanics and machine learning, and exploring pharmacogenomics and miRNA-regulated networks in psoriasis in collaboration with prestigious institutions like IIT Madras,

Georgetown University, JIPMER, CIBA, and ILS. Dr. Anandaram has published extensively in international journals on bioinformatics and systems biology. His work earned him the "Young Scientist Award" from The Melinda Gates Foundation for his research on miRNA dynamics in infectious diseases. He has reviewed over 200 manuscripts in systems biology and is currently focused on predicting novel lead molecules and biomarkers to target inflammatory pathways in infectious and autoimmune disorders using computational techniques.

B. Gayathri is an Associate Professor of Computer Science at Bishop Heber College, a position she has held since 2008. Her research focuses on Green Cloud Computing, reflecting her commitment to environmental sustainability. She has completed 46 online courses from reputable platforms like Coursera, Udemy, and Simplilearn. Dr. Gayathri has authored 8 books and published 31 journals, with 182 paper presentations. She has received 38 awards, including the Best Teacher Award from Former Governor Rossaiah. Notable recognitions include International Women Innovator of the Year and International Young Scientist of the Year. Dr. Gayathri serves as an Editorial Member of IJITRA and reviewer for esteemed journals like JENRS and ASTESJ. She holds 21 filed and published patents, including 4 granted Indian patents, one UK patent, and 2 copyright. Dr. Gayathri is a sought-after guest speaker, having spoken on 73 occasions, and has organized 30 seminars and conferences. She is a member of six professional organizations, demonstrating her dedication to academic and professional excellence.

Sudeep Biswas currently pursuing a Bachelor of Technology (B.Tech) in Computer Science and Engineering (CSE) from the Global Institute of Management & Technology, located in Krishnanagar, India. Biswas engaged in coding competitions, hackathons, and workshops to further hone their skills.

Parag Chatterjee received the B.E. degree in Computer Science & Engineering from the Karnatak University, Dharwad, India, in 1998 and received M.Tech. degree from IIEST Shibpur, Kolkata, India, in 2007. He has an academic experience of 20 years. Currently he is working as an Assistant Professor in the Department of CSE at Global Institute of Management and Technology, Krishnanagar, West Bengal, India. His current research interests include Data ware Housing, Data mining and Wireless sensor. He has 15 publication in national and international journals and 3 Patents.

D.Kirubakaran has obtained his Ph.D from Anna University in 2010 and M.E. degree from Bharathidasan University in 2000. His area of interest is AC-AC converters for induction heating & renewable energy systems. He had guided 10 Ph.D research scholars. He has published more than 60 papers on referred international journals. He is a life member of ISTE. He is having 22 years of teaching experience.

He is working as Professor & Heading EEE department at St. Joseph's Institute of technology, Chennai since 2011.

D.Mohanapriya, is currently working as Assistant Professor Department of Computer Science and Engineering in Manakula Vinayagar Institute of Technology, Puducherry. She has Published more than 25 International Journals and Conferences. Area of Interest includes CyberSecurity, Blockchain Technology and Artificial Intelligence.

Annmol Damien is a research scholar at CHRIST (Deemed to be University), Bengaluru, specializing in Finance. With an MBA in Finance and Marketing, she brings academic rigor and practical insights from two years of professional experience in the MSME sector. Her research focuses on exploring innovative solutions in finance, specializing in Neurofinance, aiming to bridge theoretical frameworks with real-world applications.

G Revathy is working in SASTRA Deemed to be University, She has published around 65 research articles and holds 10 Patents. She has been a book editor for 5 books.

Soumen Ghosh currently pursuing a Bachelor of Technology (B.Tech) in Computer Science and Engineering (CSE) from the Global Institute of Management & Technology, located in Krishnanagar, India. Ghosh I engaged in coding competitions, hackathons, and workshops to further hone their skills.

Rupam Hazra, currently persuing B.Tech(C.E.E) from Global Institute of Management & Technology, Krishnanagar, India. He is dedicated to mastering the intricacies of software development and computer systems.

Bhavana S. Karmore working as Assistant Professor and Dean at BCA and MCA Department of G. H. Raisoni University, Amravati for U.G. and P. G. Department . Total Years of Experience 15+ . She publish more than 20+ Papers in International Journals . She work as International Expert at International Conferences. She worked as a Editorial Board Members in 5+ International Journals and Reviewer in 10+ International Journals and International Conferences . Project Guided 10+ PG students and 10+ UG Students . She publish more than 10+ books of Electronics and Computer Engg. subjects. 3 Design patent, 2 copyright .

Surinder Kaur is AP in Department of Fashion Designing,Guru Nanak Girls College, Ludhiana

Manu K S is presently working as an Associate Professor in School of Business and Management, CHRIST (Deemed to be) University, Bengaluru. He received his Doctor of Philosophy (Ph.D) from University of Mysore. He possesses more than ten years of teaching, research and consultancy experience. His area of interest includes Financial Econometrics, Capital and Derivatives Markets, Business Analytics and other Financial Markets. He has published various research articles in refereed journals. He has attended and presented papers in various National and International Conferences. He has been a Financial Econometrics Resource person trainer for few Faculty Development Programs.

Shashank Kumar is an Associate Professor and currently serves as the HOD at Uttaranchal College of Health Sciences, Uttaranchal University. He is a dedicated Musculoskeletal therapist with a total of more than 9 years of work experience in academic as well as in clinical areas. With a strong background in the field of physiotherapy, he has been actively involved in academic practice, bridging the gap between research and hands-on -patient care. His contributions extend to mentoring students, conducting workshops, and implementing evidence-based rehabilitation strategies to improve treatment outcomes.

Chairman M, is currently working as Assistant Professor, Department of Electronics and communication engineering in Hindustan College of Engineering and Technology, Coimbatore. He pursuing his Ph.D. in Electronics and Communication Engineering from Anna University, Chennai, M.E. in VLSI Design in the year 2022 from Kalaignarkarunanidhi Institute of Technology, Coimbatore and B.E. in Electronics and Communication Engineering in the year 2020 from Kalaignarkarunanidhi Institute of Technology, Coimbatore. He has 2 years of teaching experience in his service. His research interests are Image Processing, VLSI, Embedded Systems, Internet of Things, and Network Security. He has published research articles in 5 reputed scopus indexed journals, granted three patents, published 7 book chapters and two books in reputed publications, and attended 10 national and international conferences.

M.Martinaa is an experienced academic professional with over 16 years in teaching and research. She serves as the Assistant professor of Computer Science Engineering at SASTRA University, specializing in Wireless Sensor Neworks, IoT, and Cloud Computing. She has published around 15 international journal papers and participated in 15 international conferences.

Nivodhini M K is an Assistant Professor in the department of CSE in K.S.R College of Engineering.Research interests include Machine Learning Techniques and Neural Networks.

M.A. Mohamed Aslam is a dedicated academic and researcher specializing in Computer Science and Engineering, with expertise in Cloud Computing, Artificial Intelligence, and Machine Learning. Currently pursuing his Ph.D. at SASTRA Deemed to be University, SRC Campus, Kumbakonam, he has established himself as a knowledgeable educator and mentor, guiding students toward academic excellence and groundbreaking research. With a Master's and Bachelor's degree in Computer Science and Engineering from Arasu Engineering College, Anna University, He has cultivated a deep understanding of emerging technologies. His passion for machine learning, Iot-based healthcare systems, and data security has led to multiple scholarly publications in Scopus-indexed and UGC Care Group 2 Journals. As an Assistant Professor at Arasu Engineering College (2021–2024), he has played a crucial role in curriculum development, student mentorship, and research initiatives. His commitment to academia is evident in his participation in faculty development programs, national workshops, and technical webinars, where he contributes to discussions on advanced computing and innovation. He has published several research papers on cutting-edge topics, such as brain pathology detection using deep learning, Iot-based healthcare solutions, and AI-driven smart farming. His work has been recognised in renowned journals, including Technische Sicherheit (Technical Security Journal), Journal of Physics: Conference Series and Book Chapter Springer. His dedication to knowledge dissemination and technological advancement makes him a significant contributor to the AI and machine learning community.

Jain Mathew, Dean of the School of Business and Management at CHRIST (Deemed to be University), Bangalore, is a distinguished academic leader with decades of experience in teaching and research in Finance and Human Resource Management. Renowned for mentoring scholars and advancing knowledge, his research spans leadership taxonomy, market volatility, and social media technology, supported by esteemed international collaborations. Dr. Mathew has earned recognition through prestigious awards and is celebrated for his impactful publications and conference presentations. His affiliations with global academic institutions and key university committees underscore his influence in the field. A passionate educator and researcher, Dr. Mathew's contributions continue to inspire and shape the next generation of scholars and professionals.

Sridevi Nair is an Assistant Professor at CHRIST University, Bengaluru, India. With a background spanning experience in the hospitality industry and academic institutions, she brings practical insights into the classroom and research. Her academic interests lie in Organizational Behaviour, Human Resource Development, Gamification, and Learning, with a focus on exploring innovative methodologies and applications in these areas. An avid researcher, she actively seeks opportunities for collaboration through interdisciplinary projects and partnerships with industry and academia. Passionate about advancing knowledge and bridging theory with practice, she is committed to fostering impactful research and contributing to the broader academic and professional community.

P. Mathivanan is an Associate Professor in KIT -CBE College. He has Published more than 25 International journals and Conferences. His areas of interest is Machine Learning, Artificial Intelligence.

Partha Paul currently pursuing a Bachelor of Technology (B.Tech) in Computer Science and Engineering (CSE) from the Global Institute of Management & Technology, located in Krishnanagar, India. Paul engaged in coding competitions, hackathons, and workshops to further hone their skills.

Jyoti Prasad Patra Is Presently The Principal, Nigam Institute of Engineering and Technology NIET, AT Govindpur PO Munduli, District Cuttack, Odisha India 754006 UG PG Diploma in engineering. He has to his credit 35 Years of experience 50 Indian utility Published Patents 12 Indian/UK Designs patents Registered 02 Indian /foreign copyright registered 01 Scopus book chapter published 03 International /National Award Received 06Text Books 02 E-Book Published 04 SCIE 11 Scopus Journal Article Published 15 IEEE Conference Articles Published 12 M.Tech Theses Guided. He Was Born 12th September 1965 and awarded Ph.D In Electrical Engineering In 2013 from SOA University Bhubaneswar which has NIRF Ranking of 14 In University 24 in Overall Category In 2024.

R.Dhivya is an Assistant Professor in the Department of Information Technology at M.Kumarasamy College of Engineering, Karur. She completed his B.Tech. in Information Technology and MTech. in Multimedia Technology with First class Distinction from Paavai Engineering College, Namakkal and Karunya University, Coimbatore. She has 10 years of teaching and research experience. She has published 15 papers in National and International Journals.Mrs. Dhivya is devoted to researching the applications of Machine Learning and Image Processing in diverse domains, including Data Science. Her research, teaching, and leadership have been instrumental in advancing the field.

Nisha Rathore, an Assistant Professor at Amity University Chhattisgarh, holds a B.Tech. from PSIT Kanpur, an M.Tech. from NIT Patna (2019), and enrolled for a Ph.D. course at IIT (ISM) Dhanbad. A GATE and UGC-NET qualifier, she has authored 3 books, published over 5 research papers, and holds 2 patents. Her research interests include brain-computer interfaces, soft computing, fuzzy logic, machine learning, and computer networks. She actively contributes as a member of IAENG and the Institutional Innovation Council. Nisha is committed to advancing computer science and engineering through innovative work. Additionally, she supports student well-being with her self-help book, The Paradise of Ultimate Happiness.

Deepali Rani Sahoo is an Associate Professor at Manipal Law School, Manipal University Rajasthan.

Nidhi Sharma is working as an Assistant Professor at Uttaranchal College of Health Sciences, Uttaranchal University, Dehradun, Uttarakhand. She graduated from College of Physiotherapy, Christian Medical College & Hospital, Ludhiana, Punjab in 2018, Master's degree with specialization in Neurological Physiotherapy from Himalayan Institute of Medical Sciences, Swami Rama Himalayan University, Dehradun, Uttarakhand in 2021, and is currently pursuing Ph.D. With a good research background, she has a good number of publications, patents, and copyrights to her credit.

Arvind Kumar Shukla has an experience of more than 18 years in teaching and Industry field. Currently, He is working as Professor & HoD at Department of Computer Applications, IFTM University, Moradabad. He has got his Ph.D Computer Science from Banasthali Vidyapith Rajasthan. He published more than 50 research papers in various reputed National/International Journals. Dr. Shukla is associated with Banasthali University, Rajasthan. Dr. Shukla has consistently been an active member of Academic Council, Executive Council,Board of Studies, Research Degree Committee of IFTM University, Moradabad. Dr. Shukla has supervised the thesis/dissertation of more than 15 students of M.Tech & Ph.D level.

Abhendra Pratap Singh is working as an Assistant Professor in HMR institute of technology and management, affiliated to GGSIPU, India. He completed M.Tech in Tool Engineering from Delhi Institute of Tool Engineering, Delhi . India and Completed B.Tech in Mechanical Engineering from Amity University, Noida Sector-125, India. He has 4 years experience of teaching. In the vast landscape of artificial intelligence in mechanical Engineering.He interests encompass Static, mechanical and Dynamics of composite structures, Honeycomb Structures, Hybrid sandwich modeling, tapered and laminated composites and Cutting. He has an impressive publication record with over 20+ research papers in Indexed high-impact factor international journals. He has attended many seminars, conferences, and training programs. Her innovative spirit is demonstrated by many filed and published patents.

Abhijeet Sudhakar is a data scientist with 3.5 years of experience in translating medical data into key strategy insights and valuable actions for better decision making. Expertise in developing, modifying and running models, performing data exploration and mining, and solving predictive and prescriptive problems, extracting key features from medical images. Hands on experience with AWS infrastructure to create pipelines, run the programs and Docker Containers to deploy the models. Good understanding of radiomics, genomics, transcriptomics, proteomics to extract useful insights from medical data like MRI Brain Tumor, Breast Cancer and produce different visualizations and graphs. Strong background in advanced mathematics, statistical and machine learning models, Python Programming Language, R language. Holistic view about Medical Domain, Data Science and Backend Development to bridge gaps between the domain and technology.

Varadharajan Bhoopathy is a Professor at Sree Rama Engineering College. He holds a Ph.D. in Information and Communication Engineering from Anna University and has over 16 years of academic experience. Specializing in wireless sensor networks and IoT, he has published numerous research papers and holds several international patents. Dr. Bhoopathy is a recipient of prestigious awards for his contributions to education and research. His commitment to academia continues to drive innovation and excellence in the field of computer science

Shraddha N. Zanjat currently serves as an Assistant Professor in the EEE Department at the School of Engineering and Technology (SOET), Sandip University, Nashik. With 10 years of teaching experience and 2 years of industry exposure, she brings a wealth of knowledge to her role. She holds a post-graduation degree, a Master of Technology in Electronics and Communication, from RTMN University. Additionally, she is pursuing her PhD at Sandip University, Nashik. Prof. Shraddha N. Zanjat has presented ten conference papers at both national and international levels and has published five papers in esteemed international journals such as IEEE, UGC, and others. Furthermore, she has obtained five Indian design patents. She is also a member of two technical association bodies. Prof. Shraddha N. Zanjat has registered four copyrights under her name and has published two book chapters in Scopus-indexed publications. Additionally, she has authored more than 50 books with international publications and has received royalties of about 20k.

Index

A

B

C

D

E

F

H

I

M

N

O

P

R

S

T

U

V

X

www.ingramcontent.com/pod-product-compliance
Lightning Source LLC
LaVergne TN
LVHW080352011026
849161LV00009B/2276